Communication in Transportation Systems

Otto Strobel
Esslingen University of Applied Sciences, Germany

Managing Director: Lindsay Johnston
Editorial Director: Joel Gamon
Book Production Manager: Jennifer Yoder
Publishing Systems Analyst: Adrienne Freeland
Development Editor: Austin DeMarco
Assistant Acquisitions Editor: Kayla Wolfe
Typesetter: Lisandro Gonzalez
Cover Design: Jason Mull

Published in the United States of America by
    Information Science Reference (an imprint of IGI Global)
    701 E. Chocolate Avenue
    Hershey PA 17033
    Tel: 717-533-8845
    Fax: 717-533-8661
    E-mail: cust@igi-global.com
    Web site: http://www.igi-global.com

Library of Congress Cataloging-in-Publication Data

Communication in transportation systems / Otto Strobel, editor.
      pages ; cm
  Includes bibliographical references and index.
  Summary: "This book bridges theoretical knowledge and practical applications of the cutting edge technologies for communication in automotive applications"-- Provided by publisher.
    ISBN 978-1-4666-2976-9 (hardcover) -- ISBN 978-1-4666-2978-3 (print & perpetual access) -- ISBN 978-1-4666-2977-6 (ebook) 1. Intelligent transportation systems. 2. Electronics in transportation. 3. Wireless communication systems. I. Strobel, Otto, 1950- editor.
  TE228.3.C6515 2013
  621.382--dc23
                            2012037375

British Cataloguing in Publication Data
A Cataloguing in Publication record for this book is available from the British Library.

All work contributed to this book is new, previously-unpublished material. The views expressed in this book are those of the authors, but not necessarily of the publisher.

To my very beloved grandchildren, Jamie and Sevi

Table of Contents

Detailed Table of Contents

Chapter 1

Otto Strobel, Esslingen University of Applied Sciences, Germany

Jan Lubkoll, Friedrich-Alexander University Erlangen-Nuremberg, Germany

Daniel Seibl, Esslingen University of Applied Sciences, Germany

The main idea of these lecture notes is to give an overview of Optical Data Buses for Automotive Application. The most important devices for fiber-optic transmission systems are presented, and their properties discussed. In particular, we consider such systems working with those basic components which are necessary to explain the principle of operation. Among them is the optical transmitter, consisting of a light source, typically a low speed LED or in case of higher demands a high speed driven laser diode. Furthermore, the optical receiver has to be mentioned. It consists of a photodiode and depending on the demands a low performance receiver circuit or a low noise high bit rate front-end amplifier. Yet, in the focus of the considerations, the optical fiber is the dominant element in optical communication systems.

Chapter 2

Piet De Pauw, BVBA De Pauw, Belgium

Communication networks in general, can be divided into 4 different classes: Ring (one directional or bidirectional), Star (active or passive), Tree, and Bus (passive). These 4 classes of communication networks are also used in transport systems. The main properties of these networks are overviewed. Then, the most important automotive networks are discussed: LIN, CAN, MOST, Ethernet, and Flexray, as well as their implementations in cars. For the MOST and Ethernet networks, different physical layer implementations are possible. The different factors determining the choice of a network are discussed. Cost is a major driver for automotive networks. For avionics networks, different standards for the network protocols and the physical layer implementations exist. In most cases physical layer implementations are proprietary, although, due to cost reduction pressure, more and more standardization is ongoing, and a tendency exist to adapt automotive network standards, and for the pressurized part of the airplane, also automotive physical layer implementations.

Chapter 3

This chapter is divided into three sections. The first section gives an overview of the different cables used in automotive communication systems. Although 6 different kind of cables and their corresponding connector systems have been qualified for use in automotive applications (3 electrical cable types and 3 optical cable types), up to now, only three types are commonly used: Unshielded Twisted Pairs (UTP), Coax cable, and 1 mm PMMA POF. GOF and multimode PCS up to now are not used for mass volume production. The second section overviews the optical connectors used in automotive communication systems. The electrical connectors used in LIN and CAN networks. The optical connectors standardized by MOST. In the early phase of MOST implementation, the number of MOST connectors has been large: long pigtails, short pigtails, micro pigtails, and these connectors are combined with electrical connections with a wide range of pin sizes. The number of connectors used in MOST – mainly under cost pressure is now drastically reducing. The third section overviews the different fiber optic transceivers used for MOST. The different packages for MOST25 and MOST150 transceivers are discussed. The most difficult step was the conversion of the fiber optic transceivers to be compatible with reflow SMD processes. This step is now taken. Also the evolution towards higher data rates is shown.

Chapter 4

The idea of this chapter is to give a complete overview on a matrix approach to describe light propagation in strongly multimode fibers such as 1-mm diameter plastic optical fibers. These large core fibers accept such a huge number of travelling modes that they can be viewed as a continuum. Thus, light propagation can be described as a power flow by a differential equation that can be more easily solved using matrices. Thus, the key of this method is the propagation matrix that is calculated from the diffusion and attenuation functions characteristic for a given fiber type. The propagation matrix has temporal frequency dependence and can be used to obtain not only angular power distributions but also temporal parameters such as pulse spread or bandwidth. This approach is flexible to introduce localized perturbations of power distribution provided they can be modeled as matrices. Thus, the effect of devices such as scramblers or connectors and also of disturbances such as curvatures and tensions can be introduced at different points in the fiber path to assess their impact on transmission properties. One of the most critical parameters when designing a network is its bandwidth and how it decreases when increasing the link reach. This dependence has been assumed to be linear when both bandwidth and length are represented in logarithms with a slope whose value provides information of the processes underlying propagation. Thus, the authors apply the model to calculate the bandwidth versus length dependence under different conditions analyzing the value of the slope and explaining previous experimental findings.

Chapter 5

Modern medium and high end vehicles are no longer imaginable without using technologies to broadcast local available data. The speed information for example is used by many well known functions: the anti blocking system, the radio, the dashboard, the cruise control, the electronic stability program, etc. Usually, this data is distributed among vehicle's electronic control units by various serial bus systems. The succeeding sections introduce the automotive communication system named FlexRay™. The development of FlexRay™ had been initialized by requirements expected for drive-by-wire systems. The content is focused on its electrical physical layer beginning with active components like bus interfaces as well as passive components like common mode filters and bus-cables. Comparisons to the state of the art systems CAN and LIN are used to support the comprehensibility

Automotive bus systems like e.g. LIN, CAN, and FlexRay™ distribute their serial data streams NRZ2 coded in the base band among the communication nodes. The nodes are interconnected by passive nets. Depending on the type of application some of these nets may consist of up to one hundred meters of different bus cables arranged in various topologies. The individual pieces of information inside the data streams are represented by voltage steps and current steps. They have to be passed among all connected nodes via the bus cables. The succeeding sections introduce the commonly known transmission line theory focused of the physical effects being relevant for automotive serial bus communication in the time domain.

This chapter reviews some of the network topologies and technologies within current vehicular systems. This is then followed by a proposal from the authors with initial viability results, into the possibility of implementing optical wireless links to either replace or complement these existing ideas. The initial motivation for this work (Green, 2010) is that there exist multiple pathways within a vehicle such as the engine compartment, within the frame of the chassis, or the internal cockpit that all lend themselves nicely to free space optical propagation. The first specialised study on the viability of optical wireless communications within the vehicles cabin was then published in (Higgins et al, 2012) which provided a further impetus to the concept. It is hoped that through the original results presented here, the reader can gain a basic understanding of the concepts compared to the current technologies, and are then able instigate their own research ideas.

Resilient optical transport networks have received much attention as the backbone for future Internet protocol (IP) networks with enhanced quality of services (QoS) by avoiding loss of data and revenue and providing acceptable services in the presence of failures and attacks. This chapter presents the principles of designing survivable Dense-Wavelength-Division-Multiplexing (DWDM) optical transport networks including failure scenarios, survivability hierarchy, routing and wavelength assignment (RWA), demand matrix models, and implementation approaches. Furthermore, the chapter addresses some current and future research challenges including dealing with multiple simultaneous failures, QoS-based RWA, robustness and future demand uncertainty accommodation, and quality of service issues in the deployment of resilient optical backbones for next generation transport networks.

Chapter 9

Joaquín Beas, Tecnológico de Monterrey, México

Gerardo Castañón, Tecnológico de Monterrey, México

Ivan Aldaya, Tecnológico de Monterrey, México

Gabriel Campuzano, Tecnológico de Monterrey, México

Alejandro Aragón-Zavala, Tecnológico de Monterrey, México

In recent years, considerable attention has been devoted to the merging of Radio over Fiber (RoF) technologies with millimeter-wave-band signal distribution. This type of system has great potential to support secure, cost-effective coverage and high-capacity vehicular/mobile/wireless access for the future provisioning of broadband, interactive, and multimedia services. In this chapter, the authors present an overview of an RoF access networks in the context of in-vehicle networks, with special attention to the figures of merit of the system and the basic enabling technologies for downlink/uplink transmission in the RoF land network, which is divided in three main subsystems: Central Station (CS), Optical Distribution Network (ODN) and Base Station (BS). The chapter first reviews the up-conversion techniques from baseband to mm-waves at the CS, and then the different BS configurations. The work finally applies these concepts to the development of an access network proposal for in-vehicle wireless application.

Chapter 10

Riccardo Scopigno, Istituto Superiore Mario Boella (ISMB), Italy

Vehicular Ad-Hoc Networks (VANETs) are wireless networks primarily meant to enforce vehicular safety. The incumbent international VANET solution is based on an adaptation of WLAN to the 5.9 GHz band and to the vehicular environment: it is universally known as IEEE 802.11p. One of the main reasons for the success of IEEE 802.11p lies on the functional requirement of a decentralized solution, that is, one able to work in the absence of infrastructure. While Filed-Operational Tests are being developed world-wide and new VANET applications, not restricted to safety, are being developed, new requisites are emerging. Some limitations of the IEEE 802.11p are coming to light as well: stakeholders must be aware of them to prevent misleading conclusions on reliability and, most importantly, improper solutions for the safety which the protocol is aimed at.

Chapter 11

Kira Kastell, Frankfurt University of Applied Sciences, Germany

Communication in transportation systems not only involves the communication inside a vehicle, train, or airplane but it also includes the transfer of data to and from the transportation system or between devices belonging to that system. This will be done using different types of wireless communication. Therefore in this chapter, first, the fundamentals of mobile communication networks are shortly described. Thereafter, possible candidate networks are discussed. Their suitability for a certain transportation system can be evaluated taking into consideration the system's requirements. Among the most prominent are the influence of speed and mobility, data rate and bit error rate constraints, reliability of the system and on-going connections. As in most of the cases, there will be no single best wireless communication network to fulfil all requirements, and in this chapter also hybrid networks are discussed. These are networks consisting of different (wireless) access networks. The devices may use the best suited network for a given situation but also change to another network while continuing the on-going connection or data transfer. Here the design of the handover or relocation plays a critical role as well as localization.

Chapter 12

Vehicular networks are deployed as hybrid networks, which consist of a cooperation of different radio access networks. Seamless communication in vehicular networks relies on proper network planning and thorough dimensioning of network protocols. Both are assessed and verified by simulation. In transportation networks, the location of the mobile devices and their pattern of movement are very important. Therefore, different mobility models suited to vehicles in transportation networks are introduced. Then, the need for location information is exemplified. Mobility models and location play an important role in the verification of handover protocols. One hybrid handover protocol is given in detail. It provides low handover latency with additional mutual authentication to allow the transfer from one radio access network to another while maintaining the network's built-in security standard. This protocol is easily extensible to include a broad variety of networks.

Chapter 13

One of the main parameters for providing traffic safety of vehicles is the knowledge of their speed. In this chapter results of research activities on a microwave radiometric sensor for the measurement of the velocity of land vehicles are presented. The work concentrates on a Radiometric Speed Sensor of Correlation Type (RM SSCT). For the analysis of design principles of the RM SSCT, the main parameters of radiometers are defined. The nature and statistical characteristics of radio thermal radiation of a terrestrial surface and objects are considered. For an estimation of the influence of parameters of the antenna system and the linear path of the receiver on parameters of the signal formed at the output of the correlator, a statistical analysis of the radiometric system of correlation type is carried out. Using a statistical model of the RM SSCT, the parameters of the antenna system were optimized as well as the radiometric receivers for various types of objects present on a terrestrial surface. Statistical results of the performance of the RM SSCT and an analysis of the basic characteristics of the SSCT for various types of objects are finally presented.

Chapter 14

The idea of this chapter is to give an overview of a relatively new technology – that of using ultrasound to transmit data at short ranges, within a room say. The advances that have made this a useful technology include the ability to utilize a sufficiently wide bandwidth, and the availability of instrumentation that can send and receive ultrasonic signals in air. The chapter describes this instrumentation, and also covers the various aspects of ultrasonic propagation that need to be discussed, such as attenuation, spatial characteristics, and the most suitable forms of modulation. Provided such details are considered carefully, it is demonstrated that ultrasonic systems are a practical possibility for in-room communications.

In recent years, driver assistance systems have become a strong trend in automotive engineering. Such systems increase safety and comfort by supporting the driver in critical or stressful traffic situations. A great variety of surround sensors with different fields of view include radar, ultrasonic, laser, and vision systems. These sensors are based on different technologies and measurement principles. They all have their specific advantages and disadvantages and range from low-cost to high-end systems. They also differ in size, mounting position, maintenance, and weather compatibility. Hence, such sensors are used in various configurations to explore the surroundings ahead, sideways, and behind a vehicle. In addition, vehicle dynamics information from speed, steering angle, yaw rate, and acceleration sensors is available. Data fusion algorithms on raw data, feature, or object levels are used to collect all this information and set up vehicle surround models. An important issue in this context is the question of data accuracy and reliability. Situation interpretation of the traffic scene is based on these surround models. Any situation interpretation has to be performed in real-time, independent of the situation complexity. Typically, the prediction horizon is a couple of seconds. Depending on the results of the driving environment analysis critical situations can be identified. In consequence, the driver can be informed or warned. Some driver assistance systems already perform driving tasks like following, lane changing, or parking autonomously. The art of designing new, valuable driver assistance systems includes many factors and aspects and is still an engineering challenge in automotive research.

Preface

In the last five decades, landline network communication has mainly been considered for application in telecom areas. The most well-known use is for high-speed, middle, and long distance systems as well as MAN and LAN networking; any last-mile application, including in-house communication to a single users' desk, needs to be connected to the rest of the world. Finally, mobile communication, in particular cell phones, more recently smart phones, tablets, tablet PCs, laptops, PCs, etc. have been developed to replace cable-based phone calls, emails and internet communication.

In the last ten years, communication in transportation systems has been more and more in demand – for communication within a vehicle, from one vehicle to another and to land-line networks, too. Development started in high-end cars with application in the infotainment area and has already reached airplanes and vessels where sensor-relevant needs were also addressed. These techniques began with low data rates. Car communication technologies for the coming decade will also include high bit-rate systems up to the region of Gbit/s.

In this book, the contributors present state-of-the-art and next-decade technologies for optical data buses in automotive applications. There are electrical, optical, and microwave solutions to meet the target. Media Oriented System Transport (MOST) an optical data-bus technology is nowadays used in cars. MOST150 is the current standard with a bit rate of 150 Mbit/s, and it is an adequate solution for optical multimedia data transmission in automobiles. MOST 150 operates with LEDs, a Polymer Optical Fibre (POF) and silicon photodiodes. However, to enable the next step towards autonomous driving, new bus systems with higher data rates will be required.

New generation aircraft covered by carbon fibre instead of conventional metal fuselages present additional challenges. In such cases, optical data transmission based on new laser types, VCSELs and Polymer Clad Silica (PCS) fibres, enables EMC compatibility and paves the way for the future.

Furthermore, in the last five years, we have learned a lot about the needs of companies producing state-of-the-art systems. Companies use practical systems in order to compete in the market of such products. They want a complete solution for their needs, and they do not care if fibre optics was part of the solution or not. Consequently, nowadays, other physical techniques have to be developed: Wireless applications open up a new field of data transmission. High-speed wireless LED transmission offers short-range data transmission without EMI/EMC problems. Visible light communication (VLC) using high power LEDs is an interesting technique. The first aim using these light sources is to light a room. However, at the same time, they can be modulated and transmit a data signal. Higher bandwidths compared to non-optical wireless applications will offer high-speed up and downloads not only in transportation systems but also in offices, labs, and private homes. Finally, non-optical techniques should be mentioned. WLAN and even radar systems can be used in automotive applications, for example, to guarantee greater safety in limited visibility situations like heavy rain, fog or snow. Moreover, combinations of optical and microwave solutions like radio over fibre (RoF) have to be considered as well.

Also non-electromagnetic microwave physical layers, like ultrasonic techniques, must be considered, as they are necessary to give the driver information about physical objects in close proximity to a car. This application is extremely useful in assisting manual parking or even automated parking as well as in dangerous situations, for example, when moving slowly through a crowded pedestrian area.

Automotive manufacturers are also interested in ready-to-integrate system solutions which offer many options, enabling a cost-effective mass production set-up adapted to the particular application. The complete product chain will be available: hard-ware components, low and high-level communication protocols implemented in soft-ware and/or hard-ware, conformance tests, measurements tools, simulations models, application development tools and, last but not least, standard soft-ware modules for ECUs. Their hardware and software interfaces will be flexible and defined clearly. Perfect compatibility is not usually required; the focus is on straight-forward portability approaches. Hardware flexibility in particular enables the finding of cost-effective solutions. However, system designers are challenged in supporting a proper development procedure; plug-and-play approaches are usually not available.

Chapter 1 mainly introduces lecture notes to give an overview of Optical Data Buses for Automotive Applications. The most important devices for fiber-optic transmission systems are presented, and their properties discussed. These systems can operate as transmission links with bit rates up to 40 Gbit/s. But communication systems are also used for recent application areas in the MBit/s region, e.g. in aviation, automobile and maritime industry. Therefore - besides pure glass fibers - polymer optical fibers (POF) and polymer-cladded silica (PCS) fibers have to be taken into account. Moreover, even different physical layers like optical wireless and visible light communication can be a solution solution as well as non-optical techniques or microwaves in Radio over Fiber (RoF) systems - even ultrasonic techniques (non-microwaves) are necessary to give the driver information about bodies surrounding the car at small distances.

In Chapter 2, the authors divide communication networks generally into four different classes: Ring (one directional or bidirectional), Star (active or passive), Tree and Bus (passive). These classes of communication networks are also used in transportation systems. The main properties of these networks are overviewed. Then the most important automotive networks are discussed: LIN, CAN, MOST, Ethernet, and Flexray, and their implementations in cars. For the MOST and Ethernet networks, different physical layer implementations are possible. The different factors determining the choice of a network are discussed.

Chapter 3 gives an overview of different cables used in automotive communication systems: Electrical cable types and optical cable types, unshielded Twisted Pairs (UTP), Coax cable and 1 mm PMMA POF. Graded Optic Fibers (GOF) and multimode PCS are in discussion. Moreover, electrical connectors used in LIN and CAN networks and the optical connectors standardized by MOST are overviewed. Furthermore, different fiber optic transceivers used for MOST, MOST25, and MOST150 transceivers are discussed.

The idea of Chapter 4 is to give a complete overview of a matrix approach to describe light propagation in strongly multimode fibers such as 1-mm diameter plastic optical fibers. These large core fibers accept such a huge number of travelling modes that they can be viewed as a continuum. Thus, light propagation can be described as a power flow by a differential equation that can be more easily solved using matrices. The key of this method is the propagation matrix that is calculated from the diffusion and attenuation function characteristics for a given fiber type. One of the most critical parameters when designing a network is its bandwidth and how it decreases when increasing the link length.

Chapter 5 presents modern vehicles being no longer imaginable without using technologies to broadcast local available data. The speed information for example is used by many well-known functions: anti blocking system, radio, dash board, cruise control, electronic stability program, etc. Usually these data are distributed among vehicle's electronic control units by various serial bus systems. The succeeding sections introduce the automotive communication system named FlexRay™. The development of FlexRay™ had been initialized by the requirements expected for drive-by-wire systems. The content is focused on its electrical physical layer. Comparisons to the state of the art system CAN are used to support the comprehensibility.

In Chapter 6 automotive bus systems like e.g. LIN, CAN, and FlexRay™ distribute their serial data streams among communication nodes. These nodes are interconnected by passive nets. Depending on the type of application some of these nets may consist of up to one hundred meter different bus cables arranged in various topologies. Common known transmission line theory focused of the physical effects being relevant for automotive serial bus communication in the time domain is discussed.

Chapter 7 reviews network topologies and technologies in automotive- and similar applications, and describe how optical wireless can be a viable technology to include. This is because today's vehicle consists of numerous interacting circuits, sensors, and many other electrical components. Thus communication is necessary amongst this circuits and functions of a vehicle. If all of electrical devices, sensors, switches, and motors in vehicles are gathered together, using current techniques, then the resulting amount of cabling and networks is considerable. Thus, optical networking provides a more effective method for complex in-vehicle communications.

In Chapter 8, the authors discuss resilient optical transport networks as backbone for future Internet protocol (IP) networks with enhanced quality of service (QoS). They avoid loss of data and revenue and provide acceptable service in the presence of failures and attacks. Moreover, this chapter presents the principles of designing survivable Dense-Wavelength-Division-Multiplexing (DWDM) optical transport networks including failure scenarios, survivability hierarchy, routing and wavelength assignment (RWA), demand matrix models, and implementation approaches.

In Chapter 9, the authors present an overview of Radio over Fiber (RoF) access networks for transportation with special attention on the figures of merit and the basic enabling technologies for downlink/uplink transmission. They review the up-conversion techniques to millimeter waves, followed by use of Base Station configurations. Finally, the authors apply these concepts to the development of a network proposal for in-vehicle wireless application.

The idea of Chapter 10 is to report on Vehicular Ad-Hoc Networks (VANETs). These networks are electrical wireless networks which are primarily meant to enforce vehicular safety. The incumbent international VANET solution is based on an adaptation of WLAN to the 5.9 GHz band and to the vehicular environment: it is universally known as IEEE 802.11p – a decentralized solution that is able to work in the absence of an infrastructure.

In Chapter 11, authors discuss that communication in transportation systems not only involves the communication inside a vehicle, train or airplane but it also includes the transfer of data to and from the transportation system or between devices belonging to that system. Fundamentals of mobile communication networks are described as well as possible future candidate networks. Moreover, also hybrid networks are discussed. The devices may use the best suited network for a given situation but also change to another network while continuing the on-going connection or data transfer. Here the design of the handover or relocation plays a critical role as well as localization. Possible system layouts and handover implementations are given.

Chapter 12 deals with vehicular networks as hybrid networks, which consist of a cooperation of different radio access networks. Seamless communication in vehicular networks relies on proper network planning and thorough dimensioning of network protocols. In transportation networks the location of the mobile devices and their pattern of movement are very important. Therefore, different mobility models suited to vehicles in transportation networks are introduced. As an example a hybrid handover protocol is given in detail. It provides low handover latency with additional mutual authentication to allow the transfer from one radio access network to another while maintaining the network's built-in security standard.

One of the main parameters for providing traffic safety of vehicles is the knowledge of their speed. In Chapter 13, the authors present results of a microwave radiometric correlation sensor for speed measurements of land vehicles. The nature and statistical characteristics of radio thermal radiation of a terrestrial surface and objects are considered. Concerning the influence of parameters of the antenna system and the linear path of the receiver on parameters of the signal formed at the output of the correlator, a statistical analysis of the radiometric system of correlation type is carried out. An optimization of antenna system and radiometric receiver parameters for various types of objects observed on a terrestrial surface has been performed, too.

The idea of Chapter 14 is to give an overview of ultrasound technology to transmit data at short ranges, The advances that have made this a useful technology include the ability to utilize a sufficiently wide bandwidth, and the availability of instrumentation that can send and receive ultrasonic signals in air. Various aspects such as attenuation, spatial characteristics, and the most suitable forms of modulation are discussed. Moreover, it is demonstrated that ultrasonic systems are a practical possibility for in-room communications.

Chapter 15 deals with driver assistance systems. They increase safety and comfort by supporting the driver in critical or stressful traffic situations. A great variety of surround sensors with different fields of view include radar, ultrasonic, laser, and vision systems. These sensors are based on different technologies and measurement principles. Hence, such sensors are used in various configurations to explore the surroundings ahead, sideways, and behind a vehicle. In addition, vehicle dynamics information from speed, steering angle, yaw rate, and acceleration sensors is available. The driver can be informed or warned in critical situations. Some driver assistance systems perform driving tasks like following, lane changing, or parking already autonomously. The art of designing valuable driver assistance systems includes many factors and aspects and is still an engineering challenge in automotive research.

This collection provides a wide-ranging overview of many extraordinary issues in communication systems for automotive application. The idea of this book is to address a broad scope of readers in order to give them an introduction to communication in transportation systems. For this reason, contributors not only present work on state-of-the-art methods, but promising techniques of the future are discussed as well. On the one hand, it is important that the key differences between optical and non-optical systems are appreciated; yet, on the other hand, similarities can be seen, too. Moreover, a combination of these different physical techniques might lead to excellent results which cannot be reached using them separately. Taking all these optical microwave and ultrasonic techniques as well as GPS in common together with a high-speed high-data processing device and software may be the old human dream of autonomous driving could be realized in a not too far future.

For readers not familiar with all these topics, there is coverage of many of the fundamentals, particularly in chapter 1. The book is intended to help undergraduate, graduate, and PhD students with basic knowledge of the subjects studying communication in transportation systems. In addition, R&D engineers in companies should also find this book interesting and useful. This is true for novices as well for experts checking certain facts or dealing with areas of expertise peripheral to their normal work.

Otto A. Strobel
Esslingen University of Applied Sciences, Germany
Autumn 2012

Chapter 1
Optical Communication in Transportation Systems including Related Microwave Issues

Otto Strobel
Esslingen University of Applied Sciences, Germany

Jan Lubkoll
Friedrich-Alexander University Erlangen-Nuremberg, Germany

Daniel Seibl
Esslingen University of Applied Sciences, Germany

ABSTRACT

The main idea of these lecture notes is to give an overview of Optical Data Buses for Automotive Application. The most important devices for fiber-optic transmission systems are presented, and their properties discussed. In particular, we consider such systems working with those basic components which are necessary to explain the principle of operation. Among them is the optical transmitter, consisting of a light source, typically a low speed LED or in case of higher demands a high speed driven laser diode. Furthermore, the optical receiver has to be mentioned. It consists of a photodiode and depending on the demands a low performance receiver circuit or a low noise high bit rate front-end amplifier. Yet, in the focus of the considerations, the optical fiber is the dominant element in optical communication systems.

INTRODUCTION

In this chapter, different fiber types are presented, and their properties explained. The joint action of these three basic components can lead to fiber-optic systems, mainly applied for data communication. These systems can operate as transmission links with bit rates of 40 Gbit/s and more. But communication systems are also used for recent application areas in the MBit/s region, e.g. in aviation, automobile, and maritime industry. Therefore - besides pure glass fibers - polymer optical fibers (POF) and polymer-cladded silica (PCS) fibers have to be taken into account. More-

DOI: 10.4018/978-1-4666-2976-9.ch001

over, even different physical layers like optical wireless and visible light communication can be a solution as well as non-optical techniques or microwaves in Radio over Fiber (RoF) systems. Just for completion also non-microwave solutions, ultrasonic techniques are necessary to give the driver information about bodies surrounding the car at small distances.

Since the beginning of the sixties, there has been a light source which yields a completely different behavior compared to the sources we had before: This light source is the LASER. The first realized laser was the bulk-optic ruby laser (Maiman 1960). Only a short time after this very important achievement, diode lasers for usage as optical transmitters had already been developed (see Figure 1) (Quist 1962). Parallel to that accomplishment in the early seventies, researchers and engineers accomplished the first optical glass fiber with sufficient low attenuation to transmit electromagnetic waves in the near infrared region (Kapron 1970).

The photodiode as detector already worked (Adams & Day 1876), and thus, systems using optoelectric (O/E) and electrooptic (E/O) components for transmitters and receivers as well as a fiber in the center of the arrangement could be developed. The main fields of application of such systems are found in the area of fiber-optic transmission and fiber-optic sensors (see Figure 2).

However, the first optical transmission is much older. Native Americans, for instance, already knew communication by smoke signals a long time ago (see Figure 3, Marstaller, personal communication 1990). Furthermore, it was a very sophisticated and modern system because it already was a digital system, consisting of "binary 1" and "binary 0" (smoke/no smoke).

Charles Kao (Kao & Hockham 1966) and Manfred Börner (Börner 1967) can be regarded as the inventors of fiber-optic transmission systems. Nowadays, their invention would not be very spectacular: Take a light source as transmitter, an optical fiber as transmission medium, and a photodiode as detector! Yet, in 1966, it was a revolution because the attenuation of optical glass was in the order of 1000 dB/km, and therefore totally unrealistic for usage in practical systems. Today's fibers achieve attenuation below 0.2 dB/km, which means that after 100 km, there is still more than 1% of light at the end of the fiber. This low value of attenuation is one of the most attractive advantages of fiber-optic systems compared to conventional electrical ones (see Figure 4). In addition, low weight, small size, insensitivity to electromagnetic interference (EMI), electrical insulation, and low crosstalk must be mentioned. Apart from low attenuation, the enormous achievable bandwidth must be pointed out. That leads to a high transmission capacity in terms of the

Figure 1. Basic arrangement of a fiber-optic system

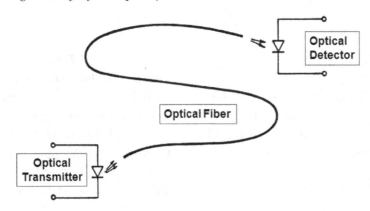

Figure 2. Application of fiber-optic systems

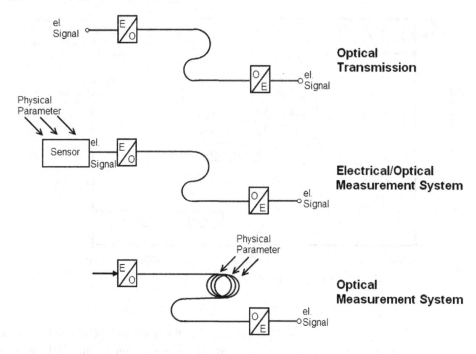

Figure 3. Digital optical transmission by use of smoke signals

product of fiber bandwidth and length. One of the most important goals is to maximize this product for every kind of data transmission. Figure 4 depicts the attenuation behavior. In particular, we observe independence from modulation frequency of fiber-optic systems in contrast to electrical ones, which suffer from the skin effect.

OPTICAL FIBERS

The most important demands on optical fibers are a proper wave guiding, low loss of optical power, and low distortion of the transmitted optical signals. The principle of operation of guiding a light wave can be explained by Snell's law (see Figure 5).

Figure 4. Attenuation of coaxial cables and optical fibers

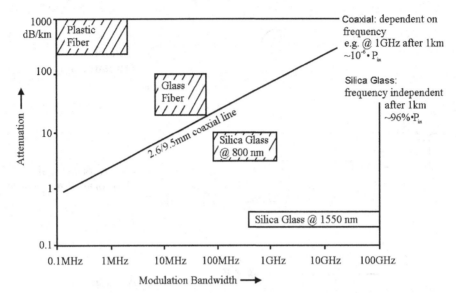

$$\frac{n_2}{n_1} = \frac{\sin\alpha}{\sin\beta} \quad (1)$$

If light is incident on an interface between two media with different refraction indices (n_1 and n_2), there is a reflected and a refracted ray in general. But for the special case that light is incident from a medium with higher refraction index ($n_2 > n_1$) as compared to the following one and furthermore the angle β exceeds a certain value (the cut-off angle β_c), there is no refraction anymore. We get reflection exclusively, the whole light is totally reflected; this effect is called "total internal reflection".

$$n_1 = n_2 \cdot \sin\beta_c = n_2 \cdot \cos\Theta \quad (2)$$

If this total internal reflection is repeated at a second interface, a waveguide is achieved (Miller et al. 1973). Figure 6 depicts this behavior. In particular, it has to be pointed out that there is no loss due to the multiple reflection because it is a total internal reflection; the coefficient $R = 1$ holds for every repeated reflection. Thus, the attenuation of the fiber is only due to losses inside the fiber.

The most important attenuation mechanisms are Rayleigh scattering and OH⁻ absorption. The scattering effect is due to inhomogeneities in the molecular structure of glass (silicon dioxide: SiO_2). Hence, statistical refraction index changes are caused. This leads to a scattering effect of the traveling light wave in the fiber, causing loss. The loss strongly depends on the wavelength of the light wave (scattered power P_S, see Figure 19). Lord Rayleigh discovered and explained that due

Figure 5. Refraction, reflection, and total internal reflection for light transition between two different media

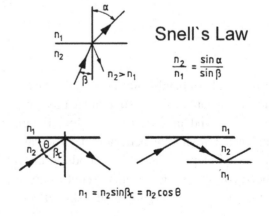

Figure 6. Total internal refraction and wave guiding

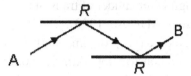

Wave is guided from A to B

to this effect, the color of the sky is blue. When we look at the sky, we see the scattered light of the white sunlight. Blue light is much more scattered than red. The same reason causes much higher losses in glass fibers for blue light than for red light (see Figure 19). Therefore, fiber-optic systems operate even beyond the red area, in the infrared (see below).

Figure 19 also depicts high attenuation peaks. These peaks are due to light absorption at undesired molecules in glass. The most important enemy in a fiber is water, which appears as OH⁻ ions in the silicon dioxide structure. The OH⁻ molecules are brought to oscillations by light waves. This effect is dominant in particular when resonance occurs at wavelengths which fit (see peaks). Hence, the energy of a light wave traveling in the fiber is absorbed, which leads to high attenuation. To achieve low fiber attenuation, the demand of purity is very high and the OH⁻ concentration must not exceed a value of 1 ppb. This is one of the reasons why it took a long time from the first idea of fiber transmission, in about 1966, to the first produced fiber in about 1972. Furthermore, it has to be mentioned that the fiber also suffers from SiO_2 self-absorption in the ultraviolet (UV) and infrared (IR) region, which in principle cannot be avoided. Whereas the UV absorption can be neglected compared to the much higher value caused by Rayleigh scattering, the IR absorption is responsible for the attenuation rise beyond 1600 nm (see Figure 19).

Besides the attenuation, there is a second cardinal problem concerning data transmission in optical fibers. Light rays in the fiber are not only traveling under one single angle. Figure 7 shows three representative existing rays (among hundreds or thousands). The existing rays are called "modes" in fibers. It is obvious that they do have different geometrical path lengths L. Yet, the determining effect for data transmission is not the geometrical but the optical path length:

$$g = n \cdot L, \qquad (3)$$

the product of the refraction index and the geometrical path length L. However, this optical path length differs for the three mentioned modes, too, because the refraction index in the fiber core is constant.

Therefore, an optical pulse travels along all the three paths in the fiber. The consequence of this is that they have different transit times and reach the fiber end at different arrival times. The three pulses superimpose and thus, we receive a broader output pulse as compared to the narrow input pulse (see Figure 7). The effect is called "pulse broadening". This behavior causes serious consequences. Since, if we want to transmit a high data rate, we have to place the second input pulse immediately after the first one. As a result, the pulses at the end of the fiber will overlap in such a manner that both pulses cannot be separated any longer. To avoid the overlap, it is necessary to place the second pulse with a greater distance

Figure 7. Pulse broadening by modal dispersion

Different transit times for different modes

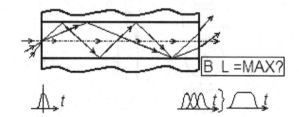

from the first one, which reduces the achievable bandwidth *B*. The second opportunity is to reduce the fiber length *L*. Both measures derogate the transmission capacity, the product of fiber bandwidth and length, the most important goal of every data transmission.

To avoid (reduce) this problem, scientists invented the graded-index fiber (see Figure 8). In contrast to the above described fiber, called step-index fiber, the refraction index is not any longer constant across the fiber (Gloge et al. 1973). The latter reveals a gradient behavior in the fiber core, whereas it still remains constant in the envelope, the cladding.

As a consequence, the optical path length $g = nL$ is now constant for each mode because in the fiber center, one can remark the shortest geometrical path length *L* and the highest refraction index *n*. In contrast to this result, one can find the longest geometrical path length linked with the lowest refraction index near the cladding. Thus, with a properly chosen index profile, we achieve a constant optical path length for all modes. At this stage, it has to be pointed out that it is not possible to achieve this goal completely. We only

obtain a good approximation and thus, there still is a certain modal dispersion left, resulting in a non-negligible remainder of transmission capacity reduction (Marcuse, 1979).

To overcome this problem, another invention was made, the construction of a monomode fiber: If we reduce the fiber core to a diameter below about ten micrometers, there will be only one ray, the ray along the optical axes transmitted, and the modal dispersion problem per se vanishes (Cohen et al. 1982). For very high data rates (about 40 Gbit/s), we have to confess that this disappearance is not completely correct due to polarization effects. Very accurate investigations lead to the result that there is a difference concerning the refraction index between two perpendicularly oriented axes in the fiber. This fact again results in different optical path lengths, and finally as described before, in the same process of pulse broadening; this dispersion is called "polarization mode dispersion" (PMD (Mahlke & Gössing 1987)).

Furthermore, another important dispersion has to be mentioned, the material dispersion (simplified chromatic dispersion) (Cohen et al. 1982). Due to the fact that there is no light source emitting at

Figure 8. Fiber types

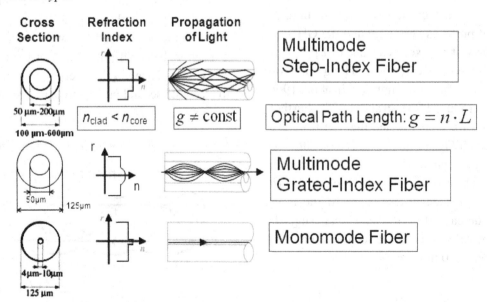

a single wavelength (see Figure 15), there is no monochromatic but always polychromatic light traveling through a fiber. Moreover, taking into account the dependence of the refraction index on the wavelength, it is obvious that we have always different refraction indices, and therefore, different optical path lengths according to different velocities of pulses traveling along the fiber.

Figure 9 shows three pulses having three representative wavelengths. They suffer from different transit times and reach the fiber end at different times of arrival.

The three pulses superimpose as described above for the modal dispersion process and thus, we again obtain a broader output pulse as compared to the narrow input pulse (see Figure 9). The result is the same as for modal dispersion, just the mechanism is different, and the effect is again pulse broadening and reduction of the transmission capacity.

Figure 10 visualizes the three common fiber types in comparison with a woman's hair. Table

1 shows an overview of fiber types depicting the most important fiber data. Furthermore, two more fiber types must be mentioned. There are also low cost applications concerning fiber-optic transmission, i.e. plastic fibers, polymer optical fibers, POF and PCS fibers, plastic cladding and silica core are used, too (see chapter 2 to 4). Their transmission capacity is much lower as compared to pure glass fibers in particular monomode fibers. However, there are applications for such fibers, e.g. if you have low data rates and only some ten meters of spacing. For example, in order to watch a machine tool in an EMI-relevant area, why not use a cheap plastic fiber transmission setup in the kHz-region?

OPTICAL SOURCES AND DETECTORS

The most important demands to optical sources are a high optical output power as well as a small

Figure 9. Pulse broadening by material dispersion

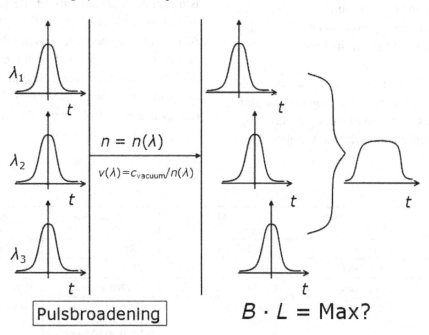

$n = n(\lambda)$

$v(\lambda) = c_{vacuum}/n(\lambda)$

Pulsbroadening

$B \cdot L = \text{Max?}$

(*n*: refractive index; λ: wavelength; c_{vacuum}: free space velocity; *v*: velocity in media)

Figure 10. Comparison of fiber types and a woman's hair

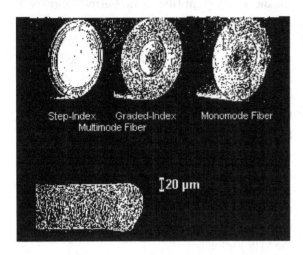

electrical input power. With regard to the fiber, the wavelengths should be in a proper range (see Figure 19). The spectral width has to be small, and for a sufficient coupling efficiency, the beam divergence should be low and the geometrical size should be small. Furthermore, a modulation capability of the injection current at high speed is favorable. To understand the principle of operation concerning optical sources and detectors, fundamental considerations about the interaction between photons and electrons have to be taken into account.

Figure 11 shows the energy band model of the semiconductor material, applied to the optical components. The lower level E_{v} is the energy level of the valence band, whereas the upper level E_{C} denotes the level of the conduction band. The difference ΔE between both levels is the energy gap E_{g}. There are three dominant effects to be considered:

From the electronic point of view, the sources and detectors are, simply put, p-n diodes following Shockley's well-known current-to-voltage characteristic curve (Burrus & Miller 1971, Panish 1976, Melchior et. al. 1970). LEDs and lasers are driven in forward direction, photodiodes in reverse voltage operation.

Absorption of Light

In single atoms, electrons may occupy only well defined "allowed" energy states. In solids, the electrons do occupy more or less broad allowed bands separated by "forbidden" gaps. A simple energy-band scheme of a semiconductor is depicted in Figure 11. At temperatures close to 0 K, the highest valence band is filled with electrons. The next higher allowed band, the so-called conduction band, is empty. The two bands are separated by an energy gap of width E_{g}. This gap is a fundamental quantity of the semiconductor and influences many properties of the material. At room temperature, a considerable number of electrons have been lifted from the valence band into the conduction band, leaving holes in the electron sea of the valence band. The necessary energy is gained from thermal movement.

Table 1. Fiber types

Type	Profile	Size	Attenuation	Bandwidth x Length
Plastic Fiber	Step Index	950/1000 µm	0,2 dB/m	< 100 MHz m
PCS Fiber	Step Index	100 – 600 µm	6 dB/km	< 10 MHz km
Multimode Glass	Step Index	> 100 µm	3 – 5 dB/km	20 Mhz km
Multimode Glass	Gradient Index	50/125 µm	2 dB/km (0.85 µm) 0,4 dB/km (1,3 µm) 0,2 dB/km (1,55 µm)	500 MHz km
Monomode Glass		5 – 10 µm		> 100 GHz km

Figure 11. Absorption and emission of photons

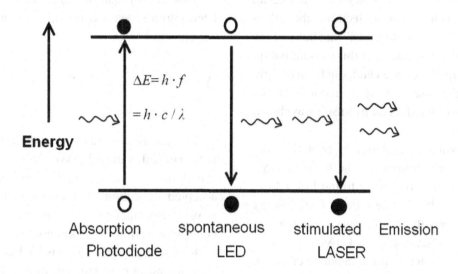

LASER: **L**ight **A**mplification by **S**timulated **E**mission of **R**adiation

A transition of electrons from the valence band into the conduction band can also take place if the required energy is supplied by radiation. According to Einstein's theory, radiation can be considered as a flow of optical quantum particles called "photons". Each photon carries a certain amount of energy which is proportional to the frequency ν of the electromagnetic wave and is given by

$$E_{\mathrm{ph}} = h\nu = h\frac{c}{\lambda}; \qquad (4)$$

h being Planck's constant, c the velocity of light, and λ the wavelength (Burrus & Miller 1971). It becomes clear that light can only be absorbed in a semiconductor, if the energy of the photons is larger than the band gap, i.e. $E_{ph} \geq E_g$. In other words: The wavelength of the light must be smaller than the wavelength λ_g that corresponds to the energy gap, E_g:

$$\lambda \leq \lambda_g = h\frac{c}{E_g}. \qquad (5)$$

If this condition for absorption is met, it appears that the optical power of the light wave Φ, is exponentially reduced while traveling through the crystal. If the power which is coupled into the crystal is denoted by Φ_0, the transmitted power that leaves a crystal of thickness d is given by

$$\Phi = \Phi_0 \exp\left(-\alpha\, d\right). \qquad (6)$$

α is called "absorption coefficient". From Equation (6), it follows that

$$\alpha = -\frac{1}{d}\ln\left(\frac{\Phi}{\Phi_0}\right). \qquad (7)$$

Emission of Light

If the conduction band is filled with more than an equilibrium distribution of electrons, then electrons can fall back spontaneously into holes of the valence band. In the course of this recombination process, energy is released either in form of light or heat. In the case of radiative recombination, one photon is emitted for each transition as shown in Figure 11.

The emission of radiation is related to the transitions of electrons from the higher energy level in the conduction band to the lower level in the valence band. The value of the energy gap E_g between these levels is distinctive for every semiconductor, e.g. for GaAs: $Eg \approx 1.43$ eV. The photon energy of the radiative emission is approximately the same as the energy of the forbidden gap:

$$E_{\text{ph}} \approx E_C - E_V = E_g \qquad (8)$$

Hence, the center wavelength of the radiation can be calculated using Equation (4):

$$\lambda \approx h \frac{c}{E_g}. \qquad (9)$$

The wavelength and the color of the emitted light consequently depend on the band gap of the semiconductor, which can be controlled by techniques of crystal growth.

There is no source which emits light at a single wavelength. It is impossible to realize pure monochromatic light, but there is always a polychromatic behavior, which can be understood as follows. The considered electrons and holes possess a thermal energy distribution with an average energy of $\frac{3}{2}kT$ each, where k is Boltzmann's constant and T the absolute temperature. Therefore, the emission bandwidth is expected to be approximately

$$\Delta E \approx 3kT \qquad (10)$$

This energy bandwidth ΔE can be recalculated into a spectral bandwidth $\Delta\lambda$, in terms of wavelengths:

$$\Delta\lambda = \frac{\lambda}{E_g} 3kT = \frac{\lambda^2}{hc} 3kT \qquad (11)$$

The quantitative description of the bandwidth is determined by the FWHM definition (full width at half maximum). The application of a generally accepted definition by international standards is very important to make results comparable. Otherwise, serious mistakes could occur.

Radiative recombination is the basic underlying principle of light-emitting diodes (LEDs) and semiconductor lasers. An LED consists of a p-n junction, where in the n-doped region, there is an excess of electrons in the conduction band whereas in the p-doped region, an excess of holes in the valence band is registered. By applying a forward current, electrons will flow from the n-type to the neighboring p-type region and holes vice versa. This process allows a huge recombination rate of electrons and holes and therefore leads to the emission of light. The energy difference ΔE is converted into a photon. In particular, the process of a following electron transition causing a second photon has nothing to do with generation of the first photon, i.e. there is a statistical behavior, named "spontaneous emission". This process occurs in an LED (Burrus & Miller 1971, Panish 1976).

In contrast to this process, the stimulated emission is completely different. The electron transition from conduction to valence band is not any longer a statistical process, but an induced or stimulated one. This process is stimulated by an already existing photon, e.g. produced by the spontaneous process. The two photons are not any longer strangers. They know each other, they are coherent, and we receive a coherent radiation caused by stimulated emission. This is the desired behavior in a laser. The two photons, now and again, cause new transitions and multiply themselves. To ensure to have enough electrons in the conduction band, the so-called first laser condition has to be fulfilled - a population inversion. Naturally, only few electrons are in the conduction band due to room temperature's heat. Thus, we need a "pumping" of electrons from the valence band to the conduction band. This can be done by optical means (first part of Figure 11). A diode laser pumps by the injection current like in the LED. The second laser condition is the necessity of an optical resonator. The light is multiply reflected by two partial reflecting mirrors. As we use GaAs or InP as semiconductor materials, the refraction index is about 3.5. Due to the fact that if light transmits between media with different refraction indices, there is not only refraction but also a reflection, the FRESNEL-Reflection (see Figure 5):

$$R = \left(\frac{n_{\text{GaAs}} - n_{\text{Air}}}{n_{\text{GaAs}} + n_{\text{Air}}} \right)^2 \approx 31 \qquad (12)$$

Thus, a reflection with a reflection coefficient R is inherently achieved by the front and the rear side of the laser chip without building an extra mirror (see Figure 13).

Figure 12. Optical power versus injection current

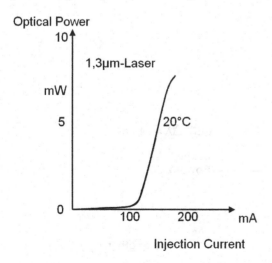

Injection Current

Figure 13. Schematic arrangement of a semiconductor laser

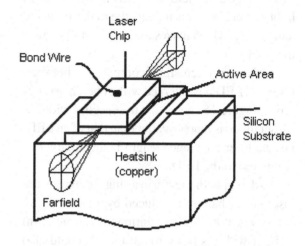

Due to the multiple reflecting process, an avalanche is produced; we receive a large amplification after exceeding a certain threshold (see Figure 12), we receive a LASER: Light Amplification by Stimulated Emission of Radiation (Panish 1976).

Figure 12 depicts the optical power versus the injection current of a semiconductor laser. Below the threshold, the laser operates as an LED. After exceeding a certain injection current, the threshold current, stimulated emission takes place. Unfortunately, there is a strong dependence of the threshold current on temperature. Hence, additional measures such as temperature control or control by a monitor diode are necessary in order to stabilize the optical output power.

Figure 13 shows the far-field distribution of a typical Fabry-Perot semiconductor laser.

Unfortunately, semiconductor lasers show beam divergence and astigmatism, whereby the divergence is different in two perpendicular directions. Both effects make the coupling into a fiber - in particular in a monomode fiber - difficult. The laser chip is mounted upside down on a silicon substrate. This mounting enables a close contact of the active area to the heat sink. The chip size is about 300 μm long, 200 μm broad and 100 μm high, the active area is in the order of 1 μm². Figure 14 visualizes a comparison between LED and laser farfield. The LED is a Lambertian light source, and thus it emits the radiation in the half sphere, which makes the fiber coupling even more difficult.

A further interesting comparison between laser and LED is concerned with coherence properties. Figure 15 shows that the spectral width of a laser diode is much smaller than that of an LED, i.e. the laser coherence length is much larger as compared to the LED.

To depict both spectra in a single diagram, the laser power had to be reduced by a factor of 50. Moreover, it has to be mentioned that the 2 nm of laser width is just a rough idea. It could also

be smaller by several decades. This results in a much better behavior concerning the material dispersion of a fiber. Thus, optical power, farfield behavior and spectral size enable the laser much more for highly sophisticated optical transmission systems, and hence exclusively lasers are used for high-speed long distance systems. In contrast to the laser, LEDs are used for low-cost, low-bit-rate and low-distance systems.

The most important demands on optical detectors (Melchior et. al. 1970, Pearsall 1981) are high sensitivity, low noise, linearity (for analog systems only), and small geometrical size. The most famous components are pin photodiodes (see Figure 16 and 17) and avalanche photodiodes (APDs). All

Figure 14. Farfield characteristics of LED and laser

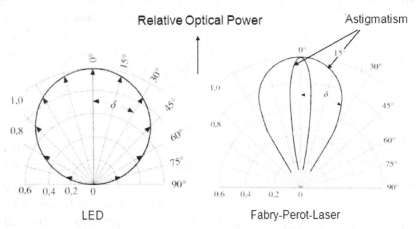

Figure 15. Spectral width of LED and laser

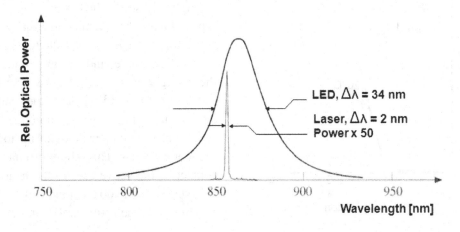

Figure 16. Photodiodes for optical systems

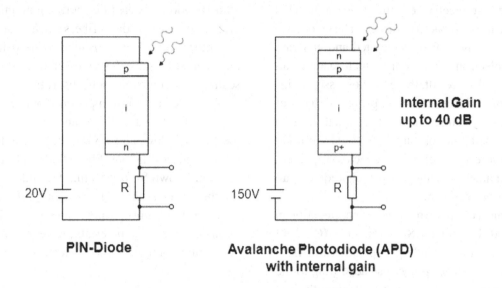

Figure 17. Schematic structure of a Ge-APD (Ebbinghaus et al. 1985)

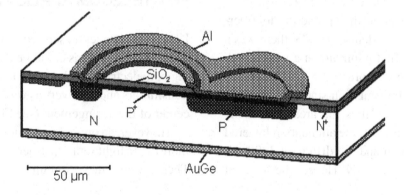

photodiodes for transmission systems are used in reverse voltage operation. This operation applies an electric field to the semiconductor material and thus an electron produced by photon absorption is experiencing a force and will be accelerated. This effect is even enlarged by introducing an intrinsic zone into a p-n diode, which results in a pin diode. This design guarantees a constant and high electric field over the whole absorption zone, whereas a simple p-n diode has only a maximum directly at the p-n junction, which leaves most part of the absorption area in a low electric field.

Therefore, the pin diode is able to operate as a high-speed component.

Figure 16 depicts an APD. This component has a highly doped p^+ zone, except for the p-n junction and the intrinsic zone. In this case, we receive a very high field at the p-n junction, besides the constant electric field in the i zone (see also voltages for comparison).

This arrangement causes such a high acceleration of charge carriers that collisions produced by the absorption process occur between the electrons and the atoms of the semiconductor material. Thus, further electrons will be freed from the atoms (lifted from the valence band into the conduction band) and we receive secondary electrons. This effect is called "impact ionization". The secondary electrons are now also accelerated and generate tertiary electrons and so on, which leads to an avalanche. As a result, the APD possesses an internal gain and thus, it is a very proper component for optical systems. The small geometrical size enables small junction capacities (see Figure 17) and therefore, high cut-off frequencies are gained. Typical diameters of high-bit-rate photodiodes are in the order or 50 μm.

The most common materials for application in photodiodes are silicon (Si), germanium (Ge), and gallium indium arsenide phosphate (GaInAsP). Figure 18 shows the proper use according to the relative responsivity versus wavelength (see also Figure 19).

Figure 19 shows a summary of the most important components and their properties for fiber-optic transmission systems: In 1973, there was a fiber featuring a first minimum at a wavelength of about 850 nm in the order of 5 dB/km. Therefore, fiber-optic transmission started at the area of this wavelength and thus, this area is called the "first window". In 1981, the attenuation lowered to about 0.5 dB/km and 0.3 dB/km at 1300 nm and 1550 nm respectively. Hence, the areas at these wavelengths are called the "second" and the "third window" where fiber-optic communication systems operate: Today's fibers reach an absolute minimum of 0.176 dB/km due to principle physical effects as described above: Rayleigh scattering and infrared self-absorption.

As optoelectronic components for light sources, we apply GaAlAs LEDs and laser diodes for the first window, and InGaAsP devices for the second and the third one. Photodiode materials are the well-known Si for 850 nm, Ge and InGaAsP for the wavelength range of about 1200 nm to over 1600 nm. Furthermore, mercury cadmium telluride (HgCdTe) materials are very promising compounds for future optical detectors (Lee 2001).

FIBER-OPTIC COMMUNICATION FOR TELECOM APPLICATIONS

Using the devices described above, fiber-optic transmission systems could be developed applying optoelectric and electrooptic components for transmitters and receivers as well as a fiber in the center of the arrangement (see Figure 1).

However, an optical communication system is more than a light source, a fiber, and a photodiode. There is a laser driver circuit necessary to provide

Figure 18. Spectral responsivity of photodiodes

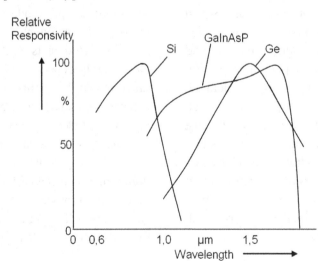

Figure 19. Spectral attenuation of optical fibers and useful wavelengths of optoelectronic devices

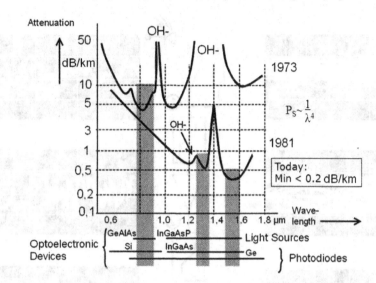

Figure 20. Optical fiber transmission principle

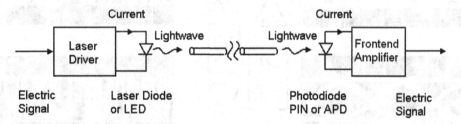

a proper high-bit-rate electric signal; this driver, combined with a laser or an LED, builds the optical transmitter. As well, the photodiode (pin or APD), together with the front-end amplifier, forms the optical detector, also called "optical receiver" (see Figure 20).

This front-end amplifier consists of a very highly sophisticated electric circuit. It has to detect a high bandwidth operating with very few photons due to a large fiber length and it is struggling with a variety of noise generators.

However, if the desired link length cannot be realized, a repeater consisting of a front-end ampli-

Figure 21. Transmitter, repeater and receiver

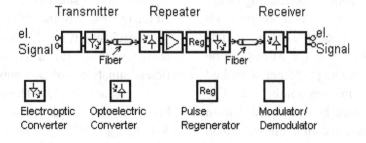

Figure 22. Eye pattern and data signal (N. Kaiser, personal communication, 1984)

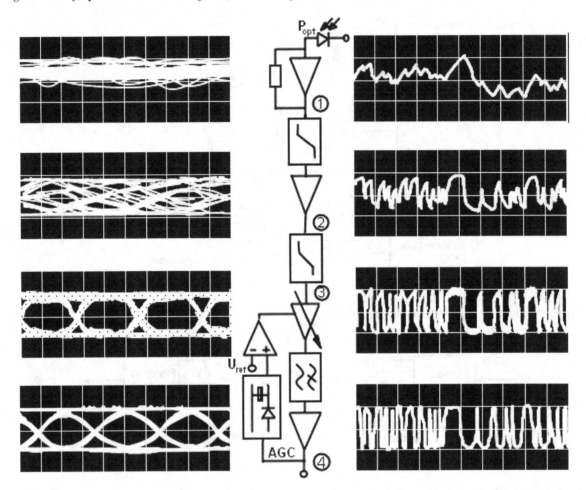

fier and a pulse regenerator will be inserted (see Figure 21). This pulse regenerator is necessary to restore the data signal before it is fed to a further laser driver followed by another laser.

Figure 22 visualizes the immense capability of the data regeneration: Directly at the front-end amplifier, (1) the eye pattern and the data signal cannot be detected as those. After a first following equalizer circuit, eye pattern and data signal are hardly recognized (2), whereas both are quite well restored after a second equalizer step (3). The non-return to zero signals at 168 Mbit/s can be seen clearly. Finally, a low pass filter is applied to suppress very high frequency noise (4).

At this point, it must be mentioned that an optical communication system is still more than

discussed in this chapter: There are further electric circuits to be taken into account, such as circuits for coding, scrambling, error correction, clock extraction, temperature power-level, and gain controls (Drullmann & Kammerer 1980).

Furthermore, until now, we have described a unidirectional system exclusively (see Figure 23), i.e. we think of a telephone link at which a person at one side of the link is able to speak. At the other side of the link a second person can listen exclusively, but the system does not operate the other way round. To overcome this insufficient situation, optical couplers on both sides of the link are inserted. Therefore, we achieve a bi-directional system (Köster 1983, Fußgänger & Roßberg 1990). The two counter propagating

Figure 23. Variety of optical transmission system

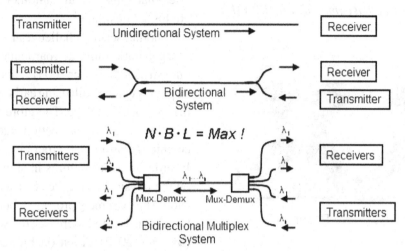

optical waves superimpose undisturbed, they separate at the optical couplers on the other side of the link and reach the according receivers. To improve the transmission capacity drastically, wavelength selective couplers are applied, called "multiplexers" and "demultiplexers". Several laser diodes operating at different wavelengths are used as transmitters; their light waves are combined by the multiplexer and on the other link end separated by the demultiplexer.

The set-up is named "wavelength division multiplex system" (WDM). If we apply this arrangement again in the two counter propagating directions, we achieve a bi-directional WDM system (Fußgänger & Roßberg 1990). The transmission capacity is risen by the number N of the channels transmitted over one single fiber.

Figure 23 depicts a scheme to describe the limits of optical transmission systems. For a single channel system, two basic limitations occur. Such systems are called "direct detection systems", consisting of one laser, one fiber, and one detector. The limitations are divided into two groups; the systems suffer from attenuation limitation and dispersion limitation: The attenuation-limited arm is governed by the transmitter power of the applied laser diode, the fiber attenuation, and the receiver sensitivity of the detector. The dispersion-limited arm is governed by the modulation bandwidth of the applied laser diode, the fiber dispersions, and the demodulation bandwidth of the detector.

Thus, for high bit rate long distance transmission systems, exclusively high-speed lasers and photodiodes will be installed as well as a monomode fiber. Figure 25 shows the eye pattern of a 43 Gbit/s data signal transmitted over a single-channel high-bit-rate system. The data rate corresponds to the cut-off frequency of about 30 GHz, which is approximately the highest frequency a single laser diode can be modulated.

Figure 24. Limits of optical transmission systems

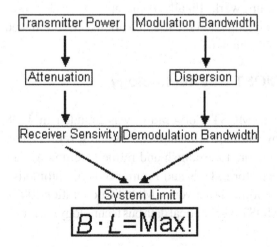

Figure 25. Eye pattern of a high-bit-rate data signal (B. Wedding, personal communication, 2001)

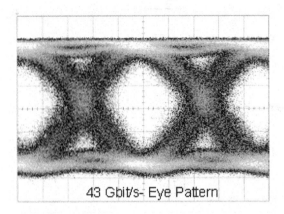

43 Gbit/s- Eye Pattern

COMMUNICATION IN AUTOMOTIVE SYSTEMS

Optical data buses in vehicles are almost exclusively used for infotainment (information and entertainment) applications. The Media Oriented Systems Transport (MOST) (MOST Cooperation 2007) is the optical data bus technology currently used in cars with a data rate up to 150 MBit/s. The development of infotainment applications in cars began with a radio and simple loudspeakers. Today's infotainment systems in cars include but are not limited to ingenious sound systems, DVD-changers, amplifiers, navigation, and video functions. Voice input and Bluetooth interfaces complement these packages. Important and basic logical links of these single components are already well-known from a simple car radio. Everybody probably knows the rise of volume in case of road traffic announcements.

However, the integration of more and more multimedia and telematic devices in vehicles led to a large increase in traffic demand - in particular for luxurious classes. A huge need for network capacity and higher complexity by integration of various applications has to be taken into account. Although MOST is the optical data bus technology currently used, alternative solutions for higher data rates that satisfy future automotive applications are highly desirable.

Another serious challenge arises in protecting new generation aircrafts, particularly against lightning strikes (Majkner 2003). This is because new airplanes will be built using carbon fiber to reduce the weight of fuselage. Therefore, these airplanes will lose a lot of protection against lightning, cosmic radiation, and other electrostatic effects. In order to avoid failures in signal transmission on the physical layer, the electrical copper wires should be extremely protected, but this solution is too expensive and increases the weight of cables (Majkner 2003). A reasonable solution is to use glass or plastic fibers as transmission medium in new airplanes. Since the FlexRay bus protocol (FlexRay Consortium 2008, also see chapter 6 and 7) is more adequate for avionic applications, it should be adapted for this transmission medium. Thus, this solution is cost-efficient and offers more safety in the aviation domain.

In this subchapter, we propose an improvement for optical data bus systems that may satisfy the requirements of future automotive applications and safety-relevant operations. First, we give an overview of MOST bus systems. Then, we present the challenges of data transmission that arise in new aircraft generations. After that, we propose two alternative solutions for optical data buses in avionic systems. Subsequently, we discuss the prototyping results and present open directions for future work. Finally, some optical wireless (see chapter 7) and radar applications are introduced (see chapter 13).

MOST Bus Technology

The MOST Corporation was founded in 1998 by automotive manufacturers and several system vendors to establish and refine a common standard for today's and future needs of automotive multimedia networks (MOST Cooperation 2007). MOST is the optical data bus technology currently used in cars with a data rate up to 150 MBit/s. This bus technology offers not only a synchronous transmission for audio and video data, but also makes available the application framework for controlling the system complexity. In particular, MOST specifies interfaces and functions for infotainment applications at a high abstraction level. As shown in Figure 26, different multimedia components can be connected in a ring topology. Furthermore, Bluetooth can be used for wireless devices and diagnosis interfaces (Grzemba 2007).

MOST technology has been developed over three generations. The first generation (MOST25) is based on a bit rate of 25 Mbit/s using an optical physical layer. A data rate of 22.58 Mbit/s can be achieved with a frame length of 512 bit. MOST25 uses a 1-mm-Polymethylmethacrylat fiber (PMMA), often called POF, as data transmission medium (see chapter 2 to 4). An LED is used as transmitter to convert the electrical signal to an optical one, using a driver circuit. The receiver converts the optical signal into an electric current using a Si-photodiode. As shown in Figure 27,

transmitter and receiver are combined in a device, called Fiber-Optic Transceiver (FOT).

The second generation (MOST50) is specified with a frame length of 1024 bit and the signal is transmitted by means of an electrical physical layer. The third generation (MOST150) was specified to satisfy large traffic demands. It is based on an optical physical layer and a data rate of up to 150 Mbit/s.

In particular, MOST specifies the physical connection between two neighboring components or Network Interface Controllers (NICs) by four specification points – SP1 to SP4 – as shown in Figure 28. These specifications are an essential criterion for the proper function of the physical layer and therefore for the whole functionality of the MOST network. SP1 and SP4 specify the interface between NICs in each subscriber. In particular, they describe the electric parameters of converters. An Electric-Optical Converter (EOC) and an Optical-Electric Converter (OEC) are used on the transmitter and receiver side respectively. SP2 and SP3 specify the connection between the device plugs and the optical fiber.

Figure 26. MOST network with ring topology (MOST Cooperation 2007)

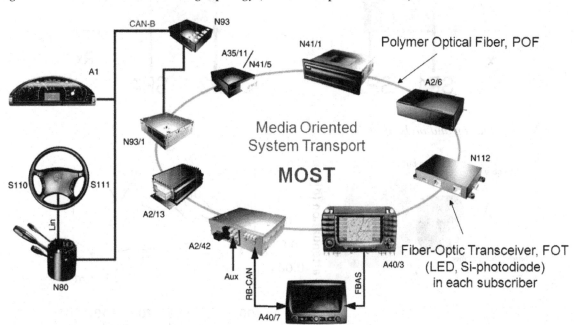

Figure 27. MOST physical layer components (MBtech Group 2008)

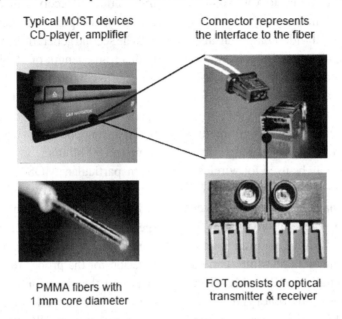

Figure 28. Location of specification points along a point-to-point link between neighboring FOT

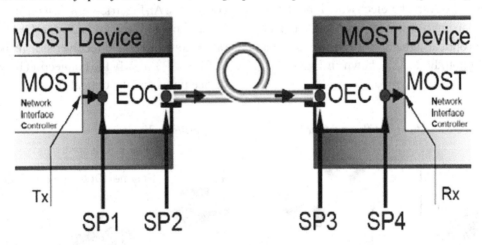

Figure 29. Attenuation and body of PCS fiber

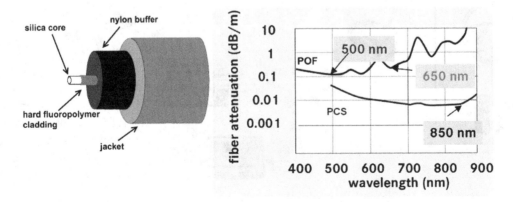

As mentioned above, the PMMA fiber is used as transmission medium for MOST bus systems. Figure 29 shows the fiber attenuation curve of PMMA fibers. Although the attenuation is lower at a wavelength of 500 nm, MOST components operate at a wavelength of 650 nm because the power of 650 nm LEDs is much higher. Taking into account the required temperature range from -40 °C to 85 °C, the wavelength of the LED varies and therefore, the worst case fiber attenuation is approximately 0.4 dB/m (Kibler et al. 2004).

However, using LEDs with 650 nm wavelength, a 1-mm-diameter POF and a large Si photodiode are sufficient for present typical traffic demands in automobile applications. In particular, short transmission distances of up to 10 m as well as a bandwidth of 100 MHz are adequate for most available automobile applications. FOT and POF are the most important components of the physical layer for MOST systems. This has been tested using a measurement setup at MB technology (Seibl et al.2008). Conventional POF/LED bus systems are capable of achieving 150 Mbit/s and will preliminarily remain the solution for cars. In particular, MOST150 is adequate for optical data transmission in cars, enabling the cross linking of onboard video cameras, laptops, GPS, and cell phones.

For higher data rates, alternative solutions are considered. Several advanced modulation techniques have been proposed recently that make this step feasible. Especially, by combining multi-carrier modulation with spectrally-efficient quadrature amplitude modulation (QAM), the first demonstration of 1 Gbit/s transmission over 100 m of SI-POF was reported (Lee et al. 2008). An alternative solution is replacing the LED as transmitter by a vertical-cavity surface-emitting laser (VCSEL) and the pure plastic fiber (POF) by a polymer-cladded silica (PCS) fiber (Strobel et al. 2007, Reiter & Schramm 2008, Lubkoll et al. 2009). Thus, the data rates can be extended into the GBit/s-region. As a result, this optical bus technology can also be used for sensor applications including safety-relevant operations like drive-by-wire, brake-by-wire, and engine management, and might finally lead to autonomous vehicle driving. If necessary also pure glass multimode or even single mode fibers would furthermore enhance the systems.

DATA BUSES IN AVIONIC SYSTEMS

In order to reduce the weight of new generation aircrafts, design engineers are going to use more and more carbon-fiber fuselage. Considered over the economic lifetime of an airplane, every saved kilogram affects a fuel economization of several thousand liters of kerosene. On one side, this technology may reduce the weight of an aircraft up to 30%, but on the other side, it introduces new safety problems and difficulties. A serious problem arises from the fact that many advantages of a closed metal fuselage get lost. An important advantage is the Faraday-cage-inherent lightning strike and cosmic radiation protection. Figure 30 shows the typical lightning strike propagation hitting an airplane. In this case, the current will propagate through the exterior skin. From Figure 30b, it can be seen that the charge channel of lightning hits the nose of the plane, travels along the skin, and leaves through the rear. In Figure 30c, the return bolt follows the charge channel. This could induce

Figure 30. Lightning strikes an airplane

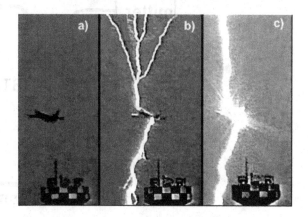

transients into wires or equipment that could be possibly disturbed, or totally damaged. However, these complications and failures can be avoided by system redundancy and special protection effects (R. Majkner 2003).

The situation would be fatal in an airplane without a completely closed Faraday cage. Lightning strikes could possibly take different paths through the plane and thus harm or even destroy several electrical components. These problems can be avoided by complex electrical protections, which cause higher costs and increase the weight of the cables. However, a reasonable and cost-effective protection method is the employment of optical wires as transmission medium used in avionic systems based on a mechanism for safety-critical systems applying a FlexRay-bus protocol (Lubkoll et al. 2009, see also chapter 5 and 6).

Application Prototypes

This section proposes a promising solution for optical data buses that can be used in avionic systems. Figure 31 shows a network where up to 8 nodes or components are interconnected by optical fibers in a star topology. Each node consists of two input/output ports, namely the optical transmitter and

optical receiver. For example, these nodes could be the landing airbrakes control units.

For simplification, all nodes apart from the transceiver are selected as optical passive/non-electrical components. Thus, the optical star offers the advantage to be used in two option modes: the first one with two fibers and a transmissive star and the second with one fiber and a reflexive star. The used data rate of the FlexRay protocol is about 10 MBit/s for a transmission distance of up to 100 m. The system should be stable between -45 °C and +85 °C. The star used in our setup comprises 8 nodes.

To satisfy the required specifications, a red LED is used as transmitter, and a step-index POF as transmission medium. Due to its lower price, robustness, and simple connector fitting, the POF seems to be suitable for avionic applications. The optical power at the receiver, including a margin of 3 dB, is about -50 dBm (Pfeiffer et al. 2001). This is due to the huge fiber attenuation caused by the link length of 100 m. The optical power budget of the POF is about 19 dB and the transmissive star has an attenuation of about 13 dB. Thus, the optical power at the receiver is too low to satisfy the necessary receiver sensitivity of -28 dBm. Hence, the standard LED/POF data bus technology cannot be used for this application.

Figure 31. Transmissive star topology

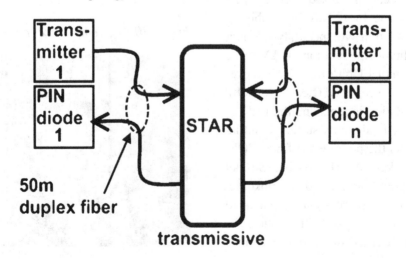

In order to reduce the total link power budget, the transmitter could be improved to reach a higher fiber input power. This would require a complex electrical circuit and special LEDs. Since a proper receiver with sufficient sensitivity is not available, the fiber and transmitter have been replaced by more sophisticated components.

Although standard silica fibers have a significant low attenuation, the POF is replaced by a 200 µm core step-index PCS fiber due to its robustness, easy connector fitting, and low price. As shown in

Figure 29, the PCS fiber combines the advantages of the POF and the glass fiber at the same time.

Since at a wavelength of 850 nm the PCS fiber attenuation is about 0.008 dB/m, the necessary link budget can be achieved. The LED is replaced by a VCSEL, which has a significantly smaller output beam divergence as shown in Figure 32. In addition, the VCSEL offers a low current consumption and a higher coupling efficiency. A light source at 850 nm also enables a gain of a higher receiver sensitivity. Figure 33 shows the receiver responsivity at 650 nm (0.47 A/W) and at 850 nm (0.63 A/W) respectively. Thus, the total link budget is 30 dB. Therefore, components with an attenuation of up to 27 dB (3 dB margin) can be used.

Based on these characteristics and advantages, we designed and implemented two prototyping solutions. The first one uses a transmissive star with two fibers (Figure 31). The second solution uses a reflexive star with a single fiber per transceiver. As shown in the upper part of Figure 34, an additional Y-coupler is used to connect transmitter and receiver. The total link budget achieved is shown in the lower part of Figure 34. It can be seen that the PCS fiber has a very low attenuation. Moreover, connectors can be easily fitted. Due to its smaller core of 200 µm, for the according length

Figure 32. VCSEL output beam (Hibbs-Brenner 2009)

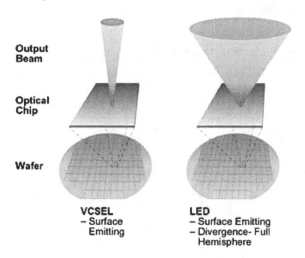

Figure 33. PIN diode responsivity

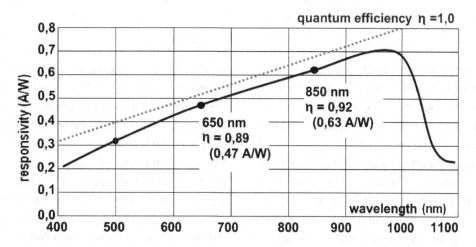

Figure 34. Optical power overview of PCS fiber

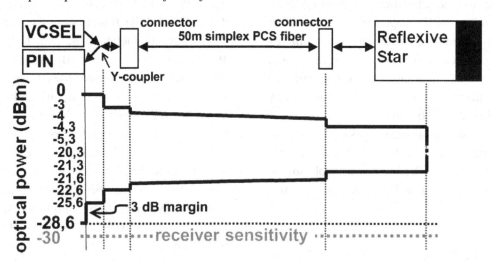

Figure 35. Demonstration of a FlexRay data communication

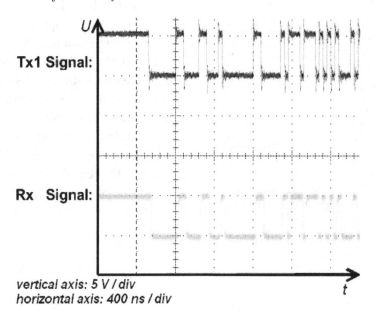

the PCS fiber provides a bandwidth significantly larger than 1 GHz, which enables the system for higher data rates. Because VCSELs and PIN diodes operate far below their limit, the distance and the number of nodes can be increased.

As shown in Figure 35 the digital input signal of the first node (Tx1) is transmitted, converted into an optical pulse, by the VCSEL. The receiver/PIN diode of the second node (Rx) hands the correct signal, reconverted into an electrical digital signal, to the front-end amplifier. Thus, the principal operational functionality for FlexRay application in the aviation domain has been demonstrated. Following investigations will have to deal with the quality description of the transmitted signal in terms of signal-to-noise ratio and bit-error rate. The upper trace shows the data signal from the driver sent to the VCSEL. Data rates up to 10 Mbit/s have been demonstrated. The lower trace shows the receiver signal.

OPTICAL WIRELESS APPLICATIONS

High-Speed Wireless LED Transmission

Optical wireless transmission offers several motivations (see chapter 7): There are no problems with electromagnetic interference (EMI), electromagnetic compatibility (EMC), and e-smog. Thus, robust use in RF-sensitive environments can be achieved. Simple shielding by opaque surfaces guarantees no crosstalk between rooms, inherent protected privacy, and insensitivity to remote sabotage. Basic configurations for wireless LED transmission are line-of-sight (LOS) and none-line-of-sight (NLOS) setups (Figure 36 and 37). Possible application areas are office labs (e.g. due to white LED panels), home area networks, mobile phones, or laptops (point-to-point data transmission for high-speed uploads/downloads), and many more (Langer, private communication 2010).

Figure 36. LOS wireless transmission

Figure 37. NLOS wireless transmission

 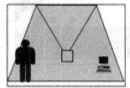

Optical Wireless Channels for Multimedia Distribution, Vehicle Control, and External Communication

An important aspect to be considered is where the optical wireless infrastructure can sit, or be used, instead of an optical fibre. Figure 38 depicts several scenarios to be investigated (Green 2010).

There might be circumstances where the use of a fibre might the better solution than optical wireless and vice versa. Doors are hollow structures, which would form a suitable enclosed space for light guiding. The engine compartment is again good as a reflective and containing space, so light can be bounced around. An important advantage is that inherently there are no problems due to temperature or petrochemical influences because it is a free space part of the communication link. Moreover light pipes might also include ducts along the body of the vehicle for air conditioning and heating.

Furthermore, the use of brake lights could provide two indications to a vehicle behind, by giving not only a visible warning of braking but also carry a warning message in the communication sense using brightness modulation, to a receiver connected to an audible warning system in the vehicle behind. The sills which form a main strengthening structure along the base of the doors also can be used as light guides. The internal cockpit of the car itself can be used as a communications space for multimedia entertainment, distributed by an optical antenna node on the dashboard and reflected from the windscreen and other surfaces. The use of optical wireless within doors and other spaces can be used to confirm that, before moving off, the doors are shut and all is satisfactory.

Figure 38. Possible optical wireless channels in a car

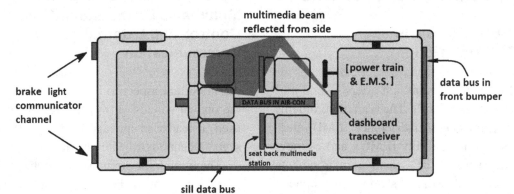

Idle-Stop Technologies Using Visible Light Communication

According to the Kyoto Protocol, the environment related automotive regulations have strengthened. The portion of renewable energy among total energy will be 1% (currently about 0.4%) in the year 2020 (Myunghee Son 2008). The necessity of energy-saving technology is growing stronger before developing alternative energy. Several lead-

ing automotive manufacturers have studied the idle-stop control device, which stops the engine while idling in traffic. To prevent unnecessary fuel consumption and exhausting emissions, the idle-stop-vehicle's engine is turned off when there is no need for propulsion or air conditioning.

Figure 39 shows a red-light-to-car communication. In this case, the red light tells a waiting driver that he will get a green sign in 50 seconds. Another example would be that a car is approach-

Figure 39. Idle-stop service scenario (Myunghee Son 2008)

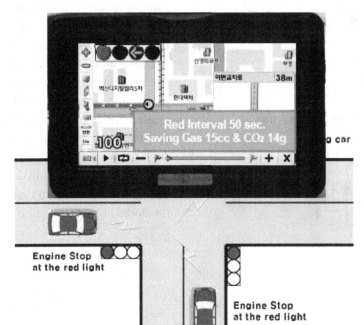

ing a red light and the driver gets the information 10 seconds before he can even see it. While the driver is waiting at the red light, the system gets the red interval from the traffic light in order to decide to stop the engine. The outcome would be a gas-mileage and CO_2 reduction of more than 5% (November 2008). Additionally, the driver does not have to fix his eyes on the traffic light permanently. These systems can predict idling intervals with accuracy to solve unnecessary engine stops and starts. Including information about the green, amber, and red time zones, the traffic volume can be regulated (Figure 40).

One of the huge advantages of traffic regulation by visible light communication is that not much new hardware is needed. LED signal lights as infrastructure are already widely provided and only the local controller would have to be updated. Additionally, receivers would have to be implemented in automobiles and the electronic control units would have to be updated. Also, car-to-car communication could be an interesting scenario. In particular safety relevant applications are of great interest (Iizuka 2008). The catch-word

is "Pre-crash safety" by VL-ISC: Visible Light Image Sensor Communication (Figure 41). Human reaction time is a problem in security, among difficult circumstances it will last up to 1.75 s. Assuming a vehicle riding with a speed of 100 km/h, it will take about 49 m before starting the breaking process. In the same situation in worst case the ICS reaction time is 0.2 s corresponding to about 5.6 m. A vehicle in front might suddenly start braking. Thus, depending on the brake pressure, immediately its stoplights will give information to a following vehicle. In critical cases then the following vehicle will automatically start emergency braking to avoid a collision or at least reduce the serious damage.

Moreover, Radar systems can be used to review conditions of limited optical visibility in case of fog, snow, rain, high smoke content, etc., (see chapter 13). Possible application areas are: control of movement of ground vehicles, search landing places for planes and helicopters and search for objects of natural and artificial origin on earth. Furthermore the system can be used for visual control or for other situations where

Figure 40. Traffic regulation by visible light communication (Myunghee Son 2008)

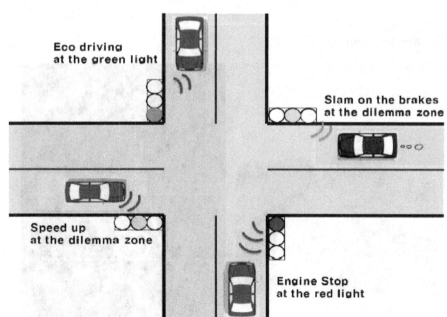

Figure 41. Car-to-car communication

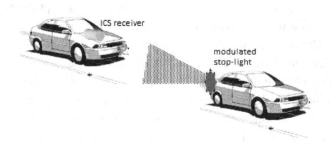

optical or IR-gauges are too difficult or impossible to use. Radar systems working in a resolved frequency range do not influence other radio-electronic devices. One practical application is the automobile radio vision system (ARVS) (Ananenkov et al. 2008). ARVS generates a radar-tracking image (RI) of road. This gives the driver an opportunity to observe precisely the road borders, cars, other subjects and obstacles within the limits of the working range even if there is no visibility (Figure 42).

Figure 43 shows a Radio over Fiber (RoF)-Systems building a bridge between fiber-optic and microwave techniques (Weinzierl 2010, see also chapter 9). Now we achieve an extension to 2 km instead of a repeater spacing of 400 m by using WLAN techniques exclusively.

CONCLUSION AND OUTLOOK

The main aim of this chapter was to give an introduction to optical communication with special regard on transportation systems. In particular fiber-optic transmission systems play an important role for this application. The reader should just be familiarized with basic components and fundamental optical techniques for such systems.

Figure 42. Radar measurement

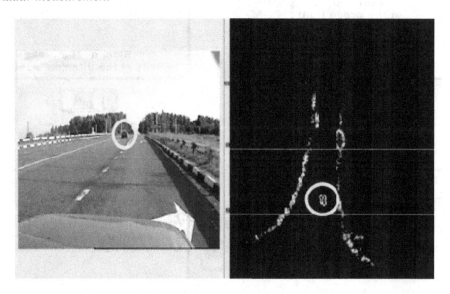

Figure 43. Trainguard MT Metro Shanghai RoF-WLAN Transmission System

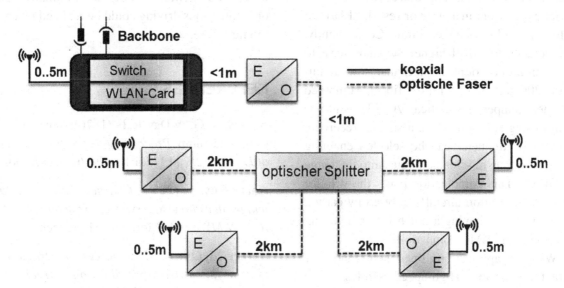

In general for more comprehensive considerations on fiber-optic communication, there is more to be dealt with (Strobel, to be published 2013), e.g. the optical amplifier for long distance systems (Payne et al. 1990, Flannery 2001, McCarthy 2001). Therefore also signal distortions have to be avoided (Kapron et al. 1970, Malyon et. al. 1991, Chbat 2000). Moreover besides point-to-point links optical LAN and MAN networks are of interest, too (Sykes 2001, Weiershausen et al. 2000, Pfeiffer et al. 2001). Also, the subject of opening up the last mile for fiber communication as well as in-house communication is of great interest. Maybe plastic fibers (Kenward 2001) like in the MOST ring for cars with sufficiently low attenuation together with LEDs could solve this problem in the near future (see chapter 2 to 4).

Concerning automotive application, we presented the state of the art and next-decade technologies for optical data buses in automotive applications. MOST is the optical data bus technology nowadays used in cars. MOST150 is the current standard with a bit rate of 150 Mbit/s and it is an adequate solution for optical multimedia data transmission in automobiles. The system consists of a ring structure; the MOST ring connects a variety of infotainment devices. It operates with a 1-mm-Polymer Optical Fiber, POF as data transmission medium (see chapter 2 to 4). An LED is used as transmitter to convert the electrical signal to an optical one. The receiver converts the optical signal into an electric current applying a Si-photodiode. Transmitter and receiver are combined in a device, called Fiber-Optic Transceiver (FOT). However, to provide the next steps taking into account more than infotainment application new bus systems with higher data rates are desirable. Safety relevant sensors and actuators like drive-by-wire and brake-by-wire will become products in the next years. There will be no steering wheel anymore - an intelligent joystick with a lot of more functions compared to a conventional steering will serve to guide a vehicle.

Additional challenges arise in new generation aircrafts. Due to safety problems in data transmission, an optical solution for data transmission is highly needed. Current standards using conventional LED/POF solutions operate with simple structures, like ring or multimedia point-to-point connections with a link length of about 10 m. For more complex or long-ranged structures like a 100 m optical network in a plane or vessel, LED/POF solutions could reach their limits. Consequently, components to fulfill higher demands have to be taken into account. In particular, low attenuation 200 μm diameter PCS-fibers, combined with less temperature-critical VCSELs could be a promising solution. Thus, a smaller receiving photodiode diameter can be selected enabling increasing transmission data rates into the range of GBit/s. This combination paves the way for the new generation aircrafts covered by carbon fiber fuselages, having a much better lightning protection and EMC compatibility.

Wireless applications open new fields for data transmission. High-speed wireless LED transmission offers short and middle range data transmission without EMI/EMC problems. Higher bandwidths than non-optical wireless applications will offer high-speed up-/downloads in office labs and private houses. Finally, non-optical techniques must be mentioned. WLAN (see chapter 10 to 12) and even RADAR systems (chapter 13) can be used in automotive applications, for example, to guarantee more safety in limited optical visibility situations. Nowadays, the user wants a complete solution for his demands and he does not ask if that solution is fiber optics or anything else. Moreover, also combinations of other physical techniques with or without fibers like Optical Wireless Communications, Optical and Non-Optical Solutions, or Microwaves in Radio over Fiber have to be developed. Just for completion also non-microwave solutions, ultrasonic techniques are necessary to give the driver information about bodies surrounding the car at small distances (see chapter 14 and 15). This application is very useful to help by manual parking or even by automated parking and in dangerous situations when moving slowly in pedestrian areas.

Taking all these optical microwave and ultrasonic techniques as well as GPS in common together with a high-speed high-data processing device and software may be the old human dream of autonomous driving could be realized in a not too far future.

REFERENCES

Adams, W. G., & Day, R. E. (1876). The action of light in selenium. *Proceedings of the Royal Society of London, 25*, 113. doi:10.1098/rspl.1876.0024

Ananenkov, A. (2008). *Characteristics of radar images in radio vision systems of the automobile. ICTON MW 2008*. Morocco: Marrakech.

Börner, M. (1967). *Mehrstufiges Übertragungssystem für Pulscodemodulation dargestellte Nachrichten*. (Patent DE1254513).

Burrus, C. A., & Miller, B. I. (1971). Small area double heterostructure AlGaAs electroluminescent diode sources for optical-fiber transmission lines. *Optics Communications, 4*(4), 307. doi:10.1016/0030-4018(71)90157-X

Chbat, M. W. (2000). Managing polarization mode dispersion. *Photonics Spectra*, 100.

Cohen, L. G. (1982). Dispersion and bandwidth spectra in single-mode fibers. *IEEE Journal of Quantiatitve Electronics, 18*(1), 49. doi:10.1109/JQE.1982.1071366

Cooperation, M. O. S. T. (2007). *MOST brand book vol. 1.1*, Aug. 2007. Retrieved December 7, 2009, from www.mostcooperation.com

Drullmann, R., & Kammerer, W. (1980). Leitungscodierung und betriebliche Überwachung bei regenerativen Lichtleitkabelübertragungssystemen. *Frequenz, 34*(2), 45. doi:10.1515/FREQ.1980.34.2.45

Ebbinghaus, G. (1985). Small area ion implanted p+n Germanium avalanche photodiodes for a wavelength of 1.3 μm. *Siemens Research and Development Report, 14*(6), 284.

Flannery, D. (2001). Raman amplifiers: Powering up for ultra-long-haul. *Fiber Systems, 5*(7), 48.

FlexRay Consortium. (2008). *Home page.* Retrieved August 7, 2009, from www.FlexRay.com

Fußgänger, K., & Roßberg, R. (1990). Uni and bidirectional 4λ x 560 Mbit/s transmission systems using WDM devices and wavelength-selective fused single-mode fiber couplers. *IEEE Journal on Selected Areas in Communications, 8*(6), 1032. doi:10.1109/49.57806

Gloge, D. (1973). Multimode theory of graded core fibers. *The Bell System Technical Journal, 52*, 1563.

Green, R. (2010). Optical wireless with application in automotives. *Proceedings of ICTON 2010*, Munich, Germany.

Grzemba, A. (2011). *MOST – The automotive multimedia network - From MOST25 to MOST150* (p. 21). Poing, Germany: Franzis Verlag.

Hibbs-Brenner, M. K. (2009). VCSEL technology for medical diagnostics and therapeutics. *Proceedings of the Society for Photo-Instrumentation Engineers, 7180*, 71800–71810. doi:10.1117/12.815307

Iizuka, N. (2008). *Some comments on merged draft from the viewpoint of the VL-ISC.* (IEEE 802.15-08-0759-02-0vlc, Project: IEEE P802.15 Working Group for Wireless Personal Area Networks (WPANs).

Kao, C. K., & Hockham, G. A. (1966). Dielectric-fiber surface waveguides for optical frequencies. *Proceedings of the IEE, 113*(7), 1151.

Kapron, F. P. (1970). Radiation losses in glass optical waveguides. *Applied Physics Letters, 17*(10), 423. doi:10.1063/1.1653255

Kenward, M. (2001). Plastic fiber homes in/on low-cost networks. *Fiber Systems, 5*(1), 35.

Kibler, T. (2004). Optical data buses for automotive applications. *Journal of Lightwave Technology, 22*, 2184–2199. doi:10.1109/JLT.2004.833784

Köster, W. (1983). Einfluss des Rückstreulichts auf die Nebensprechdämpfung in bidirektionalen Übertragungssystemen. *Frequenz, 37*(4), 87. doi:10.1515/FREQ.1983.37.4.87

Lee, S. C. J., et al. (2008). *Low-cost and robust 1 Gbit/s plastic optical fiber link based on light-emitting diode technology.* Optical Fiber Conference (OFC), San Diego, CA, USA

Lee, T. P. (2001). *Prospects and challenges of optoelectronic components in optical network systems.* Seminar on Internat. Exchange & Technology Co-operation, Sept. 22-24, Wuhan, China.

Lubkoll, J. (2009). Optical data bus technologies for automotive applications. *The Mediterranean Journal of Electronics and Communications, 5*(4), 127. ISSN 1744-2400

Mahlke, G., & Gössing, P. (1987). *Fiber optic cables* (p. 77). Berlin, Germany: Siemens AG.

Maiman, T. H. (1960). Optical and microwave-optical experiments in Ruby. *Physical Review Letters, 4*(11), 564. doi:10.1103/PhysRevLett.4.564

Majkner, R. (2003). *Overview - Lightning protection of aircraft and avionics.* Sikorsky Corp. Retrieved December 7, 2009, from http://ewh.ieee.org/r1/ct/aess/aess_events.html

Malyon, D. J. (1991). Demonstration of optical pulse propagation over 10 000 km of fiber using recirculating loop. *Electronics Letters, 27*(2), 120. doi:10.1049/el:19910080

Marcuse, D. (1979). Calculation of bandwidth from index profiles of optical fibers. *Theory Applied Optics, 18*(12), 2073. doi:10.1364/AO.18.002073

McCarthy, D. C. (2001). Growing by design. *Photonics Spectra*, 88.

Melchior, H. (1970). Photodetectors for optical communication systems. *Proceedings of the IEEE, 58*(10), 1466. doi:10.1109/PROC.1970.7972

Miller, S. E. (1973). Research toward optical-fiber transmission systems. *Proceedings of the IEEE, 61*(12), 1703. doi:10.1109/PROC.1973.9360

Panish, M. B. (1976). Heterostructure injection lasers. *Proceedings of the IEEE, 64*(10), 1512. doi:10.1109/PROC.1976.10367

Payne, D. N., et al. (1990). Fiber optical amplifiers. *Proceedings of OFC '90*, Tutorial, paper ThFl, (p. 335). San Francisco.

Pearsall, T. P. (1981). Photodetectors for optical communication. *Journal of Optical Communication, 2*(2), 42.

Pfeiffer, T. (2001). Optical packet transmission system for metropolitan and access networks with more than 400 channels. *Journal of Lightwave Technology, 18*(12), 1928. doi:10.1109/50.908792

Quist, T. M. (1962). Semiconductor Maser of GaAs. *Applied Physics Letters, 1*(4), 91. doi:10.1063/1.1753710

Reiter, R., & Schramm, A. (2008). *Verfügbarkeitsrisiko senken – Neue Physical-Layer-Spezifikation für MOST150*. WEKA Fachmedien GmbH.

Seibl, D. (2008). *Polymer-optical-fiber data bus technologies for MOST applications in vehicles. ICTON MW 2008*. Morocco: Marrakech.

Son, M., & Kang, T.-G. (2008). *Project: IEEE P802.15 working group for wireless personal area networks* (WPANs). (doc.: IEEE 802.15-08-0759-02-0vlc).

Strobel, O. (2007). *Optical data bus technologies for automotive applications. ICTON MW 2007, Sousse* (p. 1). Tunesia.

Strobel, O. A. (2013 forthcoming). *Optical communication and related microwave techniques*. Chichester, UK: John Wiley & Sons.

Sykes, E. (2001). Modelling sheds light on next-generation networks. *Fiber Systems, 5*(3), 58.

Weiershausen, W., et al. (2000). Realization of next generation dynamic WDM networks by advanced OADM design. *Proceedings of the European Conference on Networks and Optical Communication, 2000*, (p. 199).

Weinzierl, W., & Strobel, O. (2010). Simulation of throughput reduction of a WLAN over fibre system due to propagation constraints. *Proceedings of CriMiCo*, Sevastopol, Ukraine.

Chapter 2
Communication Systems in Automotive Systems

Piet De Pauw
BVBA De Pauw, Belgium

ABSTRACT

Communication networks in general, can be divided into 4 different classes: Ring (one directional or bidirectional), Star (active or passive), Tree, and Bus (passive). These 4 classes of communication networks are also used in transport systems. The main properties of these networks are overviewed. Then, the most important automotive networks are discussed: LIN, CAN, MOST, Ethernet, and Flexray, as well as their implementations in cars. For the MOST and Ethernet networks, different physical layer implementations are possible. The different factors determining the choice of a network are discussed. Cost is a major driver for automotive networks. For avionics networks, different standards for the network protocols and the physical layer implementations exist. In most cases physical layer implementations are proprietary, although, due to cost reduction pressure, more and more standardization is ongoing, and a tendency exist to adapt automotive network standards, and for the pressurized part of the airplane, also automotive physical layer implementations.

INTRODUCTION

Automotive Industries, 2005

In the past 30 years, the automotive industry has changed dramatically. The former, purely mechanical systems have been replaced by a combination of mechanical and electronic devices.

The main drivers for improvements in cars are:

- **Increase of Safety:** (Not only passive safety e.g. seat belts, but mainly by electronic

means: air bags, anti-lock braking system (ABS from German: Antiblockiersystem), Electronic stability control (ESC), Blind Spot Detection (BSD), Adaptive Cruise Control (ACC), Lane Change Assist (LCA).
- **Fuel Consumption Reduction:** Electronic Power Assisted Steering (EPAS), Replacement of belts by electronics, Lambdasonde, Motor Control Units (MCU).
- **Increase of Comfort:** MOST network.

Electronics is the key enabler for progress in the automotive industry. More than 90 percent

DOI: 10.4018/978-1-4666-2976-9.ch002

of the innovation in a car is in electronics, and electronic devices comprise more than 30 percent of the car manufacturing cost of today's luxury and mid-range cars.

Among the increasing amount of electronics in cars, data communication networks are penetrating into cars as well. Nowadays, an average car has 10 to 20 intelligent electronic devices.

Why networks have been developed? No network is needed, when devices are directly connected to each other. If there are N electronic devices in a car, the connecting each device with each other requires N x (N-1)/2 connections. Therefore, as the number of nodes increases linearly with time, the number of connections to be made increases quadratically, making networks more and more attractive. A bus network for N devices also only needs N connections. Therefore cost is the main driver for the development of automotive networks.

Because of the largely different requirements, four different categories of networks emerge: low speed, medium speed, high speed and safety critical networks.

One of the first electrical networks to be developed for cars is the controller area network (CAN), which was developed by Bosch in 1984 and made an appearance in high-end Mercedes cars in 1992. CAN has been standardized by the International Standards Organization as ISO 11898 for high and low-speed/fault-tolerant versions. High-speed CAN runs up to 1 Mbit/s and is used for engine control and power-train applications. Low-speed/fault-tolerant CAN reaches 125 Kb/s and is used for body and comfort devices. CAN is a very popular bus standard for the automotive industry, supporting in-vehicle communication and industrial automation applications.

As the data throughput reaches more than a few megabits per second, EMI issues with the current electrical automotive buses become the main concern. In addition, the introduction of hybrid cars and electrical cars has obliged car manufacturers to significantly increase the Radiative immunity

requirements from 100 to 150 V/m for cars with combustion engines to 600 V/m for hybrid and electrical vehicles.

Since the applications connected to each node of the network require electrical signals, optical networks (Optical Physical Layer or oPHY) always require electrical to optical and optical to electrical conversion. These conversions add to the cost of optical networks. On the other hand, optical connections are fully immune to electromagnetic disturbances from outside.

Although electrical networks (Electrical Physical Layer or ePHY) have difficulties coping with EMI issues, they do not need the conversions between electrical and optical signals.

This situation makes that in automotive network there is always a competition between implementations of the network as an electrical physical layer, or as an optical physical layer. The precise values of the EMI specs play an important role in the decision which physical implementation of the network to choose.

The number of network nodes is increasing year after year. In 2012 about 1.4 billion nodes are expected to be manufactured. This means the average car has about 20 nodes, while high end cars have 60 to 80 nodes.

As requirements on cars are still increasing: CO_2 emission, safety, connectivity to the outside world, infotainment... the number of nodes in automotive networks is expected to continue to rise the coming years.

As electronic drivers and receivers become complex, struggling to transfer high speed data via low bandwidth electrical wires. To overcome these problems, the automobile manufacturers have introduced fiber optic systems in addition to the networks based on copper wires. The benefits for car manufacturers are higher bandwidth, immunity to EMI, low weight, and increased transmission security (see Table 1). Compared to electrical broadband wirings, large-core optical fiber may also offer advantages in ease of handling and installation.

Table 1. Comparison between consumer and automotive transceiver ICs

	Consumer ICs	Automotive ICs
Lifetime in the field	>5 years	>20 years
Production lifetime	a few years	10 to 20 years
Production volume per chipset	10 to 50 million/yr	100k to 5 million/yr
Immunity to radiated electromagnetic disturbances	3 V/m for 80/230 MHz-1 GHz (CISPR 14-2/EN 55014-2: 1997) or 10 V/m for 80 MHz-14 Hz (IEC 61000-6-2: 1999)	150V/m (cars with combustion engines) or 600 V/m (hybrid cars and electrical cars)
Operational Temperature range	[0C, 70C]	[-40C, +85C to +125C]
Maximum failure rate in the field	<1000 ppm/year	<1 ppm/year

Standardization

Table 1 compares the characteristics and requirements for consumer and automotive IC's.

Compared to consumer markets, automotive markets are small.

Some Examples of the Consumer Market Size

The total market for optical pick-up units, to be used in CD-ROM drives, DVD-drives, and CD and DVD recorders, was estimated for the year 2000 to be 450 million units. (HIS iSupply 2011) estimates the number of large LCD panels shipped for televisions, desktop computer monitors and notebook PCs in the year 2011 to be around 740 million units. The number of Smartphones sold in 2011 is estimated to be around 420 Million units (Johnston 2011)

From the comparison of Table 1 it is clear that the requirements imposed on automotive transceivers are largely different from the requirements on commercial transceiver products. Car manufacturers are sometimes looking to leverage on the large production volumes, hence low costs of the commercial IC's. However the large differences in reliability and quality, operational temperature range, and Immunity to radiated electromagnetic fields, make that a commercial IC is difficult to be used in an automotive environment. The only exceptions are integrated circuits embedded in infotainment devices (not the transceivers of the infotainment devices). On the contrary, circuits developed for the automotive markets, as reliability and quality, EMC and operational temperature range are downwards compatible with the requirements of the industrial markets, find their way to many industrial applications. E.g. CAN transceivers are used in automation applications as an alternative for RS-485 transceivers, and promoted by the "CAN in Automation" (CiA) user organization.

By "car" we are referring to passenger cars, which are defined as motor vehicles with at least four wheels, used for the transport of passengers, and comprising no more than eight seats in addition to the driver's seat. Cars (or automobiles) make up approximately 87% of the total motor vehicle annual production in the world.

The remaining 13%, is made up by light commercial vehicles and heavy trucks, buses, coaches and minibuses.

Vehicle Manufacturers (OICA, 2011)

According to the report of Ward's Auto (http://wardsauto.com/ar/world_vehicle_popula-tion_110815/), the global number of cars exceeded 1.015 billion in 2010, jumping from

Table 2. Number of cars produced in the world every year

Year	Total number of cars produced in the world
2009	51,971,328
2008	52,940,559
2007	54,920,317
2006	49,886,549
2005	46,862,978
2004	44,554,268
2003	41,968,666
2002	41,358,394
2001	39,825,888
2000	41,215,653
1999	39,759,847

Ref: Statistics on worldwide car production issued by the *International Organization of Motor*

980 million the year before. Dividing the total number of cars in the world by the yearly production, reveals that the average car lifetime exceeds 20 years.

There is still a significant amount of room for growth in the automotive market. Ward's Auto reports that there are 1.3 people for every car in the U.S., while in China there are 17 people per vehicle, and in India 56 people per vehicle.

The world vehicle population in 2010 passed the 1 billion-unit mark 24 years after reaching 500 million in 1986. Prior to that, the vehicle population doubled roughly every 10 years from 1950 to 1970, when it first reached the 250 million-unit threshold.

Low Production Volumes per Car Model Manufacturer

For a large volume series car model, the yearly production volume is around 200 000 cars per year.

Car communication standards are first introduced in high end cars and gradually proliferate to medium class cars.

Low class cars have the least number of car communication devices, many cars have none.

As for a automotive chipset it takes about 10 years to reach the maximum production volume, the time a chipset is in high volume production is usually more than 10 years.

Figure 1. Different climate areas in a car according to SAEJ1211 (SAE 2009)

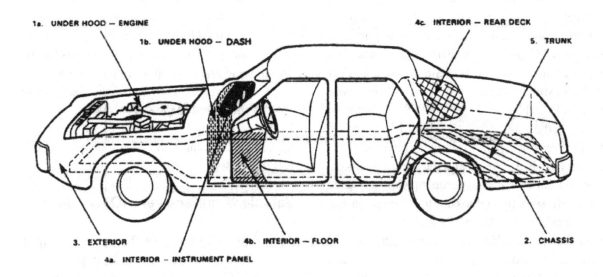

Table 3. Recommended temperature extremes for the different areas in a car from the standard SAE J1211 (SAE 2009)

	Location	Maximum Temperature
Chassis	-- Isolated Areas	+ 85 °C (+185 °F)
	-- Exposed to Heat Sources	+121 °C (+250 °F – 1200 °F)
	-- Exposed to Oils	+177 °C (+350 °F)
Exterior		+121 °C (+250 °F)
Underhood	-- Dash Panel	140 °C (285 °F)
	-- Engine (Typical)	150 °C (300 °F)
	-- Choke Housing	205 °C (400 °F)
	-- Starter Cable Near Manifold	205 °C (400 °F)
	-- Exhaust Manifold	650 °C (1200 °F)
Interior	-- Floor	85 °C (185 °F)
	-- Rear Deck	107 °C (225 °F)
	-- Instrument Panel (Top)	113 °C (236 °F)
	-- Instrument Panel (Other)	85 °C (185 °F)
Door Interior	-- No data available	
Trunk		85 °C (185 °F)
Minimum Temperature		–40 °C (–40 °F)

Table 4. Overview of different tests in order to verify electromagnetic compliance

Standard	Description
EN61000-4-2	Electrostatic Discharge
EN61000-4-3	Radiated Susceptibility Test
EN61000-4-4	Electrical Fast Transient/Burst Test
EN61000-4-5	Surge Test
EN61000-4-6	Conducted Immunity Test
EN61000-4-8	Power Frequency Magnetic Test
EN61000-4-11	Voltage Dips and Interruptions Test
EN61000-6-1	Immunity for residential, commercial and light-industrial environments
EN61000-6-2	Immunity for industrial environments
EN61547	Equipment for general lighting purposes — EMC immunity requirements
EN12016	Electromagnetic compatibility — Product family standard for lifts, escalators and passenger conveyors — Immunity

EVOLUTION OF STANDARDS

Automotive standards initially were developed by each car manufacturer for internal use, e.g., in the late 1980's when the first car networks for controlling devices and for carrying multiple signals over one conductor were implemented, all car manufacturers developed their own standards.

Mercedes developed the CAN standard, PSA (Peugeot and Citroen) and Renault introduced the VAN standard in their vehicles, BMW developed the I-bus, and the J1850 bus was developed in USA.

Figure 2. EMC compliance for automotive networks. Since the environment of a car is heavily electromagnetically disturbed, and the consequences of operational errors due to electromagnetic interference in a car might be severe, EMC requirements imposed on automotive transceivers are severe.

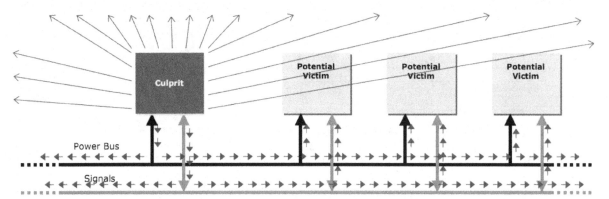

As there is little differentiation possible with a bus standard, and economies of scale allow important cost reductions, and economies of scale allowed important improvements in performance (e.g. immunity against radiated electromagnetic disturbances) the automotive world evolved over a time period of 20 years towards a single standard: CAN. Today nearly all vehicles in the world use CAN networks in order to control

Today Pre-competitive development of standards and transceivers is commonly done between the large car manufacturers.

Avionics Communication Networks

Avionics volumes are much smaller than automotive production volumes, however the number of network nodes, and hence the need for transceivers per airplane is much higher than in a car.

Up to now, each airplane manufacturer used it own standards.

Today Pre-competitive development of standards and transceivers is not done between the large avionics manufacturers.

For commercial airplanes about 90% of the nodes in a commercial airplane are in the pressurized part of an airplane, which need to fulfil the temperature interval [-40C to +85C]. This

means that there is an important potential to use circuits developed for the automotive market in commercial airplanes.

OVERVIEW OF THE DIFFERENT TYPES OF NETWORKS

Configurations

Different network configurations are used:

- Ring (One directional or bidirectional)
- Star (Active or passive)
- Tree
- Bus (Passive)

There are two types of Star networks passive and active star networks.

In an active star network the central star has a transceiver to every node connected to it. In a passive star network all connections come to a center point. This is done for optical networks having a passive optical star.

In electrical physical layer implementations, the passive star is usually replaced by a bus, which is wired from device to device.

Table 5. RTCA D0-160 environmental conditions and test procedures for airborne equipment

Section	Title	Notes
16	Power Input	115 VAC, 28 VDC and 14 VDC Power Voltage/frequency range, interruptions, surges
17	Voltage Spike	Power Leads Up to 600 V or 2x Line Voltage
18	Audio Frequency Conducted Susceptibility – Power Inputs	0.01 - 150 kHz or 0.2 - 15 kHz
19	Induced Signal Susceptibility	Interconnection Cabling E field and H Field 400 Hz – 15 kHz and spikes
20	Radio Frequency Susceptibility (Radiated and Conducted)	Conducted: 0.01-400 MHz Radiated: 0.1-2, 8 or 18 GHz
21	Emission of Radio Frequency	Power Lines: 0.15-30 MHz Interconnecting Cables: 0.15-100 MHz Radiated: 2-6,000 MHz
22	Lightning Induced Transient Susceptibility	Pin & Bulk injection, Pulse & Dampened Sine

Figure 3. Different network topologies used in vehicle networks

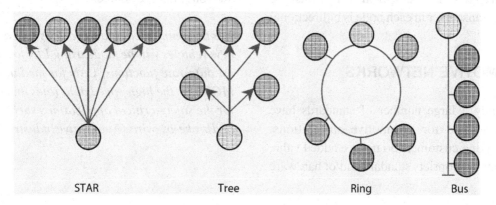

Table 6. Overview of different communication standards for vehicles and their corresponding network topology

	LIN	CAN	MOST	Firewire	Ethernet	Flexray
Ring (one directional)			X			
Ring (bidirectional)			(X)			
Daisy-chain				X		
Tree					X	
Star (active)					X	X (ePHY)
Star (passive)						X (oPHY)
Bus (passive)	X	X				ePHY

Figure 4. Passive star network versus active star network

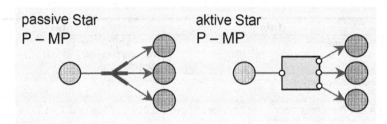

There are two types of passive optical Star networks:

1. **Transmissive Passive Optical Star:** In this case transmitter and receiver in each application are separate. In the applications (yellow), receiver (green) and transmitter (blue) are separate.
2. **Reflective Passive Optical Star:** In this case the transceiver in each node is bidirectional

AUTOMOTIVE NETWORKS

Historically a large number of standards have been developed for automotive applications. Realizing that customers do not see added value in a company propriety standard and/or hardware

implementation of this standard, while added value is perceived in system performance, a single standard is becoming dominant for each of the four automotive network niches: low speed, medium speed, high speed and safety critical. Under the heavy pressure of price, performance, quality and reliability, the transceiver components are now available on the market as standard components all from multiple sources.

It is becoming clear that regardless of carmaker, new vehicles will be made using LIN for the lowest data-rate functions, CAN for medium speed, MOST for the high-speed data rates and Flexray for the safety-critical applications such as steer- and brake-by-wire. (Automotive Industries, 2005)

Figure 5. Star network using a TRANSMISSIVE Optical Star network

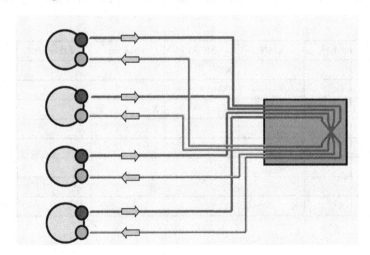

Figure 6. Communication Network with Reflective Optical Star coupler (top), and communication network with Transmissive Optical Star Coupler (bottom)

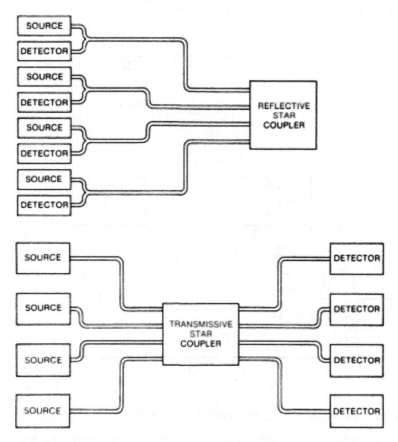

Table 7. Main types of automotive buses

	Subbus	**Event Triggered**	**Time Triggered**	**Mulitimedia**	**Wireless**
Main standard	LIN SAE-J2602	CAN	Flexray	MOST, Ethernet	Bluetooth, GSM, WLAN
Older standards	K-Line MI-bus SWCAN SbW+	VAN FTCAN SAEJ1850	Byteflight TTP TTCAN	IDB-1394 (Firewire) D2B MML	

Carrier Sense Multiple Access (CSMA) is a probabilistic Media Access Control (MAC) protocol in which a node verifies the absence of other traffic before transmitting on a shared transmission medium such as an electrical bus.

As reliability and availability of automotive networks is of crucial importance, specification

work is done thoroughly. Standardization organizations include an administration, a steering committee, a large number of associated members and an important number of working groups (physical layer, protocol, applications,.....). As the automotive networks are first implemented in high end cars, the steering committees consist

Figure 7. Comparison of the dominant automotive network standards with respect to speed and cost

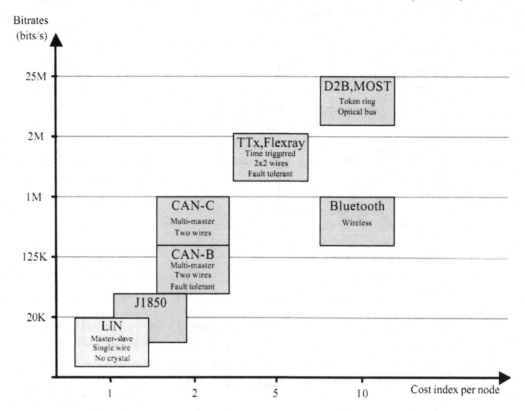

Table 8. Comparison of the main automotive network standards

Bus	LIN	CAN	Flexray	MOST
Cost/node in USD	1.5	3.0	6.0	4.0
Used in	subnets	Soft real-time	Hard real-time	multimedia
Application in car	Car body	Powertrain and Chassis	Chassis and Powertrain	multimedia
Message transmission	Synchronous	Asynchronous	Synchronous and asynchronous	Synchronous and isochronous
Message identification	Identifier	Identifier	Time slot	Place in "wagon" of bits
Architecture	Single master/up to 16 slaves	Multi-master up to 30 nodes	Multi-master up to 64 nodes	Single-master Up to 64 nodes
Access control	Polling	CSMA	TDMA	TDMA
Data Rate	20 kbps	1 Mbps	10 Mbps	150 Mbps
Physical Layer	Single wire bus	Dual wire bus	Dual Wire Star (Optical Fiber)	Optical Fiber (MOST25, MOST150) STP cable (MOST50) Coax cable (MOST150)
Babbling idiot	n/a	Not provided	Provided	n/a
Extensibility	High	High	Low	High

predominantly of high end car manufacturers, hence joining forces in a pre-competitive effort.

In order to guarantee interoperability of components made by different suppliers, all standardization consortia include a certification procedure and impose this certification to be done by independent laboratories.

For some of the older standards, e.g. sleep and wake-up procedure and status management of the transceiver are no part of the specification. In these cases, each OEM still has its own additional requirements, and an additional approval is needed by each OEM. I.e. for some of the older standards, the full interoperability is hence not valid in practice.

We now shortly discuss the most important automotive networks:

- **Low Speed:** LIN
- **Medium Speed:** CAN
- **High Speed:** MOST, Firewire (IDB-1394) and Ethernet
- **Safety-Critical:** Flexray

Box 1. Local Interconnect Network logo

Low-Speed Automotive Buses

LIN (Local Interconnect Network) is a low cost serial communication system intended to be used for distributed electronic systems in vehicles. The LIN bus is used as a cheap sub-network of a CAN bus to integrate intelligent sensor devices or actuators in today's cars. Recently LIN may be used also over the vehicle's battery power-line with a special DC-LIN transceiver.

The LIN specification is made by the LIN-consortium. In the late 1990s the Local Interconnect Network (LIN) Consortium was founded by five European automakers (Audi AG, BMW AG, Daimler AG, Volkswagen AG, Volco Car Corporation), Volcano Automotive Group and Freescale. The first fully implemented version of the new LIN specification was published in November 2002 as LIN version 1.3. In September 2003 version 2.0 was introduced to expand configuration capabilities and make provisions for significant additional diagnostics features and tool interfaces. The specification has evolved now to version 2.2.

A node in LIN networks does not make use of any information about the system configuration, except for the denomination of the master node. Nodes can be added to the LIN network without requiring hardware or software changes in other slave nodes. The size of a LIN network is typically under 12 nodes (though not restricted to this), resulting from the small number of 64 identifiers and the relatively low transmission speed. The clock synchronization, the simplicity of UART

Figure 8. LIN-bus with master and a number of slaves on a common bus

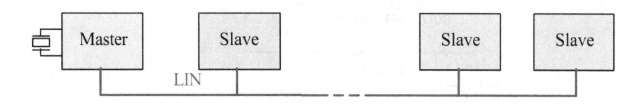

communication, and the single-wire medium are the major factors for the cost efficiency of LIN.

LIN is a single-wire serial communications protocol based on the common SCI (UART) byte-word interface. UART interfaces are available as low cost silicon module on almost all micro-controller and can also be implemented as equivalent in software or pure state machine for ASICs. The access to the bus in a LIN network is controlled by a master node so that no arbitration or collision management in the slave nodes is required, thus giving a guarantee of the worst-case latency times for signal transmission.

A particular feature of LIN is the synchronization mechanism that allows the clock recovery by slave nodes without quartz or ceramics resonator. The specification of the line driver and receiver is following the ISO 9141 single-wire standard with some enhancements. The maximum transmission speed is 20 kbit/s, resulting from electromagnetic compatibility (EMC) and clock synchronization requirements.

A LIN network comprises one master node and one or more slave nodes. All nodes include a slave communication task that is split in a transmit and a receive task, while the master node includes an additional master transmit task. The communication in an active LIN network is always initiated by the master task as illustrated in the figure below: the master sends out a message header which comprises the synchronization break, the synchronization byte, and the message identifier.

- **Master, Sync Break [13 Bits]:** Used to identify the start of the frame.
- **Master, Sync Field [Alternate 1-0 Sequences]:** Used by the slave node for clock synchronization.
- **Master, Identifier:** [6-bit long message ID and a 2-bit long parity field].
- **Master, Message ID [2, 4 or 8 data bytes]:** Optional message length information.
- **Slave Transmission ~[1-8 data bytes]:** Data bytes.
- **Slave Transmission ~[8 bit]:** Checksum (LIN consortium).

All messages are initiated by the master with at most one slave replying for a given message identifier. Therefore it is not necessary to implement a collision detection. Exactly one slave task is activated upon reception and filtering of the

Figure 9. The communication in an active LIN network is always initiated by the master (message header), and response by the intended slave or the master (message response)

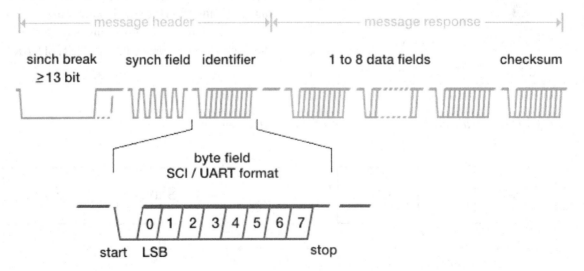

identifier and starts the transmission of the message response. The response comprises one to eight data bytes and one checksum byte. The header and the response part form one message frame.

The identifier of a message denotes the content of a message but not the destination. This communication concept enables the exchange of data in various ways: from the master node (using its slave task) to one or more slave nodes, and from one slave node to the master node and/or other slave nodes. It is possible to communicate signals directly from slave to slave without the need for routing through the master node, or broadcasting messages from the master to all nodes in a network. The sequence of message frames is controlled by the master and may form cycles including branches.

The master is typically a microcontroller, whereas the slaves are implemented as ASICs.

Current uses combine the low cost efficiency of LIN and simple sensors to create small networks. The LIN specification also handles 2400, and 9600 baud rates, and may be used as a sub-bus [sub-network] for a CANbus interface.

The main features of LIN are listed:

- Low cost single-wire implementation.
- Single master, up to 16 slaves (i.e. no bus arbitration).
- No arbitration necessary.
- Self synchronization in the slave nodes without crystal or ceramics resonator.
- Slave Node Position Detection (SNPD) allows node address assignment after power-up.
- Single wire communications up to 19,2 kbit/s at 40 meter bus length.
- Guaranteed latency times.
- Variable length of data frame (2, 4 and 8 byte).
- Configuration flexibility.
- Multi-cast reception with time synchronization, without crystals or ceramic resonators.
- Data checksum and error detection.
- Detection of defective nodes.
- Low cost silicon implementation based on standard UART/SCI hardware.
- Off-the-shelf slaves.
- Enabler for hierarchical networks.
- Operating Voltage of 12 V.

Figure 10. Example of LIN implementation (Infineon Technologies)

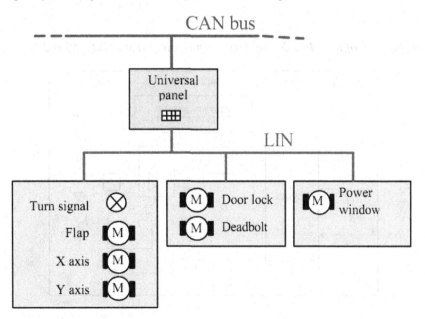

Typical applications for the LIN bus are assembly units such as doors, steering wheel, seats, climate control, lighting, rain sensor, column, switch panel, or alternator. In these units the cost sensitive nature of LIN enables the introduction of mechatronic elements such as smart sensors, actuators, or illumination. They can be easily connected to the car network and become accessible to all types of diagnostics and services.

The LIN Steering group determines the procedure for certification. The certification ensures that each LIN product is fully compliant with the LIN standard and guarantees interoperability with the other certified LIN products.

In practice, each OEM has its own additional requirements for LIN products e.g. with respect to power management (sleep and wake-up of the LIN transceiver) and status management (bus errors). Since this is not part of the LIN specification, and additional approval is needed by each OEM. The full interoperability is hence not valid with respect power management and status management. In addition, specifications with respect to EMI do not yet fully guarantee that the part in the system of the customers will pass the system EMC tests. A dedicated acceptance test might be needed in each case.

In order to save costs, more and more LIN transceivers are integrated into the application, e.g. LIN transceivers integrated with motor driver circuits, power supply voltage regulator, and microcontroller with a single package.

When the software of the application runs on the same microcontroller as the LIN software, this might give rise to difficulties in the compliance testing. Therefore in many cases the LIN software and the application software are run on a separate processor.

LIN and SAE J2602

While LIN is used in Europe, SAE J2602 is used in USA. SAE J2602 can be considered as a subset of LIN 2.2 . As a consequence products are on the market which are compatible to both LIN2.2 and SAE-J2602 specifications.

Features of SAE J2602 with respect to LIN:

- Optional requirements are removed.
- Only one speed is supported: 10.417 kbps. The speed of 19.2 kbps is not supported:
- No event triggered frames.
- Fault tolerant operation.
- Better definition of sleep–mode and go to sleep commands.

Figure 11. Block diagram of the LIN2.2/SAE J2602 transceiver from NXP (NXP 2011)

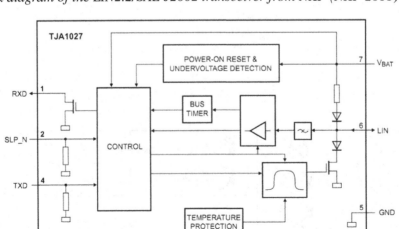

015aaa212

Figure 12. Block diagram of a LIN transceiver TH8062 with integrated voltage regulator (Melexis, 2012)

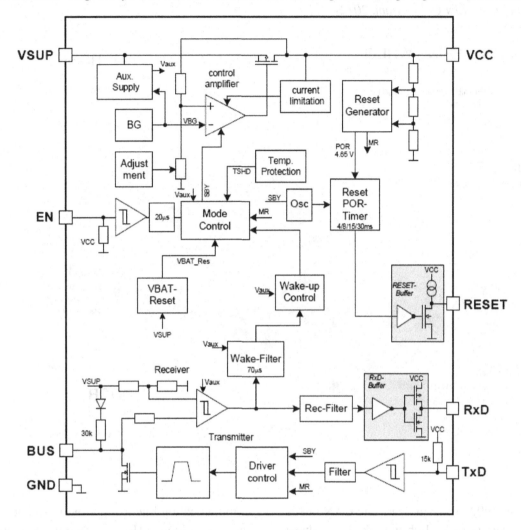

LIN applications in a car:

- **Roof:** Sensor, light sensor, light control, sun roof.
- **Steering Wheel:** Cruise control, wiper, turning light, climate control, radio.
- **Seat:** Seat position motors, occupant sensors, control panel.
- **Engine:** Sensors, small motors.
- **Climate:** Small motors, control panel.
- **Door:** Mirror, central ECU, mirror switch, window lift, seat control switch, door lock.

Medium Speed Automotive Buses: CAN

The CAN bus (Controller Area Networking) was defined in the early 1980 by a cooperation between Mercedes Benz, Bosch, and Intel. The CAN bus was initially developed for use in automotive applications

The first CAN controller was presented by Intel in 1987, quickly followed by Philips (now NXP), ST Microelectronics and others. The first volume production car with CAN came on the

Figure 13. Six different LIN networks are targeted in a car: engine, climate control, steering wheel, roof, seat, and door (LIN Consortium, 2012)

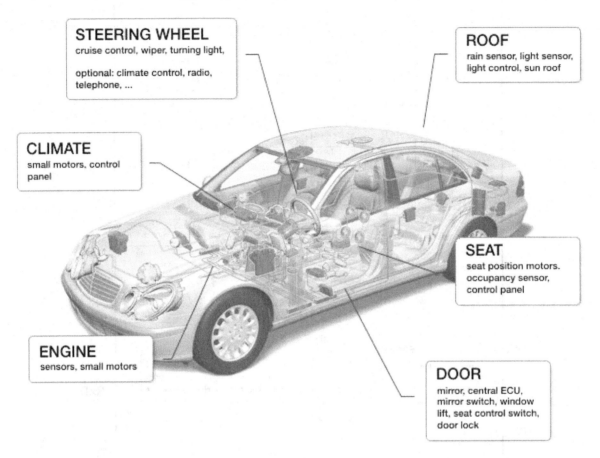

market in 1992. Today at least one CAN bus is used in every vehicle. High end vehicles can have up to 6 CAN buses.

CAN bus is one of five protocols used in the OBD-II vehicle diagnostics standard. The OBD-II standard has been mandatory for all cars and light trucks sold in the United States since 1996, and the EOBD standard has been mandatory for all petrol vehicles sold in the European Union since 2001 and all diesel vehicles since 2004.

By reducing the vehicles wiring by 2km, the vehicles overall weight was significantly reduced by at least 50kg and using only half the connectors. For the first time, each of the vehicles systems and sensors were able to communicate at very high speeds (25kbps - 1Mbps) on a single or dual-wire communication line as opposed to the previous multi-wire looms. However, the introduction of CAN Bus also increased the vehicles complexity and made after market installations even more difficult and in many cases impossible to perform.

A CAN network has the following characteristics:

- Single terminated twisted pair cable used as bus.
- Multi master.
- Maximum Data Rate: 1 Mbit/sec for a maximum cable length of 40m.
- High reliability with extensive error checking.

- Typical maximum data rate achievable is 40KBytes/sec.
- Maximum latency of high priority message <120 μsec at 1Mbit/sec.

The CAN bus is used as a fieldbus in industrial systems, e.g. medical equipment and elevators. The main driver is the low cost of CAN Controllers and processors. The CAN users for non-automotive applications are grouped in a consortium "CAN in Automation" (CiA). Bit rates up to 1 Mbit/s are possible at network lengths below 40 m. Decreasing the bit rate allows longer network distances (e.g., 500 m at 125 kbit/s).

The CAN bus is used in vehicles to connect engine control unit and transmission, and on a different bus to connect the door locks, climate control, and seat controls. The CAN BUS consists of a single twisted pair cable.

The devices that are connected by a CAN network are typically sensors, actuators, and other control devices. These devices are not connected directly to the bus, but through a host processor and a CAN controller.

Two specifications are in use:

- **CAN-B (Low speed CAN):** CAN-B is used in the comfort domain, i.e. the interior of the car (SAE classification B). Message identifiers are 11 bit and maximum data rate is 250Kbit/sec - ISO11519.
- **CAN-C (High speed CAN):** CAN-C is used in the power train of a car (SAE classification C). Message identifiers are 29 bit and maximum data rate is 1Mbit/sec - ISO 11898.

In FullCAN, each message is fully buffered in hardware. The microcontroller is informed of an interrupt when a new message has arrived. The filtering of the messages in implemented entirely in hardware. In BasicCAN, there is only one buffer for all messages arriving.

Figure 14. CAN network in a car (Canbuskit 2012)

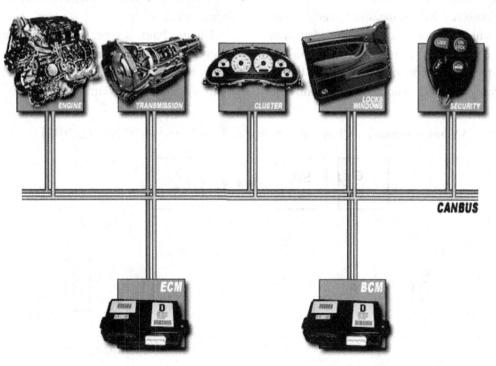

When running Full CAN (ISO 11898-2) at its higher speeds it is necessary to terminate the bus at both ends with 120 Ohms. The resistors are not only there to prevent reflections but also to unload the open collector transceiver drivers. The bus should be in all circumstances terminated correctly.

CAN is a multi-master broadcast serial bus standard for connecting electronic control units (ECUs).

Each node is able to send and receive messages, but not simultaneously. CAN uses the Provider-Subscriber Model for exchanging data. This means one participant puts data on the bus, while the message reaches all other participants. All CAN messages start with an 11-bit identifier. All participants have an identifier filter. The participants where the identifier, which has been broadcasted, passes the identifier filter are the nodes able to read the data of the CAN message.

CAN is a multi-master serial bus. This means the right of the bus access is not assigned by a central instance, but each participant can start transmitting as soon as the bus is free. If two or more nodes start transmitting at the same time, this conflict is resolved by means of an arbitration process which is based the message identifier. This is achieved by making the line drivers in the transmitter as an open collector. The line is normally high (1). When one line driver pulls the line low (0), and the other is in a high impedance state because it is send the opposite bit value (1), the line driver pulling the bus with the lowest impedance (0) wins. During the transmission a transceiver also senses the bus. As soon as a transceiver detects that an identifier it has send out (1), has been overwritten on the bus (0), it immediately stops sending. Therefore the node with the identifier with the zero always wins.

Due to the fact that this arbitration requires the transmitter to read back each bit send out, in order to detect whether it was overwritten or not, the maximum baud rate depends on the length of the bus. The signal must be able to pass through the line twice during one bit time. The maximum baud rate of 1 Mbps can only be achieved for a maximum bus line length of 40 m.

CAN is unusual in that the entities on the network, called nodes, are not given specific addresses. Instead, it is the messages themselves that have an identifier which also determines the messages' priority. Nodes then depending on their function transmit specific messages and look for specific message. For this reason there is no theoretical limit to the number of nodes although in practice it is ~64.

The mechanical aspects of the physical layer (connector type and number, colors, labels, pinouts) are not part of the specification. Each automotive electronic control unit (ECU) needs a

Figure 15. CAN bus terminated on both sides with a 120 ohm resistor (Wikipedia, 2012)

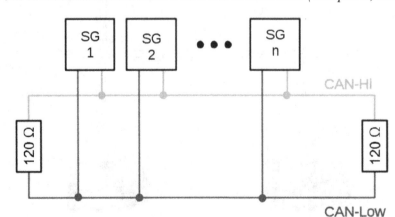

connector to connect to the CAN bus. Although several custom types of connectors are used, several de-facto standards have emerged, the most common being the 9-pin D-sub type connector with the following pin-out:

- **pin 2:** CAN-Low (CAN-)
- **pin 3:** GND (Ground)
- **pin 7:** CAN-High (CAN+)
- **pin 9:** CAN V+ (Power)

This de-facto mechanical standard for CAN requires nodes to have both male and female 9-pin D-sub connectors electrically wired to each other in parallel within the node. Bus power is fed to a node's male connector and the bus draws power from the node's female connector. This follows the electrical engineering convention that sources of power are terminated with female connectors.

MOST Automotive Networks

MOST (Media Oriented Systems Transport) is a high-speed multimedia network technology

optimized by the automotive industry. It can be used for applications inside or outside the car. The serial MOST bus uses a ring topology and synchronous data communication to transport audio, video, voice and data signals via plastic optical fiber (POF) (MOST25, MOST150) or electrical conductor (MOST50, MOST150) physical layers.

MOST technology is used in almost every car brand worldwide, including Audi, BMW, Hyundai, Jaguar, Land Rover, Mercedes-Benz, Porsche, Toyota, Volkswagen and Volvo. SMSC and MOST are registered trademarks of Standard Microsystems Corporation ("SMSC").

In order to meet the requirements of the more advanced infotainment networks (video for backseat entertainment, navigation systems,…), it became apparent that a higher data rate was needed. In order to reduce system development costs, and create a large and hence more competitive market, different companies joined together and formed in 1998 the MOST (Media Oriented System Transport) cooperation. Founders were BMW, Daimler –Benz, Becker (today Harman Automotive Division) and OASIS Silicon Systems

Figure 16. Car with CAN network in the car's comfort area, and a number of LIN networks as subnetworks of the CAN network

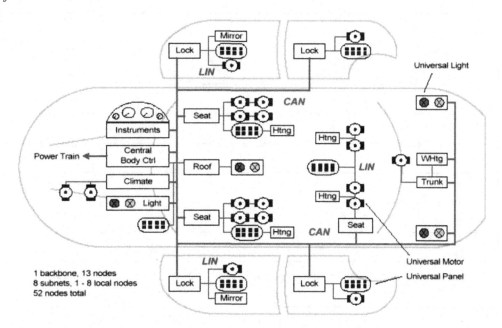

(today SMSC Inc). The MOST standard further builds on the concept of D2B, developed earlier. When the MOST cooperation became better known, the number of members quickly increased to 78, and the number of car manufacturers among the members is 16 (year 2012). The first car with a MOST network was introduced in 2001. The number of car models using MOST for their infotainment network increased year after year to 113 (year 2012). This makes MOST the dominant standard for infotainment networks.

MOST technology has been developed over three generations. The first generation (MOST25) is based on a bit rate of 25 Mbit/s using an optical physical layer. A data rate of 22.58 Mbit/s can achieved with a frame length of 512 bit. MOST25 uses a 1-mm-Polymethylmethacrylat fiber (PMMA) called POF as data transmission medium. LED's or RCLED's are used as part of the transmitter to convert the electrical signal to an optical one using a driver circuit. The receiver converts the optical signal into an electric current using a Silicon PIN-photodiode.

PRINCIPLES OF COMMUNICATION

The MOST specification defines the physical and the data link layer as well as all seven layers of the ISO/OSI-Model of data communication. Standardized interfaces simplify the MOST protocol integration in multimedia devices. For the system developer, MOST is primarily a protocol definition. It provides the user with a standardized interface (API) to access device functionality. The communication functionality is provided by driver software known as MOST Network Services. MOST Network Services include Basic Layer System Services (Layer 3, 4, 5) and Application Socket Services (Layer 6). They process the MOST protocol between a MOST Network Interface Controller (NIC), which is based on the physical layer, and the API (Layer 7).

MOST Networks

A MOST network is able to manage up to 64 MOST devices in a ring configuration. Plug-&-

Figure 17. Number of car models using the MOST protocol for their infotainment network (MOST Cooperation 2012)

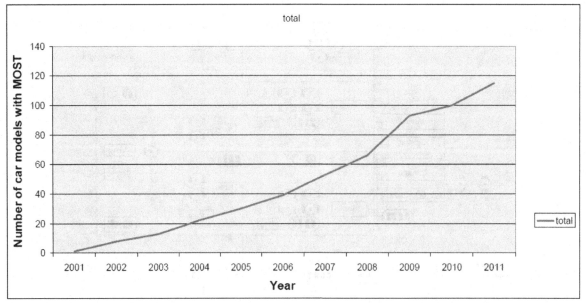

Figure 18. Ethernet protocol stacks over MOST (MOST Cooperation, 2012).

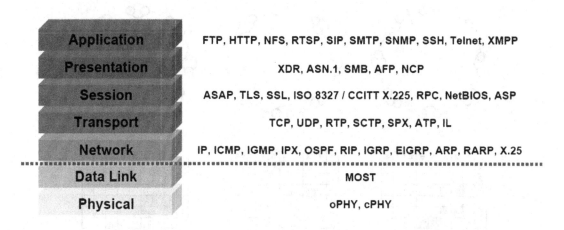

■ OSI Model

Layer	Protocols
Application	FTP, HTTP, NFS, RTSP, SIP, SMTP, SNMP, SSH, Telnet, XMPP
Presentation	XDR, ASN.1, SMB, AFP, NCP
Session	ASAP, TLS, SSL, ISO 8327 / CCITT X.225, RPC, NetBIOS, ASP
Transport	TCP, UDP, RTP, SCTP, SPX, ATP, IL
Network	IP, ICMP, IGMP, IPX, OSPF, RIP, IGRP, EIGRP, ARP, RARP, X.25
Data Link	MOST
Physical	oPHY, cPHY

Play functionality allows MOST devices to be easily attached and removed. Despite the fact that MOST is always a ring network, some car manufacturers bring all connections to and from the devices to one place, in this way making a virtual star network. In a MOST network, one device is designated the network master. Its role is to continuously supply the ring with MOST frames. A preamble is sent at the beginning of the frame transfer. The other devices, known as network slaves, use the preamble for synchronization. Encoding based on synchronous transfer, allows constant post-sync for the network slaves.

In case one node fails, the complete network fails. Therefore MOST networks implemented as a one directional ring network are only used for non safety critical networks, e.g. infotainment networks.

Safety critical MOST applications should use redundant double ring configurations. Hubs or switches are also possible in principle, but these are not well-established in the automotive sector.

The first optical network for a car was developed by Daimler around 1995. This network was based 650 nm LED's and 1 mm core PMMA Plastic Optical Fiber. The large core diameter relaxes the fiber-alignment tolerance require-

ments. The network was point to point, laid out in a ring. The simple protocol and its 5.6 Mbit/s data rate are suitable for transmitting speech and audio, and also control signals serving peripherals in an automobile. This network was called Domestic Digital Bus (D2B) and put into volume production in the S-class of Mercedes in1998. A

Figure 19. Example of a MOST network with 8 nodes. MOST network is a one directional ring network.

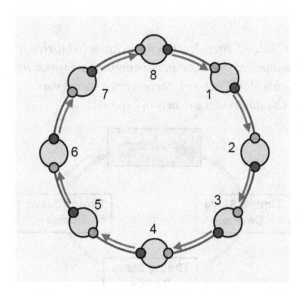

Figure 20. Consequences for a one directional ring network in case of cable fracture (left), transmitter failure (center), and controller processor failure in the application (right)

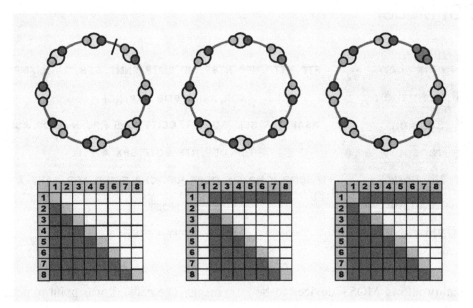

isochronous[1] bitstream of data is broadcasted in the ring network, this avoids overhead by using data labels and addresses, and all nodes of the network receive the broadcasted data quasi simultaneously. This results in high quality audio to the different speakers of the car.

Figure 21. Broadcasting a synchronous bitstream minimizes overhead in infotainment networks. In each MOST network, there is one network master. All other nodes are network slaves.

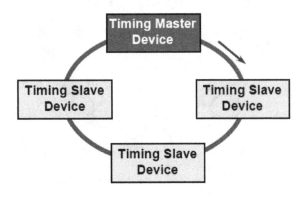

MOST Features

The strength of MOST lies in the efficiency of isochronous broadcast of time continuous data, such as uncompressed audio.

- Data is packaged in a "wagon". These wagons pass with a fixed speed through the ring network.

Therefore the MOST protocol is depicted as a train running through the ring network.

Different data types, sources of data and destinations occupy a different place and number of seats in this wagon.

This leads to the following characteristics of the MOST protocol:

- Efficient use of resources, no buffering needed into the Network controller IC's, no collisions, no sample rate conversions needed.
- Close to 100% utilization of nominal data rates, since no address labels or time labels needed for transferring a data packet

Figure 22. MOST data packaged as in a "wagon" of a train.

to its destination. The transport capacity of a MOST network with a certain data rate is comparable to the transport capacity of package switched networks such as Ethernet or Firewire, with several times the data rate of the MOST network. This explains why MOST150 and gigabit Ethernet

or Firewire S800 sometimes are presented competitors.

The rising number of infotainment applications, and the higher bandwidth required for video applications give rise to increased data rate requirements. Three generations of MOST have been developed and are on the market: MOST25, MOST50 and MOST150. The increase of data rate requirements for new infotainment networks in cars is mainly driven by the rising need in order to broadcast images, e.g. navigation system info, movies for backseat entertainment, digital television. As these data is compressed, the type of data does not need to rely critically on the efficient synchronous properties of the network. MOST150 is able to handle compressed video. In order to be able uncompressed video, which is required by some car manufacturers for safety critical applications, a fourth generation of the physical layer is needed, which is currently in preparation.

Figure 23. Comparison of a MOST25 frame (data wagon) having a fixed length of 512 bits with a MOST150 frame (data wagon) having a fixed length of 3072 bits

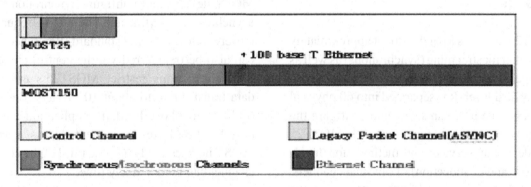

Figure 24. Composition of an Ethernet data packet

Dest Adr	Src Adr	Data	CRC
48bits	48bits	Maximum of 1506 bytes	32bits

Figure 25. The rising number of infotainment applications, and the higher bandwidth required for video applications give rise to increased data rate requirements. Three generations of MOST: MOST25, MOST50 and MOST150 (Harman-Becker 2007).

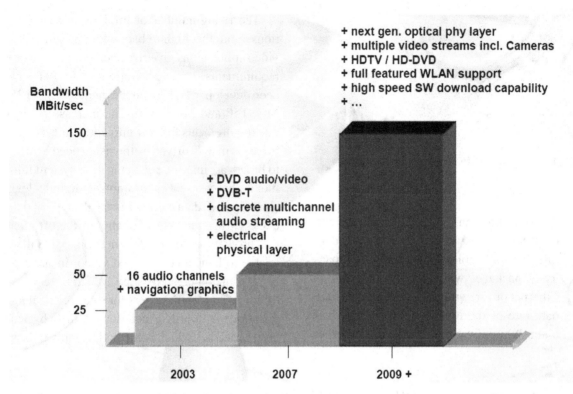

MOST25

MOST25 provides a bandwidth of approximately 23 MBaud for streaming (synchronous) as well as package (asynchronous) data transfer over an optical physical layer. It is separated into 60 physical channels. The user can select and configure the channels into groups of four bytes each. MOST25 provides many services and methods for the allocation (and deallocation) of physical channels. MOST25 supports up to 15 uncompressed stereo audio channels with CD-quality sound or up to 15 MPEG1 channels for audio/video transfer, each of which uses four Bytes (four physical channels). MOST also provides a channel for transferring control information. The system frequency of 44.1 kHz allows a bandwidth of 705.6 kbit/s, enabling 2670 control messages per second to be transferred. Control messages are used to configure

MOST devices and configure synchronous and asynchronous data transfer. The system frequency closely follows the CD standard. Reference data can also be transferred via the control channel. Some limitations restrict MOST25's effective data transfer rate to about 10 kByte/s. Because of the protocol overhead, the application can use only 11 of 32 Bytes at segmented transfer and a MOST node can only use one third of the control channel bandwidth at any time.

MOST50

MOST50 doubles the bandwidth of a MOST25 system and increases the frame length to 1024 bits. The three established channels (control message channel, streaming data channel, and packet data channel) of MOST25 remain the same, but the length of the control channel and the sectioning between the synchronous and asynchronous channels are flexible. Although MOST50 is specified to support both optical and electrical physical layers, the available MOST50 Intelligent Network Interface Controllers (INICs) only support electrical data transfer via Unshielded Twisted Pair (UTP).

MOST150

MOST150 was introduced in October 2007 and provides a physical layer to implement Ethernet in automobiles. It increases the frame length up to 3072 bits, which is about 6 times the bandwidth of MOST25. It also integrates an Ethernet channel with adjustable bandwidth in addition to the three established channels (control message channel, streaming data channel, and packet data channel) of the other grades of MOST. MOST150 also permits isochronous transfer on the synchronous channel. Although the transfer of synchronous data requires a frequency other than the one specified by the MOST frame rate, it is possible with MOST150.

One of the features of MOST150 is that it provides two new additional channels of the Ethernet channel and the Isochronous channel. Ethernet channel transmits unmodified Ethernet frames used by computer products, and brings itself into applications, by treating it as the Ethernet and the MAC address. Other Ethernet communication protocols like the TCP/IP stack and the Appletalk still communicate unchanged on the MOST.

MOST150's advanced functions and enhanced bandwidth enable a multiplex network infrastructure capable of transmitting all forms of infotainment data, including video, throughout an automobile.

3 to 5 Gbps MOST

More and more cars are equipped with cameras. So far, these cameras are not integrated into a network. However, it is expected that in the coming years most cars will be equipped with a set of cameras enabling a number of important new functions, many of these functions are seen as safety critical. Such applications are: Lane Departure Warning (LDW), Blind Spot Detection (BSD), Adaptive Cruise Control (ACC), parking aids. Cameras operating in the near infrared and far infrared, as well as radar, allow to detect obstacles not visible under fog conditions or far away in the dark. These cameras are expected to reduce the number of accidents occurring under conditions of poor visibility. At least about four cameras are expected to be placed in the car of the future: at the back of the car above the license plate, in or as a replacement of the left and the right side mirrors, and affront camera located between the rear mirror and the front window. Sending and processing these camera images allows to show the car on a dedicated display within its broader environment, allowing what is called "360 degree view", "top view" or "surround view".

In order to make use of the information of the cameras to the fullest extent and because the camera network is considered to be a safety critical network, the camera network is set-up as a star network. The star node in this network is an active star. In several cases car manufacturers want to control the camera: frame rate, position of the camera using remote control in a similar way as the side mirrors of a car are positioned, zoom in and out, control the integration times of a camera…. These requirements make that in many cases a bidirectional link between the camera and the driver assist processing unit is needed.

In order to transmit compressed HD video or uncompressed data from cameras at high frame rates (in case of a rapidly changing road environment or landscape) data rates can be 3 to 5 Gbps range are needed.

These requirements show a need for a new MOST standard for uncompressed camera images. This standardization work is now in preparation.

Figure 26. Camera network in car in addition to the MOST infotainment network

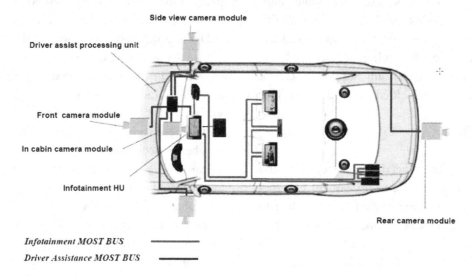

AUTOMOTIVE FIREWIRE (IDB-1394) NETWORKS

Firewire for computer industry and audio/visual devices - IEEE1394

Firewire or IEEE 1394 was developed by the computer industry as a standard for high-speed communications and isochronous real-time data transfer. FireWire is a method of transferring information between digital devices, especially audio and video equipment. Also known as IEEE 1394, FireWire is fast -- the latest version achieves speeds up to 800 Mbps. At some time in the future, that number is expected to jump to an unbelievable 3.2 Gbps when manufacturers overhaul the current FireWire cables.

One can connect up to 63 devices to a FireWire bus. Windows operating systems (98 and later) and Mac OS (8.6 and later) both support it.

Let's say you have your digital camcorder connected to your home computer. When your computer powers up, it queries all of the devices connected to the bus and assigns each one an address, a process called enumeration. FireWire is plug-and-play, so if you connect a new FireWire device to your computer, the operating system auto-detects it and asks for the driver disc. If

you've already installed the device, the computer activates it and starts talking to it. FireWire devices are hot pluggable, which means they can be connected and disconnected at any time, even with the power on.

The original FireWire specification, FireWire 400 (1394a), was faster than USB when it came out. FireWire 400 is still in use today and features:

- Transfer rates of up to 400 Mbps.
- Maximum distance between devices of 4.5 meters (cable length).

The release of USB 2.0 -- featuring transfer speeds up to 480 Mbps and up to 5 meters between devices -- closed the gap between these competing standards. But in 2002, FireWire 800 (1394b) started showing up in consumer devices, and USB 2.0 was left in the dust. FireWire 800 is capable of:

- Transfer rates up to 800 Mbps.
- Maximum distance between devices of 100 meters (cable length).

The faster 1394b standard is backward-compatible with 1394a.

Data can be sent through up to 16 hops (device to device). Hops occur when devices are daisy-chained together. Look at the example below. The camcorder is connected to the external hard drive connected to Computer A. Computer A is connected to Computer B, which in turn is connected to Computer C. It takes four hops for Computer C to access the camera.

FireWire devices can be powered or unpowered. FireWire allows devices to draw their power from their connection. Two power conductors in the cable can supply power (8 to 30 volts, 1.5 amps maximum) from the computer to an unpowered device. Two twisted pair sets carry the data in a FireWire 400 cable using a 6-pin configuration.

Some smaller FireWire-enabled devices use 4-pin connectors to save space, omitting the two pins used to supply power.

Figure 27. Firewire networks have a tree structure

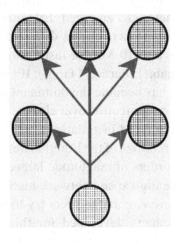

FireWire 800 cables use a 9-pin configuration. Six of those pins are the same as the six pins in the 1394a connector (shown above). Two of the added pins provide a "grounded shield" to protect the other wires from interference, and the third added pin does nothing at this time [ref].

Because FireWire 800 is backward-compatible with FireWire 400, there are a variety of adapters available to facilitate the combination of both standards on the same bus.

The IEEE 1394 interface, developed in late 1980s and early 1990s by Apple as FireWire, is a serial bus interface standard for high-speed communications and isochronous real-time data transfer. Apple first included FireWire in some of its 1999 models, and most Apple computers since the year 2000 have included FireWire ports. The interface is also known by the brand i.LINK (Sony), and Lynx (Texas Instruments). IEEE 1394 replaced parallel SCSI in many applications, because of lower implementation costs and a simplified, more adaptable cabling system. The 1394 standard also defines a backplane interface, though this is not as widely used. The High-Definition Audio-Video Network Alliance (HANA) adopted IEEE 1394 as the standard connection interface for A/V (audio/visual) component communication and control. FireWire is available in physical layer implementations: twisted pairs, wireless, fiber optic over POF, and coaxial versions using the isochronous protocols.

Promotion of IEEE1394 is done in the IEEE1394 trade Organization.

Figure 28. Assuming all of the devices in this setup are equipped with FireWire 800, the camcorder can be up to 400 meter from Computer C.

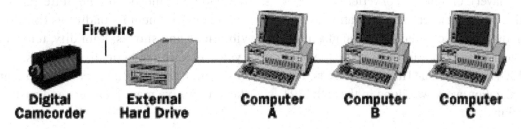

IEEE 1394 interfaces were in this market in competition with USB.

The two standards are a lot alike. Implementing FireWire costs a little more than USB, which led to the adoption of USB in computer industry as the standard for connecting most peripherals that do not require a high-speed bus.

Firewire for Automotive Applications: IDB1394

IDB-1394 has been developed as a standard for automotive infotainment networks.

Firewire is a point to point network, similar to MOST.

Firewire is like Ethernet, a data packet transmission protocol. This requires for each data packet address labels and time labels, significantly increasing the amount of overhead.

Therefore data rates in Firewire are at least 100% higher in order to obtain the same amount of net data transferred.

During the time period from 2000 to 2010, Firewire was considered by several car manufacturers as a potential candidate for implementation of the infotainment network of a car. The main driver for this was the high production volumes for Firewire transceivers from computer industry. As the production volumes are much larger for the computer industry than the potential volumes for a single car manufacturer, if these commercial Firewire devices would be used in the automotive industry, this would mean an important cost benefit.

However the stringent EMC requirements of the automotive industry require a dedicated development of the transceivers (in case of electrical physical layer), or a media converter (in case of an optical physical layer implementation.

For this purpose several car manufacturers drove the standardization for dedicated automotive IEEE1394 transceivers and media converters. This standardization work was done in the IDB1394 consortium.

Similar to MOST physical layer implementations, also for IDB-1394 several competing physical layer implementations (O/E and E/O fiber optic transceivers, connectors and media converters) have been proposed and developed as prototypes:

- Electrical over twisted pairs, electrical over coax.
- Optical over 1 mm PMMA POF, optical over 200 um core PCS.

These dedicated developments for the automotive industry, mean in reality an important barrier for a car manufacturer to commit to this standard for his infotainment system in the car. However, except from making a demo prototype, no car manufacturer so far committed to Firewire in volume production.

Automotive Ethernet

Recently a new competitor for high speed automotive networks emerged: Ethernet.

Ethernet protocols in its different generations Ethernet (10 Mbps), fast Ethernet (100 Mbps), gigabit Ethernet (1 Gbps), 10G ethernet (10 Gbps) has become the dominant standard for data communication over electric and optical cable for fixed installations.

The market size of the Ethernet transceivers is several orders of magnitude larger than the automotive high speed network market.

Automotive manufacturers try to leverage on the products developed for this market. From a standard point of view, transfer to the automotive world is straightforward. For the physical layer implementation, due to the wide automotive temperature range, and severe environmental conditions (mechanical vibrations and shocks, humidity, temperature cycles, electromagnetic disturbances, ...) and high quality and reliability requirements of the automotive manufacturers and customers, dedicated developments for most of the parts are needed for the automotive market.

Automotive Flexray Networks

The FlexRay Communication System was developed for X-by-Wire applications in vehicles that have high demands on speed, error tolerance and chronological determinism of data transmission. Flexray offers deterministic transmission of messages, and the possibility of error free communication. FlexRay is designed to be faster and more reliable than CAN, so that this network can be used for safety critical applications.

The FlexRay Communications System specifies three standard bit rates - 10 Mbit/s (corresponding to a nominal bit duration, gdBit, of 100 ns), 5 Mbit/s (corresponding to a gdBit of 200 ns), and 2.5 Mbit/s (corresponding to a gdBit of 400 ns).

In Flexray, the transmission time of each message is known within narrow specific limits. This feature allows time-triggered applications; most of these are safety-critical. In safety-critical systems, reliability of communication between the different nodes is of prime importance. In these cases, Flexray networks are set-ups as dual stars. The communication goes in parallel over the two stars.

FlexRay's prominent features are:

- High data rates
- Time- and event-triggered behavior
- Redundancy
- Fault-tolerance
- Deterministic

The Flexray standard evolved from BMW's Byteflight standard, which was earlier implemented in cars for safety critical networks. FlexRay is described as a combination of the Byte Flight protocol and TTP/C.

Implementation of the physical layer in Byteflight was an optical network, with an active optical star. The active optical star gave was needed due to limitations in the optical budget of the fiber optical transceivers, and gave rise to a relatively high implementation cost. Flexray was therefore implemented as an electrical network with a passive electrical bus, or an active electrical star.

Application in Cars

The first implementation of Flexray in series production vehicles was in 2006 in the BMW X5, enabling a new and fast adaptive damping system. Full use of FlexRay was introduced in 2008 in the new BMW 7 Series. At this moment Flexray is used in the following vehicles: Audi A6 and A8, Bentley Mulsanne, BMW X5, 7-Series and 5 Series and Rolls-Royce Ghost.

In the next years, Flexray will be used in applications where CAN has insufficient bandwidth and where a multitude of CAN buses would exceed the tolerable limit of complexity.

Consortium

FlexRay is an automotive network communications protocol developed by the FlexRay Consortium.

The FlexRay Consortium was made up of the following core members:

- Freescale Semiconductor
- Robert Bosch GmbH
- NXP Semiconductors
- BMW AG
- Volkswagen AG
- Daimler AG
- General Motors

By September 2009, there were 28 premium associate members and more than 60 associate members. The FlexRay™ Consortium has concluded its work with the finalization of the FlexRay™ Communications System specifications Version 3.0.1. The FlexRay consortium disbanded in September 2009. The FlexRay™

specifications V3.0.1 were submitted to ISO in order to be published as a standard for road vehicles. Specifications are available for download at the FlexRay website.

Networks

There are several ways to design a FlexRay cluster. It can be configured as a single-channel or dual-channel bus network, a single-channel or dual-channel star network, or in various hybrid combinations of bus and star topologies.

A FlexRay cluster consists of at most two channels, identified as Channel A and Channel B. Each node in the cluster may be connected to either or both of the channels. In the fault free condition, all nodes connected to Channel A are able to communicate with all other nodes connected to Channel A, and all nodes connected to Channel B are able to communicate with all other nodes connected to Channel B. If a node needs to be connected to more than one cluster then the connection to each cluster must be made through a different communication controller.

There are two basic topologies: a passive bus topology and active star topologies

Passive Bus Topology

The FlexRay communication network can be a single bus. In this case, all nodes are connected to this bus. In order to fulfill the high reliability requirements, the number of nodes connecting to the bus, as well as the maximum length of the connection to the bus must be limited. A 12m long bus line allows a maximum of eight stub lines with a maximum length of each stub line of 12 cm. When Flexray is implemented as a dual bus (bus A and bus B), a node can be connected to both buses or only one bus. An important argument to use a dual bus is to build in redundancy and hence increasing the reliability of the network. A short on the passive bus will cause the bus to fail.

Active Star Topology

In contrast to the bus system, a short circuit on one line, will only affect that line, and not affect the entire Flexray network. Hence the Star topology is more robust compared to the bus system. In addition the star topology supports up to 16 branches with a maximum length of 24m each. Each line is terminated perfectly, making the topology more robust with respect to signal reflections. Also here, in order to increase reliability for safety-critical applications, a FlexRay communication network will be built as a multiple star topology. There can be no more than two active stars on a network channel.

In order to increase the number of nodes n the network, or to allow larger distances between the nodes in the network, two stars can also be cascaded. Due to limitations on the maximum propagation time, the messages can only pass two stars.

Roles of a Node in a FlexRay Cluster

There are three distinct roles a node can perform.

Figure 29. Implementation of Flexray as a dual passive bus topology

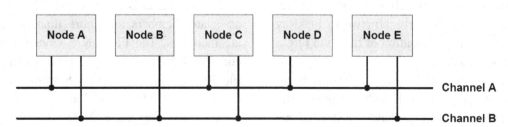

Figure 30. Implementation of a Flexray network with two star networks

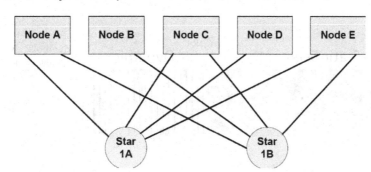

Figure 31. Cascading of Flexray Stars

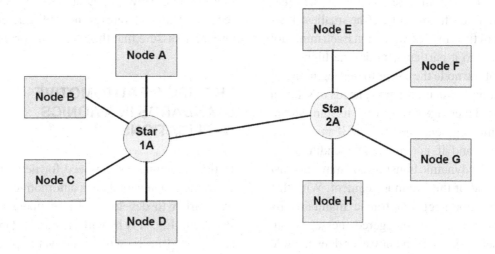

1. The role of a sync node enables a node to actively participate in the clock synchronization algorithm performed by the cluster. Sync nodes transmit sync frames that are evaluated by all nodes of the cluster to perform an alignment of clock rate, that effectively determines the cycle length, and clock offset, that effectively determines the position of the cycle start.

2. The role of a coldstart node enables a node to initiate the communication. Coldstart nodes are allowed to start transmitting startup frames in the non-synchronized state with the intent of establishing a schedule. Nodes integrate onto that new schedule by evaluating the content and timing of the received startup frames. A coldstart node is always also a sync node and a startup frame is always also a sync frame.

3. A node that is neither a coldstart node nor a sync node is referred to as non-sync node. It performs no special task.

The FlexRay communications cycle, or hyperperiod, is divided into four primary segments: static, dynamic, symbol window, and network idle time.

FlexRay provides both time-triggered and event-triggered execution models. It accomplishes this by grouping types of events into one of the cycle segments. Time-triggered communications are scheduled individually within the

Figure 32. The four parts of a Flexray cycle: static segment, dynamic segment, symbol window, and network idle time

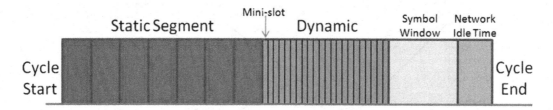

static segment. In FlexRay, event traffic must be communicated within the dynamic segment of the cycle. The dynamic segment is divided into "mini-slots", each a macrotick (the smallest practical unit of time in FlexRay) long. Each mini-slot is assigned to a particular node, the higher the priority of the node the closer to the beginning of the dynamic segment. If a node needs to send an event-based message it waits until its mini-slot in the dynamic segment, checks to see if anyone else is sending, and if not, it begins sending. The duration of a dynamic transmission may only last until the end of the dynamic segment. With this approach, time-triggered traffic maintains its determinism but event-triggered messages can also be sent. This scheme however does include the possibility of one node with an early mini-slot always sending and thus starving the remaining nodes. The symbol window segment is reserved for maintenance and identification of special cycles, such as startup and initialization. The network idle time segment is a reserved block of time used to allow nodes to adjust this clocks as part of the clock synchronization

Clocks

FlexRay system consists of a bus and processors (Electronic control unit, or ECUs). Each ECU has an independent clock. The clock drift must be no more than 0.15% from the reference clock, so the difference between the slowest and the fastest clock in the system is no greater than 0.3%.

This means that, if ECU-s is a sender and ECU-r is a receiver, then for every 300 cycles of the sender there will be between 299 and 301 cycles of the receiver. The clocks are resynchronized frequently enough to assure that this causes no problems.

THE USE OF AUTOMOTIVE STANDARDS IN AVIONICS APPLICATIONS

In the beginning, as the need for networks was much more stringent than in automotive, avionics was earlier to develop and implement network standards. E.g. "Fly by wire" is standard in avionics while "drive by wire" is not yet implemented in automotive volume production.

Avionic standards are called ARINC standards, after the company which is driving avionics network standardization. ARINC was incorporated in 1929 as Aeronautical Radio, Incorporated. The corporation's stock was held by four major airlines of the day. Through most of its history, ARINC was owned by airlines and other aviation-related companies until the sale to Carlyle.

ARINC standards today are developed by the AEEC (Airlines Electronic Engineering Committee).

- **ARINC 615:** Is a family of standards covering "data loading", commonly used for transferring software and data to or from avionics devices. The ARINC 615 standard covers "data loading" over ARINC 429.

- **ARINC 615A:** Is a standard that covers "data loading" over ARINC 664.
- **ARINC 624:** Is a standard for aircraft on-board maintenance system (OMS). It uses ARINC 429 for data transmission between embedded equipments.
- **ARINC 629:** Is a multi-transmitter data bus protocol where up to 128 units can share the same bus. It is installed on the Boeing 777.
- **ARINC 633:** Is the air-ground protocol for ACARS and IP networks used for AOC data exchanges between aircraft and the ground.
- **ARINC 653:** Is a standard Real Time Operating System (RTOS) interface for partitioning of computer resources in the time and space domains. The standard also specifies Application Program Interfaces (APIs) for abstraction of the application from the underlying hardware and software.
- **ARINC 660:** Defines avionics functional allocation and recommended architectures for CNS/ATM avionics.
- **ARINC 661:** Defines the data structures used in an interactive cockpit display system (CDS), and the communication be-tween the CDS and User Applications. The GUI definition is completely defined in binary definition files. The CDS software consists of a kernel capable of creating a hierarchical GUI specified in the definition files. The concepts used by ARINC 661 are similar to those used in user interface markup languages.

The avionics market is much smaller than the automotive market. The same functionality as in automotive is needed for avionics networks. Therefore avionics, in order to leverage on the cost, reliability and performance of the automotive applications, is using more and more automotive standards or standards based on automotive standards. Since harness lengths in avionic applications are much longer than in automotive applications, the automotive standards need a dedicated evaluation and standardization before they can be used in avionics applications.

CAN

Automotive CAN is used in avionics industry as ARINC 825 and CANaerospace.

Controller Area Network (CAN) found its way into aerospace applications because of its cost

Table 9. Overview of ARINC Avionics networking standards (Wikipedia 2012)

ARINC 664 defines the use of a deterministic Ethernet network as an avionic databus in modern aircraft like the Airbus A380, Sukhoi Super Jet 100 and the Boeing 787 Dreamliner.
ARINC 665 This standard defines standards for loadable software parts and software transport media.
ARINC 801 through 807 define the application of fiber optics on the aircraft.
ARINC 811 provides a common understanding of information security concepts as they relate to airborne networks, and provides a framework for evaluating the security of airborne networked systems.
ARINC 812 is a standard for the integration of aircraft galley inserts and associated interfaces
ARINC 818 defines a high-speed digital video interface standard developed for high bandwidth, low latency, uncompressed digital video transmission.
ARINC 822 is the standard for Gatelink.
ARINC 823 is a standard for end-to-end datalink encryption.
ARINC 825 is a standard for Controller Area Network bus protocol for airborne use.
ARINC 826 is a protocol for avionic data loading over a Controller Area Network bus.
ARINC 834 defines an aircraft data interface that sources data to Electronic Flight Bags, airborne file servers and the like.
ARINC 840 defines the Application Control Interface (ACI) used with an Electronic Flight Bag (EFB)
ARINC 841 defines Media Independent Aircraft Messaging
ARINC 842 provides guidance for usage of digital certificates on airplane avionics and cabin equipment.

Table 10. Examples of avionics network based on automotive standards (Klueser, 2012)

Protocol	Description
AFDX® / ARINC 664	AFDX®(Avionics Full DupleX Switched Ethernet) will serve as the next generation aircraft data network. It is already being implemented in the A380, A350, B787 and other programs. (AFDX® is an Airbus' registered trademark)
ARINC 810 / 812	ARINC 810 and 812 specify communication between modules in the on-board galley (Galley Master, Galley Inserts). The focus here is on power management.
ARINC 825	ARINC 825 specifies both the fundamental communication within CAN-based subsystems and between CAN subsystems, which for example are interconnected by AFDX. It offers addressing mechanisms, communication mechanisms, a service structure, profile descriptions and much more.
ARINC 826	ARINC 826 specifies Software Data Load over CAN. The mechanisms of ARINC 615A were adapted and optimized for CAN here.
CANaerospace	CANaerospace was developed by the company Stock Flight Systems. Key protocol applications are in engineering simulators, simulation cockpits and especially in the Italian field on drones (UAVs).
FlexRay	FlexRay is a scalable, flexible high-speed communication system that fulfills growing safety-related requirement in the automobile.

effective and efficient networking capability for systems to share data across a common media. The ability of CAN to transmit data across a shared shielded twisted pair cable, has advantages in terms of weight savings compared to existing avionic standard implementations. In addition, the CAN physical layer protocol specification provides additional functionality which make this data bus standard attractive to aviation applications: error recovery and protection mechanisms. Newer commercial airplanes like Airbus A380 and Boeing 787 already make use of CAN networks for all sorts of functions including flight deck systems, engine control and flight control systems.

The Airlines Electronic Engineering Committee initiated the development of ARINC Specification 825. The goal of ARINC 825 is to ensure interoperability and to simplify interoperation of CAN subsystems with other airborne networks.

Wikipedia gives following use of CANaerospace: "CANaerospace supports airborne systems employing the Line-replaceable unit (LRU) concept to share data across CAN and ensures interoperability between CAN LRUs by defining CAN physical layer characteristics, network layers, communication mechanisms, data types and aeronautical axis systems. CANaerospace is an open source project, was initiated to standardize the interface between CAN LRUs on the system level. CANaerospace is continuously being developed further and has also been published by NASA as the Advanced General Aviation Transport Experiments Databus Standard in 2001. It found widespread use in aeronautical research worldwide. A major research aircraft that employs several CANaerospace networks for real-time computer interconnection is the Stratospheric Observatory for Infrared Astronomy (SOFIA), a Boeing 747SP with a 2.5m astronomic telescope. CANaerospace is also frequently used in flight simulation and connects entire aircraft cockpits (i.e. in Eurofighter Typhoon simulators) to the simulation host computers. In Italy CANaerospace is used as UAV data bus technology. Furthermore, CANaerospace serves as communication network in several general aviation avionics systems."

FLEXRAY

The aeronautic industry and its suppliers show increasing interest in utilizing the automotive FlexRay protocol for their applications, more than ever since an opening of the standard for all industries and field of applications. With its combination of deterministic and flexible communication and data rates up to 10 Mbit/s on a single twisted wire pair, FlexRay is a promising candidate for future system developments and the modernization of CAN based systems. Signal integrity has been demonstrated and validated on topologies which include 90m harness length at a data rate of 10 Mbit/s to prove the suitability of FlexRay for aeronautic applications. The FlexRay/ByteFlight databus is used in Avidyne's FlightMax Entegra line of integrated avionics.

REFERENCES

CAN in Automation. (2012). *About CAN in Automation*. Retrieved from http://www.cancia.org/index.php?id=aboutcia

Canbuskit 2012. (n.d.). Retrieved from http://www.canbuskit.com

Consortium, L. I. N. (2012). LIN specification package. Retrieved from http://www.linsubbus.de/

Cooperation, M. O. S. T. (2012). *MOST specifications*. Retrieved from http://www.mostcooperation.com/publications/specificationsorganizationalprocedures/index.html

Firecomms. (2003). *Firecomms launches RCLED-Based S200 solution for IEEE 1394 POF applications*. Press release. Retrieved from http://pofto.com/articles/FC200R%20S200%20PR.pdf

Flexray. (2003). *Avidyne Corporation joins Flexray Consortium*. Retrieved from http://www.flexray.com/news/Avidyne_pr.pdf

Flexray. (2008). *FlexRay in the avionic context*. FlexRay Product Day, 2008. Retrieved from

Flexray. (2010). *Specifications version 3.0.1*. Retrieved from http://www.flexray.com/index.php?sid=9c03bd01bb2a03269f772bb99935a742&pid=94&lang=de

Grzemba, A. (2011). MOST: The automotive multimedia network. In *From MOST25 to MOST150*. Poing, Germany: Franzis Verlag GmbH. ISBN 9783645650618

Grzemba, A., & von der Wense, H. C. (2005). *LIN-Bus –Systeme, Protokolle, Test von LIN Systemen, Tools, Hardware, Applikationen*. Poing, Germany. *Franzis Verlag GmbH.*, *ISBN13*, 9783772340093.

HarmanBecker. (2007). *Migration from 1st to 2nd generation MOST systems*. MOST All members meeting, Frankfurt, 27th of March 2007.

Heller, C. (2010). FlexRay on aeronautic harnesses, 2010. *Institut für Luftfahrtsysteme.*, *ISBN3868537112*, 9783868537116.

Heller, C., & Reichel, R. (2009). *Enabling FlexRay for avionic data buses*. IEEE/AIAA 28th Digital Avionics Systems Conference, DASC '09, EADS Innovation Works, Munich, Germany.

HIS iSupply. (2011). *LCD panel market growth slows in 2011*. Retrieved from http://www.isuppli.com/DisplayMaterialsandSystems/MarketWatch/Pages/LCDPanelMarketGrowthSlowsin2011.aspx

http://arstechnica.com/gadgets/news/2011/07/2011willhit420millionsmartphonesalessamsunghits433percentgrowthsamsungsees300growthinsmartphonemarketshare.ars

http://www.eurtd.com/moet/PDF/EADSIW%20(Presentation)%20%20FlexRay%20Product%20Day.pdf

International Organization of Motor Vehicle Manufacturers (OICA). (2011). Statistics on worldwide car production. Retrieved from http://www.worldometers.info/cars/

Johnston, C. (2011). *Samsung sees 300% growth in smartphone market share: IMS Research finds that 420 million smartphone sales will have been made by the end of 2011*. Retrieved from.

Klueser, J. (2012). *CAN based protocols in avionics Version 1.1*. (20120412, Application Note ANION10104) Vector Informatik GmbH. Retrieved from https://www.vector.com/portal/medien/cmc/application_notes/ ANION10104_CANbased_protocols_in_Avionics.pdf

Melexis. (2012). *TH8062 voltage regulator with LIN transceiver*. Datasheet of the LIN transceiver TH8062. Retrieved from http://www.melexis.com/LINSBC'sandBUSTransceivers/(general)/TH8062583.aspx

NXP. (2011). *Datasheet TJA1027, LIN 2.2/SAE J2602 transceiver*. Retrieved from http://www.nxp.com/products/automotive/transceivers/lin_transceivers/series/TJA1027.html

(Product Code: J1211: 20090407). Retrieved from http://standards.sae.org/j1211_200904/

Robert Bosch Gmb, H. (1991). *CAN specification*, ver. 2.0. Stuttgart, Germany: Robert Bosch GmbH. Retrieved from http://www.semiconductors.bosch.de/media/pdf/canliteratur/can2spec.pdf

Robert Bosch Gmb, H. (2011). *Automotive handbook* (8th ed.). Plochingen, Germany: Robert Bosch GmbH.

SAE. (2009). *Handbook for robustness validation of automotive electrical/electronic modules*.

Ward's Auto. (2012). *World vehicle population*. Retrieved from http://wardsauto.com/ar/world_vehicle_population_110815/

Wikipedia. (2012). *ARINC*. Retrieved from http://en.wikipedia.org/wiki/ARINC

Wikipedia. (2012). *CAN bus*. Retrieved from http://en.wikipedia.org/wiki/Controller_area_network

Wikipedia. (n.d.). *ARINC825*. Retrieved from http://en.wikipedia.org/wiki/ARINC_825

Wikipedia. (n.d.). *CANaerospace*. Retrieved from http://en.wikipedia.org/wiki/CANaerospace

Zimmermann, W., & Schmidgall, R. (2011). Bussysteme in der Fahrzeugtechnik: Protokolle. *Standards und Softwarearchitektur.*, *ISBN13*, 9783834809070.

ENDNOTES

[1] Isochronous is a word derived from the Greek iso (equal) and chronos (time). It literally means regularly, or at equal time intervals. In general English language, it refers to something that occurs at a regular interval, of the same duration; as opposed to synchronous which refers to more than one thing happening at the same time. The term is used in different technical contexts, but often refers to the primary subject maintaining a certain interval, despite variations in other measurable factors in the same system.

APPENDIX: ABBREVIATIONS

- **CAN:** Controller Area Network
- **D2B:** Domestic digital bus
- **LIN:** Local Interconnect Network
- **Mbps:** Megabits per second
- **MOST150:** 150 Mbit/s version of MOST
- **MOST25:** 25 Mbit/s version of MOST
- **MOST50:** 50 Mbit/s version of MOST

Chapter 3

Physical Layer Implementations of Communication Standards in Automotive Systems

Piet De Pauw
BVBA De Pauw, Belgium

ABSTRACT

This chapter is divided into three sections. The first section gives an overview of the different cables used in automotive communication systems. Although 6 different kind of cables and their corresponding connector systems have been qualified for use in automotive applications (3 electrical cable types and 3 optical cable types), up to now, only three types are commonly used: Unshielded Twisted Pairs (UTP), Coax cable, and 1 mm PMMA POF. GOF and multimode PCS up to now are not used for mass volume production. The second section overviews the optical connectors used in automotive communication systems. The electrical connectors used in LIN and CAN networks. The optical connectors standardized by MOST. In the early phase of MOST implementation, the number of MOST connectors has been large: long pigtails, short pigtails, micro pigtails, and these connectors are combined with electrical connections with a wide range of pin sizes. The number of connectors used in MOST – mainly under cost pressureis now drastically reducing. The third section overviews the different fiber optic transceivers used for MOST. The different packages for MOST25 and MOST150 transceivers are discussed. The most difficult step was the conversion of the fiber optic transceivers to be compatible with reflow SMD processes. This step is now taken. Also the evolution towards higher data rates is shown.

CABLES USED FOR AUTOMOTIVE NETWORKS

Six different kind of cables and their corresponding connector systems have been qualified for use in automotive applications, three electrical and three optical cable types. Up to know only three types

are commonly used: Unshielded Twisted Pairs (UTP), Coax cable and PMMA Polymer, Optical Fibers (POF), details are shown in Table 1.

The main reason to limit the number of different cable types and connector systems is the high tooling and qualification costs for bringing into volume production a new cable and connector system. This poses a severe difficulty for bringing a new optical cable and the correspond-

DOI: 10.4018/978-1-4666-2976-9.ch003

Table 1. Different cables (and their corresponding connector systems) qualified for automotive networks

```
Electrical Physical Layer (ePHY)
• Unshielded Twisted Pairs (UTP)
• Shielded Twisted Pairs (STP)
• Coax cable
Optical Physical Layer (oPHY)
• 1 mm PMMA POF
• multicore Glass Optical Fiber (GOF)
• 200 um core Plastic Cladded Silica fiber (PCS)
```

ing connector system into automotive volume products, especially for optical physical layers implementations.

As can be seen from Table 2, each cable solution has at least one red box, i.e. no cable system performs excellent on all required performances.

Electrical Physical Layer (ePHY)

Every electrical physical layer implementation has the following advantages over all optical physical layer implementations:

• No need for an optical to electrical and electrical to optical conversion in every node of the network. This leads to reduced costs.

• A limited electrical power supply over the cable is possible, eliminating the need for a separate electrical power supply wiring in some nodes with small electrical power consumption (sensors e.g. in cameras).

• Implementation of bidirectional transmission full duplex can be done at low cost in the transceivers at both ends of the cable.

• Can be used for high temperatures, also for cabling in the engine compartment.

Unshielded Twisted Pair (UTP) Cable

• Advantages:
  ◦ Thin, flexible cable that is easily handled.
  ◦ Lowest cost cable material.
  ◦ Lowest cost connectors.
• Disadvantages:

Table 2. Comparison of different cables for automotive applications

Cable Type	Fiber		Copper	
Characteristics	POF	PCS	UTP	STP
Cost down potential of technology				
Temperature stability >> 85°C				
Active components (FOT, Magnetics) for >> 85°C				
Bandwidth				
Bending radius, ease of installation				
Power budget/Signal Quality/EMC@Bandwidth				
Diagnosis, Fault detection, Repair				
Scalability, Usability-over-cost in other automotive applications				
Over all sum				

Source; MOST Interconnectivity conference, "Experience and Future scenarios" by BMW AG

Figure 1. Unshielded Twisted Pair (UTP) cable (Wikipedia)

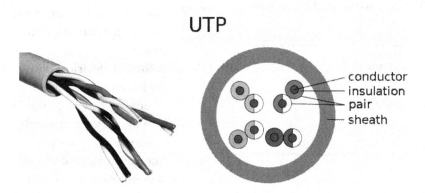

Twisted pair's susceptibility to electromagnetic interference greatly depends on the pair twisting

schemes (usually patented by the manufacturers) staying intact during the installation. As a result, twisted pair cables usually have stringent requirements for maximum pulling tension as well as minimum bend radius. This relative fragility of twisted pair cables makes the installation practices an important part of ensuring the cable's performance.

- o Limited bandwidth.
- o For high bandwidth cable, minimum bending radius is a constraint for cabling the car. The higher the bandwidth of the cable the tighter the restrictions on the minimum bending radius. A minimum bending radius requires precautions for the cabling of a car.
- o Large cross section, making it more difficult to draw a cable when there is a lack of space, e.g. a CAT-5 cable has a diameter of about 6 mm. The cross section of this cable is about 28 mm^2.
- o Limited EMI immunity.
- o Requires use of matched transformers at both ends of each UTP.

- **Poor Cable Impedance Control:** Not fully symmetric cable. Relatively broad variation of the impedance over the cable and from cable to cable. Impedance changes with bending.
- In video applications, sending information across multiple parallel signal wires, twisted pair cabling can introduce different signal delays from pair to pair known as skew. The skew occurs because twisted pairs within the same cable often use a different number of twists per meter so as to

Figure 2. Connection of twisted pair cables to transformers in order to meet EMC requirements (SMSC)

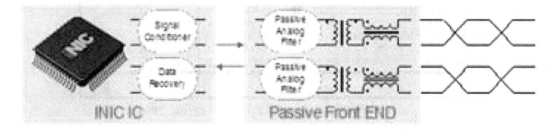

Table 3.Overview of different UTP cables (Wikipedia)

Category	Type	Frequency Bandwidth	Applications
Cat1		0.4 MHz	Telephone and modem lines
Cat2		4 MHz	Older terminal systems, e.g. IBM 3270
Cat3	UTP[6]	16 MHz[6]	10BASE-T and 100BASE-T4 Ethernet[6]
Cat4	UTP[6]	20 MHz[6]	16 Mbit/s[6] Token Ring
Cat5	UTP[6]	100 MHz[6]	100BASE-TX & 1000BASE-T Ethernet[6]
Cat5e	UTP[6]	100 MHz[6]	100BASE-TX & 1000BASE-T Ethernet[6]
Cat6	UTP[6]	250 MHz[6]	1000BASE-T Ethernet
Cat6e		250 MHz (500 MHz according to some)[who?]	
Cat6a		500 MHz	10GBASE-T Ethernet
Cat7	S/FTP[6]	600 MHz[6]	Telephone, CCTV, 1000BASE-TX in the same cable. 10GBASE-T Ethernet.
Cat7a		1000 MHz	Telephone, CATV, 1000BASE-TX in the same cable. 10GBASE-T Ethernet.
Cat8		1200 MHz	Under development, no applications yet.

prevent common-mode crosstalk between pairs with identical numbers of twists.

Screened Twisted Pair (ScTP or F/TP) Cable

ScTP cabling offers an overall sheath shield across all of the pairs. ScTP cable is different form

Shielded Twisted Pair (STP) cable. In a STP cable, each twisted pair has its own shield.

Advantages and disadvantages of ScTP are identical to UTP, except for the following:

- Advantages:
  - Excellent EMI immunity.

Figure 4. Cross section of a ScTP cable (Wikipedia)

Figure 3. Screened twisted pair (ScTP or F/TP) cable (Wikipedia)

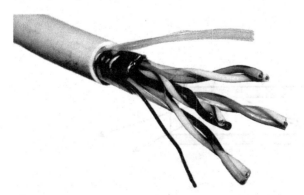

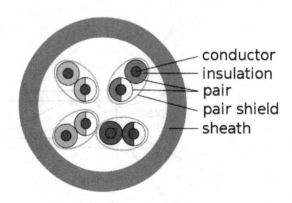

- Disadvantages:
  - Expensive connectors: require excellent connection to cable shield.
  - Connection of the cable with a plug needs to be done manually, which is expensive.

Coax Cable

Coax cable has one inner wire and one shield. Coax cable used in cars is RG-174. Compared to other coax cables of larger diameter (e.g. RG-58) the RG-174 coax cable has a higher DC resistance of its center conductor and a lower bandwidth.

Connectors shall have an impedance of 50Ω and be compliant with the mechanical and electrical properties of DIN 72954, ISO 2860-1 and SAE/USCAR 17-1 and 18-2, e.g. Fakra SMB.

Figure 5. Cross section of a coax cable

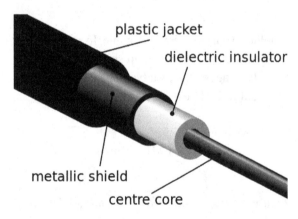

Figure 6. Optical fiber definitions

Comparison of the performance of different coax cables:

- Advantages:
  - Excellent EMI immunity.
  - Low cost cable qualified for automotive use.
  - Automated, hence low cost, assembly of cable.
  - Well controlled impedance due to full symmetry of the cable.
- Disadvantages:
  - Small diameter cable has a limited bandwidth, requiring dispersion compensation for high data rates.
  - Minimum bending radius requires precautions for the cabling of a car.

Optical Physical Layer (oPHY)

Three types of optical fiber have been qualified for automotive applications:

- 1mm PMMA POF
- 200um Plastic Cladded Silica Fiber
- 1 mm multicore Glass Optical Fiber (GOF)

The two fibers at the right side of Figure 7 are qualified for automotive applications: Hard Clad Silica Fiber (230.200 um) and PMMA POF (1000/980 um).

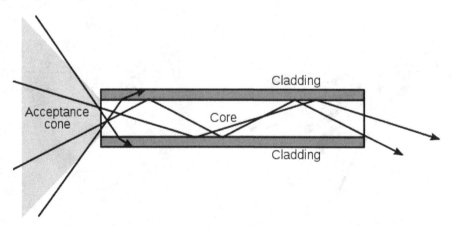

Figure 7. Comparison of different optical fibers

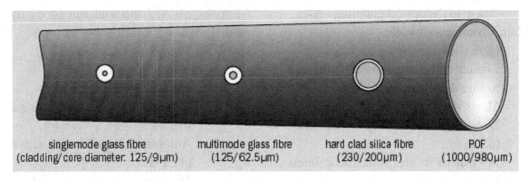

| singlemode glass fibre (cladding/core diameter: 125/9μm) | multimode glass fibre (125/62.5μm) | hard clad silica fibre (230/200μm) | POF (1000/980μm) |

Box 1. Automotive Key Requirements for Optical cable

- Based on present 85°C requirement "Test Guidelines for fiber Optic Cable / Prufrichtlinie fur LWL-Meterware"
  - Heat resistance:105°C/3,000hrs. @650nm: < 200 dB/km
  - High humidity resistance:85°C/95%/3,000hrs @650nm: < 200 dB/m
    - According to the above "Test Guidelines", PG 6 Climatic Reistance"
  - Stripping force (Inner-jacket from fiber):   >50 N
  - Pistoning @105°C/24hrs: <±30 μm
  - Thermal shrinkage @105 °C/24hrs   <0.6 %
  - Material
    - Bending stiffness/Handling while harness fabrication
    - Non-flammability

For all optical cables:

- Advantages:
  - All optical cables are fully immune to electromagnetic disturbances. This is the main driver in order to implement automotive networks with optical fiber.
  - Small diameter of the cable making it easier to put the cable in areas where there is a lack of space (usually the case in a car). The jacket of most automotive fiber optic cable has a diameter of 2.3 mm (cross section of about 4 mm²). This means that the cross section of a fiber optic cable is about 7 times smaller than the cross section of a cat-5 cable.

- Disadvantages:
  - Optical fibers need an optical to electrical and electrical to optical conversion in every node of the network. This increases costs.
  - Electrical power supply over the optical cable is not possible.
  - Implementation of bidirectional transmission full duplex requires two different wavelengths, doubles the cost of the converters, and an additional wavelength filter is needed. The additional wavelength filter results in

a significant cost increase and an increase in the optical coupling losses.

- Main advantages of POF:
  - High data rate transmission.
  - Lighter and more flexible compared to shielded electric data lines.
  - Meets strict EMC requirements.
  - Does not cause any interference radiation.
  - Insensitive to electromagnetic interference irradiation.
  - Small diameter of the cable.

1 mm PMMA POF

By using a black inner sheet of PA12, a ferrule in PA12 material can be easily laser welded to the

optical fiber. By using a black PA12 buffer layer, a ferrule in PA12 material can be easily laser welded to the optical fiber buffer layer.

- Advantages:
  - Excellent EMI immunity.
  - Low cost cable qualified for automotive use.
  - Automated, hence low cost, assembly of cable.
  - The large core diameter allows larger tolerances, which give rise to low connector prices.
- Disadvantages:
  - The large numerical aperture (NA = 0.5) of the optical fiber gives rise to a limited bandwidth, requiring for

Figure 8. Dimensions of PMMA POF used for automotive applications (Mitsubishi Rayon)

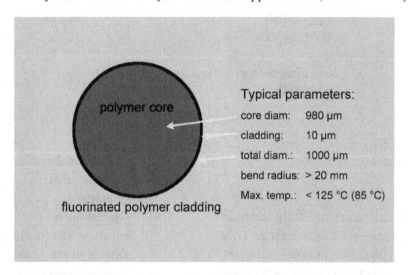

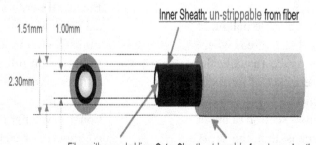

Figure 9. 1 mm PMMA plastic optical fiber terminated with a ferrule

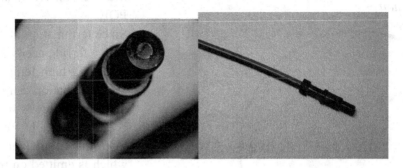

high data rates (automatic) dispersion compensation.

- ◦ Minimum bending radius of 25 mm requires precautions for cabling of a car (e.g. the use of molded bending protectors).
- ◦ The large numerical aperture (NA = 0.5) of the optical fiber gives rise to a limited bandwidth, requiring (automatic) dispersion compensation for high data rates. Bandwidth is insufficient for uncompressed camera links.
- ◦ 1 mm PMMA qualified for automotive applications is limited to a maximum ambient temperature of 95°C. Recently, improvements in the outer

jacket material of the POF made it possible to have PMMA POF qualified up to 105°C. This maximum temperature limitation for the use of PMMA POF limits the use of PMMA POF in cars, since the rooftop area requires a temperature resistance? of at least 107°C and the engine compartment requires a temperature of minimum 125°C.

Multicore Glass Optical Fiber (GOF)

- • Advantages:
  - ◦ Temperature range up to 300°C

Figure 10. Multicore glass optical fiber (GOF): Schott

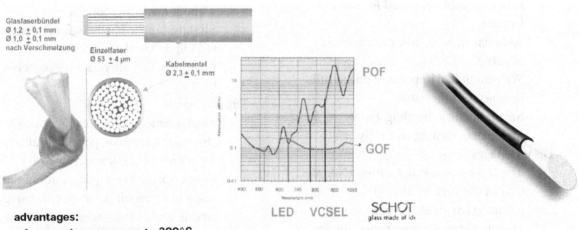

Figure 11. Cross section of multicore glass optical fiber (GOF): (Schott)

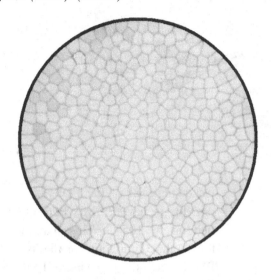

Table 4. Technical data of multicore glass optical fiber (GOF)

Fiber Core: Ø 53 ± 4 µm
Fiber Cladding: 3 µm
Fiber Bundle: 380 fibers
Fiber Bundle Diameter: Ø1 mm
Numerical Aperture: 0.5
Attenuation at 650 nm: < 0.25 dB/m
Temperature Range: 40°C to +125°C
Minimal Bending Radius: 5 mm

- ◦ Broad wavelength range: 500 nm to 1000 nm
- ◦ Low cost cable qualified for automotive use
- ◦ Automated, hence low cost, assembly of cable
- ◦ Well controlled impedance due to full symmetry of the cable.
- ◦ Small minimal bending radius of 5 mm, makes cabling relatively easy.
- • Disadvantages:
  - ◦ The large numerical aperture (NA = 0.5) of the optical fiber gives rise to a limited bandwidth, requiring (automatic) dispersion compensation for high data rates. Bandwidth is insufficient for uncompressed camera links.

- ◦ More expensive than 1mm PMMA POF.
- ◦ Since termination of the optical fiber with a ferrule needs dedicated equipment and high temperature processing, termination can only be done at the manufacturer's site.
- ◦ By coupling light from a monocore-POF into a multicore-GOF, power which is emitted on the optical cladding of each single fiber is absorbed. This brings an additional coupling loss of 0.5dB.

200 um Core Plastic Cladded Silica fiber (PCS) of Tyco Electronics (Engel 2005)

By using a black PA12 buffer layer, a ferrule in PA12 material can be easily laser welded to the optical fiber buffer layer.

- • Advantages:
  - ◦ Low cost cable qualified for automotive use.
  - ◦ Automated, hence low cost, assembly of cable.
  - ◦ Small minimum bending radius of 9 mm allows routing a cable in a car without many precautions.
  - ◦ Slightly higher bandwidth compared to POF due to the lower NA.
  - ◦ Lower optical attenuation compared to POF.
- • Disadvantages:
  - ◦ Small diameter cable requires smaller alignment tolerances, potentially affecting connector and transceiver costs.
  - ◦ Connectorization requires cutting the silica fiber, requiring more expensive cutting tools (diamond cutter or laser cutter) compared to POF.
  - ◦ Bandwidth is still insufficient for uncompressed camera links.

Figure 12. Comparison of transmission characteristics between PMMA POF and GOF (Schott)

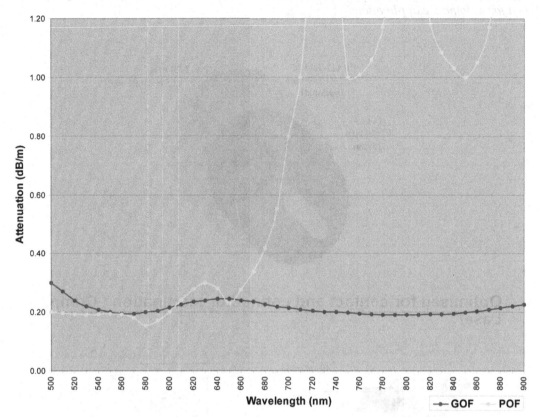

Figure 13. 0.5 dB additional coupling losses when coupling into GOF. (Schott)

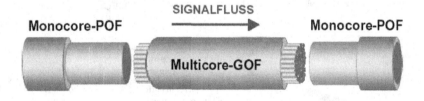

CONNECTORS

Electrical Connectors

LIN

The mechanical aspects of the physical layer (connector type and number, colors, labels, pin-outs) are not part of the LIN specification. A variety of connectors is used by the automotive manufacturers.

CAN

The mechanical aspects of the physical layer (connector type and number, colors, labels, pin-outs) are not part of the CAN specification. Each automotive electronic control unit (ECU) needs a connector to connect to the CAN bus. Although several custom types of connectors are used, several de-facto standards have emerged, the most common being the 9-pin D-sub type connector with the following pin-out:

Figure 14. Dimensions and cross section of 200 um core Plastic Cladded Silica fiber (PCS) of Delphi developed for automotive applications

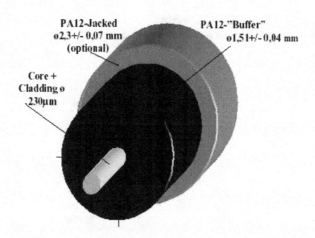

Optimised for contact and connector termination / Crimp, Laser

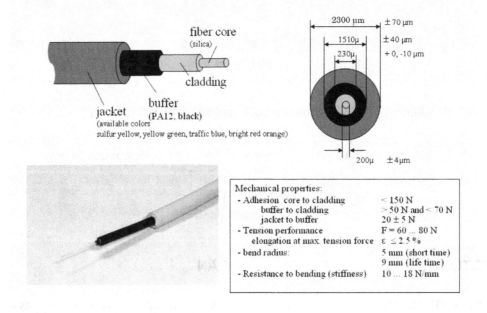

Table 5. Main advantages of using optical fiber in cars

Light Weight: Lower weight compared to UTP and ScTP electrical cable
High Bandwidth: Optical fiber offers a higher bandwidth in comparison to electrical twisted paired wires. This permits somewhat higher data rates without the need for dispersion compensation.
Immunity to EMI: Light is not susceptible to any electromagnetic fields surrounding it. It also does not interfere with other devices using electricity.
Low Cost: Optical fiber is a cost-efficient alternative to shielded electrical wires and can reduce the wiring handling cost.
Data Security: Optical fiber does not radiate electromagnetic fields, and tapping into a fiber cable without being detected is difficult, therefore data security is improved.

- **pin 2:** CAN-Low (CAN-)
- **pin 3:** GND (Ground)
- **pin 7:** CAN-High (CAN+)
- **pin 9:** CAN V+ (Power)

This de-facto mechanical standard for CAN requires nodes to have both male and female 9-pin D-sub connectors electrically wired to each other in parallel within the node. Bus power is fed to a node's male connector and the bus draws power from the node's female connector. This follows the electrical engineering convention that sources of power are terminated with female connectors.

The following connectors are used for CAN.

FLEXRAY

Optical Connectors

MOST Connectors

MOST defines one optical MOST connector. Such a connector is depicted in Figure 15 together with a 2+0 connector interface. In order to protect the connector, connectors are delivered with a dust cap.

Table 6. 9-Pin D, CAN Bus Pin Out

9 Pin (male) D-Sub CAN Bus PinOut		
Pin #	Signal Names	Signal Description
1	Reserved	Upgrade Path
2	CAN_L	Dominant Low
3	CAN_GND	Ground
4	Reserved	Upgrade Path
5	CAN_SHLD	Shield, Optional
6	GND	Ground, Optional
7	CAN_H	Dominant High
8	Reserved	Upgrade Path
9	CAN_V+	Power, Optional

Table 7. 9-Pin D, CAN Bus Pin Out

10-Pin Header CAN Bus PinOut		
Pin #	Signal Names	Signal Description
1	Reserved	Upgrade Path
2	GND	Ground, Optional
3	CAN_L	Dominant Low
4	CAN_H	Dominant High
5	CAN_GND	Ground
6	Reserved	Upgrade Path
7	Reserved	Upgrade Path
8	CAN_V+	Power, Optional
9	Reserved	Upgrade Path
10	Reserved	Upgrade Path

According to the CANopen connector pin assignment [CiA draft DR-303-1]; pins 9 and 10 are reserved supporting direct connection to a 9-pin D connector.

Table 8. RJ-Style, CAN Bus Pin Out

RJ-45 connector

RJ10, RJ45 CAN Bus PinOut			
RJ45 Pin #	RJ10 Pin #	Signal Name	Signal Description
1	2	CAN_H	Dominant High
2	3	CAN_L	Dominant Low
3	4	CAN_GND	Ground
4	-	Reserved	Upgrade Path
5	-	Reserved	Upgrade Path
6	-	CAN_SHLD	CAN Shield, Optional
7	-	CAN_GND	Ground
8	1	CAN_V+	Power, Optional

MOST Connector Interfaces

MOST defines five connector interfaces. The general name is x+y connector interface, with x being the number of optical connections made in the connector interface, and y the number of electrical connections made in the connector in-

Table 9. 7-Pin Open Style, CAN Bus Pin Out

7-Pin Open Style CAN Bus PinOut		
Pin #	Signal Names	Signal Description
1	CAN_GND	Ground
2	CAN_L	Dominant Low
3	CAN_SHLD	Shield, Optional
4	CAN_H	Dominant High
5	CAN_V+	Power, Optional

4-pin Open Style Connectors either use pins 1-4 (Version A) or pins 2-5 (Version B). 3-pin Open Style Connectors use pins 2-4. The bus node provides the male pins of the connector.

terface. These connector interfaces are identical for MOST25 as well as for MOST150.

The connection between the connector interface and the fiber optic transceiver can be achieved through a micro pigtail or a long pigtail.

In MOST, FOT's are offered in two types of package: a Single in Line package (SIL) and a Surface Mount (SMD) package. The connection between the MOST connector interface and the FOT is different for the SIL and the SMD packages.

Figure 15. MOST connector of Yazaki with dust cap, and a 2+0 MOST connector with FOT pair connected with a micro pigtail

Connection from Connector Interface to FOTs in SIL Package

There are three FOT MOST25 suppliers: Avago (using the cavity as interface package developed by Infineon), Hamamatsu, and Firecomms (both using a clear molded package). As the FOT package was not standardized for MOST25, a plastic

Box 2. FLEXRAY

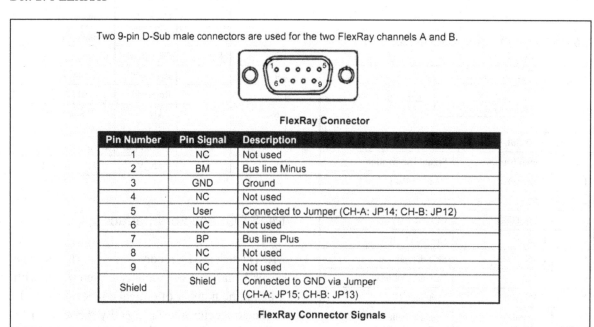

Two 9-pin D-Sub male connectors are used for the two FlexRay channels A and B.

FlexRay Connector

Pin Number	Pin Signal	Description
1	NC	Not used
2	BM	Bus line Minus
3	GND	Ground
4	NC	Not used
5	User	Connected to Jumper (CH-A: JP14; CH-B: JP12)
6	NC	Not used
7	BP	Bus line Plus
8	NC	Not used
9	NC	Not used
Shield	Shield	Connected to GND via Jumper (CH-A: JP15; CH-B: JP13)

FlexRay Connector Signals

Table 10. Overview of MOST connector interfaces (MOST standard)

"Nick name"	Number of optical contacts	Number of	electrical	contacts
		PIN = 0,63 mm	PIN = 1,5 mm	PIN = 2,8 mm
2+0	2	·	·	·
2+4	2	4	·	·
2+12	2	12	·	·
2+20	2	18	2	·
4+40	2 x 2	2 x 12	·	2 x 8

Figure 16. 2+0 and 2+4 MOST micro pigtail headers of Yazaki

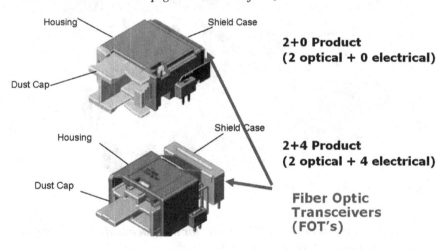

2+0 Product
(2 optical + 0 electrical)

2+4 Product
(2 optical + 4 electrical)

Fiber Optic Transceivers (FOT's)

Figure 17. MOST150 2+0 connector of Tycoelectronics interfacing with a MOST150 FOT pair in single in line package

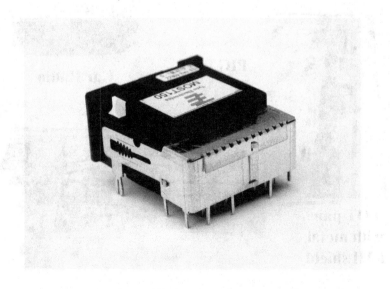

coupling structure is needed in order to fix the micro pigtail or long pigtail to the FOT. In order to reduce radiated emission form the FOT leads, a metal shield is placed around the FOT.

Connection from connectoriInterface to FOT's in SMD package. Only long pigtails are used in order to connect between the connector interface and the SMD package. As optical fiber for the pigtails, a dedicated "pigtail" fiber is used, qualified for automotive use. Since the optical pigtail fiber has a high rigidity, and in order to keep the bending radius of the optical fiber under control, a 90 degree bending protector is used.

Figure 18. 2+0 MOST connector of Tycoelelectronics interfacing with a Fiber Optic Transceiver pair though micro pigtails. The Fiber optic Transceivers are surrounded by a metal shield in order to reduce radiated emission of magnetic disturbances (Engel 2005).

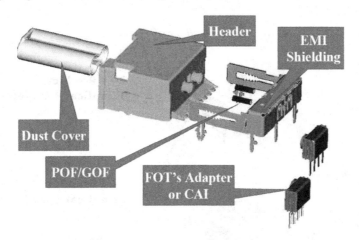

Figure 19. Connection between a standard MOST connector interface and a fiber optic transceiver using a long pigtail. Implementation in a car radio in BMW.

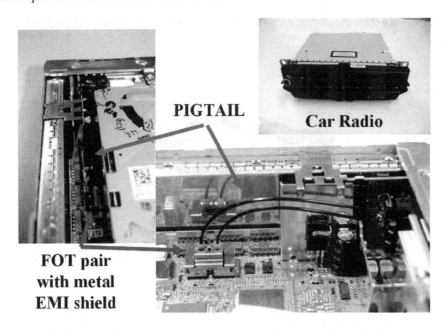

Figure 20. Connection between a standard MOST connector interface and a fiber optic transceiver using a long pigtail

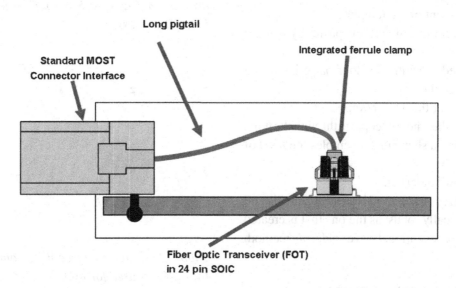

Note that the SMD has an integrated ferrule clamp. The ferrule clamp is multi-pluggable.

The single in line package is an insertion component. Electronics assembly industry is moving away from insertion components towards surface mounting technology. The main driver for this is cost. SMD components can be placed by an automatic pick and place machine. Insertion components must be fitted with their leads through narrow holes in the PCB. Due to relatively large tolerances on the leads position, the fitting of insertion components into the BCP holes is mainly done manually. Investments in and operational costs of wave soldering machines are relatively higher than for SMD reflow machines. Therefore SMD components are attractive form an assembly point of view.

During recent years, the standardized MOST connector interfaces as well the FOTs have been converted form insertion components to SMD components. As temperature peaks during reflow reach 260°C and hence the temperatures are significantly higher for the body of the components than during wave soldering processes, this requires a major reengineering of the connector interfaces as well as for the FOT package. Although FOTs

packaged in single in line packages and been made resistant to the SMD reflow temperatures, cannot be called real SMD components, since they still have to be inserted, they might be used in SMD assembly by a technique called "Pin-in-paste".

Figure 21. Pigtails ended with dedicated MOST150 ferrules (different for receiver pigtail and transmitter pigtail) inserted into an SMD MOST150 transceiver of Melexis

1. Printed circuit board with a plated-through drill hole.
2. Alignment of the template.
3. Application of solder paste by screen printing.
4. Desired amount of solder paste is pushed through the hole.
5. Fitting of the THR component.
6. Pin pushes the solder paste through the hole, forming a shape that resembles the head of a match.
7. Reflow soldering.
8. Connection achieved is as mechanically and electrically robust as the one that is created by the conventional wave soldering method.

TRANSCEIVERS USED FOR AUTOMOTIVE NETWORKS

Market Size

End of 2004, Infineon (now Avago) reported that they had shipped 11 million transmitter/receivers pairs to BMW and in total over 25 Million MOST transmitter/receivers. In November 2011, Hamamatsu reported they have shipped 50 million MOST transmitter/receivers pairs, and they serve about 60% of the MOST market. The MOST25 market in 2011 is estimated to be about 18 million transmitter/receivers pairs per year.

Every MOST transceiver has a receiver and a transmitter. When MOST is used in infotainment networks, a ring network is made as shown in Figure 23.

Packages

MOST150 Transceivers Packaged in Single Inline Package

Development of an optical package which withstands the high temperatures of SMD processes is very challenging. The main reason is the differ-

Figure 22. The Pin–in–paste technique is used for integrating through hole components, which are resistant against SMD reflow temperatures in SMD processes.

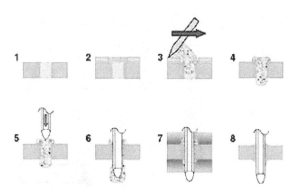

Table 11. Overview of cabling connectors and transceivers used for MOST

MOST25
OPHY 650 nm LEDs or RCLEDs 1mm PMMA POF
(FOT) Transceiver technology in 4 pin side looker package:
● Cavity as interface (Infineon)
● Clear molding (Hamamatsu, Firecomms)
● Clear molding over molded with black molding compound (Sharp)
(FOT) Transceiver technology in 24 pin SMD package:
● Cavity molding (Melexis)
● Pre molded package (Hamamatsu)
Headers:
● **Micro Pigtail:** 2+0, 2+4, 2+12, 2+20,
● **Long Pigtail:** 4+40
aOPPHY 850 nm VCSEL 200 um core PCS
(FOT) Transceiver technology in 4 pin side looker package:
● Clear molding (Hamamatsu)
● Cavity molding (Melexis)
● Pre molded package (ASTRI & Acasia technologies, Finisar)
MOST50
ePHY UTP cable with 2 twisted pairs and line transformers
MOST150
OPHY 650 nm LEDs or RCLEDs 1mm PMMA POF
(FOT) Transceiver technology in 7 pin side looker package:
● Clear molding (Hamamatsu, Avago)
● Cavity molding (Melexis)
(FOT) Transceiver technology in 24 pin SMD package:
● Cavity molding (Melexis)
● Pre molded package (Hamamatsu)
Headers:
● **Micro Pigtail:** 2+0 (side looker package),
● **Long Pigtail:** 4+40 (SMD package)
ePHY coax cable bidirectional transceiver on both sides electrical transceiver EQCOLOGIC
MOST3 to 5 Gbps
aoPHY 850 nm VCSEL small diameter silica fiber
(FOT) Transceiver technology in 24 pin SMD package:
● Pre molded package (Hamamatsu)

Figure 23. MOST transceivers connected in a ring network

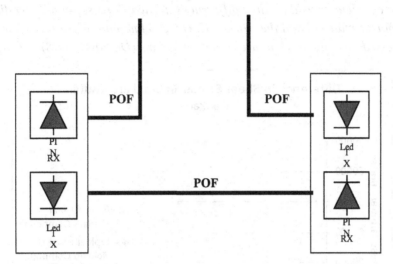

ences in thermal expansion coefficients between three materials: the silicon, the lead frame where the silicon is mounted upon, and the plastic material of the package. Three optical packaging technologies have been developed which were able to withstand the temperatures of the SMD reflow processes:

- Pre molded cavity packages.
- Cavity molding technique.
- Clear molding.

Pigtail insertion into SMD FOTs is mistake proof (Poka-Yoke). Receiver and transmitter have two times a different diameter so that receiver ferrule cannot be inserted in the transmitter coupling hole, and the transmitter ferrule cannot be inserted into the receiver coupling hole of the fiber optic transceiver package.

Table 12. Overview of the different thermal expansion coefficients of different package materials.

Material	TCE below Tg (for epoxies)
Silicon	2.6 ppm/C
Cu Leadframe	16.5 ppm/C
Pure epoxy without filler	64 to 67 ppm/C (*)
Black Epoxy (standard amount of filler)	Depending on the molding compound: alfa1 = 7 to 20 ppm/C
Clear Epoxy with micron size glass filler	52 ppm/C (30% filler) (**) 43 ppm/C (50% filler) (**) 35 ppm/C (70% filler) (**)
Clear Epoxy with nanopartcles filler	58 ppm/C (20% filler) 52 ppm/C (30% filler =max)

Figure 24. As shear forces rise with a third power from the center of the package, small differences in thermal expansion coefficients lead to large differences in shear forces, possibly leading to delamination between the plastic material and the silicon IC. Once delamination occurs, the wire bonds on the IC surface are stressed, leading to IC failures after some time. (De Pauw 2008)

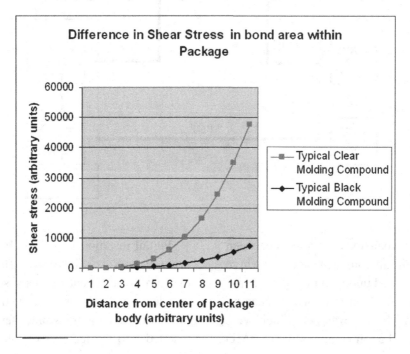

Figure 25. MOST150 fiber optic transceivers packaged into two 7-pin single in line packages

Figure 26. MOST150 transceiver in SMD package of Melexis

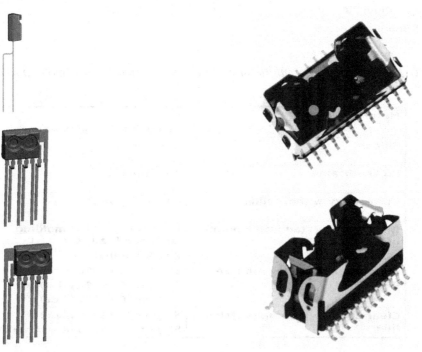

Figure 27. Pigtail insertion into SMD FOTs is mistake proof (Poka-Yoke)

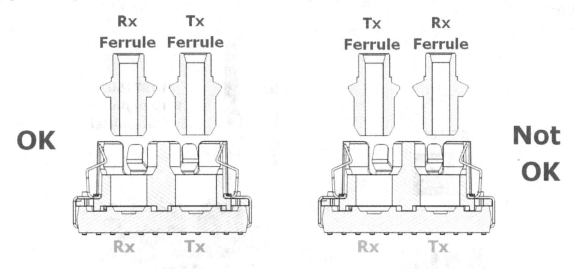

State of the Art Description: 150 Mbps MOST

Melexis provides, in a single plastic SMD component, a fully integrated fiber optic transceiver. The fiber optic transceiver also includes the optical connector. For the receiver a dual PIN photo detector was implemented in order to obtain maximum EMC immunity. The receiver die is fully monolithic. Sensitivity for an extinction ratio of minimum 10 and a BER below 2 E-9 is -23.5 dBm minimum for 650 nm light over the entire temperature range -40°C to +95°C.

Transmitter

The optical power coupled into an optical fiber is at minimum -6.5 dBm for an extinction ratio of minimum 10 dB for 650 nm light over the entire temperature range -40°C to +105°C. Hence a minimum optical budget of 17 dB can be guaranteed. Modulation and peaking currents can be fully programmed, as well as their temperature

Figure 28. Fully integrated 150 Mbps receiver die MLX75603, placed into the MOST150 transceiver of Melexis (De Pauw 2008)

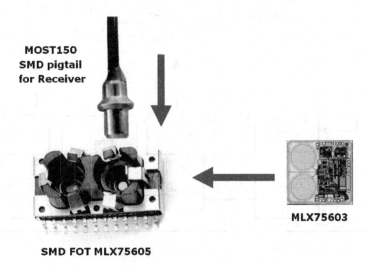

Figure 29. 150 Mbps LED driver die MLX75604 and LED, placed into the MOST150 transceiver of Melexis (De Pauw 2008)

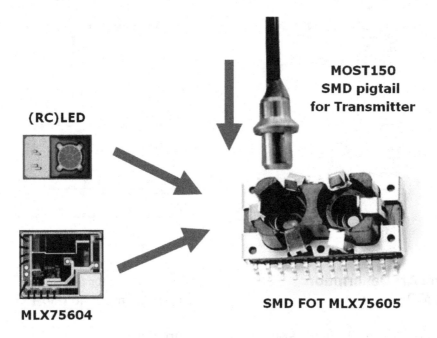

Figure 30. Transmitter circuit of Melexis MLX75604

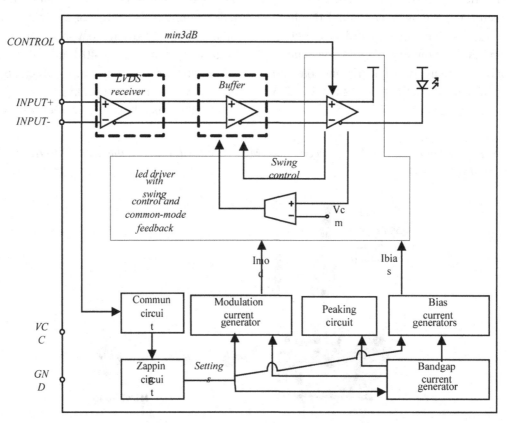

dependence. Using the appropriate temperature coefficient for the LED modulation current, the MLX75604 die can be trimmed so that the optical power variation of the LED over the temperature range [-40°C, +95°C] can stay within a 2 dBo wide band. The LED is mounted on top of the transmitter die. This minimizes radiated EM disturbances.

In order to make a 150 Mbps transmitter one can choose between two low cost light sources: LEDs and RCLEDs. LEDs are slow, hence they need peaking current in order to reduce the rise time. The use of LEDs, in contrast to the use of last RCLEDs, has two advantages: Higher optical power at high temperatures and higher reliability. Indeed, using LEDs, the transmitter can be qualified to be used in ambient temperatures up to 105°C, exceeding the requirements in an important way.

LEDs and RCLEDs might degrade over the product lifetime. An important aging mechanism is the generation of crystal defects (dislocations) within the active region where the electrons and holes recombine. Crystal defects give rise Shockley-Read-Hall recombination, which competes with the radiative recombination, hence generating heat instead of photons. The superior reliability and high temperature performance of LEDs is attributed to the significantly lower current density in the active region of the device.

Figure 31 gives the performance of the LED during a 55X accelerated life test, and figure 32 shows the predicted time to 10 ppm and 50% cumulative failure rate points, versus the requirements. Note the predicted reliability exceeds the requirements.

Connector Performance

Mechanical stress in an automotive environment is very severe, making the connector design chal-

Figure 31. Performance of the LED of Melexis transmitter during a 55X accelerated high temperature life test. Failure criterion is light output variation of more than 1 dB. (De Pauw 2008).

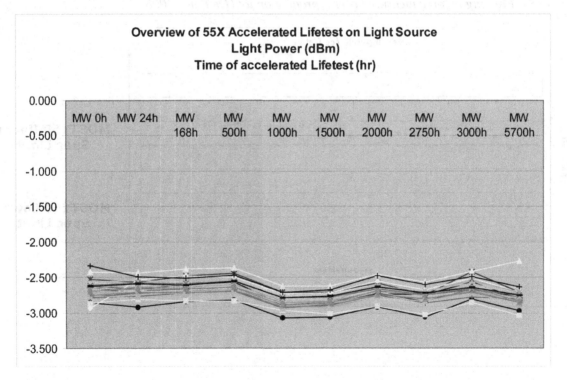

Figure 32. Predicted time to 10 ppm and 50% cumulative failure rate points, versus the requirements (De Pauw 2008)

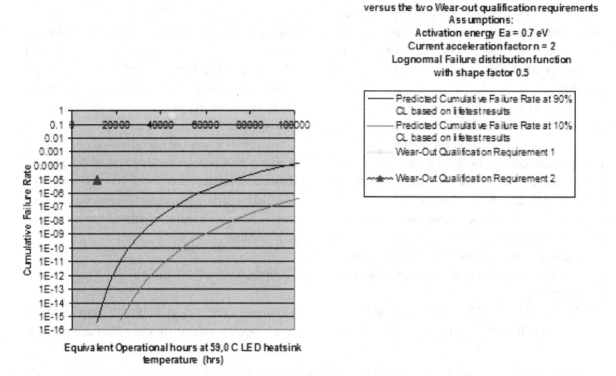

Figure 33. Pressing in force measured over a production lot (De Pauw 2008)

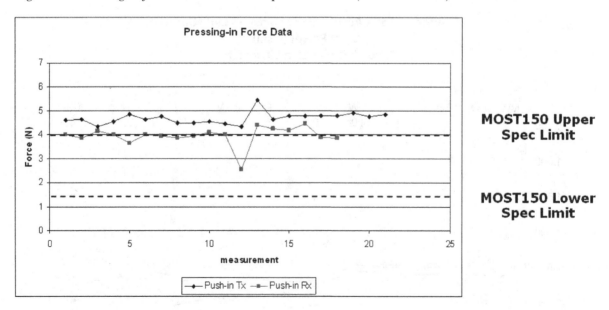

Figure 34. Push-out force measured over a production lot (De Pauw 2008)

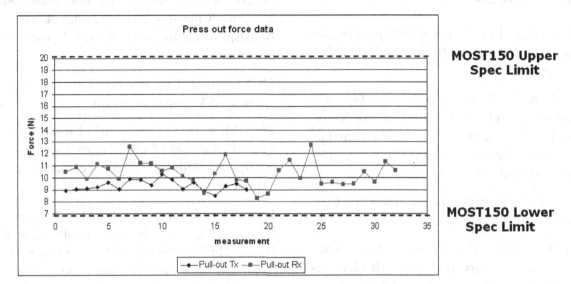

lenging. The connector is a multiple pluggable connector, and meets the low pressing-in force and high push-out force requirements of automotive applications.

Assembly of the Fiber Optic Transceiver

With MLX75605, Melexis provides a low cost fiber optic transceiver, able to withstand lead-free SMD processes. In order to obtain this performance, they applied a cavity molding process.

Figure 35. Cavity molding technique (De Pauw 2008)

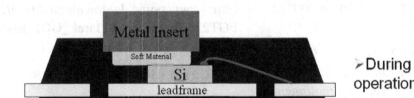

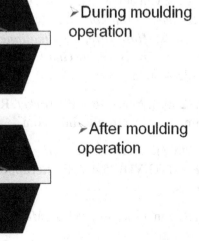

REFERENCES

Consortium, L. I. N. (n.d.). *LIN-specification package*. Retrieved from http://www.lin-subbus.de/

De Pauw, P. (2008). *Melexis new fiber optic ttransceivers*. ITG VDE 5.4.1, Fachgruppentreffen Krems. Retrieved from http://www.pofac.de/downloads/itgfg/fgt25/FGT25_Krems_De-Pauw_Melexis-Transceivers.pdf

Delphi. (2003). *MOST Interconnectivity Conference, Frankfurt*.

Engel, A. (2005). *Tyco Electronics next generation MOST networking*. MOST All Members Meeting, Frankfurt, 5th of April 2005.

Flexray. (2010). *Flexray specifications version 3.0.1*. Retrieved from http://www.flexray.com/index.php?sid=9c03bd01bb2a03269f772bb99935a742&pid=94&lang=de

Group, I. G. I. (2009). Plastic optical fiber (POF) selected reprint series, Vol. 1. Retrieved from http://igigroup.net/osc3/product_info.php?cPath=22&products_id=174

Grzemba, A. (2011). *MOST: The automotive multimedia network: From MOST25 to MOST150*. Poing, Germany: Franzis Verlag GmbH.

Grzemba, A., & von der Wense, H.-C. (2005). *LIN-Bus –Systeme, Protokolle, Test von LIN-Systemen, Tools, Hardware, Applikationen*. Poing, Germany. *Frazis-Verlag GmbH.*, *ISBN-13*, 978–3772340093.

Liburdi, F. (2005). *New IDB-1394 socket*. Retrieved from www.electronic-links.com

Melexis. (2008). *Melexis new fiber optic ttransceivers*. ITG VDE 5.4.1, Fachgruppentreffen Krems.

Mitsubishi Rayon. (2009) *MOST All Members Meeting*, Frankfurt.

MOST. (n.d.). *Specifications*. Retrieved from http://www.mostcooperation.com/publications/specifications-organizational-procedures/index.html

MOST. (n.d.). *Cooperation*. Retrieved from www.MOSTCooperation.com

POSESKA. (n.d.). *Technical characteristics of POF*. Retrieved from http://www.pofeska.com/pofeskae/download/pdf/technical%20data_01.pdf

Robert Bosch Gmb, H. (1991). *CAN specification*, v 2.0. Stuttgart, Germany: Robert Bosch GmbH. Retrieved from http://www.semiconductors.bosch.de/media/pdf/canliteratur/can2spec.pdf

Robert Bosch Gmb, H. (2011). *Automotive handbook* (8th ed.). Plochingen, Germany: Robert Bosch GmbH.

Schott. (n.d.). *Spec sheet*. Retrieved from http://www.schott.com/korea/korean/download/specsheet_datacom_e&a_2_5.pdf

Sklarek, W. (2006). *High data rate capabilities of multicore glass optical fiber cables*, (p. 22). Fachgruppentreffen der ITG-FG 5.4.1, Optische Polymerfasern, Oct 2006. Retrieved from http://www.pofac.de/downloads/itgfg/fgt22/FGT22_Muenchen_Sklarek_GOF-Buendel.pdf

SMSC. (n.d.). Retrieved from www.smsc.com/

Tyco Electronics. (2003). *MOST Interconnectivity Conference*, Frankfurt.

Warrelmann, J. (2003). *Glasfaserbündel für Datenkommunikation*, (p. 15). Fachgruppentreffen der ITG-FG 5.4.1, Optische Polymerfasern, Offenburg 25- 26, März 2003. Retrieved from http://www.pofac.de/downloads/itgfg/fgt15/FGT15_Offbg_Warrelmann_GOF-Buendel.pdf

Wikipedia. (n.d.). *Unshielded twisted pair*. Retrieved from http://en.wikipedia.org/wiki/Unshielded_twisted_pair#Unshielded_twisted_pair_.28UTP.29

Ziemann, O., Krauser, J., Zamzow, P., & Daum, W. (2008). *POF handbook: Optical short range transmission systems* (2nd ed.). Springer.

Zimmermann, W., & Schmidgall, R. (2011). Bussysteme in der Fahrzeugtechnik: Protokolle, Standards und Softwarearchitektur. ATZ/MTZ-Fachbuch. Auflage, Germany: Vieweg+Teubner Verlag. ISBN-13: 978-3834809070

APPENDIX: ABBREVIATIONS

- **ABS:** Anti-lock Braking System
- **ACC:** Adaptive Cruise Control
- **aoPHY:** advanced optical physical layer
- **CAN:** Controller Area Network
- **D2B:** Domestic digital bus
- **DAB:** Digital Audio Broadcasting
- **FOT:** Fiber Optic Transceiver
- **GOF:** Glass Optical Fiber bundle
- **HVAC:** Heating, Ventilating, and Air Conditioning
- **IDB:** ITS Data Bus
- **LED:** Light-Emitting Diode
- **LIN:** Local Interconnect Network
- **Mbps:** Megabits per second
- **MOST150:** 150 Mbit/s version of MOST
- **MOST25:** 25 Mbit/s version of MOST
- **MOST50:** 50 Mbit/s version of MOST
- **PD:** PhotoDiode
- **PIN:** diode Positive-Intrinsic-Negative diode
- **PMMA:** Polymethyl Metacrylate
- **POF:** Plastic Optical Fiber
- **RCLED:** Resonant Cavity Light-Emitting Diode
- **RX:** Receiver
- **TIA:** Trans-Impedance Amplifier
- **TX:** Transmitter
- **UI:** Unit Interval
- **VCSEL:** Vertical Cavity Surface Emitting Diode

Chapter 4
Modeling of Polymer Optical Fibers

Javier Mateo
Universidad de Zaragoza, Spain

Ángeles Losada
Universidad de Zaragoza, Spain

Alicia López
Universidad de Zaragoza, Spain

ABSTRACT

The idea of this chapter is to give a complete overview on a matrix approach to describe light propagation in strongly multimode fibers such as 1-mm diameter plastic optical fibers. These large core fibers accept such a huge number of travelling modes that they can be viewed as a continuum. Thus, light propagation can be described as a power flow by a differential equation that can be more easily solved using matrices. Thus, the key of this method is the propagation matrix that is calculated from the diffusion and attenuation functions characteristic for a given fiber type. The propagation matrix has temporal frequency dependence and can be used to obtain not only angular power distributions but also temporal parameters such as pulse spread or bandwidth. This approach is flexible to introduce localized perturbations of power distribution provided they can be modeled as matrices. Thus, the effect of devices such as scramblers or connectors and also of disturbances such as curvatures and tensions can be introduced at different points in the fiber path to assess their impact on transmission properties. One of the most critical parameters when designing a network is its bandwidth and how it decreases when increasing the link reach. This dependence has been assumed to be linear when both bandwidth and length are represented in logarithms with a slope whose value provides information of the processes underlying propagation. Thus, the authors apply the model to calculate the bandwidth versus length dependence under different conditions analyzing the value of the slope and explaining previous experimental findings.

DOI: 10.4018/978-1-4666-2976-9.ch004

INTRODUCTION

Plastic optical fibers (POFs) are being considered for high-performance fiber links at very short distances because of their ductility, light weight and ease of connection as compared to glass optical fibers in spite of their narrower bandwidths. Pulse broadening in multimode fibers is mainly due to modal dispersion, that in PMMA based POFs seems to be related mainly to optical power diffusion. High NA POFs are characterized by propagating a huge number of modes that can be approximated to a continuum of modes, each mode characterized by its propagation angle. Thus, the power exchange between modes can be viewed as power diffusion whose physical origin is either Rayleigh scattering by impurities in the bulk material or Mie scattering produced by irregularities in the core-cladding interface (Bunge el al., 2006). As power propagates through the length of the fiber, diffusion induces power transfer between neighboring angles which is translated as mode mixing, an effect that is much stronger than that occurring in multimode glass fibers. Bends, curvatures and other macroscopic strains also act as diffusive agents transferring power to other further separated angles or modes. All these diffusive effects intrinsic and extrinsic to the fiber have an impact on the fiber's modal dispersion which consequently determines its temporal response and bandwidth.

Power diffusion changes the angular power distribution as light propagates throughout the fiber by transferring power from some angles towards others. These changes affect the optical paths and produce delays which have an impact on temporal properties and, in particular, on bandwidth. In general, mode coupling tends to reduce modal dispersion in some conditions leading to a dependence of bandwidth with fiber length different from the expected linear rate, which is not necessarily the square-root dependence found for glass fibers after they have reached equilibrium. The strong diffusion in POFs how-ever, does not guarantee a rapid achievement of steady conditions, which can only be reached after 75-100 m propagation (Djordjevich & Savović, 2008). Since the main use of current POFs is in short-range applications, POF links will be mostly working in non-stationary conditions. In these distance ranges, the launching conditions have a strong impact on the fiber performance and localized disturbances such as curvatures, bends and stress have also a strong influence in propagation properties (Losada et al., 2002, Fuster & Kalymnios, 1998). Their effect is not simply power loss but also power transfer from one mode to the others which are further away from their nearest neighbors. These changes in power distribution due to mode mixing also alter the modes' relative temporal delays and have a direct relation to bandwidth (Kalymnios, 1999). On the other hand, for under-filled launching conditions, the effect of diffusion can increase modal dispersion and, in addition, significantly changes output-field properties that can degrade beam quality with possible consequences for power delivery and sensory systems. Thus, a model that accounts for power transfer among modes is necessary to describe the behavior of the fiber in such conditions.

Another source of changes in the angular power distribution is the differential attenuation, different from the overall attenuation that reduces the whole propagating power as an exponential function of the fiber length. Differential attenuation is a function of the angle, as power following longer paths characterized by higher propagation angles suffer higher losses than power confined near the fiber center. The mechanisms behind attenuation are not so different from those causing diffusion. For attenuation, however, the power that leaves a mode is not transferred to another mode but it is radiated and lost. Both diffusion and attenuation act jointly to change the angular power distribution but their relative influence changes depending on several factors. Usually, diffusion determines the initial power spread as power is launched into the fiber, while attenuation is responsible for the reshaping of the power distribution when it is closer to the steady state distribution. In general, diffusion has a stronger impact for short fiber lengths and differential attenuation dominates at longer lengths, but it also depends on the launching conditions (Losada et al., 2007). External factors can also alter differential attenuation for the same reasons as in diffusion (Arrue et al., 2001, Durana et al. 2003, Losada et al. 2003). Thus, a model able to incorporate angular diffusion and attenuation to account for differential power losses and power interchange is necessary to obtain an accurate picture of the fiber behavior. In addition, this model has to be flexible to study the propagation properties for different launching conditions and has to be able to integrate different disturbances to quantify their effects over propagation.

A direct way to describe mode mixing is provided by the power flow differential equation introduced by Gloge (1972) in the context of POFs. The solution of this equation can be found using different approaches as described in (Breyer et al., 2007), but in the next section we apply the finite-difference method to solve the temporal generalization of the power flow equation in the frequency domain (Mateo et al., 2009). The resulting equations are re-written in matrix form that provides a fast and robust approach to describe power propagation., This matrix approach is flexible to introduce the effects of localized disturbances providing they can be described as matrices as described in the following section. Next, we illustrate the model potential to obtain different spatial and temporal parameters related to fiber transmission, and how useful it is to predict them for conditions where its measurement is difficult or unpractical. The last chapter offers a discussion of the bandwidth in POFs.

PROPAGATION MATRIX MODEL

Here, we describe a fast and robust matrix approach of the finite-difference method to solve the temporal generalization of the power flow equation in the frequency domain that provides the space-time evolution of the optical power when it is transmitted throughout the fiber. The obtained 3D propagation matrix that fully characterizes the fiber behavior gives a complete description of the power distribution at a given fiber length that can be used to derive parameters with strong impact over fiber transmission capabilities, giving a full insight of fiber behavior.

Gloge Diffusion Equation in Matrix Form

We start from Gloge's power flow differential equation to describe the evolution of the modal power distribution as it is transmitted throughout a POF where different modes are characterized by their inner propagation angle with respect to fiber axis (θ), which can be taken as a continuous variable (Mateo et al., 2006) due to the high number of modes propagating through the 1-mm core POF. We make no assumptions about the angular diffusion and attenuation which are described as functions of θ, $d(\theta)$ and $\alpha(\theta)$ respectively. Following the procedure described by Gloge (1973) to introduce the temporal dimension, the partial derivative of the optical power, $P(\theta,z,t)$, with respect to z, we get the following equation:

$$\frac{\partial P(\theta,z,t)}{\partial z} = -\alpha(\theta)P(\theta,z,t)$$
$$-\frac{n}{c\cos\theta}\cdot\frac{\partial P(\theta,z,t)}{\partial t}$$
$$+\frac{1}{\theta}\frac{\partial}{\partial\theta}\left(\theta\cdot d(\theta)\cdot\frac{\partial P(\theta,z,t)}{\partial\theta}\right). \quad (1)$$

Then, we take the Fourier transform at both sides of Equation (1) and use the Fourier derivation property to obtain the following simplified equation:

$$\frac{\partial p(\theta, z, \omega)}{\partial z} = -\left(\alpha(\theta) + \frac{n}{c \cos \theta} \cdot j\omega\right) p(\theta, z, \omega)$$
$$+ \frac{1}{\theta} \frac{\partial}{\partial \theta} \left[\theta \cdot d(\theta) \cdot \frac{\partial p(\theta, z, \omega)}{\partial \theta}\right],$$

$$(2)$$

where $p(\theta, z, \omega)$ is the Fourier transform of $P(\theta, z, t)$.

To solve this differential equation we implement a finite-difference method where we use a forward difference for the first z derivative, and first and second-order central differences for the first and second angular derivatives, respectively. Thus, the power at angle θ and distance $z + \Delta z$ is obtained as the linear combination of the power at the same angle and the two adjacent angles $(\theta + \Delta\theta, \theta - \Delta\theta)$ for a distance z as shows the following equation in Box 1.

Equation (3) can be expressed in a more compact representation in matrix form. In fact, the differential changes in the angular power distribution at each Δz step are given by a simple matrix product. Thus, given the angular power at an initial length z_1, the power distribution at a longer length z_2 can be calculated with the following matrix equation:

$$\mathbf{p}(z_2, \omega) = \left(\mathbf{A}(\omega) + \mathbf{D}\right)^m \cdot \mathbf{p}(z_1, \omega),$$

$$(4)$$

Table 1. Attenuation γ and $Q_N(\theta)$ parameters for the best fits to the SSD of the three fibers

	γ			$Q_N(\theta)$ [a]				
Fiber	*np/m*	*dB/m*		σ_1	θ_1	σ_2	θ_2	*RMSE*
GH	0.0331	0.1438		6.5327	0.2689	4.0744	0.2689	2.479×10^{-3}
HFB	0.0344	0.1496		6.1991	0.2651	3.9867	0.2651	5.517×10^{-3}
PGU	0.0404	0.1754		6.4475	0.3275	5.1740	0.1920	2.578×10^{-3}
[a] Defined in Equation (12).								

Box 1.

$$\begin{aligned} p(\theta, z + \Delta z, \omega) \; &= \left[1 - \left(\alpha(\theta) + \frac{n}{c \cos \theta} \cdot j\omega\right)\Delta z\right] p(\theta, z, \omega) \\ &+ \frac{\Delta z}{2 \cdot \Delta\theta}\left(\frac{d(\theta)}{\theta} + d'(\theta)\right)\left(p(\theta + \Delta\theta, z, \omega) - p(\theta - \Delta\theta, z, \omega)\right) \\ &- \frac{2d(\theta)\Delta z}{\Delta\theta^2} p(\theta, z, \omega) \\ &+ \frac{d(\theta)\Delta z}{\Delta\theta^2}\left(p(\theta + \Delta\theta, z, \omega) + p(\theta - \Delta\theta, z, \omega)\right). \end{aligned}$$

$$(3)$$

where $\mathbf{p}$ is a vector where each component k is the power at the discretized propagation angle $\theta = k \cdot \Delta\theta$, and $m = \dfrac{z_2 - z_1}{\Delta z}$ is an integer that can be found for any pair of lengths, $z_2 > z_1$ providing we choose a small Δz.

This method we propose to solve Equation (2) by calculating multiple matrix powers is more efficient than performing the same number of iterations, particularly when using MatLab®. Even more, as will be explained in the next paragraph, matrices $\mathbf{A}$ and $\mathbf{D}$ are sparse which makes it even more efficient. In addition, it is not necessary to re-calculate these matrices when changing the initial condition, as they only depend on the fiber diffusion and attenuation.

Propagation Matrix

In the Equation (4), the term: $(\mathbf{A}(\omega) + \mathbf{D})$ is a three dimensional matrix depending, not only on input and output angles, but also on frequency ω, that is called "Propagation matrix", $\mathbf{M}$. This matrix carries all space-time information concerning power propagation through the fiber and thus, gives a complete description of the fiber as a transmission system. $\mathbf{M}$ is calculated from two matrices: $\mathbf{A}$ and $\mathbf{D}$ that describe the changes in power distribution due to power attenuation and diffusion when power propagates through an infinitesimal fiber length.

$\mathbf{A}$ is a diagonal matrix that describes power propagation without diffusion. Its elements are obtained from Equation (3) as

$$A_{k,k}(\omega) \approx 1 - \Delta z \cdot \alpha \left(k \cdot \Delta\theta\right) - \Delta z \cdot \frac{n}{c \cos(k \cdot \Delta\theta)} \cdot j\omega$$

(5)

which is the first order approximation of

$$A_{k,k}(\omega) = \exp\left(-\Delta z \cdot \alpha \left(k \cdot \Delta\theta\right) - \Delta z \cdot \frac{n}{c \cos(k \cdot \Delta\theta)} \cdot j\omega\right)$$

(6)

that is the exact solution of the Equation (2) in the absence of diffusion. We used this later expression to ensure that the matrix elements are always positive which solves stability problems for large α values and permits the use of greater Δz steps.

Notice that $\mathbf{A}$ is the only frequency dependent term in Equation (4). Iteration over the values of ω gives the complete spatial and temporal evolution of the optical power in the fiber. The complex values of $A_{k,k}(\omega)$ are obtained by sampling the angular frequency ω as required for a precise calculation of the inverse discrete Fourier transform of $p(\theta, z, \omega)$ to obtain $P(\theta, z, t)$.

The matrix $\mathbf{D}$ is a tri-diagonal matrix which accounts for diffusion along the fiber. Its elements for $k > 0$ are:

$$D_{k,k-1} = \left(d\left(k \cdot \Delta\theta\right) - \frac{1}{2} \frac{d\left(k \cdot \Delta\theta\right)}{k} - \frac{1}{2} d'\left(k \cdot \Delta\theta\right) \Delta\theta\right) \frac{\Delta z}{\Delta\theta^2}$$

$$D_{k,k} = -2d\left(k \cdot \Delta\theta\right) \frac{\Delta z}{\Delta\theta^2}$$

$$D_{k,k+1} = \left(d\left(k \cdot \Delta\theta\right) + \frac{1}{2} \frac{d\left(k \cdot \Delta\theta\right)}{k} + \frac{1}{2} d'\left(k \cdot \Delta\theta\right) \Delta\theta\right) \frac{\Delta z}{\Delta\theta^2}.$$

(7)

These matrix elements describe power diffusion through a differential fiber length giving the fraction of the power that flows out from a given angle and that drifting to this angle from the adjacent ones.

To obtain the boundary condition at $k = 0$, corresponding to $\theta = 0$, we use the approximation proposed in (Mateo et al., 2009), and the fact that $P(\theta)$ is an even function of the angle, obtaining

$$D_{0,0} = -4d\left(0\right) \frac{\Delta z}{\Delta\theta^2} \qquad D_{0,1} = 4d\left(0\right) \frac{\Delta z}{\Delta\theta^2}.$$

(8)

The other boundary condition at the maximum $k = N$ is given by Equation (7). Thus, so that there are no losses due to diffusion, the value of $D_{N,N-1}$ must compensate for the absence of the term $D_{N,N+1}$ such that the sum of terms is zero, resulting in

$$D_{N,N-1} = 2d(N)\frac{\Delta z}{\Delta\theta^2} \qquad D_{N,N} = -2d(N)\frac{\Delta z}{\Delta\theta^2}. \tag{9}$$

Experimental Diffusion and Attenuation Functions

In this section, we describe our procedure to characterize power propagation in POFs by obtaining the diffusion and attenuation functions.

For large z values, when the angular power distribution has reached its Steady State Distribution (SSD), the solution of the equation can be expressed as the product of two functions of independent variables:

$$P_{SSD} = Q(\theta)e^{-\gamma z}. \tag{10}$$

The function $Q(\theta)$ describes the shape of the SSD profile that depends only on the propagation angle, while the dependence on fiber length z is given by a decreasing exponential function which accounts for the power decrease due to the fiber attenuation γ. Introducing this solution into Equation (1), an equation can be obtained relating: $\alpha(\theta)$, $Q(\theta)$ and $D(\theta)$. Thus, $\alpha(\theta)$ can be expressed in terms of the others as follows:

$$\alpha(\theta) = \gamma + \frac{1}{Q(\theta)\theta}\frac{\partial}{\partial\theta}\left[\theta D(\theta)\frac{\partial Q(\theta)}{\partial\theta}\right] \tag{11}$$

In this way, a shape for the attenuation function does not have to be assumed as it can be directly calculated provided γ, $Q(\theta)$ and $D(\theta)$ are known.

The fiber attenuation may be calculated from the total optical power as a function of length being the slope of the curve $\ln\left(P_T(z)\right)$ versus z the value of the attenuation γ.

In order to obtain the angular attenuation using Equation (11), we have to provide an analytical expression for $Q(\theta)$ and for its derivatives. Thus, the normalized SSDs for the three fibers were fitted by a product of two sigmoid functions of the squared inner angle, given by:

$$Q_N(\theta) = \frac{\left(1 + e^{-\sigma_1^2\theta_1^2}\right)\left(1 + e^{-\sigma_2^2\theta_2^2}\right)}{\left(1 + e^{-\sigma_1^2\left(\theta_1^2-\theta^2\right)}\right)\left(1 + e^{-\sigma_2^2\left(\theta_2^2-\theta^2\right)}\right)} \tag{12}$$

This function has several characteristics that make it very suitable for our purposes: It has a flat asymptotic behavior at the origin and at infinity, it decreases monotonically and it is even and continuously derivable. In addition, the four free parameters that characterize these functions give sufficient flexibility to model the different slopes of the head and the tails of the FFPs.

The steady state is obtained as the radial profile of a far field image obtained for a long fiber. Figure 1 shows optical power versus length and Figure 2 shows the normalized SSD, given by $Q_N(\theta)$ for three fiber types. These fibers are poly methyl methacrylate (PMMA) optical fibers of similar properties (1-mm diameter and high NA) from three different manufacturers: ESKA PREMIER, GH4001 (GH) from Mitsubishi, PGU-FB1000 (PGU) from Toray and HFBR-RUS100 (HFB) from Avago. The steady state distributions are very similar, for all fibers even though each was measured at a different fiber length, as the distance that the light has to propagate before reaching the steady state depends on the fiber characteristic diffusion and attenuation.

The values of γ and the parameters that best fit the SSDs are given in Table 1, as well as the

Figure 1. Total power versus length for the three tested fibers

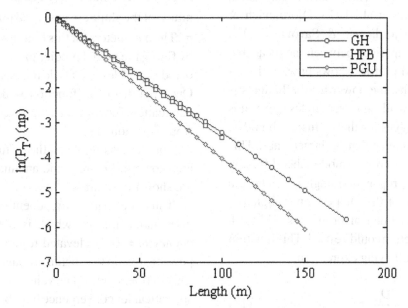

Figure 2. SSD radial profiles for the three tested fibers. Data symbols represent the raw data from the radial profiles and the lines give the best-fit to Equation (12).

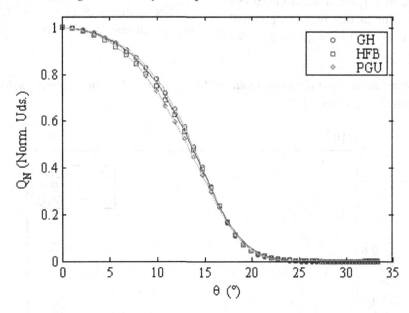

root mean square error (RMSE) for the three fibers. The goodness of the fits can be observed in Figure 2 where they are displayed as lines along with the original data as symbols for the three fibers tested. Notice the presence of power beyond the

inner critical angle (19°) for all fibers whose SSDs are very similar which is not surprising as they have similar optical configuration.

To estimate $D(\theta)$, in the first approach the diffusion function is modeled by a constant D_c

as it is usually assumed (Rousseau & Jeunhomme, 1977, Jeunhomme et al., 1976, Djordjevich & Savović, 2000, Djordjevich & Savović, 2004). However, the comparison between the measured and the predicted profiles shows how, in the intermediate lengths, the power at small angles is underestimated, while at larger angles power it is overestimated suggesting that diffusion should be higher at small angles than at larger ones. This fact justifies our proposal to model fiber diffusion as a function of propagation angle. We chose a sigmoid function of the squared inner angle because of its mathematical properties, although other similar function could be used. This function is given by the following expression:

$$D(\theta) = D_0 + \frac{D_1}{1 + D_2 e^{\sigma_d^2 \theta^2}} \tag{13}$$

where D_0, D_1, D_2, and σ_d are free parameters of our model. The value of this function at small angles tends to $D_0 + D_1/(1 + D_2)$ while for larger angles tends to D_0. The position and magnitude of the slope are governed jointly by D_2 and σ_d. The parameters for this function were obtained by fitting the model predictions to measurements of radial profiles of far field images for long fibers. Once the diffusion function is determined, the attenuation function can be directly calculated from Equation (11).

For these same fibers, the diffusion functions are shown in Figure 3 and attenuation functions are shown in figure 4:

Thus, the spatial component of the complete propagation matrix, which is obtained for frequency $\omega = 0$, is elevated to the corresponding power to obtain the spatial propagation matrix for a given fiber length. The values of Δz and $\Delta \theta$ that are critical for convergence have been determined according to the required precision. In the calculations presented here we have used $\Delta z = 0.001$ m and $\Delta \theta = 0.005$ rad obtaining accurate results.

The spatial propagation matrix is a two dimensional matrix that can be represented as an

Figure 3. Diffusion functions for the three modeled fibers obtained by modeling diffusion with the sigmoid function given by Equation (13)

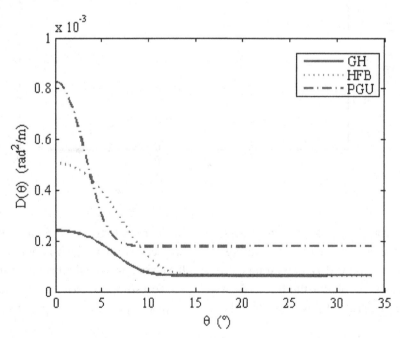

Figure 4. Attenuation functions for the three modeled fibers given by Equation (11)

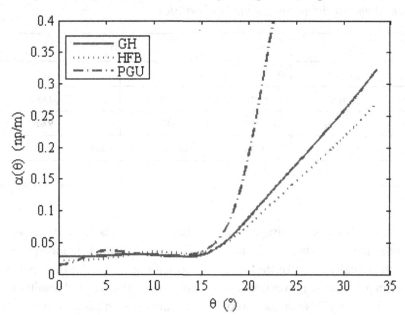

image. Figure 5 shows the propagation matrix for a particular fiber (GH from Mitsubishi) at three different lengths (25 meters, 50 meters and 100 meters, from left to right), calculated from the attenuation and diffusion functions obtained using the parameters given in Table 2. The vertical and horizontal axes are angles in degrees and the highest values are shown in red and the lowest in blue.

The figure reveals the effects of propagation on spatial distribution of optical power. As light travels through the fiber, power spreads towards higher angles as shows the widening of the images for longer lengths. The higher losses for higher angles are indicated by the shortening of

the diagonal. The matrices have been individually normalized to their respective maxima. Otherwise, the images for 50 and 100 meters will be hardly visible due to overall attenuation that gives the power loss with fiber length.

LOCALIZED DISTURBANCES IN POF

As mentioned in the introduction, power diffusion can also be produced by macroscopic agents that disturb power propagation over a small length of fiber. This intense and localized diffusion can be caused by different devices such as scramblers,

Figure 5. Spatial propagation matrix of GH fiber for three different lengths: 25m, 50 m, and 100 m

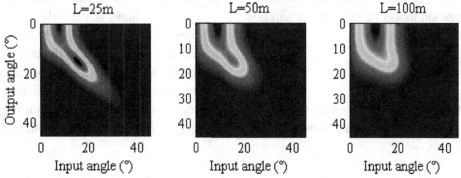

Table 2. Parameters for the constant diffusion and sigmoid diffusion functions that minimizes the error between experimental and model-predicted far field profiles

Fiber	D_c		$D(\theta)$ [a]				
	rad^2/m	RMSE	D_0	D_1	D_2	σ_d	RMSE
GH	1.171×10^{-4}	22.7×10^{-3}	6.356×10^{-5}	2.338×10^{-4}	0.321	**11.66**	16.4×10^{-3}
HFB	1.649×10^{-4}	17.1×10^{-3}	6.196×10^{-5}	6.513×10^{-4}	0.466	**9.626**	8.45×10^{-3}
PGU	2.271×10^{-4}	12.9×10^{-3}	1.775×10^{-4}	1.561×10^{-3}	1.403	**16.54**	8.92×10^{-3}
[a] Defined in Equation (13).							

tappers, connectors, etc., or it can arise in the vicinity of curvatures and other kinds of strains, intentionally or unintentionally placed on the fiber. The abrupt changes in paths cause power transfer from some angles to other very far from their next neighbors. The agents that produce diffusion are stationary and do not have a direct effect over the temporal component, but they alter spatial power distribution which affects ray paths, introducing temporal delays and thus, changing the impulse spread function, and thus, the frequency response and bandwidth. These localized disturbances can be introduced as matrices into the model framework and their effects on the transmission properties can be evaluated, provided the device has been characterized as a matrix. Here, we will first describe the injection matrix that is introduced to account for the strong diffusion produced when power is launched into the fiber, and that was necessary to get an agreement of experimental and simulation results. Then, we will describe an experimental procedure developed to characterize the effects of any spatially localized device as a matrix, and show how it can easily be used in the propagation model framework, as will be illustrated with an example.

Injection Matrix

As large NA POF useful reach is relatively short compared to other fiber types, the power distri-

bution cannot achieve its steady state before the detector and thus, launching conditions have a strong impact on the fiber performance. In fact, previous experimental results suggest that the input distribution has a strong impact on bandwidth changing the balance of diffusion and differential attenuation. We found a disagreement when comparing model predictions of radial profiles for short fibers (Losada et al., 2007) and also, when validating the model using experimental frequency responses (Losada et al., 2008), suggesting diffusion at the power launching stronger than the slow diffusion that takes place when power propagates. To get a better agreement to the experimental measurements, it is necessary to introduce another matrix, which was called injection matrix, and that accounts for the strong diffusion produced when launching power into the fiber (Mateo et al., 2009). The injection matrix is different for each fiber type and can be estimated by extracting the radial profiles from experimental far field patterns (FFPs) for short fibers obtained launching a He-Ne laser beam at different angles (Losada et al., 2007, Mateo et al., 2009).

Figure 6 shows the angular scans of radial profiles for the three PMMA POFs whose characteristics are described in the last section. This representation of the matrix of radial profiles versus input angle permits the compact visualization of the whole scan of launching angles. In the images, each column represents the radial profile for the corresponding input angle as a function of the output angle on the vertical axis. The mirror images of the profiles, corresponding to negative output angles, are also shown for the sake of symmetry. The profiles for negative injection angles have also been measured and are not exactly the same as those obtained for the positive angles, revealing the deviations from the ideally symmetric response. The injection matrix is an average of the four quadrants shown in the images because its elements represent power transfer only between non-negative angles, as those of the propagation matrix shown in Figure 3. The matrix product of the injection matrix, $\mathbf{J}$, and the source angular power distribution, $\mathbf{p}(z = 0_-)$, gives the vector describing the angular power distribution just after entering the fiber $\mathbf{p}(z = 0_+)$ as:

$$\mathbf{p}(z = 0_+) = \mathbf{J} \cdot \mathbf{p}(z = 0_-) \tag{14}$$

Introducing this matrix into the model calculations, we found a better agreement between the model predictions and the experimental results (Mateo et al., 2009, Losada et al., 2011). In addition, the injection matrix diminishes the impact of the input power distribution of the source, equalizing the effects of sources of different widths. The effect of the injection matrix for narrow sources is to diffuse power towards higher angles, while for wider sources is to filter out power at the highest angles.

Device Characterization

In the previous subsection, the potential of the matrix approach to characterize and to incorporate the power spread at the fiber input, which was validated by comparison of model predictions and experimental measurements, has been described. However, power disturbances can be located at any position along the fiber and can have many different origins. Thus, it is necessary to devise a method to obtain the characteristic matrix for any device or defect, provided they have a linear behavior. Thus, an experimental procedure was designed to obtain the characteristic matrix for a linear localized disturbance and was tested for a corrugated scrambler shown in Figure 7 (Fuster & Kalymnios, 1998).

The method consists on isolating the effects of the disturbance from propagation effects by comparing experimental measurements obtained without the device and with the device near the fiber output end. These measurements were radial profiles extracted from output FFPs measured at different launching angles following a similar procedure as that devised to obtain the injection matrix.

The effects of these disturbances are linear in the sense that the power at a given angle in the output distribution can be obtained as a linear combination of the power at different input angles. As this combination is different for different angles,

Figure 6. Injection matrices for GH, HFB, and PGU fibers

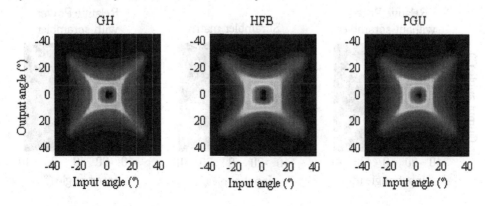

Figure 7. Corrugated scrambler

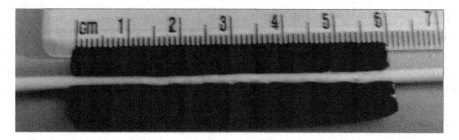

the system is spatially (or angularly) variant. The device is then modeled as a multiplicative matrix, **S**. Thus, given the input power profile as a column vector, $\mathbf{p}_i$, where each vector element is the power at a given angle, the output power vector, $\mathbf{p}_o$ can be obtained by the matrix product of $\mathbf{p}_i$, by the characteristic matrix **S**, or $\mathbf{p}_o = \mathbf{S} \bullet \mathbf{p}_i$. The radial profiles for the whole angular scan can be arranged in a matrix form, where each column is a power vector whose horizontal index represents the input angle. Thus, the effects of the device over the whole data set can be directly calculated as the matrix product:

$$\mathbf{P}_o = \mathbf{S} \cdot \mathbf{P}_i \qquad (15)$$

where $\mathbf{P}_o$ and $\mathbf{P}_i$ are matrices built as column power vector aggregates. Images in Figure 8 show the scan of radial profiles for a fiber inserting a cor-

rugated scrambler to illustrate the procedure with an example. The image on the left shows the radial profiles before inserting the device and while the image on the right shows the effects of inserting the scrambler on the radial profiles. The scales are the same for the left and right images, so those measurements obtained with the scrambler are dimmer indicating power loss. Matrix **S**, shown in the center, is the characteristic matrix for the scrambler. In this representation, each column indicates the spread of the power in a given input angle due to the device. Each element in the column gives the relative power transferred to the angle indicated by the row index.

Once the matrix **S** is obtained to fit the experimental results, the effects of the device can be introduced at any position in the fiber link. As an example, a device is inserted first right after the input end of the fiber and then, before its output end. Thus, the characteristic matrix has to

Figure 8. Effect of a scrambler over the spatial distribution of optical power. The middle image shows the characteristic matrix for the scrambler.

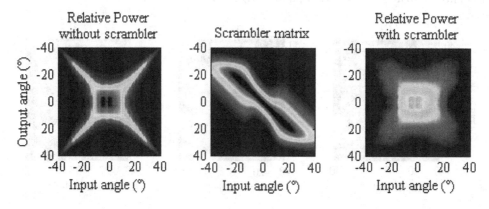

be introduced as the right or left matrix product of **S** by the fiber propagation matrix, $\mathbf{M}(\omega)$, respectively, as described in the following equations:

Device at input end, $\quad \mathbf{p}(z_2, \omega) = \mathbf{M}(\omega)^m \cdot \mathbf{S} \cdot \mathbf{J} \cdot \mathbf{p}(z_1, \omega)$

Device at output end, $\mathbf{p}(z_2, \omega) = \mathbf{S} \cdot \mathbf{M}(\omega)^m \cdot \mathbf{J} \cdot \mathbf{p}(z_1, \omega)$

$$(16)$$

where **p** are the power vectors as defined in Equation (4) and Equation (15), **S** the localized disturbance characteristic matrix, and **J** the injection matrix described before,. This approach was used to simulate the dependence of fiber bandwidth on the scrambler position. Simulations were in good agreement with experimental measurements of bandwidth versus fiber length showing how the scrambler near the detector increases the system bandwidth acting as a spatial filter while it degrades the performance if placed near the transmitter end of the fiber (Kalymnios, 1999).

SPACE-TIME POWER DISTRIBUTION

Once the propagation and injection matrices have been determined, it is possible to calculate the output power distribution for a given input source with a known angular output distribution using the next equation:

$$\mathbf{p}(L, \omega) = \left(M(\omega) \right)^{L/\Delta z} \cdot \mathbf{J} \cdot \mathbf{p}_s(0, \omega) \qquad (17)$$

where $\mathbf{p}(L, \omega)$ is a column vector, with each of its elements representing the power at a given angle for each frequency ω, and $\mathbf{p}(0, \omega)$ is a vector obtained by concatenating the source spatial distribution multiplied by the value of the source frequency response at each frequency. Thus, Equation (15) provides the evolution of the space-time optical power distribution with fiber length from which angular power distribution, attenuation, bandwidth and pulse spreading can be derived. The frequency response at a given length $z = L$

can be calculated as the total power for all angles at a given frequency:

$$H(L, \omega) = \int p(\theta, L, \omega) \sin \theta \, d\theta \qquad (18)$$

whose inverse Fourier transform gives the pulse temporal spreading at this length, $h(L, t)$. As an optical fiber may be modeled as a linear system, we could assume that if we obtain the frequency response for a fiber length, it should be possible to use this function to obtain the frequency responses of lengths multiples of the basic length by obtaining the powers of the initial function. However, an important aspect revealed by the triple dependence of power with propagation angle, length and time is that to obtain the frequency response at one given length L, it is not enough to know the frequency response at any shorter length, $H(L_0 < L, \omega)$. To compute the total acquired delay at a given angle and fiber length it is necessary to know the previous path followed by the power reaching that angle, which implies to know either $\mathbf{p}(L_0 < L, \omega)$. Thus, to be able to calculate the frequency response at any given length, it is necessary to know the angular power distribution right at the fiber input: $\mathbf{P}(z = 0, t = 0)$ or $\mathbf{p}(z = 0, \omega)$, where there is no propagation acquired temporal delay. The path changes produced by power diffusion are the origin of the intertwined spatial and temporal dependences. Thus, to have further knowledge of fiber behavior, the output power distribution can be shown as a joint function of output angle and time at any fiber length by taking the inverse Fourier transform of the calculated frequency responses for each output angle. This bidimensional function is called the space-time power distribution: $ST(\theta, t)$ and can be represented as an image as in Figure 8, where the central graph shows the power at the output of a 100-m HFB fiber as an example. Time is shown on the horizontal axis in nanoseconds and output angle in degrees on the vertical axis. In the image, each row represents the temporal pulse arriving at a given output angle, while the image columns are the radial profiles of the spatial power distribution at fixed times.

For a given fiber length, the space-time power distribution conveys all the useful information as the integrated power over the output angle results in the temporal pulse spread obtained by:

$$h(z,t) = \int ST(\theta,t)\sin\theta \, d\theta \qquad (19)$$

which is shown normalized on the right of the figure. The pulse spread is very asymmetric with power rising first very steeply to its maximum and then, slowly decreasing with a long tail extending up to 30 ns. Its Fourier transform is the frequency response of the fiber at that length and for the launching specified by the initial condition.

In addition, the integrated power over time of $ST(\theta, t)$ gives the radial profile of the FFP represented on the left graph as normalized power on the horizontal axis versus angle in degrees on the vertical axis is given by:

$$P(z,\theta) = \int ST(\theta,t)dt \qquad (20)$$

On the image, the superimposed solid blue line joins the angular positions at which the maximum power reaches the fiber end at each temporal delay. The dashed magenta line shows the delay, $\delta(\theta)$,

obtained without diffusion which is given by the ray-theory inverse cosine law:

$$\delta(\theta) = L \cdot \frac{n}{c}\left(\frac{1}{\cos\theta} - 1\right) \qquad (21)$$

where L is the fiber length and n the refraction index of PMMA. Figure 9 shows how optical power that exits the fiber over a cone from $0°$ to $8°$ is concentrated over a relatively short time slot. Above this angular range, pulses have a wider time spread and their peaks increase with the output angle as shows the blue line in the image. However, the power peaks are reached at lower times than for the cosine prediction indicating noticeable shorter delays than those that would be obtained in the absence of diffusion. In other words, diffusion improves fiber transmission capability. The horizontal time shift at the lowest angles does not affect the fiber behavior as it is an overall delay. This shift is consistent with the strong diffusion at launching described by the injection matrix that spreads power over a range of angles.

For a given length, the image representation of the ST function helps to visualize the power distribution over space and time at the fiber output and to understand the relationship between the angular power distribution and the pulse spreading through modal diffusion. In the next

Figure 9. The central graph is the image representation of the space-time power distribution at the output of 100 m of the HFB fiber. On the right, the graph shows the overall pulse spread obtained as the integral of power over output angle. The power integral over time renders the radial profile of the FFP shown on the left.

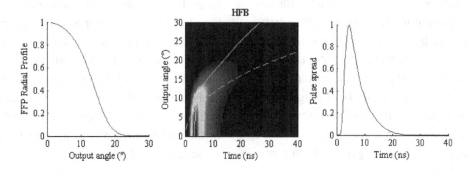

Figure 10. The images give the representation of the space-time power distribution at the output of 75, 100, and 150 meters of the PGU fiber, respectively.

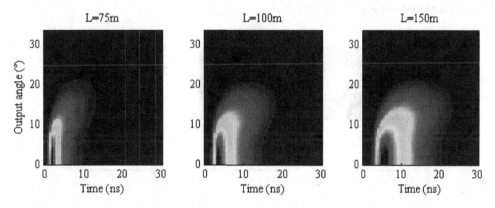

figure, the ST power distribution is shown at three fiber lengths (75, 100 and 150 m) to see how it changes through propagation by the joint effects of diffusion and attenuation.

The widening of the image with increasing fiber length illustrates the temporal spread acquired as power is diffused with propagation, and it is not noticeably greater for the higher angles. The effects of differential attenuation are revealed by the absence at longer lengths of the power initially present at the highest angles that have higher losses than angles below 15°. The power exiting the fiber at the highest angles is also that with the longest delays which suggests an easy way to improve fiber capacity by spatial filtering out of the tail at the higher angles. We found experimental evidence of bandwidth increase by spatial filtering when inserting a scrambler before the receiver (Losada et al., 2011), when using a photo-detector with a small effective area (Mateo et al., 2003) and when withdrawing the fiber end from the detector (Heredia et al., 2007). As most power is confined in a range of lower angles, filtering out the power at the highest angles will imply small power loss while producing a narrower overall impulse response. As an example, the ST power distribution for three 50-m long fibers is shown in the upper row of Figure 11. The red dashes show the limit imposed by a spatial filter with a cut-off chosen as a compromise to

increase bandwidth with small power loss. The corresponding frequency responses shown below are calculated from the impulse response obtained using Equation (19), with and without spatial filtering. The ST power distributions show noticeable differences among fibers. For example, the HFB fiber has a narrower distribution that indicates that it has less diffusion than the other two fibers. Thus, the knowledge of the characteristics of the fiber type can be used along with the model to design the best filter to that particular fiber at a given length.

POF BANDWIDTH VERSUS LENGTH

Fiber bandwidth can be obtained from the frequency response as the frequency where half of its maximum value is reached. Bandwidth as a function of fiber length in logarithmic coordinates has been usually fitted with straight lines. The slope of these lines is known as the concatenation factor and is related to mode coupling. In the absence of mode coupling and differential attenuation, the concatenation factor is one and therefore, a slope higher or lower than one is evidence of diffusive effects (Mateo et al., 2003). In multi-mode glass fibers whose typical working lengths guarantee that the measurements are performed when equilibrium has been reached, slopes lower than one

Figure 11. In the upper row, the ST power distributions of the three fibers for 50 meters are shown as images. The red lines mark the filter cut-off at 15°. The lower row shows the frequency responses obtained with the filter (lines) and without the filter (dashes).

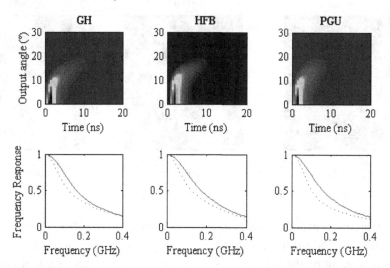

are usually found. In fact, it is usual to assume a slope of 0.5 in Step-Index fibers that reveals selective attenuation of higher order modes. In POFs, slopes both higher and lower than one have also been reported when measuring bandwidth under different experimental conditions. SI-POF bandwidths have been measured changing the aperture of the source to assess the effect of launching on its dependence with length (Garito et al., 1998, Ziemann et al., 2008). Bandwidth versus length has also been measured introducing devices that modify the power distribution, both at the trans-

mitter or at the receptor end (Losada et al., 2011), applying a spatial filter at the receptor (Mateo et al., 2003), and also using fibers that had previously been subject to strain (Losada et al., 2004). The propagation model is able to reproduce these experimental results and, can be used to extend its predictions where it would be difficult or impractical to measure, so it is a useful tool to explain the reasons for a given behavior (Mateo et al., 2009). Figure 12 shows a schematic representation of the effects of diffusion predicted using the model for three different conditions. The upper row shows

Figure 12. Schematic representation of the effects of diffusion described by the model. The upper row shows an under-filled injection. On the second and third row, the effects of a scrambler near the launching and detector ends of the fiber, respectively, are illustrated.

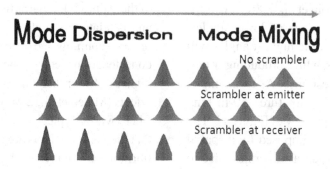

the case of an under-filled injection where the power is quickly diffused right after launching. The second row describes how using a scrambler near the launching end of the fiber has an equalizing effect on power distribution that implies less abrupt variations with length. Finally, the lower row illustrates the filtering effects of the same scrambler when it is close to the detector.

An important factor for bandwidth measurements is the way light is launched into the optical fiber. Bandwidth versus length measurements show that slopes are higher when using lower aperture sources (Garito et al., 1998, Ziemann et al., 2008). In fact, using conventional commercial laser sources, that imply an under-filled launch, we found two distinctive regions in the log-log plot of bandwidth versus length as shows Figure 13. Here, the measurements with no scrambler show that, for shorter fiber lengths, bandwidth decreases with length more steeply than for longer lengths. The upper row of Figure 12 illustrates the power distribution changes for an under-filled launch. We argue that for under-filled launch most power is coupled at low angles where the dominant effect is modal dispersion caused by strong diffusion (Ritter, 1993). Thus, the power in more confined (inner) modes is transferred to the pe-

ripheral (outer) modes, quickly increasing the power spread and degrading the frequency response (Losada et al., 2002). This rapid power spread produces a decrease in bandwidth with length at higher rates than the case without diffusion characterized by a slope of one. The lower row shows the effects of a scrambler placed just before the detection that acts as a spatial filter as we explained before making the bandwidth decrease even steeper than without it. On the other hand, in over-filled conditions or for longer fiber lengths, concatenation factors can be less than one. These conditions can be also achieved by using a scrambler near the transmitter end of the fiber as shown in Figure 13, where the slope for the data measured with the scrambler near the launching end is 0.7 for fibers longer than 60 meters. In these cases, the starting power distribution is closer to that attained after propagation through the transition length, where diffusion and differential attenuation are balanced (Djordjevich & Savović, 2008) and variations in power distribution are less abrupt than when starting with a narrower distribution. In these overfilled conditions, diffusion causes power transfer between modes that averages out the different delays, acting as an equalizer. Thus, mode mixing, jointly

Figure 13. Representation of the bandwidth versus length measurements for a fiber in three conditions: with a scrambler at the launching (diamonds) and detector (squares) ends and without the scrambler (circles). The data were fitted using two straight lines whose slopes are also shown.

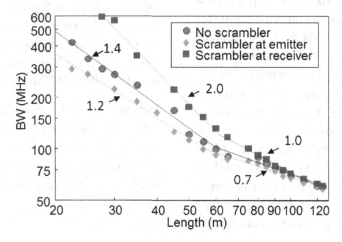

with differential attenuation that attenuates the higher flank of the impulse response, explains how for longer fiber lengths the concatenation factor can be less than one. These effects are illustrated in the middle sequence of Figure 12 that shows the smoother changes with length in the power distribution.

Slopes lower than one are hard to measure experimentally as very long fiber lengths are needed to reach the transition length that depends on the amount of diffusion. For fibers with very intense diffusion, such as the fiber shown in Figure 13, it can be around 60 meters. These fibers display even higher slopes for the first segment, suggesting a very intense diffusion, while there is a shallower segment, with a slope less than one that indicates how their strong diffusion makes them reach the transition length sooner than other fibers. Using the model to extend predictions to longer fiber lengths it was shown that all fibers exhibit a similar behavior. Slopes lower than one have also been found in fibers that have been suffering a continuously applied strain along their whole lengths. The strain causes an increase of their effective diffusion and modifies their standard behavior.

CONCLUSION

Here we have described an approach to model light propagation in POFs based on the power flow differential equation that was simplified using the finite-difference method in the frequency domain. The resulting equations were re-written in matrix form providing a fast and robust method to describe power propagation. The propagation matrix, that characterizes a given fiber type, was calculated from its diffusion and attenuation functions that can be determined from experimental data. In addition, this approach is also flexible to introduce localized perturbations of power distribution as matrices. The model provides the space-time evolution of the angular power distribution when it is transmitted throughout the fiber which gives a detailed picture of the POFs capabilities for information transmission. This space-time function can be used to derive parameters with strong impact over fiber transmission capabilities and to extend model predictions where it is difficult or unpractical to measure them. Thus, the model gives a full insight of fiber behavior that can be applied to improve the performance of actual POF networks.

REFERENCES

Arrue, J., Zubia, J., Durana, G., & Mateo, J. (2001). Parameters affecting bending losses in graded-index polymer optical fibers. *IEEE Journal on Selected Topics in Quantum Electronics*, 7(5), 836–844. doi:10.1109/2944.979345

Breyer, F., Hanik, N., Lee, S. C. J., & Randel, S. (2007). Getting the impulse response of SI-POF by solving the time-dependent power-flow equation using the Crank-Nicholson scheme. In Bunge, C. A., & Poisel, H. (Eds.), *POF modelling: Theory, measurement and application*. Norderstedt, Germany: Verlag Books on Demand GmbH.

Bunge, C. A., Kruglov, R., & Poisel, H. (2006). Rayleigh and Mie scattering in polymer optical fibers. *Journal of Lightwave Technology*, 24(8), 3137–3146. doi:10.1109/JLT.2006.878077

Djordjevich, A., & Savović, S. (2000). Investigation of mode coupling in step index plastic optical fibers using the power flow equation. *IEEE Photonics Technology Letters*, 12(11), 1489–1491. doi:10.1109/68.887704

Djordjevich, A., & Savović, S. (2004). Numerical solution of the power flow equation in step-index plastic optical fibers. *Journal of the Optical Society of America. B, Optical Physics*, 21, 1437–1438. doi:10.1364/JOSAB.21.001437

Djordjevich, A., & Savović, S. (2008). Coupling length as an algebraic function of the coupling coefficient in step-index plastic optical fibers. *Optical Engineering (Redondo Beach, Calif.)*, *47*(12), 125001. doi:10.1117/1.3049858

Durana, G., Zubia, J., Arrué, J., Aldabaldetreku, G., & Mateo, J. (2003). Dependence of bending losses on cladding thickness in plastic optical fibers. *Applied Optics*, *42*(6), 997–1002. doi:10.1364/AO.42.000997

Fuster, G., & Kalymnios, D. (1998). Passive components in POF Data Communications. *Proceedings of the 7th International Conference on Plastic Optical Fibres and Applications*, (pp. 75-80).

Garito, A. F., Wang, J., & Gao, R. (1998). Effects of random perturbations in plastic optical fibers. *Science*, *281*, 962–967. doi:10.1126/science.281.5379.962

Gloge, D. (1972). Optical power flow in multimode fibers. *The Bell System Technical Journal*, *51*(8), 1767–1783.

Gloge, D. (1973). Impulse response of clad optical multimode fibers. *The Bell System Technical Journal*, *52*, 801–816.

Heredia, P., Mateo, J., & Losada, M. A. (2007). Transmission capabilities of large core GI-POF based on BER measurements. *Proceedings of the 16th International Conference on Plastic Optical Fibres and Applications*, (pp. 307-310).

Jeunhomme, L., Fraise, M., & Pocholle, J. P. (1976). Propagation model for long step-index optical fibers. *Applied Optics*, *15*, 3040–3046. doi:10.1364/AO.15.003040

Kalymnios, D. (1999). Squeezing more bandwidth into high NA POF. *Proceedings of the 8th International Conference on Plastic Optical Fibres and Applications*, (pp. 18-24).

Losada, M. A., Garcés, I., Mateo, J., Salinas, I., Lou, J., & Zubía, J. (2002). Mode coupling contribution to radiation losses in curvatures for high and low numerical aperture plastic optical fibres. *Journal of Lightwave Technology*, *20*(7), 1160–1164. doi:10.1109/JLT.2002.800377

Losada, M. A., Mateo, J., Garcés, I., Zubía, J., Casao, J. A., & Pérez-Vela, P. (2004). Analysis of strained plastic optical fibres. *IEEE Photonics Technology Letters*, *16*(6), 1513–1515. doi:10.1109/LPT.2004.826780

Losada, M. A., Mateo, J., Martínez, J. J., & López, A. (2008). SI-POF frequency response obtained by solving the power flow equation. *Proceedings of the 17th International Conference on Plastic Optical Fibres and Applications*.

Losada, M. A., Mateo, J., & Martínez-Muro, J. J. (2011). Assessment of the impact of localized disturbances on SI-POF transmission using a matrix propagation model. *Journal of Optics*, *13*, 055406. doi:10.1088/2040-8978/13/5/055406

Losada, M. A., Mateo, J., & Serena, L. (2007). Analysis of propagation properties of step index plastic optical fibers at non-stationary conditions. *Proceedings of the 16th International Conference on Plastic Optical Fibres and Applications*, (pp. 299-302).

Mateo, J., Losada, M. A., & Garcés, I. (2006). Global characterization of optical power propagation in step-index plastic optical fibers. *Optics Express*, *14*, 9028–9035. doi:10.1364/OE.14.009028

Mateo, J., Losada, M. A., Garcés, I., Arrúe, J., Zubia, J., & Kalymnios, D. (2003). High NA POF dependence of bandwidth on fibre length. *Proceedings of the 12th International Conference on Plastic Optical Fibres and Applications*, (pp. 123–126).

Mateo, J., Losada, M. A., & Zubia, J. (2009). Frequency response in step index plastic optical fibers obtained from the generalized power flow equation. *Optics Express, 17*(4), 2850–2860. doi:10.1364/OE.17.002850

Ritter, M. B. (1993). Dispersion limits in large core fibers. *Proceedings of the International. Conference on Plastics Optical Fibres and Applications,* (pp. 31-34).

Rousseau, M., & Jeunhomme, L. (1977). Numerical solution of the coupled-power equation in step-index optical fibers. *IEEE Transactions on Microwave Theory and Techniques, 25,* 577–585. doi:10.1109/TMTT.1977.1129162

Ziemann, O., Zamzow, P. E., & Daum, W. (2008). *POF handbook: Optical short range transmission systems.* Springer Verlag.

Chapter 5
FlexRay™ Electrical Physical Layer:
Theory, Components, and Examples

Jürgen Minuth
Esslingen University of Applied Sciences, Germany

ABSTRACT

Modern medium and high end vehicles are no longer imaginable without using technologies to broadcast local available data. The speed information for example is used by many well known functions: the anti blocking system, the radio, the dashboard, the cruise control, the electronic stability program, etc. Usually, this data is distributed among vehicle's electronic control units by various serial bus systems. The succeeding sections introduce the automotive communication system named FlexRay™. The development of FlexRay™ had been initialized by requirements expected for drive-by-wire systems. The content is focused on its electrical physical layer beginning with active components like bus interfaces as well as passive components like common mode filters and bus-cables. Comparisons to the state of the art systems CAN and LIN are used to support the comprehensibility.

ELECTRICAL PHYSICAL LAYER FLEXRAY™

Scope

The section "Electrical Physical Layer FlexRay™" introduces FlexRay™ focused on the electrical physical layer specifications and application notes version 3.0.1. All main features are taken into consideration. Dedicated detail information is skipped; examples for them are: bus-guardian, wake-up procedures, implementation variants, timing constrains, timing clusters, gateways, system design rules, and EMC tests.

INTRODUCTION

Flexray™ System

FlexRay™ had been specified by an industrial consortium consisting of American, European and

DOI: 10.4018/978-1-4666-2976-9.ch005

Japanese car makers and their suppliers. The basic requirements were reasoned by the expected communication requirements of safety relevant drive by wire applications end of the last millennium. On one hand the protocol and the physical layer were inspired additionally by the known CAN system and on the other hand by a time triggered approach of a university research laboratory and by a proprietary automotive bus implementation. FlexRay™ specifications are available in the version 3.x. Rausch (2008) explained how to understand and use FlexRay™ focused on protocol's view. The succeeding sections summarize the published properties from physical layer's point of view[1].

The FlexRay™ communication is based on synchronized distributed clocks (relative time). Wake-up, start-up and synchronization procedures are included. A final accuracy in the range of few 100 ns is achievable. Additional properties are (the succeeding list points out some important properties only):

- Up to 64 nodes can participate in the system.
- Three baud rates are supported: 2,5 Mbit/sec, 5 Mbit/sec and 10 Mbit/sec.
- A 1st channel interconnects all nodes. It consists of passive lines and optionally of an active star.
- An active star (AS) may be extended by a communication node optionally.
- A 2nd channel may be used optionally enabling redundant communication or increasing the gross data rate.
- The communication scheme shall be fixed during compile time; however flexible multiplexing procedures are possible.
- The communication scheme contains a single communication cycle.
- The communication cycle contains a static part and/or a dynamic part.

- The static part contains several timing slots; each of these slots may contain one FlexRay™ frame transmitted by one of the nodes.
- The dynamic part enables an event driven communication comparable approximately with CAN's arbitration procedure.

All protocol procedures are implemented in a communication controller (CC). The connection between the communication controller and the passive bus lines is done by the bus driver (BD).

The basic range of application for FlexRay™ can be seen on one hand in (safety relevant) distributed closed loop control systems and on the other hand in communication systems which require a gross data rate not achievable with CAN. Ongoing developments like automotive Ethernet or CAN FD[2] may modify this state.

The FlexRay™ specifications are done in a way which gives maximum flexibility to the system designer. However the system designer cannot follow a straight forward plug and play approach. A lot of necessary components are specified very loosely. From the electrical physical layer point of view it is specified in detail: bus-driver, active stars and procedures to cover and evaluate the system behavior; however passive components like connectors, bus-cables, common mode chokes, electro static discharge protection elements, printed circuit board layouts or termination circuits are specified more or less roughly.

Electrical Physical Layer

The ISO OSI seven layer model defines as level 0 a so called "physical layer". This layer applied to FlexRay™ (or CAN) contains all hard-ware components which are involved in transferring protocol information (e.g. frames, messages, symbols, in-frame bits etc.) among the mathematically perfect working digital protocol machines.

Figure 1. Overview FlexRay™ system and communication

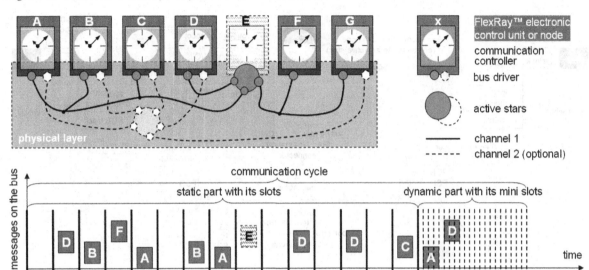

The border to the physical layer crosses each communication node internally. The following effects and components are included.

- Timing effects of crystals like e.g. accuracy, temperature dependencies and aging drift.
- Timing effects of oscillator circuits, PLLs and clock-scalers inside the µC like edge jitter, duty cycle variations, frequency variations etc.
- Timing effects of µC's I/O-pins like dynamic input/output characteristics, detection thresholds etc.
- Supply voltages of the different hardware components; e.g. two connected circuits could be driven by different supply voltages by what levels and thresholds could vary non-correlated.
- Track on the printed circuit boards (PCB) including their explicitly mounted termination and their implicitly termination caused by e.g. parasitic components like input impedances.
- Transceivers (CAN, LIN), bus drivers (FlexRay™) and optionally active stars[3] (FlexRay™) are included.
- Additionally components for electro static discharge (ESD) protection, common mode

chokes (CMC), bus cables, connectors and inline connectors, stub cables, branch cables, termination networks, splices etc. are included.

Depending on the communication protocol, the selected communication speed and the selected topology[4] a subset of the mentioned components should be used to enable the communication. Detailed information is available in the actual specifications (CAN, FlexRay™) and the data sheets (e.g. CAN low-speed and high-speed transceivers, FlexRay™ bus drivers and active stars).

Typical FlexRay™ Set-Up

The succeeding figure visualizes a typical FlexRay™ set-up. A one channel active star topology with two branches is chosen. The optionally allowed 2nd channel is indicated by the dotted lines.

Various FlexRay™ components are offered on the market. They are developed by taking the available specifications into account. Conformance test services are available to check the matching between specification and product.

Figure 2. Boundary of the physical layer (FlexRay™ wording selected

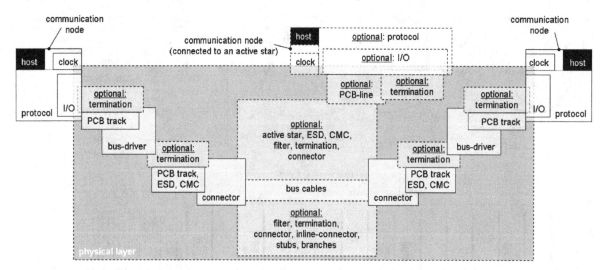

Figure 3. Components of a FlexRay™ communication system

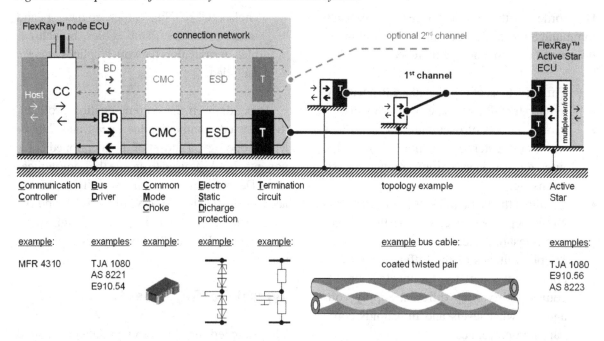

Typical CAN Set-Up

The succeeding figure visualizes a typical CAN set-up. A passive bus topology with two stubs is chosen. An active star topology is not supported.

Typical LIN Set-Up

The succeeding figure visualizes a typical LIN set-up. A passive bus topology with two stubs is chosen. An active star topology is not supported.

Physical Representations of the Communication Information

From physical layer's point of view automotive serial communication systems distinguish three pieces of information which have to be coded in states inside the communication media. Depending on the implementation various physical representations are possible.

These three pieces of communication information are coded in CAN (and LIN):

- Two communication media states are defined.

Figure 4. Components of a CAN communication system

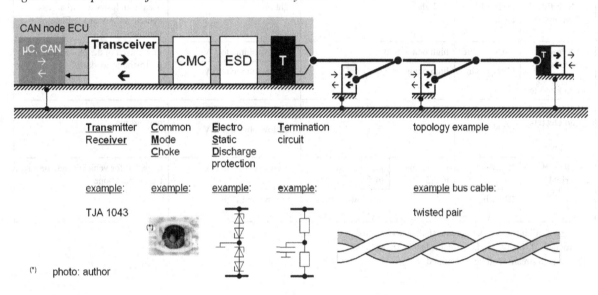

Figure 5. Components of a LIN communication system

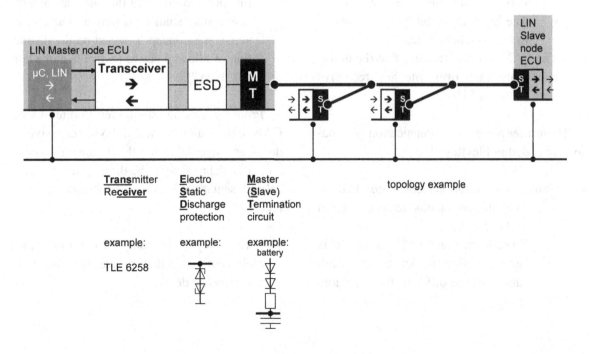

Table 1. Examples of various interpretations when coding communication information.

		Pieces of Information to be Coded			
	Media	Bus Free(***)	"1" bit(*)	"0" bit(**)	Remark
LIN	-	recessive		dominant	protocol view
CAN	-	recessive		dominant	protocol view
FlexRay™	-	idle	active		protocol view
			data_1	data_0	
LIN electrical	LIN wire	high ohmic to the battery		pull low-ohmic to 0 V	standard for vehicle's LIN networking
CAN optical ≤ 1 Mbit/sec	optical fiber cable	no light		light on	used in industry e.g. to access vehicle's CAN inside EMC chambers
CAN low speed electrical ≤ 125 kbit/sec	CAN_{high} wire	high ohmic to 0 V		pull low-ohmic to 5 V	rarely used in the passenger compartment networking
	CAN_{low} wire	high ohmic to 5 V		pull low-ohmic to 0 V	
CAN high speed electrical ≤ 1 Mbit/sec	CAN_{high} wire	high ohmic to 2,5 V		pull low-ohmic to 5 V	standard for vehicle's CAN networking
	CAN_{low} wire			pull low-ohmic to 0 V	
FlexRay™ electrical ≤ 10 Mbit/sec	BP wire	high ohmic to 2,5 V	pull low-ohmic to 5 V	pull low-ohmic to 0 V	expected for vehicle's networking, e.g. as backbone
	BM wire		pull low-ohmic to 0 V	pull low-ohmic to 5 V	

(*) logical "1" inside the data stream shall be transmitted
(**) logical "0" inside the data stream shall be transmitted
(***) no data stream shall be transmitted

- ○ The information bus free and bit "1" is converted to the recessive state.
- ○ The information bit "0" is converted to the dominant state.
- The CAN protocol requires that the dominant state shall overwrite any recessive state (ditto for LIN).

These three pieces of communication information are coded in FlexRay™:

- Three communication media states defined.
  - ○ The information bus free is converted to the idle state.
  - ○ The information bit "1" and bit "0" is summarized to the active state. Both sub-states are different from the idle state.

- FlexRay™'s electrical physical layer specification requires that the idle state and the active state shall be different. The chosen active star concept is based on the distinction of idle and active (active is named often "busy").

From the message timing level point of view CAN distinguishes two delays: recessive to dominant edge delay and the dominant to recessive edge delay. Due to the three different physical representations FlexRay™ distinguishes delay types:

- The delay of the complete communication element[5] is called propagation delay or symmetric delay.

Figure 6. Physical bus signals: Flexray™ and CAN

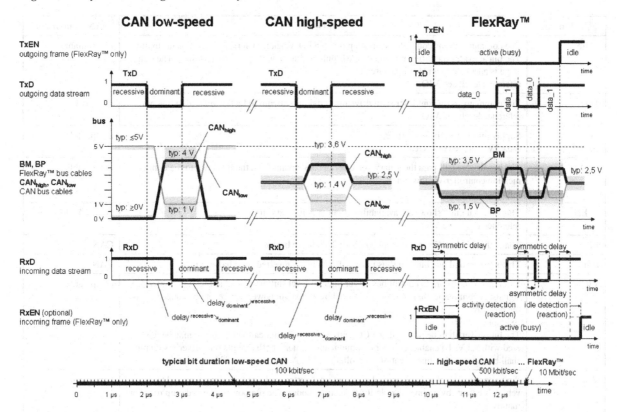

- The delay until a receiver detects the communication element just started is called activity detection.
- The delay until a receiver detects that the communication element still ended is called idle detection.
- The difference in the symmetric delay between the data_1 to data_1 edge and the data_0 to data_0 edge is called asymmetric delay. The asymmetric delay generates lengthened or shortened bits.

Active Components

Node Bus Driver

The first key element of a FlexRay™ network is the bus driver mounted inside a node. Its task is to pass protocol information like frames and symbols between the communication controller and the cables inside the wiring harness. However it is not able to detect the type of protocol information[6]. The bus-driver offers two interfaces to support the serial data stream passing:

- Communication Controller Interface
  - Unidirectional transmitting interface from the communication controller.
  - Unidirectional receiving interface to the communication controller.
- Bus Interface
  - Bidirectional differential mode interface to two-wire bus cables.

Additionally interfaces to control the power-modes of the device and the complete node are available. The succeeding table introduces all relevant pins and describes their function.

Table 2. Pins of the bus driver and their functions

Pin	Function	CAN comparison
TxEN	The communications controller shall keep this pin low when transferring a serial data stream. The pin activates the differential mode push-pull output stage to drive the bus lines. The output stage is adequate to an H-bridge circuit. reference level: see VIO	not available
TxD	The communication controller shall drive the serial data stream to be transmitted to this pin. reference level: see VIO	comparable
VCC; GND	Supply pins to run the bus-driver. typ.: 5V shall be connected	comparable
RxD	The bus driver shall pass the binary voted serial data stream from the bus lines to the communication controller via this pin. reference level: see VIO	comparable
RxEN	This optional pin[*] shows the availability of an ongoing data stream. reference level: see VIO	not available
BP	That's the first bus-line pin (bus plus) driven by a halve bridge.	CAN_{low}
BM	That's the second bus line pin (bus minus) driven by an inverted used halve bridge.	CAN_{high}
VIO	Reference voltage pin to control bus driver internal level shifters placed at the digital interface signals. A communication controller powered by a separate power source can be connected to the bus driver. Example: The bus driver shall be powered at VCC with 5 V. A communication controller may be powered with 3,3 V. To enable a proper communication among both circuits bus driver's VIO pin shall be connected communication controller's 3,3 V. A common ground is required.	comparable
VBat	Pin to connect the electronic control unit's internal battery voltage (usually connected to the vehicle's battery by an inverse polarity protection diode). The pin powers the wake-up receiver mainly.	comparable
e.g. STB; EN; ERR	Pins (stand-by, enable, error) to control the operating modes of the bus driver and to be informed about asynchronous events like e.g. under-voltage detected. An implementation by SPI is possible as well. reference level: see VIO	comparable
INH	Output pin to control an external voltage regulator.	comparable

[*] The function to drive the optional pin is mandatory internal the device.

Figure 7. Block chart of a node bus driver

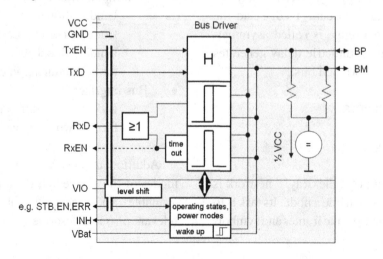

Transmitting Path

The transmitting path contains an H-bridge typically build by n-channel and p-channel MOS-FET plus an additional diode in series. The diodes suppress reverse currents e.g. in case of an unsupplied bus driver while ongoing bus communication. The push-pull stages are used inverted and generate a differential voltage on the bus. The H-Bridge is driven directly by the RxD signal and enabled by the RxEN signal. While being enabled the bus levels are generated by high ohmic resistors being connected to 50% of the supply voltage between VCC and GND.

Receiving Path

The receiving path contains two sub-paths: one sub-path generates the data stream RxD out of the bus signals by using a fast differential Schmitt trigger; one sub-path detects the availability of serial data streams by using a slow differential window comparator. Optionally the detected availability is offered as RxEN signal. Both modules use identical threshold; the specified values vary between ±150 mV and ±300 mV with a matching of maximum ±30 mV. The window comparator output signal is used to mask the Schmitt trigger output signal to the final RxD signal.

Operating States and Power Modes

Usually bus drivers support two main power modes based on the ability to control an external voltage regulator by an INH pin. The host may command, via its control interface, the bus driver to switch off the voltage regulator. If the wake-up receiver detects the presence of wake-up pattern the bus driver switches the voltage regulator on again autonomously. Ongoing communication may wake up the bus driver too – usually it does. Dedicated symbols are available guaranteeing a proper wake-up during communication. The

wake-up detector is connected to the electronic control unit internal battery voltage. In case the voltage regulator is activated communication is supported.

Functional Classes

The modules and interfaces of the bus driver are allowed to be implemented in several arrangements specified as functional classes. These classes support various fields of application.

Active Star

The second key element of a FlexRay™ network is the active star. It consists of several bus driver functionalities interconnected by a multiplexer or router. Its main task is to exchange the protocol information (e.g. frames and symbols) among all connected branches. Implementations based on a single chip per branch as well as single chip implementations to connect several branches are available. Transmitting and receiving paths are equivalent to the bus driver's paths.

An active star printed circuit board typically used in laboratories is shown by the figure. Each branch connection is designed identically consisting of bus-drivers, common mode chokes and termination; an electro static protection circuit is not implemented.

Routing

While there is no serial data stream on the bus (idle phase) the active star listens to all branches. If detecting activity (differential window comparator; delay app. 100 - 150 ns) due to the beginning of a serial data stream on the bus at one of the branches this branch remains in the listen or receive mode; the other branches are switched to the transmitting mode and the serial data stream is going to be passed from the receiving branch (differential Schmitt trigger) to all transmitting

Figure 8. Active star block diagram and its laboratory set-up (example)

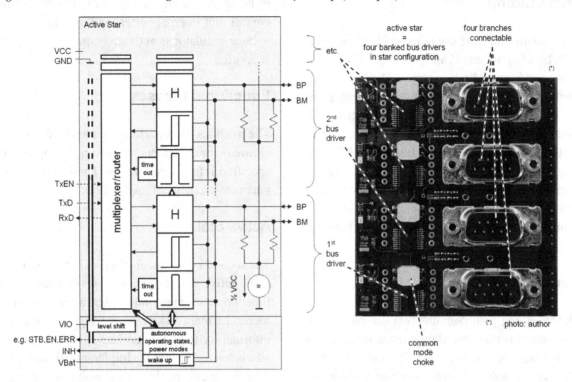

branches (H-bridge output stages). At the end of the serial data stream the receiving branch detects idle (differential window comparator; delay app. 100 - 150 ns) and the transmitting branches are switched back to the listen mode again.

The active star enables the implementation of large topologies with e.g. 64 communication nodes connected to one channel by avoiding any signal integrity problem. The active star routes the communication elements without having any information about the communication relationships. The routing is based on real-time analog principles only.

Operating States and Power Modes

Usually active stars support two main power modes autonomously based on the ability to control an external voltage regulator by an INH pin. The active star switches the voltage regulator off autonomously if there is no communication

during a dedicated time-out available. Optionally a host can command, via its control interface, the active star to switch off the voltage regulator. The active star switches the voltage regulator on again autonomously, if the wake-up receiver detected the presence of wake-up pattern. Ongoing communication may wake up the active star too – usually it does. Dedicated symbols are available guaranteeing a proper wake-up during communication. The wake-up detector is connected to the electronic control unit internal battery voltage. In case the voltage regulator is activated communication is supported.

Functional Classes

The modules and interfaces of the active star are allowed to be implemented in several arrangements specified as functional classes. These classes support various fields of application.

Block Charts and Operating Modes

The hardware block chart of a basic FlexRay™ electronic control unit (node) is straightforward. Four main components are necessary: a microcontroller (μC), a communication controller, a bus-driver and a voltage regulator. Usually standardized software modules with well-defined application programming interfaces (API) are embedded.

- The μC is connected via some control lines to the bus driver (e.g.: STB and EN).
- The communication controller (integrated inside the μC optionally) is connected via some data lines to the bus driver (RxD, TxD and TxEN).
- The bus driver is connected via a control line to the voltage regulator (INH).
- The voltage regulator is connected to the battery and offers the stabilized voltage (VCC).
- The bus-driver is powered by the stabilized voltage VCC and the battery.
- The μC and the communication controller are powered by the stabilized voltage VCC.

The bus driver offers several operating modes to control on one hand the operating modes of the complete electronic control unit and on the other hand the operating modes of the complete networked topology.

Bus Driver Modes

- **PowerOff Mode:** The bus driver is powered negligible. It behaves high-ohmic. Noactive operation is supported.
- **PowerDown Mode:** The bus driver is powered below its normal level. It behaves high-ohmic. Monitoring functions are supported. Neither communication nor wake-up detection is supported.
- **PowerOn Mode:** The bus driver is powered above its normal level. All active functions are supported.
- **(PowerOn Mode) Normal Mode:** The bus driver receives and transmits communication elements; the external voltage supply is activated by the INH output of the bus driver.
- **(PowerOn Mode) Standby Mode:** The bus driver neither receives nor transmits

Figure 9. Basic FlexRay™ ECU's schematic and bus driver's states (simplified)

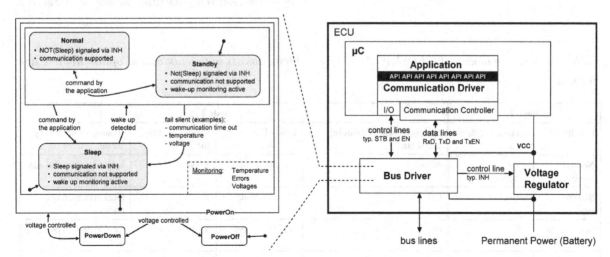

communication elements, but it is sensitive for wake up events; the external voltage supply is activated by the INH output of the bus driver.

- **(PowerOn Mode) Sleep Mode:** The bus driver neither receives nor transmits communication elements, but it is sensitive for wake up events; the external voltage supply is de-activated by the INH output of the bus driver.

ECU Modes

From a general topology point of view only a few operating modes are relevant (limp home modes and partial networking are not taken into consideration here):

- Net-wide communication in with each node participates. All nodes are in their Normal mode.
- Local operation of one node at least. The communication is de-activated. The bus lines are biased to ground. The nodes are in Standby mode or in Sleep mode.
- All nodes and the communication are not active. The nodes are in their Sleep mode. The bus lines are biased to ground.

ECU's Power Mode Controlling

The power modes of an ECU are controlled by the ECU itself and by the networked system. If the application is active it is able to command the hardware deactivating the voltage regulator. Usually the right to give the local command is based on a successful net-wide negotiation e.g. by OSEK[7] network management. The command is passed, via the bus driver (transceiver in case of CAN), to the voltage regulator. If the voltage regulator is not active, the application is not active usually as well; however the bus driver (transceiver in case of CAN) is able to filter wake-up events from the bus-lines. When such a wake up event is detected the bus-driver activates the voltage regulator. With the availability of the regulator output voltage the application is able to work again.

ACTIVE STAR'S POWER MODE CONTROLLING

The active star is able to control its power modes autonomously (if there is no host added). During an ongoing communication the active star keeps the voltage controller active (by the INH signal). The absence of the communication is monitored by a time-out. When the time-out elapses the active

Table 3. Operating modes of a FlexRay™ topology which consists of two nodes and optionally one active star

node 1			node 2		
bus driver	voltage regulator	communication (including active star[*])	bus driver	voltage regulator	topology's operating mode
Normal mode	active	Active	Normal mode	active	net-wide operation and communication
Standby mode	active	Off	Standby mode	active	local operation two nodes
Sleep mode	off	Off	Standby mode	active	local operation one node
Sleep mode	off	Off	Sleep mode	off	net-wide sleep mode

[*] The active star supports autonomous operating modes.

Figure 10. ECU's power mode switching (Sleep ⇔ Normal)

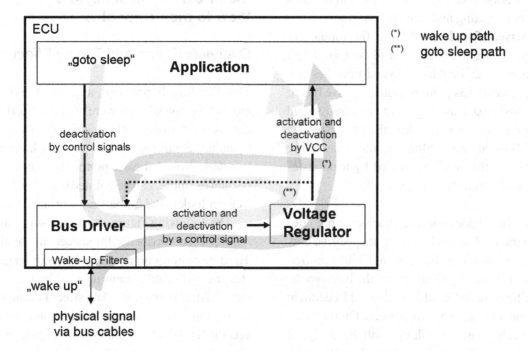

(*) wake up path
(**) goto sleep path

star de-activates the voltage regulator. When the voltage regulator is not active the active star is yet able to filter wake-up events from the bus-lines. When such a wake up event is detected the active star activates the voltage regulator. With the availability of the regulator output voltage the active star is able to pass communication elements again.

Topologies

FlexRay™ offers a broad range of various topologies. In principle each topology is allowed which meets the timing requirements and the signal shape requirements. Each topology shall be proofed against these requirements separately.

- **Point-to-Point:** A single line connects two bidirectional communication access points. Nodes and active stars are allowed. The line is terminated at both ends. The termination shall match with the line impedances.

- **Passive Bus with Stubs:** Adding short stubs with communication access points (nodes or active stars) to a point-to-point connection generates a passive bus. The stubs are terminated high-ohmic.

- **Passive Star with Branches:** Connecting all stubs of a passive bus at one point generates a passive star. Additional passive damping elements are recommended to limit reflections.

- **Active Stars with Branches:** The active star is a module with autonomous signal multiplexing and routing. Knowledge of the communication relations is not required. It is a multi-directional amplifier which refreshes the signal level. The bit-timing is not touched[8]. A communication signal received from one of n branches is going to be transmitted to the other n-1 branches. If a host is not connected the module works autonomously. In case a host is connected the module is able to work like a communication node in parallel to the autonomous multiplexing and routing behavior.

- **Mixed Topology:** Any of the topologies pointed out are allowed to be mixed. An active star may have several branches implemented as point-to-point, passive bus or passive star as long as the required signal properties are met. Only the baud rates 2,5 Mbit/sec and 5 Mbit/sec allow cascading two active stars connected by a point-to-point connection, additionally.

FlexRay™ does not specify which topologies are guaranteed to work perfect. It specifies how topologies should be build up and which properties must be met by them and by the bus signals. This allows on one hand a maximized freedom to combine a broad range of options. On the other hand each chosen topology shall be analyzed individually whether the specified and/or individual requirements are met.

Electrical Physical Layer's View to the Protocol

Overview Frame and Timing Effects

The FlexRay™ protocol and the FlexRay™ electrical physical layer were not developed from each other independently. Technical and physical boundary conditions of the physical layer were transferred into dedicated protocol properties. Important examples can be pointed out when having a closer look to the frame structure and handling.

The following figure illustrates some of the main effects described in the succeeding sections. The transmitting communication controller generates the TxD data stream masked by the TxEN signal. After passing the data stream through the channel the receiving communication controller gets the RxD data stream. Propagation delays are neglected in the figure, the received data stream RxD is triggered by the falling byte start sequence edges of the transmitted data stream TxD.

Figure 11. Topologies

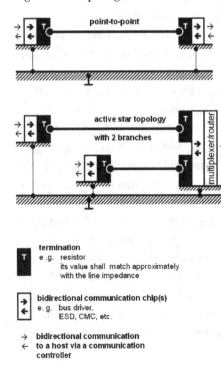

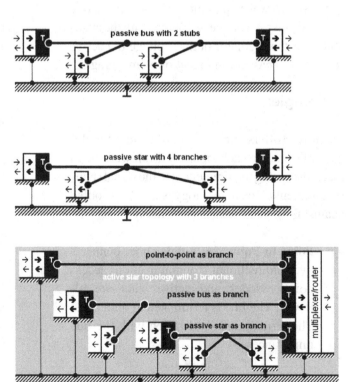

Figure 12. Effects appearing at the data streams when passing the physical layer

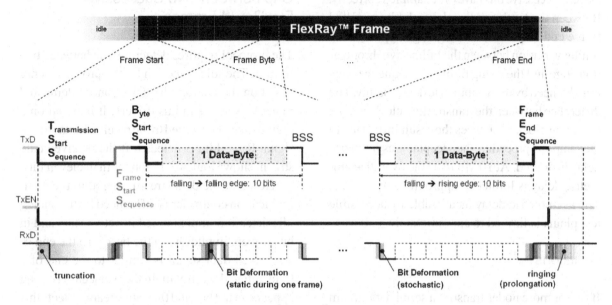

From receiver's point of view bit-deformations, truncation and optional prolonging ringing appear. Between two identical edges (rising to rising or falling to falling) a little stochastic jitter appears. The granulation of 100 ns (10 MBit/sec) is achieved with high accuracy. Between two different edges (rising to falling or falling to rising) a static bit-deformation overlaid with a little jitter appears. The static bit-deformation varies from frame to frame usually. Irradiation may increase the stochastic edge jitter primarily.

Availability of a Transmission Start Sequence (TSS)

The active star concept effects that each communication element shall be allowed to be truncated at its start. The decoder (as a part of the communication controller) shall still be able to get the message faultless. A so called transmission start sequence (TSS) heads each frame. It initializes the routing procedure inside an active star. It is allowed to be truncated down to a single bit-time. The length of the transmission start sequence can be adjusted by a protocol parameter.

Availability of a Byte Start Sequence (BSS)

Although FlexRay™ requires crystals with a high accuracy a re-synchronization of the asynchronous sampling and decoding is necessary. Each transferred byte is headed by a fix two bit byte start sequence. It allows a resynchronization triggered by the edge between the two bits.

Stuff-bits known by CAN are not beneficial in FlexRay™: the optional shortening of frames is not able to increase the gross data rate in a system based on constant time slots; the implementing of a stuff bit procedure requires a more complex state machine.

Limitations of the Asymmetric Delay

The FlexRay™ protocol samples each bit at 62,5% of the bit from receiver's clock point of view. The sampling is going to be re-synchronizes every ten bits with the falling edge in the byte start sequence. The worst case scenery is represented by the very last byte of the FlexRay™ frame. Based on the falling byte start sequence re-synchronization eleven successive bits have to be sampled correctly. Between the tenth and the eleventh bit (so called frame end sequence) a rising edge appears. The timing variation between the falling synchronization edge and the rising frame end sequence edge is influenced by the asymmetric delay mainly. The difference between the transmitting clock and the receiving clock influences the result little. Due to the sampling procedure the acceptable asymmetric delay is limited. At 10 Mbit/sec the overall asymmetric delay is limited to approximately ±37 ns. The asymmetric delay measurable by accessible test plains is limited to approximately ±30 ns.

Collisions

If two or more nodes transmit a serial data stream at the same point in time collisions are going to take place. This may happen somewhere on the communication paths. During the start up collisions are expected and allowed. During the normal communication collisions are not allowed. The correctness of frames is ensured by several cyclic redundancy checks (CRC). In case of any error the frame is going to be refused. Error frames and frame repetitions known by CAN are not specified.

The superposition of FlexRay™ data-streams produce unpredictable resulting data streams. Depending on the properties of the involved circuits and depending on the location of the collisions several variations are possible:

- One data stream is broadcasted inside the complete topology,
- One part of the topology receives the one data stream and another part of the topology receives the other data stream,
- Each node inside the topology receives a different data stream
- Etc.

In contrast the superposition of CAN frames is specified by the states dominant and recessive.

Gap between Two Successive FlexRay™ Frames

The protocol requires detecting idle between two frames. The idle detection by the protocol is not based on the analog idle detection implemented in active stars or in bus drivers. It is based on a digital time-out applied to the received data stream signal RxD. Idle is detected if the received data stream shows logical "1" longer than eleven bits.

However the data stream offered to the communication controller is influenced by the analog idle detection implemented in active stars and in bus drivers implicitly. In the best case the end of a received data steam is similar to the end of a transmitted data stream. In the worst case ringing appears on the bus and the data stream reflects this ringing. It may happen that the data stream gets stack at logical "0" until the analog idle detections masks the data stream to logical "1". The digital idle detection inside the protocol gets delayed by these effects. The gap between two frames shall be long enough to include all idle detection effects; additionally protocol clock accuracies shall be taken into account – an adjustable protocol parameter exists.

Accuracy of the Slot Timing

FlexRay™'s slot timing is based on distributed clocks synchronized by protocol's clock correction procedures. The achievable accuracy is proportional to the maximum propagation delay inside a FlexRay™ channel.

Two different types of clocks have to be distinguished. The sampling procedure is based on an 80 MHz clock generated from a crystal directly or by PLL. These clocks work asynchronous among each other and they are responsible for the bit-timing and the sampling of the transmitted data stream. The slot-timing however depends on a distributed clock synchronized among all FlexRay™ nodes which participate in this procedure. This clock is available in all FlexRay™ nodes. The protocol allows selecting whether a node participates in the distributed clock synchronization or overtakes an external clock.

Sampling and Glitch Filtering

Usually FlexRay™ topologies do not show any glitches – neither in laboratory set-ups nor inside vehicles. However dedicated stress conditions may enable glitches visible on the received RxD stream. Vehicle manufacturers EMC draft specifications describe different module tests which are able to generate glitches:

- Stress the bus-line by putting them under a strip line. Transmit standardized test-pulses (e.g. pulses[9] 3a and 3b with minimum edge time of 5 ns and amplitude up to approximately 100 V) to the strip line while a communication is running.

These test pulses are characterized by a rise (or fall depending on the polarity) time of 5 ns. The strip-line couples a common mode transient pulse (up to approximately 50% of the test pulse peak amplitude) to the bus lines. The common mode pulses can be able to stress the common mode ratio of the connected components; common mode chokes cannot damp them completely. On the other hand the common mode pulses can be transferred to differential mode pulses if the set-up is not balanced to reference ground perfectly. The differential mode pulses would be superposed to the data signals.

At the end glitches on the received RxD stream may appear. Cross-talk effects directly in the printed circuit board are able to generate glitches as well.

The communication controller samples the data stream signal eight times per bit cyclically. A successive glitch filter is specified to suppress faulty samples. After passing the glitch filter the fifth sample is always used as bit-value. The glitch filter generates the binary voted average of five previous samples and offers therefore some properties:

- The data stream is delayed by 3 sample periods;
- A single faulty sample value is going to be suppressed;
- Two successive faulty samples are going to be suppressed.

The succeeding figure shows a sampling and glitch filtering example.

One stand-alone sample or two combined stand-alone samples can be voted as glitches comprehensibly. They are going to be suppressed by the filter. If a glitch appears nearby an edge several scenarios could took place:

- **Target Pattern:** xxx11110000xxx
- **Glitch Scenery 1:** xxx1111**0**100xxx
- **Glitch Scenery 2:** xxx1111**1**000xxx
- **Glitch Scenery 3 (Two Glitches):** xxx11**1**0**1**000xxx
- **Glitch Scenery 4:** xxx111**00**000xxx
- **Glitch Scenery 5:** xxx11**0**1**0**000xxx

Potential glitches are marked in bold. If the target pattern is unknown (usually from receiver's point of view) the glitch filter cannot detect and correct the faulty sample(s). It has to smooth the edge comparable to a low pass filter (realized by the three of five voting). From physical layer's point of view glitches are able to increase the asymmetric delay on the pass from the transmitting protocol to the receiving protocol.

Communication Example in the Time Domain

Physical layer's properties can be shown easily when analyzing an example. Six communication nodes marked with the integer numbers 1 ... 6 are interconnected by a passive bus. Two nodes are terminated; four nodes are connected via stubs.

Figure 13. Sampling and glitch filtering example

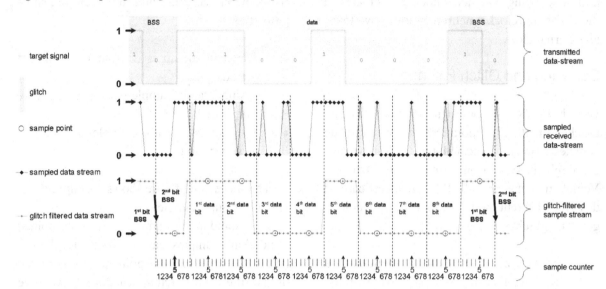

- The passive bus consists of five partial lines. Their lengths are varying between two meters and six meters. The four stub lines have a length of one meter each.
- The termination matches with the line impedance.
- CMC, ESD protection, block capacitor, power lines and parasitic components are considered.
- The CMC has a small stray inductance due to a coupling factor of 99%. (see: common mode choke section)
- Two nodes number 1 and number 2 send simplified identical dummy frames at 10 Mbit/sec.

All effects introduced in the previous sections are visible. Each receiver gets its message delayed and truncated. Each single bit is visible clearly. A little asymmetric delay appears (not recognizable in the chosen scaling).

Some nodes recognize ringing at the end of the frame. By random, the ringing gets stack to logical "1" at some nodes (see e.g. node number 4). By random, the ringing gets stack to logical "0" at some other nodes (see e.g. nodes number

2 and 5) until the successful idle detection masks the data stream to logical "1".

The communication is expected to work perfect. The truncation and the ringing can be taken into account when calculating the protocol parameters.

Requirements to the Signal Shapes

BP and BM on the Bus-Cables

Overview

Each circuit and each interface between two circuits generates a portion of the symmetric delay (propagation delay) and a portion to the asymmetric delay (bit-distortion). The symmetric delay influences the slot timing; protocol parameters allow an adaptation in a very broad range up to several μs. The asymmetric delay is limited to ±three sample periods (due to eight samples per bit while the fifth sample is used as bit-value always); a 10 Mbit/sec communication requires maximum ±37.5 ns therefore.

The FlexRay™ electrical physical layer specification and its application notes describe three different procedures validating whether the requirements to signals are met:

Figure 14. FlexRay™ communication inside a passive bus with six nodes

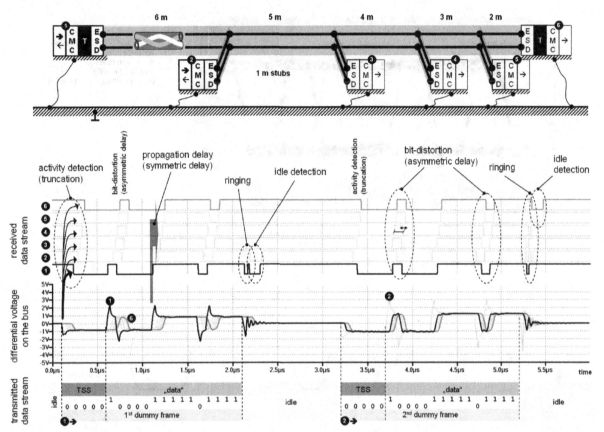

- **Mask Test:** The mask tests specify the requirements to FlexRay™ bus drivers transmitting stages when driving a standardized load. The mask test checks the asymmetric delay and levels of single bits.
- **Eye Chart Test:** The eye chart tests describe the minimum signal shapes which must be detected correctly by a FlexRay™ bus driver's receiving stage. The eye chart test checks the asymmetric delay and levels of a superposition of all bits of one message at least.
- **Signal Integrity Voting:** The signal integrity voting describes a procedure to check whether a measured or simulated FlexRay™ bus signal can be detected by a FlexRay™ bus driver's receiving stage correctly in principle.

Masks and Eyes

The masks represent the output capability of a bus driver when driving a standard load (40 Ω ∥ 100 pF) and the eye charts represent the minimum input capability of a bus driver. Between both the passive network with damping effects as well as reflections or refractions is placed.

The FlexRay™ electrical physical layer specification defines several masks and the application notes propose one eye chart per gross data rate. Individual eye charts may be defined to reflect dedicated system requirements. An example is included inside the succeeding overview figure.

Some oscilloscope manufactures just implemented the FlexRay™ eye chart test inside their products. The procedure is triggered by the falling BSS edge.

Figure 15. Masks and eye charts

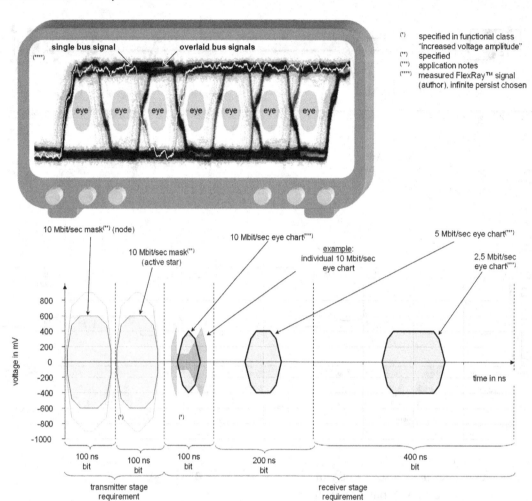

Signal Integrity Voting

A lot of perfect operating topologies would fail in an eye chart test. Usually an eye chart test requires that the signals do not have big under shoots nearby the centre of a bit. On the other hand these under shoots would be suppressed by each bus driver's receiver stage due to its Schmitt trigger behavior. The signal integrity voting introduced by the FlexRay™ electrical physical layer application notes covers the behavior of a receiver stage by defining a kind of worst case generic receiver.

- The speed limit of the receiver stage is modeled by a low pass filter 1st order with a cut off frequency of 14 MHz.

This module approximates the behavior of a worst case slow receiver. However implementations in silicon usually have a non-linear filtering behavior.

- The signal integrity voting shall be applied to detected single bits: $xxx_{000}{}^{1}{}_{0}xxx$ or $xxx^{111}{}_{0}{}^{1}xxx$.

The detection of single bits represents the hardest requirements to receiver stages.

- Sixteen Schmitt triggers cover the complete specified range of data detection thresholds. The thresholds are specified to vary within ±150 mV up to ±300 mV. They shall match within ±30 mV.

Several properties are derived from the single bit inside each detected bit-stream sequence:

- Shortest and longest bit;
- Minimum or maximum level (depending on the bit value);
- Duration passing the idle area (rising edge and falling edge).

The sixteen results are voted separately. Only if all results meet the specified requirements the signal integrity voting is passed:

- The shortest bit shall be longer then approximately 70 ns.
- The maximum additional asymmetry (durations longest bit – shortest bit) shall be less 7 ns.
- The edges shall pass the idle area maximum in 50 ns (avoiding idle detection during an ongoing data stream).
- The levels shall be 30 mV bigger than the maximum threshold of 300 mV at least (a little margin is required).

The succeeding figures illustrate the operating of the signal integrity voting. A noisy 10 MHz/sec FleyRay™ signal example with an undershot during the bit is going to be voted. The low pass filter smoothes the noisy signal. The visualized threshold areas show implicitly the effect of the asymmetric delay. Three values are derived from the low pass filtered signal directly:

- **Maximum Level Inside the Bit:** $\approx 1{,}2$ V $>> 330$ mV ✓
- **Idle Dwell Time (Rising Edge):** 7,6 ns $<< 50$ ns ✓ (duration passing the idle detection area)
- **Idle Dwell Time (Falling Edge):** 7,6 ns $<< 50$ ns ✓ (duration passing the idle detection area)

The effect of the asymmetric delay is shown directly by the superposition of 16 detected data streams according 16 defined threshold combinations.

Each of the 16 Schmitt triggers has two own thresholds and detects individual bit durations:

- **Detected Minimum Bit Duration:** 101,3 ns > 70 ns ✓
- **Detected Maximum Bit Duration:** 102,4 ns < 130 ns ✓
- **Detected Average Bit Duration:** n 101,85 ns
- **Maximum Additional Asymmetry:** $\pm0{,}55$ ns $<< 7$ ns ✓ $\pm\frac{1}{2}$ (maximum bit duration – minimum bit duration)

Figure 16. Signal integrity test and voting

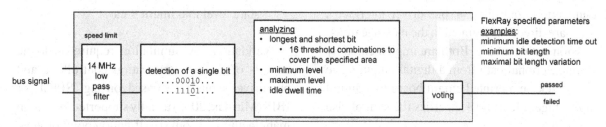

Figure 17. Signal Integrity test "bit-length detection" applied to a measured FlexRay™ signal example

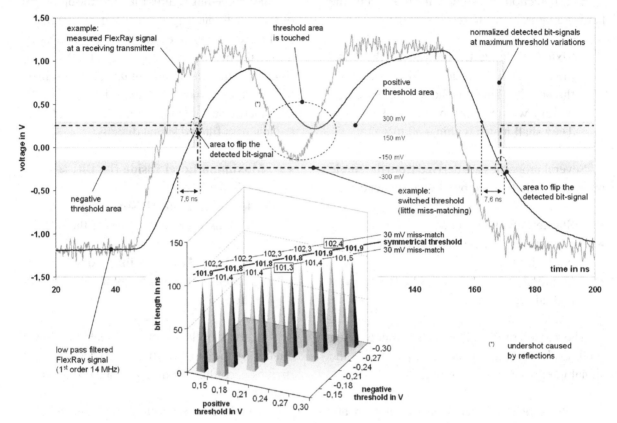

Although the big undershoot during the high-level of the differential signal would fail in an eye chart test, the corresponding bit will be detected clearly. The detected properties are within their specified range clearly. From signal integrity point of view a perfect signal shape is available.

TxD and RxD on the Printed Circuit Board

The signals TxD and RxD represent the "communication controller ⇔ bus driver interface". They are directly integrated in the message transmission path's timing. Both are implemented by a unidirectional path from a digital output stage via a track on a printed circuit board to a digital input stage. FlexRay™ specifies the sum of rise and fall time in a special case which represents a typical implementation directly:

- RxD/TxD output followed by a 1 ns 50 Ω micro-strip line and terminated by a 10 pF capacitor. The capacitor represents the maximum input capacitance of a digital receiver stage.

- The voltage at the capacitor is voted by its sum of rise and fall time detected between the thresholds at 20% at 80% of the steady state level maximum 9 ns

Validations by simulation require analogue models of the communication controller's and bus-driver's interface based on e.g. SPICE or IBIS (Mirmak, 2005) usually supported by silicon manufacturers. From specification point of view generic output and input stages can be used to do validation. The succeeding figure introduces an example based on the generic switch (see corresponding section). The values are chosen to show available physical effects.

- Impedance of the generic output stage high-side switch ∞ down to 100 Ω
- Impedance of the generic output stage low-side switch 100 Ω up to ∞
- Pin capacitance of the output stage 10 pF
- **Bit Transmitted by the Output Stage:** 30 ns (The short timing is chosen to visualize the effects best)
- **Open Load Rising Edge:** 2 ns
- **Open Load Falling Edge:** 3 ns
- **Output Stage's Impedance:** 100 Ω (See section SPICE compliant switch)
- **Pin Capacitance of the Input Stage:** 10 pF (Ideal) Substrate Diodes Available
- **Micro-Strip Line Impedance (Lossless):** 50 Ω
- **Micro-Strip Line Propagation Delay:** 1ns
- **Speed of the Input Stage:** 500 MHz (Negligible)
- **VCC Related Input Stage Thresholds (No Hysteresis):** 40%, 50%, 60%

Results

Although the propagation delay of the micro-strip line is shorter than the open load edge timings several reflections appear. The signals show steps

combined superposed with an exponential function typically for capacitive charging.

The detected bit duration and the asymmetric delay depend on the detection threshold.

- **Bit Durations Detected by the CMOS Receiver Stage:** 31,8 ns; 30,65 ns; 29,6 ns (no hysteresis)
- **Rising Edge at the Receiver Stage:** 4,6 ns
- **Falling Edge at the Receiver Stage:** 5 ns

The shown example represents approximately the worst case according FlexRay™ specification. Typical samples are going to show smaller values.

Asymmetric Delay Budget

FlexRay™ specifies an asymmetric delay budget. It describes aspects of the physical layer which leads to timing derivations between two alternating edges on the path from a transmitter to a receiver. Two aspects are covered:

- Maximum derivation of two alternating edges between ten bits: "ten bit asymmetric delay".

The timing is measured from the falling edge of the last byte start sequence to the rising edge of the frame end sequence. That's the worst case scenery from sampling point of view.

Figure 18. Example: Generic Model of the CC ⇔ BD interface

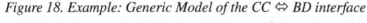
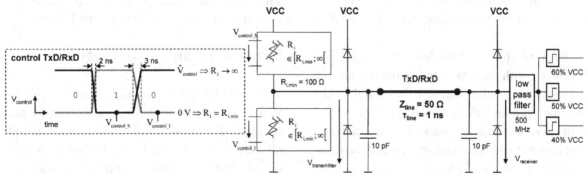

Figure 19. Result CC ⇔ BD interface

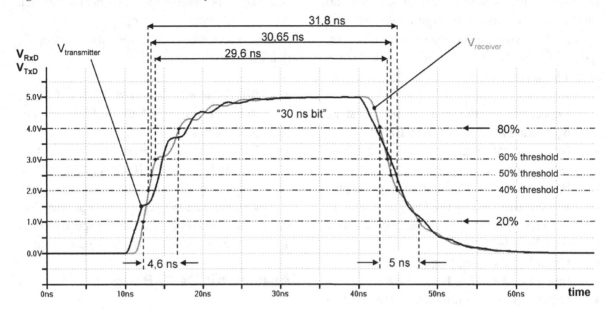

- Maximum shortening of a single bit.

That's worst case from bus driver's receiving capability point of view.

The ten bit asymmetric delay depends on the topology and the location of the communication partners inside the topology. A typical example topology "active star with passive buses" explains the effects. The specification and the application notes distinguish two behaviors:

- Static asymmetric delay on the longest path (timing budget path) which is constant during a frame but varies from frame to frame.
- Stochastic asymmetric delay on the longest path (timing budget path) which varies from edge to edge in each frame. This effect is mainly caused by irradiation (EMC).

Each of the involved hard ware modules beginning with the crystal inside the transmitter via bus drivers and ending with the crystal inside the receiver adds a differing small portion to the entire asymmetric delay. Some of these values are specified as required properties, some of them are specified as educated guess.

- The maximum static asymmetric delay in this topology (constant during a frame) is worst case specified to about ±30 ns at 10 Mbit/sec (sum of specified and educated guess values).
- The maximum stochastic asymmetric delay in this topology (varying from edge to edge) is estimated to about ±12 ns (mainly educated guess values due to EMC).

The worst case limit is given by the properties of the sampling/decoding procedures inside the communication controller: ±3 sampling periods = ±37,5 ns at 10 Mbit/sec. A contradiction seems to appear because the communication controller requirement is not met: 30 ns + 12 ns > 37,5 ns. The succeeding sections discuss potential approaches for interpreting this contradiction.

Random Sample Experiment

The asymmetric delay of a FlexRay™ topology shall be always within the limits given by the timing budget. One way to ensure this is to add all corresponding worst case values according the specification. However the simple addition could be far away from any reality and it is hard

Figure 20. Timing budget path example: Active star topology two passive buses as branches at least

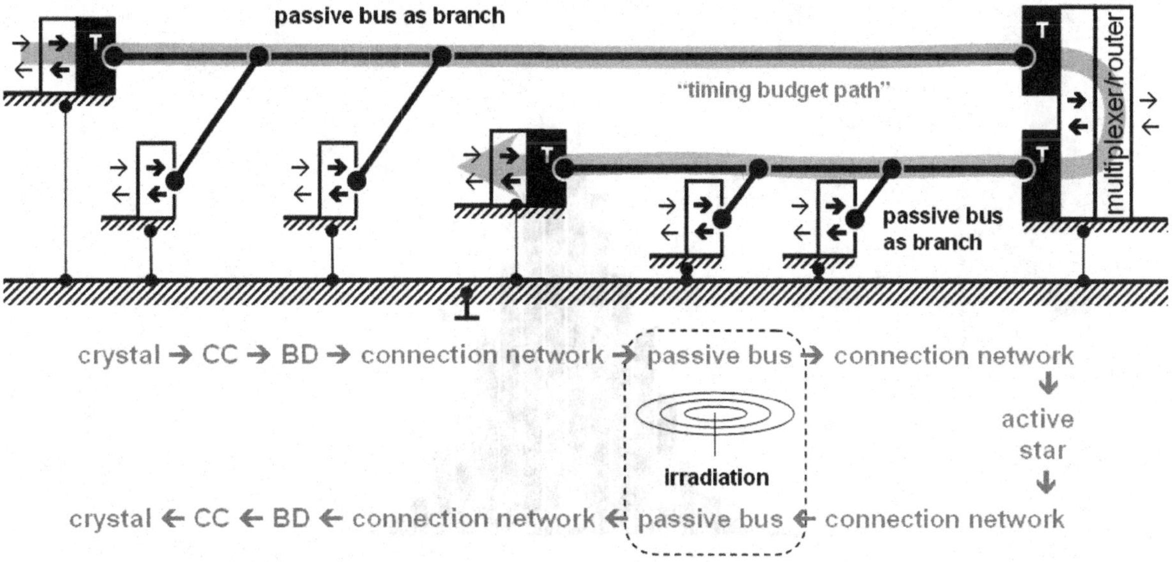

to get reliable data particular with regard to EMC influences. From the application point of view the asymmetric delay distribution inside a topology can be interesting to estimate e.g. a bit-error rate. The specification and the application notes provide a system designer with two values for the active star topology with two or more passive buses (±30 ns and ±12 ns). Both values are the sum of several more or less uncorrelated portions. Due to the lack of field data, random sample experiments shall help to estimate probability distributions. An example is shown in the succeeding figure:

- Each portion of the static delay and the asymmetric delay is varied uniformly.
- The asymmetric delay portions are uncorrelated among each other.
- Ten thousand experiments were taken and arranged in 2 ns intervals.

The experiment confirms the assumption as expected: although each portion is distributed uniformly the final distribution looks like the Gaussian bell curve. The worst case will be reached practically "never". The shown experiment tries to give an idea how to take probability distributions into account eventually. It can be used to pre-estimate the distribution of the timing budget in various FlexRay™ topologies.

Theory

One of the basic rules in statistics can be applied to the FlexRay™ parameter "timing budget". The rule points out, that each parameter which consists of sufficient enough statistically distributed single parameters ends in a Gaussian (or normal) distribution according the central limit theorem (Crilly (2008), Adamek (2010), Bronstein (1981) and others):

Adding k parameter to one parameter:

$$y = \sum_{i=1}^{k} y_i \qquad (1)$$

with their expectations μ_i and their standard deviations $\tilde{A}_i^2$

Figure 21. Distribution if taking 10.000 random samples at the topology example

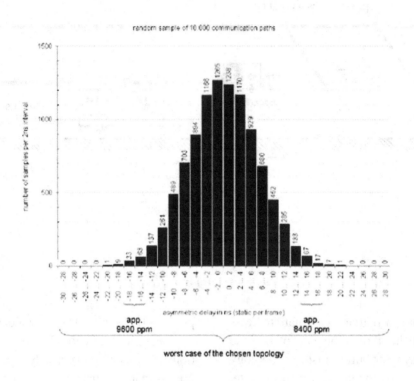

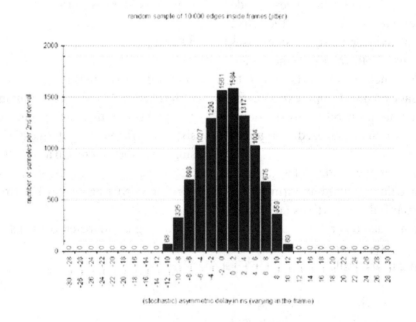

Results in a probability distribution:

$$y = \frac{1}{\sqrt{2\pi\sigma^2}} e^{-\frac{(x-\mu)^2}{2\sigma^2}} \qquad (2)$$

with the expectation:

$$\mu = \sum_{i=1}^{k} \mu_i \qquad (3)$$

and the standard deviation:

$$\sigma^2 = \sum_{i=1}^{k} \sigma_i^2 \qquad (4)$$

When applying this rules to the FlexRay™ timing budget effect (according the shown experiment),

Uniform distributed parameter y_i:

$$y_i = \begin{cases} 0 & \text{for} & x < -c_i \\ \dfrac{1}{2\,c_i} & \text{for} & x \in \left[-c_i; +c_i\right] \\ 0 & \text{for} & x > c_i \end{cases} \qquad (5)$$

with their expectations $\mu_i = 0$ and their standard deviations:

$$\sigma_i^2 = \int_{-c_i}^{c_i} \frac{1}{2c_i} x^2 dx = \frac{c_i^2}{3}$$

Results in a probability distribution (bell-shaped curve):

$$y = \frac{1}{\sqrt{2\pi\sigma^2}} e^{-\frac{x^2}{2\sigma^2}} \qquad (6)$$

with the expectation:

$$\mu = 0 \qquad (7)$$

and the standard deviation:

$$\sigma^2 = \frac{1}{3} \sum_{i=1}^{k} c_i^2 \qquad (8)$$

This result can be used to pre-estimate the distribution of the timing budget in various FlexRay™ topologies as well. All estimations are based on a preliminary approach which is sufficient so long as results from experiences in the field are not available:

- Each asymmetric delay parameter which delivers a portion of the complete asymmetric delay is characterized by a uniform probability distribution (non-correlated among each other).

Conclusion

Chapter 3.5.1 of the FlexRay™ electrical physical layer application notes hint probability calculations applied to the asymmetric delay. The previous section illuminated this from a pragmatic and a theoretic point of view as first approach how to approximate these effects in principle. Ongoing experiences in mass products will improve the examinations.

Passive Components

Bus Cables

Specified Properties

The FlexRay™ specifications describe requirements and properties of FlexRay™ compliant cables[10] very roughly: their design (two wires with an optional shield) as well as their differential mode impedance, differential mode propagation delay and differential mode attenuation.

The FlexRay™ bus driver and active star are specified for applying differential mode line impedances between 80 Ω and 110 Ω. However nor the common mode line impedance neither its frequency dependent losses are specified. The differential mode impedance range can be achieved with various line implementations; examples are: shielded and unshielded as well as coated twisted pair lines. Flat lines are not included. Values between 90 Ω and 100 Ω represent available cables.

Pulses Traveling along Lines

In the lossless case (e.g. premium coaxial BNC cables) the behavior of signals (e.g. pulses) along lines is very simple:

- Voltage pulse and current pulse have similar shapes; the ohmic line impedance appears as proportionality factor between them.
- While traveling along the line the shape remains unchanged. Neither damping nor edge smoothing takes place.

The behavior changes when passing a pulse into a lossy line (e.g. standard twisted pair). Several basis effects appear:

- The line impedance behaves like a frequency dependent complex resistor. However its ohmic amount dominates.
- The first corner of a rising or falling edge remains unchanged nearly because of the causal behavior.
- The second corner of a rising or falling edge gets smoothened because of the frequency dependent losses due to the skin effect and the loss angle of the isolator between the wires.
- The steady state level of long pulses gets damped because of the direct current resistor of the wires primary. The direct current resistor of the isolator can be neglected mostly.
- Short pulses are converted into shark dorsal fin shaped pulses.

The described effects of lossy lines can be measured (see example in the corresponding section). Time domain simulation tools offer two types of transmission line models usually: lossless and constant current losses. Both models are not able to cover all described effects.

Figure 22. Traveling of pulses along lossless and lossy lines

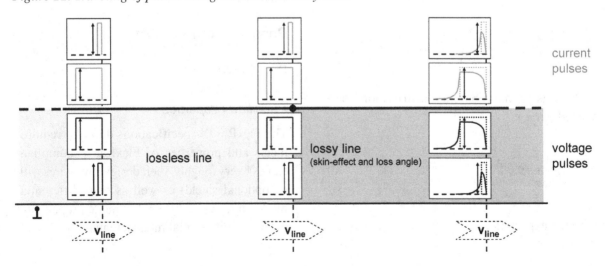

Line Parameters and their Dependencies

Lines consist of approximately parallel wires with an isolator between them. Four different parameters characterize lines: the conductance of the isolator, the resistance of the wires, the capacitance and the inductance between the wires. Each of these parameters varies with the frequency: some more, some less. The common known Skin effect influences the inductance and the resistor mainly. Dielectric losses in the isolator influence the conductance mainly. Various proportionalities appear: $\sim f$, $\sim \sqrt{f}$, $\sim \frac{1}{\sqrt{f}}$. These dependencies can be seen most in frequency dependent losses. E.g. Meinke and Gundlach (1986) describe losses in three steps: up to some 100 kHz direct current losses are dominant; for higher frequencies wire losses proportional to the square root of the frequencies are dominant (Skin effect); for very high frequencies losses proportional to the frequency are dominant (isolator losses).

Standard simulation tools in the time domain support two types of lines usually: lines without any losses and lines with constant losses. Particularly lines with constant losses are used very often analyzing automotive communication topologies. Anyhow frequency dependent losses are absence the achievable accuracy meets most engineers' requirements often; in particular up to the communication speed of FlexRay™. Some dedicated use cases suggest the consideration of frequency dependent losses although. Examples are long cables, wires with small diameters, isolators with high loss angles or calculations covering the worst case.

Measuring of Pulses Traveling along a Cable

A generator drives a signal into a wire above reference ground. Three voltages are measured. The equivalent circuit of the set-up consists of:

- **Pulse Generator:**
  - Transmitting a 4 V / 100 ns pulse and a 4 V / 50 ns pulse, impedance 50 Ω.
  - Rise and fall time of a few ns (perfect rectangle signals from an engineer point of view).
- Two identical lines in a row:
  - **Length:** 48 m
  - **Propagation Delay:** $\approx$ 215 ns
  - **Line Impedance:** $\approx$ 48 Ω
- 50 Ω termination at the end of the 2nd line
  - The voltages V_1, V_2 and V_3 according to the succeeding figure were measured with a 200 MHz oscilloscope. The used measuring setup avoided any reflections. Assuming the lines are perfectly lossless the oscilloscope should show practically rectangle signals at all three test plains. Lossless simulations show exactly this behavior. Smoothed pulses with overshoots and undershoots appear on the oscilloscope.

Detected properties of the acquired signals:

- Differences between "constant losses simulation" and measurement appear clearly. The differences are approximately proportional to the cable length; in other words, they increase with the cable length.
- The upper corner of the rising edge gets smoothened more and more.
- The lower corner of the falling edge gets smoothened more and more.
- The achievable level gets damped more and more.
- Rectangle pulses convert more and more to a shark fin shape when traveling along the line.
- The slew-rate of the edges decreases along the line.

Figure 23. Transmitting of rectangle pulses (4V / 100 ns; 4V / 50ns)

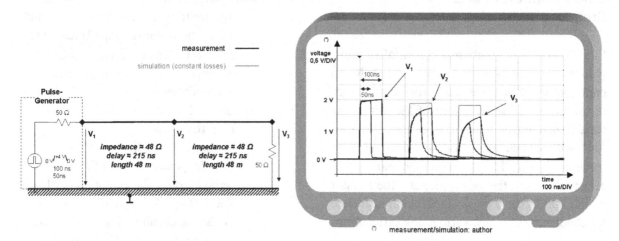

Mainly the measurement results are caused by the two physical effects yet pointed out: the loss angle of the isolator and the skin effect. The usage of direct current losses only generates clear differences. However many topologies (especially when using shorter lines) do not require frequency dependent losses to evaluate them.

Unshielded Twisted Pair Cables

Unshielded twisted pair cables can be described as superposition of three transmission lines: twice wire ⇔ ground and once wire ⇔ wire (according VDE 0472 part 516). The symmetry enables its characterization by two line impedance values only. A typical example shows the succeeding figure.

Impedance of one wire to ground $Z = 300\ \Omega$
$$(9)$$

Impedance between the wires $Z_{12} = 120\ \Omega$
$$(10)$$

Common mode impedance $Z_{com} = Z\|Z = 150\ \Omega$
$$(11)$$

Differential mode impedance $Z_{dif} = Z_{12}\|(Z+Z)$
$= 100\ \Omega$
$$(12)$$

From simulation point of view the superposition is described e.g. by Hoefer and Nielinger (1985): two separate lines to ground with the impedance Z each and a line with the impedance Z_{12} connecting the first and the second line. The propagation delays of the three lines shall be identical. From data communication point of view

Figure 24. Unshielded twisted pair example and its operating modes

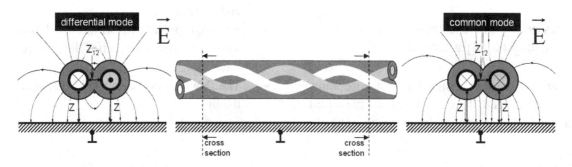

it is recommended choosing the differential modes speed. In most cases the common mode propagation delay is a little bit less than the differential mode delay.

Termination Networks

Termination networks are necessary additions to bus systems. They fulfill various tasks, some of them are obvious and some others are not. Some termination networks do not support each task:

- Terminate the bus lines to avoid reflections of the communication signals.
- Offer an ohmic direct current load for the transceiver or bus-driver.
- Terminate the bus lines to avoid reflections from radiation point of view.
- Terminate the bus lines to avoid reflections from irradiation point of view.
- Bias the bus-lines.

The succeeding section introduces some typical termination networks and their properties.

Split Termination

The FlexRay™ electrical physical layer specification recommends implementing a split termination (used in CAN too). When being connected to the end of a typically twisted pair cable, best results can be achieved.

- **Differential Mode Impedance:**

$$R_{T,dif} = R_{T1} + R_{T2} \tag{13}$$

(Direct current and radio frequency)

- **Common Mode Impedance:**

$$R_{T,com} = R_{T1} \| R_{T2} + R_{T3} \tag{14}$$

(Radio frequency)

Some general design rules can be stated by using the defined twisted pair line impedances (see corresponding sections):

- **Differential Mode Impedance:**

Figure 25. Split termination

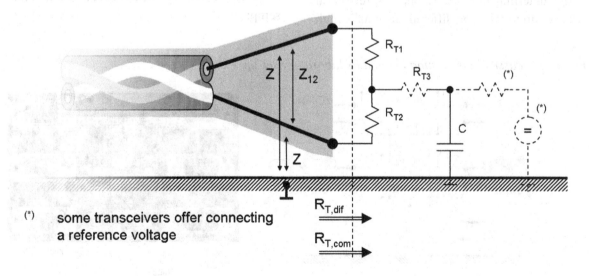

(*) **some transceivers offer connecting a reference voltage**

$$R_{T,dif} \overset{!}{\approx} Z_{line,dif} \quad (= Z_{12} || 2Z) \tag{15}$$

- **Common Mode Impedance:**

$$R_{T,com} \overset{!}{\approx} Z_{line,com} \quad \left(= \frac{Z}{2}\right) \tag{16}$$

Recommended matching $=> 99\%$ $R_{T1} \overset{!}{\approx} R_{T2}$
$$\tag{17}$$

(avoiding differential to common mode converting; being balanced to ground)

$$\text{Capacitance} = C \approx 4.7 - 10 \text{ nF} \tag{18}$$

When following the recommended rules an unshielded twisted pair bus cable is terminated practically perfect (common mode and differential mode use case). Just available parasitic components produce little ringing effects.

Passive Star Termination

The FlexRay™ electrical physical layer application notes recommend implementing a passive star with a R||L circuit per bus wire (used in CAN too). This termination is able to suppress reflection of data communication differential signal's high frequency parts. It is obvious to add split termination optionally (named if implemented: central termination). Reflections can be suppressed and the direct current level can be controlled.

The dimensioning is similar to the split termination:

$$R_{T,dif} \overset{!}{\gg} Z_{line,dif} \quad (= Z_{12} || 2Z)$$

$$R_{T,com} \overset{!}{\gg} Z_{line,com} \quad (= \frac{Z}{2})$$

The following guide line can be used to achieve the dimensioning above approximately:

$$R = R_{T1} + R_{T2} = \frac{Z_{line,dif}}{2} \tag{19}$$

$R_{T1} + R_{T2}$ represent the parallel circuit of the usually available two terminations; each with $Z_{line,dif}$. $R + R$ represents the high frequency relevant termination of each branch assuming that the parallel circuit of the other branches and the split termination end in a value $<< Z_{line,dif}$.

The value of the inductivity L depends on the cable length, on the bit duration and on the required ringing limit.

The effect of the passive star termination can be shown easily by the succeeding simulation setup:

Figure 26. Passive star termination and its laboratory set-up

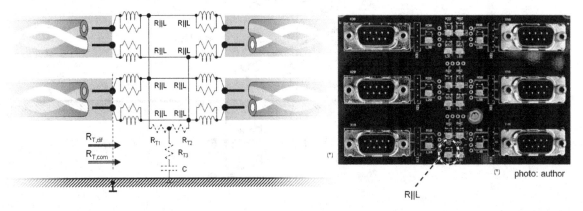

Figure 27. Example of symmetric passive star implementations

use case: four branches + central termination

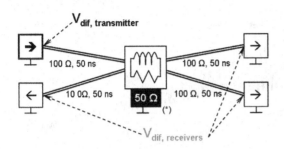

use case: six branches (two of them: terminated)

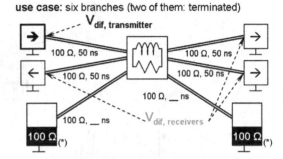

(*) two terminated 100 Ω lines are approximately similar to a single 50 Ω resistor

- Four (or six: see figure and the hint added to the next topic) identical lossy lines (differential mode line impedance 100 Ω, propagation delay 50 ns each).
- Central termination of 50 Ω ($R_{T1} + R_{T2}$)

Hint

The central termination of 50 Ω is nearly equivalent to two additional branches connected to the passive star, if this branches are terminated perfectly at their transceiver side.

R_{T3} and C do not appear – only the differential mode is relevant here.

- One transmitter (CAN high-speed compliant) transmits the states recessive - dominant - recessive.

Hint

FlexRay™ bus-drivers generate similar effects when transmitting idle – active (or busy) - idle.

- High-ohmic receivers with parasitics and biasing.
- Common mode chokes with a low stray inductance.

- The resistor value R of the part "R‖L" is chosen to 50 Ω; according to the recommendation.
- The inductance value L of the part "R‖L" is chosen to 0 µH; 1 µH and 10 µH

The result of the simulation is very clear:

- The inductance L controls the ringing appearing at the edges.

Two behaviors have to be distinguished per bus system:

A. Output stage switches from off to on (or high ohmic to low ohmic).
B. The output stage switches from on to off (or low ohmic to high ohmic).

Explanation

A. Represents in case of CAN: recessive to dominant edge.
B. Represents in case of CAN: dominant to recessive edge.
A. Represents in case of FlexRay™: idle to active (or busy) edge.
B. Represents in case of FlexRay™: active (or busy) to idle edge.

Figure 28. Signals at the symmetric passive star example

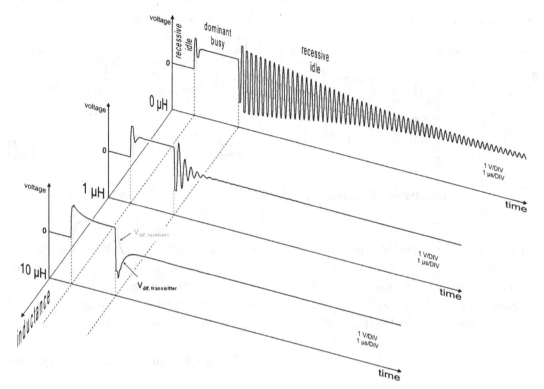

1. The most challenging case is when the stage gets passive. The inductance L controls the damping constant of the ringing (caused by reflections on the lines). Even in the best case a big undershot appears.

2. The frequency appearing in the ringing is inverse proportional to the propagation delay of the line.

The example shows that the inductance shall be chosen somewhere nearby 10 μH. On one hand this value shall be chosen proportional to the length of the lines and on the other hand the remaining undershot (app. 500 ns in the example) should be shorter than the bit timing. A communication speed less than 1 Mbit/sec is achievable.

Passive star setups can be used in low-speed CAN applications very easily. Passive stars in FlexRay™ applications are possible only if the length of the lines are limited.

FlexRay™'s Direct Current Bus Load

When driving data_0 or data_1 in the steady state the differential mode load between the bus lines is specified as direct current bus load. The minimum value is 40 Ω; the maximum value is 55 Ω. All effects shall be included: especially termination resistors and parasitic impedances of the connected components. The value is available per passive net inside any topology.

Common Mode Chokes

Common mode chokes are used in automotive ECUs usually. They are placed between the bus driver and the optional termination or ECU's connector to the bus lines. Two main requirements are:

- The bidirectional communication in the differential mode shall be influenced marginally.

Figure 29. Examples of common mode chokes

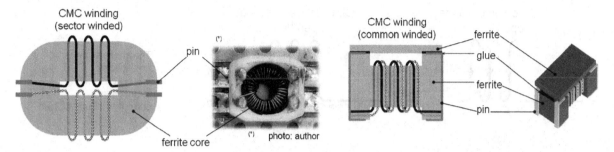

- Common mode signals shall be suppressed.

Explanation

The bus driver (and the ECU) may not be stressed by common mode signals coming from outside the ECU (e.g. via ECU's wiring harness).

ECU's wiring harness may not be stressed by common mode signals coming from inside the ECU.

Common mode signals can be generated by several sources.

- Continuous wave signals, amplitude modulated signals or transient signals which are injected to the bus cables (against ground or chassis).
- Transmitting FlexRay™ communication elements via (little) asymmetric bus driver output stages. Asymmetries can be found as offset errors or/and edge errors. Offset errors are caused by e.g. high- and low-side switches with various impedances. Edge errors are caused by e.g. none simultaneously working high- and the low-side switches (e.g. various slew rates or various propagation delays).
- Transmitting regular differential mode communication signals via set-ups not balanced to ground are able to convert them to common mode signals.

Design and Properties

The set-up of a common mode chokes is straightforward. The basis set-up is a ferrite bead which covers all wires of a cable and which is usually placed near to the connector (examples: monitor cable, USB cable). Applied to a two-wire bus cable both cables are winded in the same direction round a magnetic closed ferrite. The windings can be arranged separately on the ferrite (sector winded) or they can be arranged together (common winded). Due to various production procedures little air gaps may appear.

When running a common mode choke with a differential mode signal the magnetic fields caused by the two windings compensate each other. The stray inductance caused by the non-perfect magnetic coupling of the two coils influences the result a little. In case of FlexRay™ damped ringing effects can appear at the change from active to idle (comparable to the CAN change from dominant to recessive).

When running a common mode choke with a common mode signal the magnetic fields caused by the two windings accumulate. The main inductance dominates the result and damps the signal clearly.

Equations

The equivalent circuit of a common mode choke is built by two coupled coils mainly. Assuming symmetry, both coils have the same inductance L and a coupling factor near to one. From data com-

Figure 30. Operating of common mode chokes

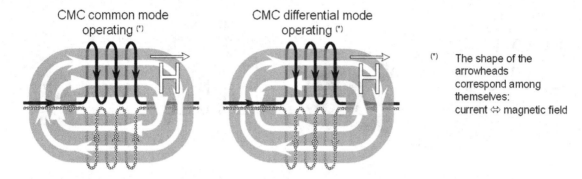

Figure 31. Definition of CMS's differential mode inductance L_{dif} and common mode inductance L_{com}

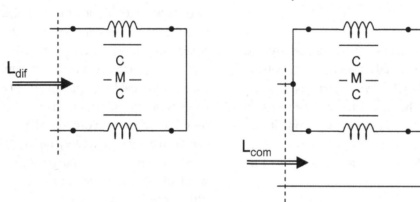

munication point of view (differential mode) the inductance L_{dif} is active. From radiation point of view (common mode) the inductance L_{com} is active.

- **Inductance of One Coil:** L in H
- **Magnetic Coupling of the Two Coils:**

$$k \in [0;1[\qquad (20)$$

- **Differential Mode Inductance:**

$$L_{dif} = 2L(1-k) \qquad (21)$$

(proportional to the stray inductance $\ll$ L)

- **Common Mode Inductance:**

$$L_{com} = L\frac{1+k}{2} \qquad (22)$$

- **Coupling Factor:**

$$k = \frac{4 - \dfrac{L_{dif}}{L_{com}}}{4 + \dfrac{L_{dif}}{L_{com}}} = 1 - \frac{2L_{dif}}{4L_{com} + L_{dif}} < 1 \qquad (23)$$

- **Inductance of One Coil:**

$$L = L_{com} + \frac{L_{dif}}{4} \qquad (24)$$

Special Case k=1 (Perfect Coupling, Not Producible):

- **Differential Mode Inductance:**

$$L_{dif} = 0 \qquad (25)$$

- **Common Mode Inductance:**

$$L_{com} = L \qquad (26)$$

Special Case k=0 (No Coupling, Two Separate Coils):

- **Differential Mode Inductance:**

$$L_{dif} = 2L \qquad (27)$$

- **Common Mode Inductance:**

$$L_{com} = \frac{L}{2} \qquad (28)$$

Example

A typical representative of a sector winded common mode choke is chosen to point out the main properties. This choke is used in CAN applications often. The usage in FlexRay™ applications is possible but not recommended - the stray inductance is higher than specified in the FlexRay™ electrical physical layer specification.

Measurements in the time-domain (see succeeding sections) enable the calculation of the CMC's main values:

- **Differential Mode Inductivity:**

$$L_{dif} \approx 3.9 \mu H \qquad (29)$$

- **Common Mode Inductivity:**

$$L_{com} \approx 43 \mu H \qquad (30)$$

- **Coupling Factor:**

$$k = 1 - \frac{2L_{dif}}{4L_{com} + L_{dif}} \approx 95{,}6\% \qquad (31)$$

- **Inductance of One Coil:**

$$L = L_{com} + \frac{L_{dif}}{4} \approx 44 \mu H \qquad (32)$$

From data communication's signal integrity point of view the common mode choke works like a 3,9 µH inductor along the signal path approximately.

Generic Simulation Model

Usually common mode chokes' simulation models are offered by their manufacturers (e.g. TDK). A simple generic model (applicable for signal integrity analysis) can be generated by using three main values: inductance L, coupling factor k and the serial resistance of one coil. E.g. internal resonances or magnetic losses are not covered.

Figure 32. Sketch to get the electrical behavior of a CMC

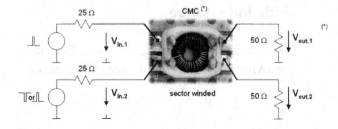

[*] The housing of the shown CMC is opened (photo: author). The internal design with the ferrite core and the windings is visible. The CMC is mounted on a strip board (1/10 inch grid).

Figure 33. Basic simulation model of a CMC and its SPICE syntax

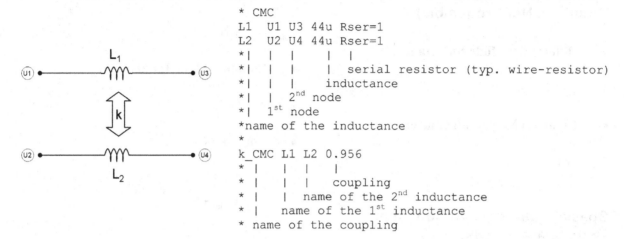

```
* CMC
L1   U1 U3 44u Rser=1
L2   U2 U4 44u Rser=1
*|   |  |   |   |
*|   |  |   |   | serial resistor (typ. wire-resistor)
*|   |  |   | inductance
*|   |  2^nd node
*|   1^st node
*name of the inductance
*
k_CMC L1 L2 0.956
* |   |  |   |
* |   |  |   coupling
* |   |  name of the 2^nd inductance
* |   name of the 1^st inductance
* name of the coupling
```

Hints

- The resistor(-value) Rser can be added separately if the available SPICE parser does not support Rser as a parameter of an inductor.
- Depending on the required accuracy additional resistors and capacitances parallel to the coils are recommended (e.g. to deal with losses of the magnetic material or with internal resonances).
- Common mode chokes show little (app. ±20% regarding the inductance in the usual operating area) non-linear behavior which is not covered by the linear generic model.

Operating Modes

The differential mode operating represents the serial data communication which should not be influenced by the availability of the common mode choke usually. The time domain results correspond with the expected effects. The common mode choke works as being a stand-alone inductivity.

The oscilloscope image allows to derivate some characteristic values:

- **Time Constant:**

$$\tau = \frac{L_{dif}}{150\ \Omega} \approx 26\ \text{ns} \tag{33}$$

(differential mode inductance $L_{dif} \approx 3{,}9\ \mu\text{H}$)

The common mode operating represents the radiation of common mode disturbances sent by the bus driver. The time domain results correspond with the expected effects. The common mode choke works as being a stand-alone inductivity nearly.

The oscilloscope images allow to derivate some characteristic values:

- **Time Constant:**

$$\tau = \frac{L_{com}}{37{,}5\ \Omega} \approx 1150\ \text{ns} \tag{34}$$

(common mode inductance $L_{com} \approx 43\ \mu\text{H}$)

Impacts of the Common Mode Inductance

On one hand common mode chokes are first class improving EMC behavior; on the other hand de-

Figure 34. CMC in differential mode operating at 150 Ω load

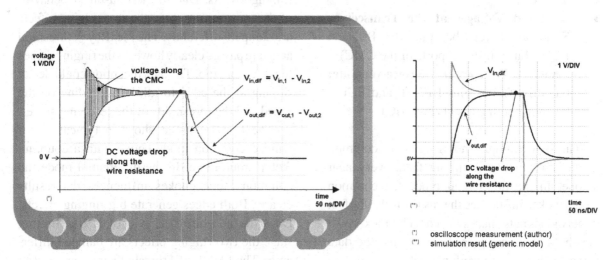

Figure 35. CMC in common mode operating at 37,5 Ω load

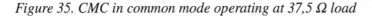

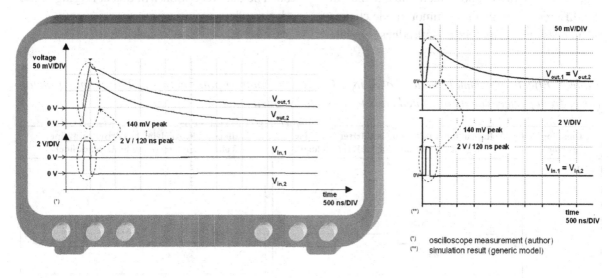

pending on its properties the signal shapes in the differential mode are influenced. A basic point to point connection visualizes the main effects. The set-up is typically for FlexRay™ active star branches:

- A four meter transmission line;
- A generic driver stage with 20 ns switching time and ±700 mV levels;
- Typical parasitic components (biasing, pin-capacitances);

- Perfect matching termination (start and end of the line);
- Each coil of the CMC has an inductance of 100 μH;
- The magnetic coupling factor varies: 99,99%; 99,5%; 98%;
- Stimulation of a1 μs FlexRay™ communication element (10 bits at 10 Mbit/sec logical zero) marked as "symbol" in the succeeding figure.
- **TxD:** ...$^{1111}_{0000000000}$ ^{1111} ...

- **TxEN:** $\ldots^{1111}{}_{0000000000}{}^{1111}\ldots$
- **Analyzed Voltages at the Transceiver Side:** bus driver's bus pins BP; BM and ECU's bus pins (both ports of the CMC)
- **Analyzed Voltages at the Receiver Side:** bus driver's bus pins BP; BP and ECU's bus pins (both ports of the CMC)

The first example shows a perfect common mode choke with a coupling factor very near to one. In the differential mode the common mode chokes influences the results little. Perfect trapezoid signals shapes appear. The FlexRay™ threshold areas are passed clearly and the data reception is going to work perfect.

The second example shows a typical common mode choke with a coupling factor near to one. In the differential mode the common mode chokes influences the results a little. Both edges generate ringing effects. Due to the non-linear behavior of the transmitting stage the two ringing effects are damped differently. The FlexRay™ threshold areas are passed clearly however the ringing passes the threshold too. The data reception could detect ringing at the edge to idle – depending on the speed and the thresholds of the receiving stage.

The third example shows a typical sector winded common mode choke with a coupling factor around one. In the differential mode the common mode chokes influences the results clearly. Both edges generate big ringing effects. Due to the non-linear behavior of the transmitting stage the two ringing effects are damped differently. The FlexRay™ threshold areas are passed clearly however the ringing passes the threshold, too. The data reception will detect ringing at the edge to idle.

Table 4. Effects of common mode chokes in a FlexRay™ point-to-point network. The coupling factor varies from perfect common winded coils up to sector winded coils.

Coupling Factor	Voltage at the transmitting bus driver (gray line) Voltage at the transmitting ECU (black line)	Voltage at the receiving bus driver (gray line) Voltage at the receiving ECU (black line)
99.99%		
99.5%		
98%		

The ringing on the bus lines is not only caused by the coupling factor of the common mode choke or rather its stray inductance. A lot of additional properties increase or decrease the ringing result. Most of these properties are not specified clearly and educated guess values have to be assumed:

- Losses inside the ferrite of the common mode choke;
- Parasitic capacitances nearby the common mode choke, e. g: pin capacitances;
- Losses inside the transmission lines: skin effect and isolator's loss angle;
- Reflections due to parasitic capacitive loads at the lines; e.g. pin capacitances of bus drivers, electro static discharge protection circuits and the common mode choke itself;
- Reflections when using e.g. passive buses or passive buses with stub lines mainly caused by their line impedances.

Applying a similar simulation to CAN generates comparable results.

ESD Protection

The usage of ESD protection elements connected to the bus-lines and ECU's ground is recommended. Their equivalent capacitance should be less than 20 pF. A maximum mismatch of 2% is recommended. Available products are implemented on a single silicon die placed inside a three pin package.

Measurement Examples

FlexRay™ at a Passive Star

The example in this section points out some typical application-oriented properties of the FlexRay™ electrical physical layer. A passive network usually applied in conformance tests is selected. A dummy

frame combining different bit-patterns checks the longest path inside the chosen topology.

The measurement set-up is characterized by:

- Three networked ECUs with bus drivers, common mode chokes and ESD protection components (infinite termination).
- Two additional networked ECUs with bus drivers, common mode chokes, ESD protection components and a practically perfect termination.
- A passive star with five branches and filtering elements according the FlexRay™ specification is used.
- Coated twisted pair lines are used (the branch lengths vary from 1 meter up to 11 meters).
- A 10 Mbit/sec FleyRay™ communication is stimulated.
- A 100 bit dummy frame is used combining various bit patterns.
- A four channel 500 MHz oscilloscope was used to get the signal shapes.

The dummy frame is transferred among the two terminated ECUs. The path between both contains more than 10 m bus-cables. Approximately 20 m bus-cables are used to build up the complete passive star. The measurement results can be characterized by:

- The FleyRay™ dummy frames are transferred from TxD/TxEN via the topology to RxD faultless.
- The differential mode signals (named uBus) are acceptable. The signal shapes of transmitted single bits are near to the (worst case) limit.
- Little ringing is visible on the differential bus signal.
- The filtering elements mounted inside the passive star damp reflections successfully.
- Bit deformations appear: a transmitted single bit (...00010... or ...11101...) is

Figure 36. Measurements at a passive star topology used for a 10 Mbit/sec FlexRay™ communication

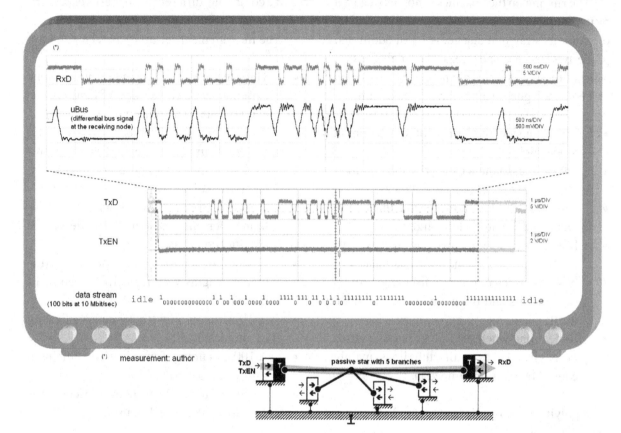

shorter than the bits inside a homogenous pattern (…101010…).

- Reflections caused by an edge appear with a delay longer then the bit duration. In other words the signal shape of a transferred bit depends on the history bit pattern before the considered bit.

The used topology is an element of the FlexRay™ conformance test which has to be passed by each bus-driver type. The received RxD signal is faultless here although the bus-signals are far away from being perfect.

CAN with Ground Shift

The example in this section points out some typical application-oriented properties of the CAN electrical physical layer. The results visible here appear in FlexRay™ as well.

To visualize the transient behavior best the bit-duration should match with the time constants during the recessive or idle phase especially. The effects shown in CAN easily, can be shown in FlexRay™ only if varying the time resolution dramatically – the clarity of the images would be lost.

The example points out the correlations between common mode and differential mode caused by any asymmetry. A lot of different effects are able to generate asymmetries. Examples are:

- Recessive (idle) level differs from 2,5 V;
- The ground shifts differs among the ECUs (constant or temporarily);

- The high-side output stages do not match with the low-side output stages perfectly (delay, slew rate, transient impedance, steady state impedance);
- The line impedances are not balanced to reference ground perfectly (reference ground: usually the chassis inside the car or a metal sheet inside laboratory set-up);
- The termination circuits are not balanced perfectly to reference ground;
- The ECUs and their power lines are not balanced perfectly to reference ground
- Etc.

Plenty additional more or less important asymmetry effects exist. Applying a ground shift only to a single ECU enables the visualization of the basic properties appearing in asymmetric used networks. The selected ground shift scenery generates bus signals which can be measured in automotive applications usually.

The chosen measurement set-up is characterized by:

- Five networked ECUs with high-speed transceivers, common mode chokes and ESD protection (high ohmic termination).
- A passive star with five branches, summarized termination and filtering circuits are used. This passive star is implemented according the FlexRay™ specification.
- Twisted pair lines (branch lengths app. 4 meters) are used.
- A 182 kBit/sec CAN communication is running.
- A single ECU is applied with a ground shift of app. 2,1 V.
- A four channel 500 MHz oscilloscope is used to get the signal shapes (channel 1 and 2 connected to CAN_{high} and CAN_{low}, two mathematically operation are applied generating the common mode signal and the differential mode signal).

Some definitions and statements enable easy reading and point out the interconnection between CAN and FlexRay™:

- Voltages are measured to reference ground.
- The recessive level in CAN is similar to the idle level in FlexRay™.
- The CAN common mode voltage (voltage at CAN_{high} + voltage at CAN_{low}) is similar to the FlexRay™ common mode voltage (voltage at BM + voltage at BP).
- The dominant common mode voltage in CAN is similar to the common mode voltage during activity in FlexRay™.
- The CAN transceiver is similar to the FlexRay™ bus driver here:
  - **CAN:** TxD 1111**data_stream**1111
  - **FlexRay™:** TxD 11111111111111111111
  - TxEN 1111**data_stream**1111

Each ECU transmits CAN messages. Each branch of the passive star has approximately the identical length. An example of two successive CAN messages is chosen to visualize the asymmetry effects best. The measurement results can be characterized by:

- The CAN communications works perfect among all ECUs.
- The differential mode signals are perfect (neither ringing nor bit-deformation appears at any message). Each message shows the expected in-frame acknowledge (ACK) bit enlargement.

Hint

This effect does not appear in FlexRay™. FlexRay™ does not support in-frame acknowledgement. Collisions are handled different from physical point of view.

Figure 37. Measurements at a passive star topology (182 kbit/sec CAN)

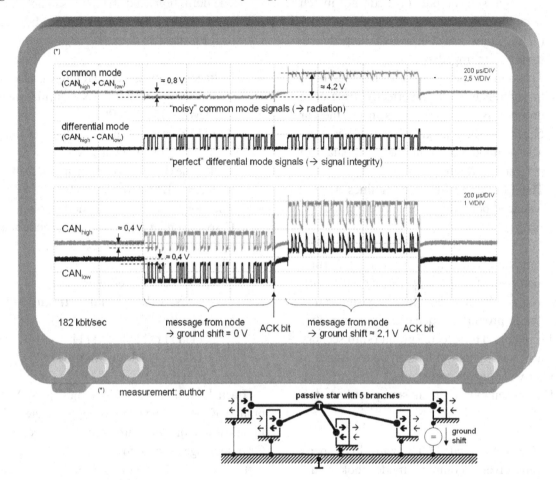

- The common mode signals are very noisy; the radiation is expected being influenced negatively.
- The filtering circuits mounted inside the passive star damp reflections successfully.

The scenery with a ground shift applied to one of five ECUs can be analyzed very easily[11]:

- **[Bus Driver's] Transceiver's Supply Voltage:** VCC ≈ 5V (35)
- **[Idle] Recessive Level Without Any Ground Shift:** ≈ ½ VCC ≈ 2,5 V (36)
  [Idle] Recessive Level in the Use Case: 1 of n ECUs with ground shift

$$\frac{\text{VCC}}{2} + \left.\frac{V_{\text{ground shift}}}{n}\right|_{\substack{\text{here} \\ n=5}} \approx 2,5 \text{ V} + \frac{2,1 \text{ V}}{5} \approx 2,9 \text{ V}$$

(37)

Explanation

Four ECUs try to generate 2,5 V and one ECU tries to generate 2,5 V + 2,1 V; each via some 10 kΩ.

The expected recessive common mode voltage shall reach app. 5,8 V.

- [Activity or data_0] Dominant common mode level of an ECU without ground shift.

$$2 \cdot \frac{VCC}{2} \approx 5V \qquad (38)$$

- [Activity or data_0] Dominant common mode level of an ECU with ground shift.

$$2 \cdot \left(\frac{VCC}{2} + V_{ground\ shift} \right) \approx 2 \cdot (2{,}5V + 2{,}1V) = 9{,}2V \qquad (39)$$

- Common mode step from "recessive" [idle] to "message without ground shift".

$$5{,}8\ V \rightarrow 5\ V = 800mV \text{ step down} \qquad (40)$$

$$CAN_{high} \text{ and } CAN_{low} \text{ each with } 400\ mV$$

Common mode step from "message without ground shift" to "message with ground shift".

$$5\ V \rightarrow 9{,}2\ V = 4{,}2\ V \text{ step up} \qquad (41)$$

- The measured common mode voltage shows triangular peaks during each recessive bit. The effect appears at each message.

Explanation

During a [idle] recessive phase the common mode voltage tries to reach the [idle] recessive common mode level.

The measurement shows plenty of effects appearing in case of ground shift. On one hand the common mode voltage jumps up and down (driven by messages and bits); on the other hand the differential mode voltage is practically perfect. The CAN communication works perfect, radiation effects increase.

Signals in the Time and Frequency Domain

For the sake of completeness differential mode and common mode measurements in the frequency domain are added (ground shift here 0 V). They were measured and calculated by a 200 MHz 8 bit oscilloscope with an internal FFT[12] function. A dedicated message is used to visualize a line spectrum best: identifier 5 55 (hexadecimal) and data 55 55 55 55 (hexadecimal).

- **Communication Speed:** 182 kbit/sec
- **Average Voltage Differential Mode:** app. 1 V
- **Average Voltage Common Mode:** app. 10 mV
- **Offset Between Common Mode and Differential Mode Spectra:** app. 40 dB
- **Distance Between Two Spectral Lines:** 182 kHz
- **Distance Between Two Spectral Zero Points (Due to 50% duty cycle):** 182 kHz

If the frequency and dynamic requirements are not too high, even a FFT which is built in most oscilloscopes is able to show good results; particularly if the time domain signal is acquired by using an over-sampling averaging function. The spectrum example is available from direct current up to 2 MHz; the dynamic range covers more than 50 dB.

Generic Bus-Driver Model

This section introduces a generic bus driver model optimized for minimum simulation time and compliant to each SPICE derivate available on the market. The model shall not replace silicon manufacturer's bus-driver or active star model. However it enables an easy way analyzing any passive net whether it is able to run FlexRay™ signals. The generic model takes the relevant FlexRay™ parameter into account.

Figure 38. CAN time and frequency domain measurement

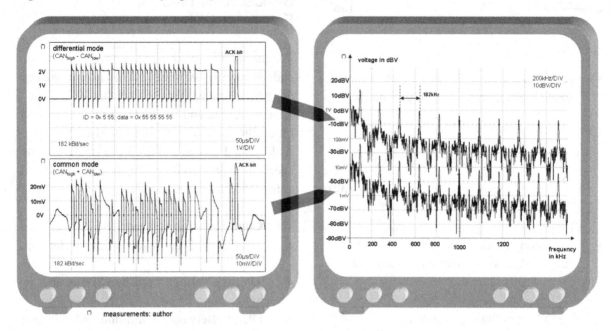

The model consists of three blocks:

- A transmitter stage to drive the bus-signals into the bus-lines via the pins BP and BM.
- A receiver stage to generate the data stream at the RxD pin and to generate the frame signal at the RxEN pin.
- Some biasing and parasitic components.

The generic bus driver is a compromise between accuracy and simulation speed represented by the requirements which are met:

- Parameters relevant for the signal integrity shall be taken into account:
  ○ Impedances of the transistors driving the bus-lines;
  ○ Edge timings;
  ○ Data detection and frame detection thresholds (voltage domain and time domain).
- Properties of bus drivers implemented in silicon shall be taken into account:
  ○ Reverse current protection;
  ○ Capability to reflect signals from the bus-lines;

Figure 39. Generic bus driver block chart

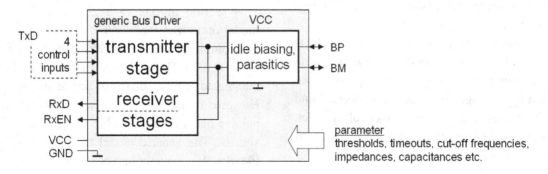

○ Capability to pass the effect of reflections to the power pins;

○ Capability the pass variations of the power supply to the bus lines;

○ Capability to work with ground shifts;

○ Parasitics.

Further parameter which are covered by silicon manufacturers models usually are out of the scope here. Examples:

- Propagation delay or symmetric delay;
- Operating modes like e.g. standby, sleep and wake up;
- ESD clamping;
- Current limitations of the transistors inside the transmitter stage;
- Operational common mode range limitations;
- Dependencies based on temperature variations, ageing etc.

Idle Biasing and Parasitic Components

Each FlexRay™ bus-driver (as well as CAN high-speed transceivers) biases the two bus-lines in the similar way: both lines are connected to 2,5 V by high-ohmic impedances. In parallel the receivers (wake up, data and frame) and the transmitter stages load the bus with their impedances (high-ohmic voltage dividers and parasitic capacitances). These effects can be modeled by two voltage dividers (e.g. 20 kΩ range) connected between VCC and GND and additional capacitances (e.g. 10 pF range). Details are defined in the corresponding data sheets, some limits are defined in the electrical physical layer specification. Although the signal integrity is influenced little by these values it is recommended to use them.

Generic Transmitter Stage (H-Bridge)

The generic transmitter stage is built by an H-bridge based on SPICE switches (see corre-

sponding section). Each switch is controlled by an own voltage $V_{control...}$. This enables an easy way implementing timing asymmetries if common mode disturbances shall be provoked. The value $R_{i,min}$ shall be selectable per switch individually.

The reverse current protection behavior is simulated by voltage sources or diodes. Applying an ohmic load to the bus pins shows the expected trapezoid results.

Generic Receiver Stage

The generic receiver stage is a mathematical implementation representing the FlexRay™ specified target behavior and the FlexRay™ application notes. Both were influenced implicitly by the state of feasibility nowadays visible by products available on the market.

The generic receiver stage meets some basic requirements:

- Receiver stage's speed limit shall be taken into consideration so that a short peak gets suppressed.
- The Schmitt trigger behavior to detect the data stream shall be available.
- The window comparator behavior to detect frames shall be supported. Hereby the activity detection time-out and the idle detection time-out shall be active.
- Schmitt trigger and window comparator shall work with identical thresholds.
- At the beginning of a frame the effect of truncation shall be supported.
- At the end of a frame the effect of ringing and idle detection shall be supported..

The generic receiver stage is based on the signal integrity voting extended by activity and idle detection circuits:

- Receiver stage's speed limit modeled by a low pass filter 1ˢᵗ order (minimum cut-off

Figure 40. Generic FlexRay™ transmitter stage (H-bridge)

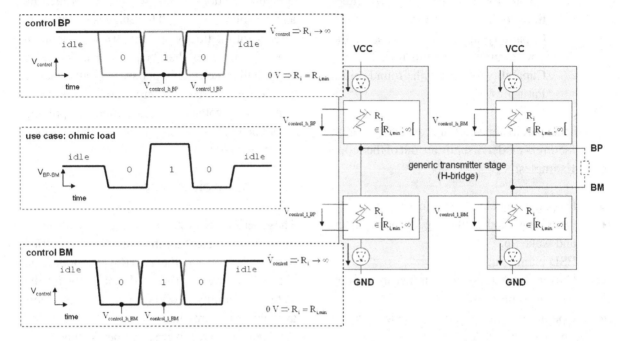

frequency according FlexRay™ electrical physical layer is 14 MHz).

- A Schmitt trigger with two thresholds to detect the data steam (e.g. ±150 mV)
- A window comparator with two Schmitt trigger thresholds to detect the frame (e.g. ±150 mV).
- The activity detection timeout and the idle detection timeout are modeled by a second low pass filter followed by a second Schmitt trigger (e.g. 2 MHz and 20%/80% thresholds). The output of this Schmitt trigger is the RxEN signal.
- The RxEN signal is used to mask the output signal of the first Schmitt trigger to the RxD data signal.

The previous figure illustrates a working example. The bus signal of an example FlexRay™ frame (two single bits applicable to the signal integrity voting) is available.

- The transmission start sequence is simulated by five zero-bits at the beginning. The frame finishes by four one-bits followed by a little ringing. Example FlexRay™ frame:
  - **TxD:** …$^{1111}_{00000\ 0000}\ ^{1}_{0}\ ^{11111\ 11111111}$ …
  - **TxEN:** …$^{1111}_{00000000000000000000}\ ^{1111}$ …
- The low path filter smoothes the bus signal a little however ringing is still available.
- The Schmitt trigger converts the filtered bus signal into a preliminary RxD signal. The RxD signal gets stuck at zero here due to the remaining ringing. The Schmitt trigger works like a flip-flop; its lower threshold is passed a last time and the output shows permanently zero (cancelled by the RxEN masking).
- The window comparator converts the filtered bus signal into a preliminary RxEN signal. The short idle pulse while the bus signal passes the idle are going to be suppressed in the following two modules. The low pass and the successive Schmitt trigger

Figure 41. Generic FlexRay™ receiver stage

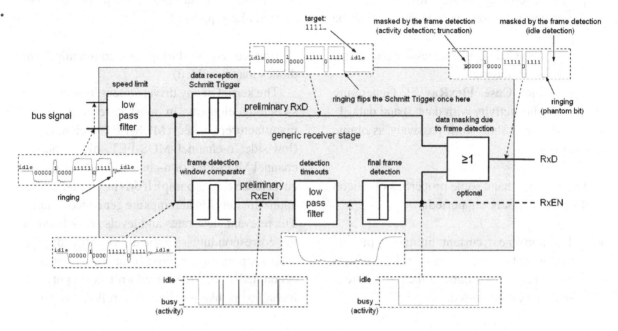

realize idle detection and activity detection to the final RxEN signal.

Finally the generic receiver stage converts the differential bus signal into a data stream signal and into a frame signal. Due to the Schmitt trigger and the window comparator timing and threshold properties effects like truncation, bit-distortion or frame end prolongation appear.

SPICE Compliant Generic Switch (Inverting Design)

Switching output stages are necessary for the simulation of serial automotive communication. This could be a CAN or a FlexRay™ along cables inside the wiring harness as well as the digital RxD/TxD interface between a protocol chip or a communication controller and a transceiver or a bus driver. The evaluation of topologies often requires the adjustment of two parameters at least: impedance and switching speed or slew rate of a driver stage. Implementing output stages by FETs

available in libraries do not offer the manipulation of these parameters. The generic SPICE switch introduced in the succeeding section bridges this gap, in other words this switch allows implementing adjustable non-linear driver stages.

Boundary Conditions and Requirements

From an abstract point of view various requirements are to be met by the generic SPICE switch:

- The SPICE switch shall be applicable for the implementation of non-linear driver stages.
- Any SPICE compliant tool shall be able to use the switch for signal integrity simulations in the time domain.
- The switch shall be independent from any transceiver or bus-driver model implementation.
- The key-values (e.g. levels, timing and slew-rates) shall be adjustable easily.

- The operating mode which supports data transmission and reception shall be possible.
  ○ **Use Case CAN:** Recessive and dominant states
  ○ **Use Case FlexRay™:** Generating idle, activity with data_0 and data_1
- The switch shall work passively as ohmic resistor.

On the other hand some properties of silicon switches shall not be supported:

- The non-linear current limitation properties of real output stages (e.g. FET effects when being used outside their saturation) shall not be supported.

- Reverse current protection properties shall not be supported.

(If required yet during data communication: diodes may be added)

The kernel of any driver state is its non-linear switching element. In most cases the silicon manufacturers select MOS-FET technologies (low-side: n-channel MOS-FET; high side: p-channel MOS-FET or n-channel MOS-FET with an included charge pump). If driving an ohmic load approximately linear ramps are generated usually. The relevant slew-rates and levels are defined in the corresponding specifications. A quadripole shall support the required behavior. A galvanic decoupled input port shall control the output port. It shall work like a voltage controlled resistor.

Table 5. Properties of the switching element

Interface	Pins	Description
Voltage controlled input $V_{control}$	Positive terminal: in1 Negative terminal: in2	**Test Set-Up:** • Out1 connected to a voltage source • Out2 connected via an ohmic resistor to ground • V_{out2} approximately (*) proportional $V_{control}$ • The element behaves like a voltage controlled ohmic resistor R_i
Ohmic resistor output	Positive terminal: out1 Negative terminal: out2	$V_{control} = \hat{V}_{control} \quad \Rightarrow \quad R_i \to \infty; \quad V_{out2} = 0V$ $V_{control} = 0V \quad \Rightarrow \quad R_i = R_{i,min}; \quad V_{out2} > 0V$ • The switching duration is given by the $V_{control}$-ramp.

(*) A perfect proportionality cannot be supported by the chosen SPICE compliant implementation. On the other hand the implementation enables fast simulation speed and best simulation stability.

Figure 42. Voltage controlled resistor R_i implemented by using a voltage controlled current source I_{Ri}

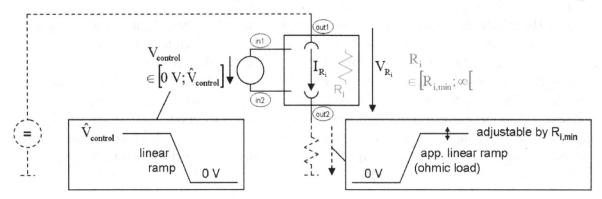

The SPICE compliant switching element can be implemented by using a non-linear voltage controlled current source second order.

- **2ⁿᵈ Order Voltage Controlled Current Source:**

$$I_{R_i} = \frac{V_{R_i}}{R_{i,min}} - \frac{V_{R_i} \cdot V_{control}}{R_{i,min} \cdot \hat{V}_{control}} \quad (42)$$

- **Passive Property of the Source:**

$$R_i = R_{i,min} \frac{\hat{V}_{control}}{\hat{V}_{control} - V_{control}} \quad (43)$$

SPICE Polynomial and Syntax

Hoefer and Nielinger (1985) show how SPICE uses polynomials higher order to define non-linear behaviors. A second order polynomial uses two input variables (here: $V_{control}$ and V_{Ri}) and one output variable (here: I_{Ri}). The required equation can be transformed straightforward to the polynomial (see Box 1).

The SPICE syntax is practically identical to the coefficients of the polynomial. Just some reference information shall be added in front (see Box 2).

SPICE Compliant Generic Switch (Non-Inverting Design)

Following the boundary conditions and the requirements of the inverting design a change of the polarity to a non-inverting design can be done easily.

The SPICE compliant switching element can be implemented by using a non-linear voltage controlled current source second order.

- **2ⁿᵈ Order Voltage Controlled Current Source:**

Box 1.

$$f(\;x,\quad y\quad) = p_0 + \quad p_1 \quad x + \; p_2\; y + \; p_3 \; x^2 + \quad p_4 \quad xy + \; p_5 \; y^2$$

$$V_{R_i} \quad V_{control} \qquad 0 \qquad \frac{1}{R_{i,min}} \qquad 0 \qquad 0 \qquad \frac{-1}{R_{i,min} \cdot \hat{V}_{control}} \qquad 0 \qquad (44)$$

Box 2.

```
Gname out1 out2 poly(2) out1 out2 in1 in2 0 {1/Ri_min} 0 0 {-1/Ri_min/Vcontrol_max} 0
*  |     |    |       |    |    |   |  |  |         |        | |            |               |
*  |     |    |       |    |    |   |  | p0       p1       p2 p3           p4              p5
*  |     |    |       |    |    |   |  |
*  |  output pins     |    |    |   |  |
*  |          second order |    |   |  |
*  |             polynomial |    |   |  |
*  |                 1st control pins | |
*  |                          2nd control pins
* voltage controlled current source (name selectable)
```

Figure 43. Voltage controlled resistor R_i implemented by using a voltage controlled current source I_{Ri}

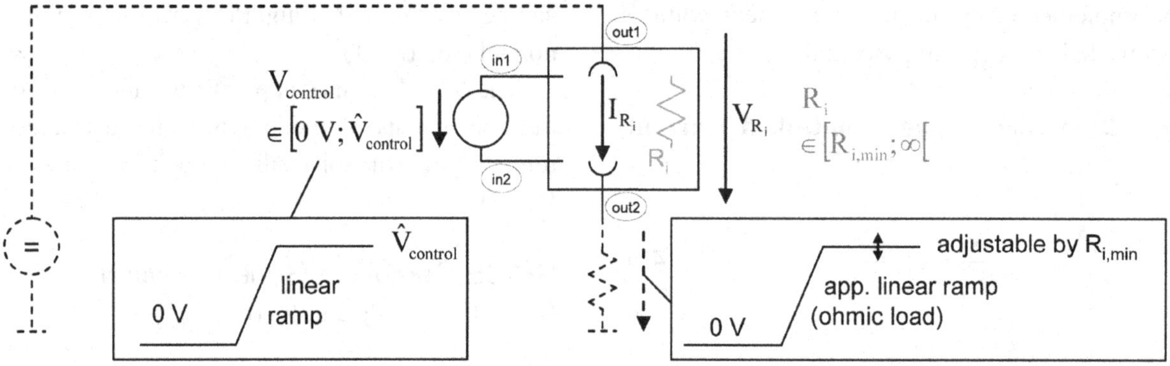

$$I_{R_i} = \frac{1}{R_{i,min} \cdot \hat{V}_{control}} V_{R_i} \cdot V_{control} \qquad (45)$$

- **Passive Property of the Source:**

$$R_i = R_{i,min} \frac{\hat{V}_{control}}{V_{control}} \qquad (46)$$

SPICE Polynomial and Syntax

The required equation can be transformed straightforward to a second order polynomial with two input variables (here: $V_{control}$ and V_{Ri}) and one output variable (here: I_{Ri}) (see Box 3 and 4).

Box 3.

$$
\begin{array}{ccccccccc}
f(& x, & y &) = & p_0 + & p_1\ x + & p_2\ y + & p_3\ x^2 + & p_4\ xy + & p_5\ y^2 \\
\uparrow & \uparrow & & & \uparrow & \uparrow & \uparrow & \uparrow & \uparrow & \uparrow \\
V_{R_i} & V_{control} & & & 0 & 0 & 0 & 0 & \dfrac{1}{R_{i,min} \cdot \hat{V}_{control}} & 0
\end{array}
\qquad (47)
$$

Box 4.

```
Gname out1 out2 poly(2) out1 out2 in1 in2 0 0 0 0 {1/Ri_min/Vcontrol_max} 0
* |      |    |       |    |    |   |   | | | |          |                 |
* |      |    |       |    |    |   | | p0 p1 p2 p3       p4                p5
* |      |    |       |    |    |   | |
* |   output pins     |    |    |   | |
* |           second order |    |   | |
* |           polynomial   |    |   | |
* |                 1st control pins | |
* |                        2nd control pins
* voltage controlled current source (name selectable)
```

Application: Recessive⇔Dominant Differential Mode Stage

The recessive⇔dominant differential mode stage represents two typical operating modes of automotive bus communication:

- CAN transceiver when transmitting serial data
- FlexRay™ bus-driver when starting and finishing the transmitting of a serial data stream frame or when transmitting e.g. wake-up symbols

The stage is built by two generic switches and two voltage sources which simplify the availability of reverse current protection diodes. If necessary the sources can be replaced by diodes; however longer simulation duration will appear. The switches are controlled by two separate control sources; their equivalent resistor varies between 100 Ω and infinite. Two 50 Ω resistors build a symmetric load. The bias level is driven by two 10 kΩ resistors. Two 2,5 V sources supply the set-up. The stage is symmetrical to 2,5 V.

The generic switches are driven by the control voltages simultaneously. Three states are generated: dominant state → recessive state → dominant state (FlexRay™: idle → active → idle). The switching time is 20 ns, the recessive state takes 100 ns.

The simulation shows the expected behavior:

- The trapezoid control voltages generate an inverted differential mode output signal at the load.
- The edges of the output signal are approximately linear.
- The edge duration and recessive duration are available with high accuracy.
- The steady state levels are based on voltage dividers.
- Asymmetries can be applied optionally: various switching speed (high/low), various delays and various stead state levels.

The simulation demonstrates that the introduced switch can be used designing recessive⇔dominant output stages for CAN data transmitting or FlexRay™ frame and symbol transmitting (idle ⇔ active).

Figure 44. Set-up of a recessive⇔dominant differential mode stage by using two generic switches

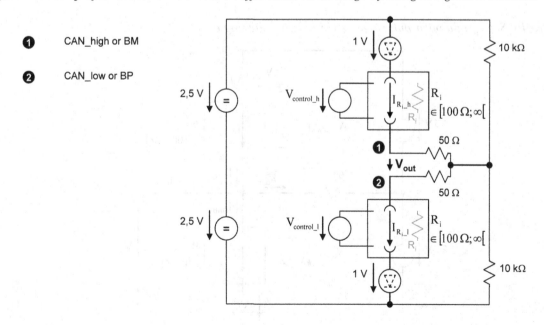

Figure 45. Simulation result of the recessive⇔dominant stage

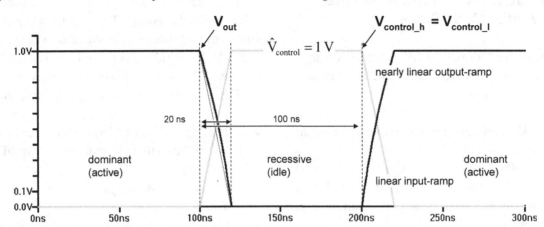

Application: Push-Pull Stage

The push-pull stage represents a typical operating mode of automotive bus communication:

- FlexRay™ bus-driver when transmitting serial data (e.g. BP)

The stage is built by two generic switches and two voltage sources which simplify the availability of reverse current protection diodes. If necessary the sources can be replaced by diodes; however longer simulation duration will appear. The switches are controlled by two separate control sources; their equivalent resistor varies between 100 Ω and infinite. A single 50 Ω resistor builds a symmetric load. Two 2,5 V sources supply the set-up. The stage is symmetrical to 2,5 V.

The switches are driven by the control voltages simultaneously but inverted. Three states are generated: high state → low state → high state. The states high and low correspond to the bit-states data_0 and data_1 in the serial data stream. The

Figure 46. Set-up of a push-pull stage by using two switching elements

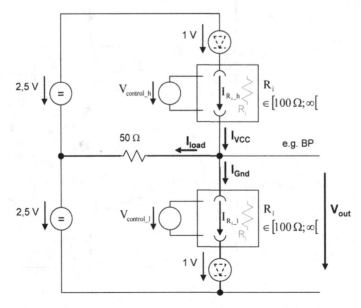

Figure 47. Simulation results of the push-pull stage

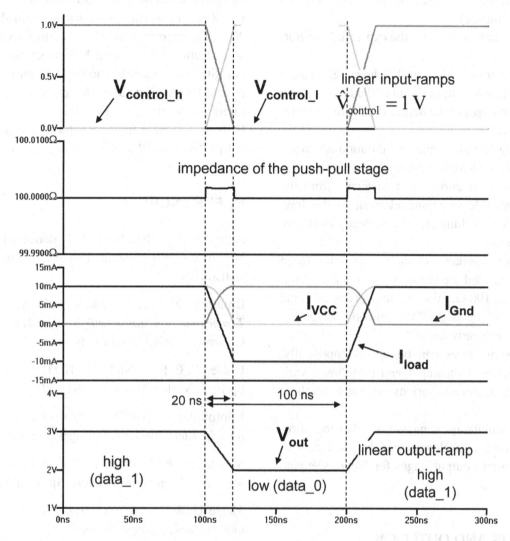

Figure 48. Active suspension of BMW's model X5

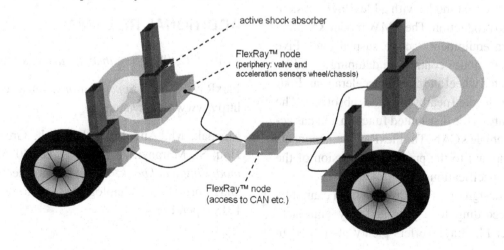

switching time is 20 ns; the low state takes 100 ns (10 Mbit/sec).

The simulation shows the expected behavior:

- The trapezoid control voltages generate a trapezoid output signal at the load.
- The edges of the output signal are perfectly linear.
- Edge duration and low duration are available with high accuracy.
- The load current is transferred from the high-side switching element to the low-side switching element perfectly (and vive versa).
- The impedance of the two switches (seen from load's point of view) is nearly constant 100 Ω, minor variations appear at the switching edges: "the push-pull stage behaves nearly linear ".
- Asymmetries can be applied optionally: various switching speed (high/low), various delays and various stead state levels.

The simulation demonstrates that the introduced switching element can be used designing push-pull output stages for FlexRay™ data transmission.

STATUS AND OUTLOOK

In 2006 the first model with a FlexRay™ system went into production. The BMW model X5 offers the extra equipment "active suspension". Five FlexRay™ nodes control the damping based on the detected wheel and chassis accelerations. Four nodes are located nearby the shock absorbers. The fifth controls the distributed function and has access to vehicle's CAN. The FlexRay™ components are compliant to the predecessor version of the current specifications.

The design of a FlexRay™ system cannot be done according to a plug-and-play approach. However FlexRay™ offers plenty of support in designing an optimized and robust system. The FlexRay™ consortium work had been finished in 2010. The current version 3.0.1 of the FlexRay™ specifications where published in October 2011. An ISO taskforce works in building an international FlexRay™ standard. Several car manufacturers are working on FlexRay™ integrations. Perhaps – in near future – vehicles are driving with a "*FlexRay™ inside*" logo.

REFERENCES

Adamek, T. (2010). Statistik 1. Retrieved from http://www.uni-stuttgart.de/bio/adamek/numerik/statistik1.pdf

Bronstein, N. I., & Semendjajew, K. A. (1981). *Taschenbuch der Mathematik*. Thun und Frankfurt, Germany: Verlag Harri Deutsch.

Hoefer, E. E. E., & Nielinger, H. (1985). *SPICE*. Springer Verlag. ISBN 3-540-15160-5

Küpfmüller, K. (1973). *Einführung in die theoretische Elektrotechnik*. Springer Verlag.

Meinke, H., & Gundlach, F. W. (1986). *Taschenbuch der Hochfrequenztechnik*. Springer Verlag.

Tietze, U., & Schenk, C. (1993). *Halbleiterschaltungstechnik*. Springer Verlag.

ADDITIONAL READING

Crilly, T. (2008). *50 mathematical ideas*. Quercus.

FlexRay™. (n.d.). *Specifications*. Retrieved from http://www.flexray.com/

Mirmak, M., Angulo, J., Dodd, I., Green, L., Huq, S., Muranyi, A., & Ross, B. (2005). *Ibis modelling cookbook*. Government Electronics and Information Technology Association and The IBIS Open Forum.

Rausch, M. (2008). *FlexRay*. Carl Hanser Verlag. ISBN 978 3-446-41249-1

Zimmermann, W., & Schmidgall, R. (2007). *Bussysteme in der Fahrzeugelektronik*. Vieweg und Teubner. doi:10.1007/978-3-8348-9188-4

KEY TERMS AND DEFINITIONS

Active: Status of the communication in case of FlexRay™ when a message or serial data stream is available (see "busy").

AS: Abbreviation of active star: the active star enables connecting several bus lines. Active multi-directional 1 to n-1 amplifiers are included.

API: Abbreviation of application programming interface.

BD: Abbreviation of FlexRay™'s bus driver: the bus driver converts logical data streams to their physical representation on the bus lines and vice versa. In CAN systems the similar circuit is the transceiver.

BSS: Abbreviation of byte start sequence: each byte inside a FlexRay™ frame is headed by this two bit sequence. The included falling edge re-synchronizes the asynchronous bit sampling every ten bits.

Busy: Status of the communication in case of FlexRay™ when a message or serial data stream is available (see "active").

CAN: Abbreviation of controller area network: an automotive serial bus.

CMC: Abbreviation of common mode choke: the common mode choke suppresses common mode signals which may be overlaid to differential mode signals.

Dominant: Status of the communication in case of CAN or LIN when a logical zero bit is available.

ECU: Abbreviation of electronic control unit.

FES: Abbreviation of frame end sequence: each FlexRay™ frame is ended by this two bit sequence.

FlexRay™: A made-up word for an automotive serial bus.

FSS: Abbreviation of frame start sequence: after the transmission start sequence each FlexRay™ frame is continued by this one bit sequence.

Idle: Status of the communication in case of FlexRay™ when not any message or serial data stream is available.

LIN: Abbreviation of local operating network: an automotive serial bus.

PCB: Abbreviation of printed circuit board.

PLL: Abbreviation of phase lock loop. That is a procedure controlling the clock speed inside silicon devices.

Prolongation: The end of each FlexRay™ frame ($\rightarrow$ FES) or symbol will prolonged optionally on the way from the transmitting node to the receiving node.

PS: Abbreviation of passive star: the passive star enables connecting several bus-lines without using any active components.

Recessive: Status of the communication in case of CAN or LIN when a logical one bit is available.

SPICE: Simulation tool for passive and active electric circuits. Spice represents the simulations kernel typically; the graphical front-ends vary among the suppliers.

Truncation: The beginning of each FlexRay™ frame ($\rightarrow$ TSS) or symbol will be cut on the way from the transmitting node to the receiving node.

TSS: Abbreviation of transmission start sequence: each FlexRay™ frame is headed by this sequence consisting of several identical bits. Their number can be selected during the system design.

ENDNOTES

[1]   source of all measurements, simulations and photos: author

[2]   CAN with flexible data rate; published by Robert Bosch GmbH in 2012

[3]   The wording "active star" is used in FlexRay™ twice. An active star chip is an

integrated circuit which supports an active star behavior. Several of these chips can be banked. An active star electronic control unit contains one active star chip at least and optionally a complete communication node.

4 FlexRay™ offers the option to use two channels connected to one protocol chip in parallel. Either a doubling of the gross date rate or a redundant transmission is supported. From physical layer's operating principles point of view a limitation to the consideration of a single channel is sufficient.

5 The wording communication element summarizes FlexRay™ frames and symbols here.

6 Exception: bus drivers are able to detect wake-up pattern (and distinguish them from "noise")

7 Abbreviation of a German phrase. In English: "open systems and their interfaces in automotive electronic systems" consisting of an operating system, communication and network management.

8 Bit reshaping procedures (time domain) are not specified

9 see "automotive electrical stress" according ISO 7637-3:1995

10 The words "cable" and "line" shall describe identical components here:
on one hand from wiring point of view and on the other hand from data communication point of view

11 The following syntax is used if CAN and FlexRay behaves similar.
without brackets: CAN naming;
with brackets: FlexRay naming
Example: recessive [idle] level

12 Fast Fourier transformation

APPENDIX

Symbols

- ‖: Used when two components are placed parallel to each other (used making equations clearly arranged)
- **V:** Voltage
- **I:** Current
- **R:** Resistor or resistance
- **C:** Capacitor or capacitance
- **L:** Inductivity or inductance
- **,:** The comma is used as decimal comma according ISO. Example: one dot five is written as 1,5

Chapter 6
Transmission Lines for Serial Communication:
Theory and Practice

Jürgen Minuth
Esslingen University of Applied Sciences, Germany

ABSTRACT

Automotive bus systems like e.g. LIN, CAN, and FlexRay™ distribute their serial data streams NRZ[1] coded in the base band among the communication nodes. The nodes are interconnected by passive nets. Depending on the type of application some of these nets may consist of up to one hundred meters of different bus cables arranged in various topologies. The individual pieces of information inside the data streams are represented by voltage steps and current steps. They have to be passed among all connected nodes via the bus cables. The succeeding sections introduce the commonly known transmission line theory focused of the physical effects being relevant for automotive serial bus communication in the time domain.

TRANSMISSION LINES

The transmission line theory is common knowledge described e.g. by Küpfmüller (1973) or Meinke and Gundlach (1986).

Motivation

Communication among several nodes requires communication media and unidirectional or bidirectional interfaces which connect each of them. In the electrical use case[3] transceivers and

bus-drivers work as interfaces while (bus) cables work as media. As long as the distance between the interfaces is very short, (bus) cables between them can almost be interpreted as perfect short circuits. If the distance increases, various equivalent circuits based on discrete components are allowed. Their selection depends on the application. If the distance increases more the interpretation of (bus) cables as transmission lines is recommended.

The transmission line theory is introduced to almost all electrical engineering students during their study programs. Effects like propagation delay, damping, refraction and reflection appear. A clear threshold to decide whether the transmission

DOI: 10.4018/978-1-4666-2976-9.ch006

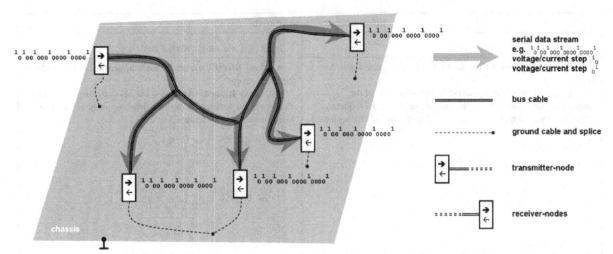

Figure 1. Serial communication among distributed nodes via bus cables

line theory shall be used is not defined. In case of NRZ coding in the base band it depends on the communication speed and on the edge timing or rather on the expected or required accuracy.

Usual automotive serial communication based on electrical cables like CAN as well as FlexRay™ use NRZ coding in the base band. They require the interpretation of their (bus) cables as transmission lines. Their communication signals can be seen as superposition of voltage or current steps representing ${}^1\!\backslash_0$ and ${}_0\!/^1$ edges in the time domain. However, automotive serial communication based on LIN can be seen as exception; its communication speed is slow enough to allow replacing (bus) cables by discrete capacitances, often.

Scope

Transmission lines are built up by two more or less parallel wires (or strands)[4]. Voltages between the wires and currents through the wires depend on the position of the test plain on the line. The theory describes an approach and solutions covering the measurable behaviors. Idealizations enable the theory deriving straight forward solvable differential equations[5]. The theory is based on distributed circuits or rather on putting an infinite number of infinitesimal short circuits in a row. Each of these circuits is build by discrete components.

DIFFERENTIAL EQUATIONS AND THEIR SOLUTIONS

For example, following the Maxwell equations (respectively Faraday's law of induction and Ampère's law) results in two differential equations which describe voltages and currents along the line in the time domain.

- **Voltage:**

$$
\frac{d^2v(x,t)}{dx^2} = L'C'\frac{d^2v(x,t)}{dt^2}
$$
$$
+ \left(R'C' + G'L'\right)\frac{dv(x,t)}{dt} + R'G'\,v(x,t)
$$

(5)

- **Current:**

$$
\frac{d^2i(x,t)}{dx^2} = L'C'\frac{d^2i(x,t)}{dt^2}
$$
$$
+ \left(R'C' + G'L'\right)\frac{di(x,t)}{dt} + R'G'\,i(x,t)
$$

(6)

The equations describe waves including their attenuation. They are valid for any signal shape

Table 1. Electrical effects at cables (or rather bus lines)

Effect when measuring into the ...	
... Input port at an open ended cable	... Input port at a short circuited cable
• **A Very Low Direct Current Flows:** Losses inside the isolator appear • **A Lossy Capacitor Appears:** A slight decrease of the capacitance for higher frequencies (mostly negligible). Frequency dependant losses which can be described approximately by aconstant loss angle. $C \approx const$ $G \approx const + \underbrace{\langle \sim f \rangle}_{\substack{\text{amount}\\\text{propotional}\\\text{to the frequency}\\\text{(isolator's loss angle)}}}$	• **A Low Direct Current Resistor appears:** Losses inside the wires appear • **A Lossy Inductor Appears:** The skin effects generates an increase of the resistor and a slight decrease of the inductance when increasing the frequency $R \approx const + \underbrace{\langle \sim \sqrt{f} \rangle}_{\substack{\text{wire's resistance}\\\text{(Skin effect)}}}$ $L \approx const + \underbrace{\langle \sim 1/\sqrt{f} \rangle}_{\substack{\text{wire's inner inductance}\\\text{(Skin effect)}}}$
Common Usage Definitions	
Conductance per length $\quad G' = \dfrac{G_{line}}{\ell} = \dfrac{\text{conductance}}{\text{length of the line}}$ (1)	Resistance per length $\quad R' = \dfrac{R_{line}}{\ell} = \dfrac{\text{resistance}}{\text{length of the line}}$ (2)
Capacitance per length $\quad C' = \dfrac{C_{line}}{\ell} = \dfrac{\text{capacitance}}{\text{length of the line}}$ (3)	Inductance per length $\quad L' = \dfrac{L_{line}}{\ell} = \dfrac{\text{inductance}}{\text{length of the line}}$ (4)

Figure 2. Transmission line with the length ℓ and its infinitesimal representation

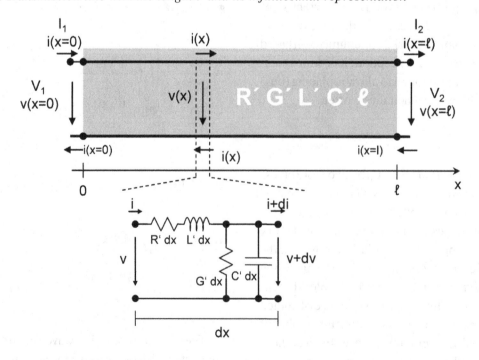

in the time domain (examples: sine wave signals, steps, pulses etc.); the steady state and the transient state is covered. Describing voltages and currents by complex numbers (steady state of sine wave signals) enables the derivation of two linear differential equations in the frequency domain, one for the complex voltage $v(x)$ and one for the complex current $i(x)$.

- **Complex Voltage:**

$$\frac{d^2v(x)}{dx^2} = \left(R' + j\omega L'\right)\left(G' + j\omega C'\right)v(x)$$

(7)

- **Complex Current:**

$$\frac{d^2i(x)}{dx^2} = \left(R' + j\omega L'\right)\left(G' + j\omega C'\right)i(x)$$

(8)

From communication's point of view two basic solutions are important:

- General solution for lossy lines in the frequency domain.

The reverse Fourier transformation enables an easy way back to the time domain.

- Special solution for lossless lines in the time domain to transmit e.g. steps or pulses.

General Solution for Lossy Lines in the Frequency Domain

The commonly known approach $y(x) = const \cdot e^{\lambda x}$ results in a solution with two constants. Their final values depend on the boundary conditions and the initial conditions given by the application.

- **Solution for the Complex Voltage:**

$$v(x) = \underbrace{V_f \cdot e^{-\gamma x}}_{\substack{\text{forward traveling} \\ \text{voltage wave} \\ \rightarrow \text{positive x-direction}}} + \underbrace{V_r \cdot e^{\gamma x}}_{\substack{\text{reverse traveling} \\ \text{voltage wave} \\ \leftarrow \text{negative x-direction}}}$$

(9)

with the propagation coefficient[6]

$$^3 = \sqrt{\left(R' + j\omega L'\right)\left(G' + j\omega C'\right)}$$

(10)

Its real part describes the attenuation of sine wave-form signals traveling along the line.

The two differential equations are depreciated by using a single schematic. Therefore (and due to the law of induction) the complex current $i(x)$ is proportional to the complex voltage $v(x)$ always. The constant of proportionality is the line impedance Z_{line}.

- **Line Impedance:**

$$Z_{line} = \sqrt{\frac{R' + j\omega L'}{G' + j\omega C'}}$$

(11)

- **Solution for the Complex Current:**

$$i(x) = \underbrace{\frac{V_f}{Z_{line}} \cdot e^{-\gamma x}}_{\substack{I_f \\ \text{foreward running} \\ \text{current wave} \\ \rightarrow \text{positive x direction}}} - \underbrace{\frac{V_r}{Z_{line}} \cdot e^{^3 x}}_{\substack{I_r \\ \text{reverse running} \\ \text{current wave} \\ \leftarrow \text{negative x direction}}}$$

(12)

Special Solution for Lossless Lines in the Time Domain

Lossless lines offer a simple differential equation and a simple solution for signal transmission. Any signal shape f(x) or g(x) can be transmitted along lines without changing its shape. The Ohmic line impedance appears as a proportional constant of the quotient voltage-shape/current-shape.

- Characterization of a Lossless Line
  - $R' = 0 \; \Omega/_m$
  - $G' = 0 \; ^S/_m$
  - $L' \neq 0 \; ^H/_m$
  - $C' \neq 0 \; ^F/_m$

- **Differential Equation:**

$$\frac{d^2 v(x,t)}{dx^2} = L'C' \frac{d^2 v(x,t)}{dt^2} \qquad (13)$$

- **Solution:**

$$\left.\begin{array}{c} v(x,t) \\ Z_{line} \cdot i(x,t) \end{array}\right\} = \underbrace{f\left(x - v_{line}t\right)}_{\substack{\text{foreward running} \\ \text{wave} \\ \rightarrow \text{ positive x-direction}}} \left\{\begin{array}{c} + \\ - \end{array}\right\} \underbrace{g\left(x + v_{line}t\right)}_{\substack{\text{reverse running} \\ \text{wave} \\ \leftarrow \text{ negative x direction}}}$$

$$(14)$$

- **Line Impedance:**

$$Z_{line} = \sqrt{\frac{L'}{C'}} \quad in \quad [\Omega] \qquad (15)$$

Basis Equations for Lossless Lines

The equations are available only when using lossless lines. Lines with little losses can be handled as well, if minor deviations are accepted. Often these deviations are hidden by the measuring tolerances.

- **Line Impedance (Field View):**

$$Z_{line} = \sqrt{\frac{L'}{C'}} = \sqrt{\frac{L_{line}}{C_{line}}} \overset{typ.}{\in} [10\Omega \; ; \; 300\Omega] \overset{typ.}{=} 100\Omega \qquad (16)$$

- **Line Impedance (Signal View):**

$$Z_{line} = \frac{\hat{V}}{\hat{I}} = \frac{\text{"amplitude voltage wave"}}{\text{"amplitude current wave"}} \qquad (17)$$

- **Propagation Velocity:**

$$v_{line} = \frac{1}{\sqrt{L'C'}} = \frac{c_0}{\sqrt{\varepsilon_r}} = \frac{\ell_{line}}{\tau_{line}} \overset{typ.}{=} 20 \frac{cm}{ns} \qquad (18)$$

- **Distributed Line Capacitance:**

$$C' = \frac{1}{Z_{line} \, v_{line}} \overset{typ.}{=} 50...100 \frac{pF}{m} \qquad (19)$$

- **Distributed Line Inductance:**

$$L' = \frac{Z_{line}}{v_{line}} \overset{typ.}{=} 0,5...1 \frac{\mu H}{m} \qquad (20)$$

- **Line Capacitance:**

$$C_{line} = C' \ell_{line} = \frac{\tau_{line}}{Z_{line}} \qquad (21)$$

Line Inductance:

$$L_{line} = L' \ell_{line} = Z_{line} \, \tau_{line} \qquad (22)$$

Pulses, Cables, and Lines

Lines' Cross Sections and Impedances

Line impedances can be derived from the geometric cross section data (cylindrically symmetric design). Inductive and capacitive amounts are coupled by the Maxwell equations. The wire-internal inductance is assumed to be zero, in other words the operational frequency is so large that the current flows at the surface of the wire only (due to the skin effect); however losses are neglected.

Each of these equations contains an arcosh(x) or ln(x) function. Thereby the variable x represents the quotient of wire-diameter and wire-distance. Commonly used technical cross sections generate

Table 2. Line cross section and their line equations according to e.g. Meinke and Gundlach (1986)

Cross Section	Line Impedance	
Wire above a metal sheet: 	$Z_{line} = \dfrac{60\Omega}{\sqrt{\varepsilon_r}} \, ar \cosh \dfrac{2a}{d}$	(23)
Two wires: 	$Z_{line} = \dfrac{60\Omega}{\sqrt{\varepsilon_r}} \, ar \cosh \dfrac{4a^2 - d_1^2 - d_2^2}{2 d_1 d_2}$	(24)
	Special case: $d_1 = d_2 = d$ $\quad Z_{line} = \dfrac{60\Omega}{\sqrt{\varepsilon_r}} \, ar \cosh \dfrac{a}{d}$	(25)
Micro strip line[*] (typically used on printed circuit boards): $\varepsilon_r \le 16, \quad \dfrac{w}{h} \ge 0.05$ 	$\varepsilon_{r,eff} = \dfrac{\varepsilon_r + 1}{2} + \dfrac{\varepsilon_r - 1}{2} \begin{cases} \dfrac{1}{\sqrt{1 + 12\frac{h}{w}}} + 0.04\left(1 - \dfrac{w}{h}\right)^2 & for \dfrac{w}{h} \le 1 \\[12pt] \dfrac{1}{\sqrt{1 + 12\frac{h}{w}}} & for \dfrac{w}{h} \ge 1 \end{cases}$	(26)
	$Z_{line} = \dfrac{120\pi\,\Omega}{\sqrt{\varepsilon_{r,eff}}} \begin{cases} \dfrac{\ln\left(\frac{8h}{w} + \frac{w}{4h}\right)}{2\pi} & for \dfrac{w}{h} \le 1 \\[12pt] \dfrac{1}{\frac{w}{h} + 2.46 - 0.49\frac{h}{w} + \left(1 - \frac{h}{w}\right)^6} & for \dfrac{w}{h} \ge 1 \end{cases}$	(27)

[*] micro strip line equations are based on empiric

values of x in the interval [1,5;10] approximately. This ends in a factor in the interval [1;3]. Isolators offer relative permittivity in the interval [1;10], usually. This value influences the result of Z_{line} valued by the square root.

The succeeding figure shows typical numbers for an example line: one wire with the diameter of 0,4 mm above metal sheet – covered by an isolator with $\varepsilon_r = 4$.

Summarizing standard line configurations cause line impedances in the range of approximately 10 Ω up to 300 Ω. This range covers flat lines up to two wire lines with a distance of a few centimeters. Overland lines (e.g. high voltage power lines) result in significantly higher line impedances. However a typical value applicable to data communication lines is 100 Ω.

Lines' Frequency Dependant Parameters

Two physical main effects cause frequency dependant line parameters. On one hand the isolators between the wires show a loss angle which is constant in the frequency range, approximately. On the other hand the current flow through a wire is going to be pushed to the surface of the wire for higher frequencies. This effect is known as skin effect, caused by the law of induction.

The constant loss angle generates a line conductance value G' proportional to the frequency. The skin effect has two coupled consequences. First, the line resistance value R' is proportional to the square root of the frequency. Second, an inner inductance of the wire L'_{inner} is reciprocally proportional to the square root of the frequency. It appears in the direct current application (typically 50 $^{nH}/_m$ per wire) and disappears for higher frequencies. Depending on the line's diameter, the inner inductance disappears at frequencies higher than a few ten kHz.

The line capacitance C' and the line impedance L' can be seen as frequency independent, approximately.

When simulating lossy lines its parameters shall not be replaced by simple approximation

Figure 3. Line impedance, capacity and inductance of a wire above ground

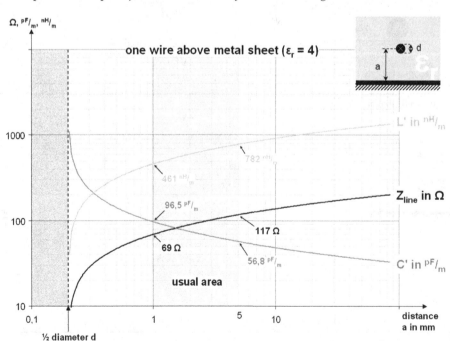

equations which try to build up f ½, f -½ and the f proportionality beginning with the direct current behavior. If doing this, the simulation and measuring results will not match. A straight forward way to enable simulations with frequency dependant effects is described e.g. by Bidyut and Wheeler (1998) and Elco and Burke (2003). The approaches are based on approximating the frequencies' dependencies by R-L und R-C circuits. Other sources describe designing frequency dependant losses based e.g. on rational polynomial function approximations of the propagation function and the line impedance. The effects of the frequency dependant losses are going to be shown in the following section.

Measuring

A generator drives a signal into a wire above reference ground. Three voltages are measured. The equivalent circuit of the set-up consists of:

- **Pulse Generator:**
  - Impedance 50 Ω,
  - Transmitting three different pulses with an amplitude of 4 V each: 50ns, 100ns and 3µs,
  - Rise and fall time of a few ns (perfect rectangle signals from an engineer point of view).
- **Two Identical Lines in a Row:**
  - **Length:** 48 m

  - **Propagation Delay:** ≈ 215 ns
  - **Line Impedance:** ≈ 48 Ω
- 50 Ω termination at the end of the 2nd line

The voltages V_1, V_2 and V_3 were measured with a 200 MHz oscilloscope. The used measuring setup avoided any reflections. Assuming the lines are perfectly lossless, the oscilloscope should show practically rectangle signals at all three test plains. Lossless simulations show exactly this behavior – the measurement results differ clearly.

Detected properties of the acquired signals (3 µs pulse – duration > line propagation delay):

- The steady states of all pulses are horizontal.
- The steady state at the test plain V_2 (centre between both lines) is 50% of pulse generator's open circuit voltage.
- The direct current voltage drop (direct current losses) along the 1st line is identical to the direct current voltage drop of the 2nd line.
- The voltage at the pulse generator shows a little over shoot after the rising edge and a little under shoot after the falling edge (which passes the 0V line briefly).
- The starting edge of the pulse remains nearly rectangular.
- The following edges get smoothened.
- The slew-rate of the edges decrease along the line.

Figure 4. Equivalent circuit for transmitting pulses

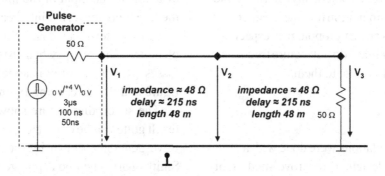

Figure 5. Transmitting rectangle 4V pulses (50ns, 100ns, and 3µs)

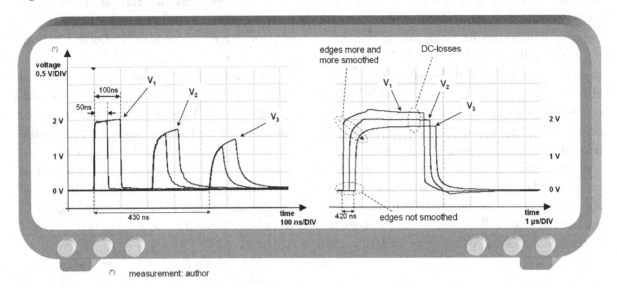

When the 50 ns and 100 ns pulses travel along the line, three main effects appear (duration < line propagation delay):

- The upper corner of the rising edge and the lower corner of the falling edge get smoothened more and more.
- The achievable level gets damped more and more.
- Rectangle pulses convert more and more to a shark fin shape when traveling along the line.

Mainly, the measurement results are caused by the two physical effects yet pointed out: the loss angle of the isolator and the skin effect. The usage of lossless lines or even direct current's losses only (supported by practically each available SPICE compliant simulation tool) generates clear differences. However, many topologies (especially when using shorter lines) do not require frequency dependant losses to evaluate them.

Simulation

The following simulations were done with a non-commercial tool designed[7] to prove models of lossy lines. The simulation works in several steps:

- The 4 V / 3 µs time domain pulse gets transferred into the frequency domain.
- The frequency dependant line losses are defined by suitable networks inspired by Bidyut and Wheeler (1998) and Elco and Burke (2003).
- The simulated spectrum gets transferred into the time domain as pulse answer.
- The pulse answer gets integrated to the step answer.

Simulation and measuring results match practically perfect. Skin effect and loss angle are responsible for the measured smoothing, under-shot, and over-shot effects. The spectrum shows the main effect (measuring in the time domain shows them implicitly): the high frequency portions available at the input of the line are absorbed by the line (and converted into heat) when traveling along to the terminated end. For the sake of completeness, the result when using direct current losses (R'=const) only is added in the previous figure and the succeeding figure.

The succeeding figure shows the simulation result gotten at the output terminals of a 50 Ω sine wave generator drawn into a 50 Ω normalized Smith chart. The line is plugged to the generator (open end - infinite termination at the end). The

Figure 6. Transmission of a 3 µs pulse via a lossy line

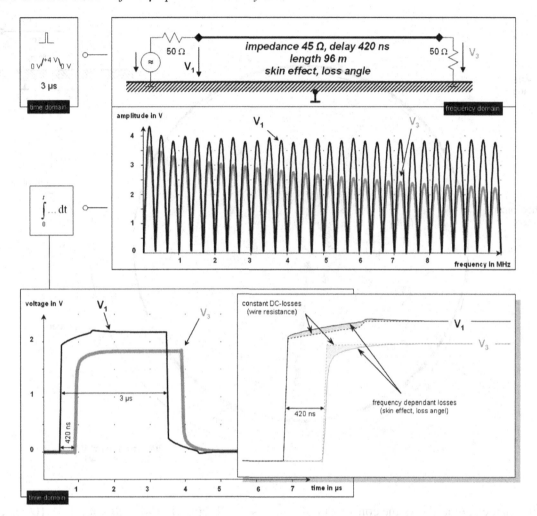

Smith chart offers the complex reflection-factor and the complex input resistance of the line (measured as load at the generator ports). Two results are pointed out: line with direct current losses (R' = const) only and line with the frequency dependant losses (skin-effect and loss-angle).

- Both curves start at the open circuit point (frequency = 0 Hz) due to the open end.
- The direct current losses' curve has circles far away from the centre. The absolute value of the reflection-factor sticks at approximately 0,85. The line behaves alternately high ohmic, capacitive, low ohmic, inductive, high ohmic etc. Low damped

high frequency resonances could appear in dedicated scenarios.

- The skin-effect and loss-angle curve turns spirally in clockwise direction to a single centre at an ohmic resistor of 45 Ω. This ohmic value corresponds with the line impedance. The centre is located nearby the 50 Ω centre of the Smith chart. In other words, for higher frequencies an open ended 45 Ω line behaves like being terminated; the reflections factor trends to -0,05.

Potentially available high frequency resonances are going to be damped.

Figure 7. Smith chart applied to input port of a line (open end)

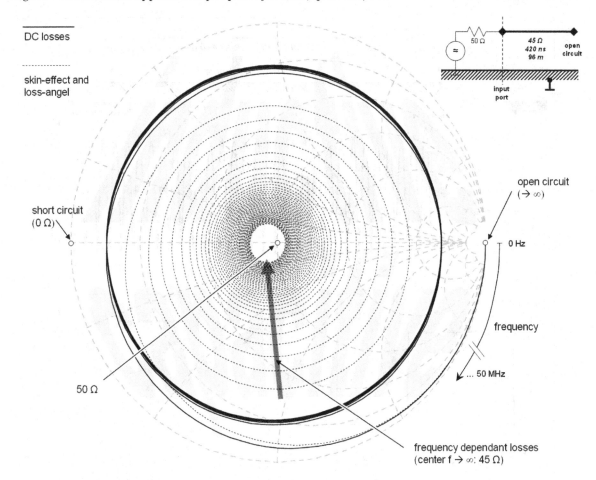

The final example shows the conversion of a rectangle pulse into a shark fin shape when traveling along the lossy line. Simulation and measuring results match practically perfect here, too.

Pulse Shapes

The deformation of pulses when traveling along cables depends on the one hand on frequency dependant cable-losses and on the other hand on the pulses' spectral distribution. The succeeding section offers a view to three different basic pulse shapes. Their line spectra are commonly known when repeating them periodically. In addition the standard statement "The radiation trend of a switching output stage increases, when their rising or falling edge timings decrease." is going to be reflected.

- **Rectangle Pulse Shape:** T_i = 100ns, amplitude $\hat{V}$ = 1 V (28)
- **Repetition Period:** T = 1 µs (29)
  Direct Current Component:

$$20 \log \frac{\hat{V} \cdot \dfrac{T_i}{T}}{1\mu V} = 100 \text{dB}\mu V \qquad (30)$$

- **Triangular Pulse Shape:** Characterization according to the rectangle pulse.
- **cos² Pulse shape:** Characterization according to the rectangle pulse.

On one hand when bringing the edge timings in ascending order (20%/80% is chosen to define the timings):

Figure 8. Transmission of a 100ns pulse via a lossy line

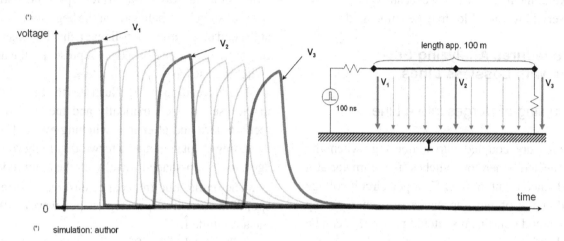

simulation: author

Figure 9. Periodically repeated pulse shapes in the time domain and their line spectra

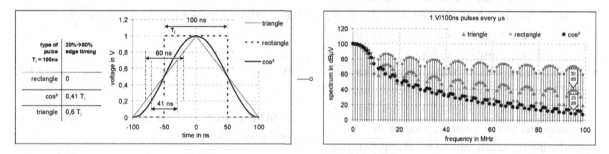

- **Rectangle:** 0ns → **Triangular:** 41ns → cos²: 60ns

On the other hand when bringing the radiation relevant line spectrum amplitudes into an order (chosen area: app. 90 – 100 MHz):

- **Rectangle:** ≈ 70 dBµV → **cos²:** ≈ -30 dB less → **Triangular:** ≈ -25 dB less

Although the edge of the cos² pulse is approximately 50% faster then the triangular edge, the radiation relevant part of the cos² pulse is approximately 25 dB smaller compared to the triangular pulse. The edge timings (rise and fall) do not correspond with the high frequency portions of the signals in every case. The radiation relevant spectral lines are not directly visible in

edge timings; they are hidden in the maxima of the 2nd derivation (or in turning points of the 1st derivations).

It is obvious that the deformation-rate of pulses depends on their spectral distribution – in particular because the damping of lines increases with the frequency. Bringing the deformation-rate to be expected into an order:

- **Rectangle:** (Big Deformation) → **Triangular:** (Medium Deformation) → **cos²:** (Little Deformation)

Controlling pulse shapes enables two optimization approaches in serial communication: reduction of radiation and reduction of shape deformation. The shown effects and statements shall not be seen as recommendations for future developments directly; however an expanded view to serial bits on cables may be stimulated.

Procedures Analyzing Set-ups with Lossless Lines

Coupling of Surges into a Line

Surges are coupled into a line e.g. when the connected generator switches its parameter at a dedicated point in time. The open circuit voltage and/or the ohmic impedance may change; any linear and non-linear switching operation can be realized:

- **Switching with a Linear Behavior:** The open circuit voltage changes its value only; the duration is assumed to be approximately zero ($\ll$ any line delay)
- **Switching with a Non-Linear-Behavior:** The ohmic impedance changes its value at least ($R(t=t_0) \neq const$). The non-linearity happens at the switching point in time t_0; the duration is assumed to be approximately zero ($\ll$ any line delay).

A waveform generator represents the linear switching use case only; a CAN transceiver represents a non-linear switching use case only.

However a FlexRay™ bus driver represents both types of switching behavior; at the beginning and at the end of a frame a non-linear behavior exists. Inside the frame the bit-edges represent a linear switching use case, approximately.

The succeeding figure illustrates the behavior in the use case schematically and the state directly before and after the switching event. The switching event generates a forward running voltage surge (step-shaped wave) with the amplitude $\hat{V}$. The ohmic line impedance ensures a corresponding current surge (step-shaped wave) with the amplitude $\hat{I}$.

Following Kirchhoff's voltage law results in three equations:

- **Voltages and Currents at $t < t_s$:**

$$V = V_0 - R_0 \cdot I \tag{31}$$

with the current I at $t < t_s$ (e.g. caused by termination resistors)

- **Voltages and Currents at $t > t_s$:**

$$V + \hat{V} = V_1 - R_1\left(I + \hat{I}\right) \tag{32}$$

Figure 10. Switching operation to inject surges on lines

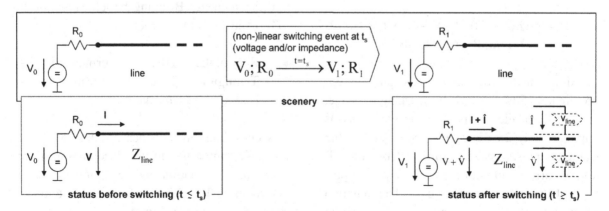

- **Line Impedance:**

$$Z_{line} = \frac{\hat{V}}{\hat{I}} \tag{33}$$

The three equations allow to derivate a solution for each pair of the four unknown values (I, V, $\hat{I}$ and $\hat{V}$). Two solutions are pointed out: $\hat{V}(I)$ and $\hat{V}(V)$.

- **Knowing I:**

$$\hat{V}(I) = R_1 || Z_{line} \cdot \left[\frac{V_1 - V_0}{R_1} + I \cdot \left(\frac{R_0}{R_1} - 1 \right) \right] \tag{34}$$

- **Knowing V:**

$$\hat{V}(V) = R_1 || Z_{line} \cdot \left[\frac{V_1 - V}{R_1} + \frac{V - V_0}{R_0} \right] \tag{35}$$

Special Case: Surge Generated by a Pulse Generator

That is the standard laboratory use case when connecting generators (arbitrary, pulse, burst, functional) to any line set-up. The impedance is constant and the voltage steps from a first to a second value. The transmitting of FlexRay™

bus driver's data_1 to data_0 edges and data_0 to data_1 edges is similar to this special case.

- **Switching Operation:**

$$\underbrace{\begin{array}{c} V_0 \\ R_0 = R_i \end{array}}_{t < t_s} \xrightarrow{\quad t=t_s \quad} \underbrace{\begin{array}{c} V_1 = V_0 + \Delta V \\ R_1 = R_i \end{array}}_{t > t_s} \tag{36}$$

- **Derivation:**

$$\hat{V}(I) = R_1 || Z_{line} \cdot \left[\frac{V_1 - V_0}{R_1} + I \underbrace{\left(\underbrace{\frac{R_0}{R_1} - 1}_{\substack{0 \\ (R_0 = R_1)}} \right)} \right]$$

$$= R_i || Z_{line} \cdot \left[\frac{V_0 + \Delta V - V_0}{R_i} \right] \tag{37}$$

- **Result:**

$$\hat{V} = \Delta V \cdot \underbrace{\frac{Z_{line}}{Z_{line} + R_i}}_{\substack{\text{transmission} \\ \text{factor}}} \tag{38}$$

Figure 11. Surge generated by a pulse generator

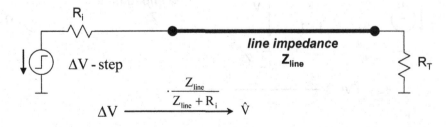

The result is surprisingly simple and allows some general interpretations:

- A voltage divider covers the input behavior completely.
- The result is not influenced by any offset applied to the source; the principle of superposition is visible.
- The equivalent circuit for a line in the use case "coupling a surge into a line" is an ohmic resistor. Its value is equal to the line impedance.
- The termination R_T does not influence on the result in the used scenario.

Special Case: Surge Generated by Closing a Switch

That is the standard case when connecting a bus-driver or transceiver to a passive net. Closing a switch represents the recessive to dominant edge in case of CAN and the idle to active (busy) edge in case of FlexRay™.

- **Switching operation:**

$$\underbrace{\begin{matrix} V_0 \\ V = 0V \\ \underbrace{R_0 \to \infty} \end{matrix}}_{t<t_s} \quad \xrightarrow{t=t_s} \quad \underbrace{\begin{matrix} V_1 = VCC \\ R_1 = R_i \end{matrix}}_{t>t_s} \qquad (39)$$

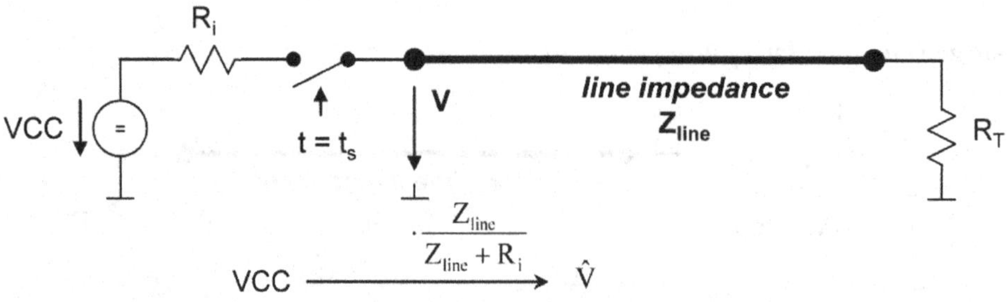

- **Derivation:**

$$\hat{V}(V) = R_1||Z_{line} \cdot \left[\frac{V_1 - V}{R_1} + \underbrace{\frac{V - V_0}{R_0}}_{\substack{0 \\ (R_0 \to \infty)}} \right]$$

$$= R_1||Z_{line} \cdot \left[\frac{VCC}{R_1} \right] \qquad (40)$$

- **Result:**

$$\hat{V} = VCC \cdot \underbrace{\frac{Z_{line}}{Z_{line} + R_i}}_{\substack{transmission \\ factor}} \qquad (41)$$

The result is surprisingly simple, compliant to the example "surge generated by a pulse generator" and allows some general interpretations:

- A voltage divider covers the input behavior completely (see example "surge generated by a pulse generator").
- The equivalent circuit for a line in the uses case "coupling a surge into a line by closing a switch" is an ohmic resistor. Its value

Figure 12. Surge generated by closing a switch

Figure 13. Surge generated by opening a switch (stub line)

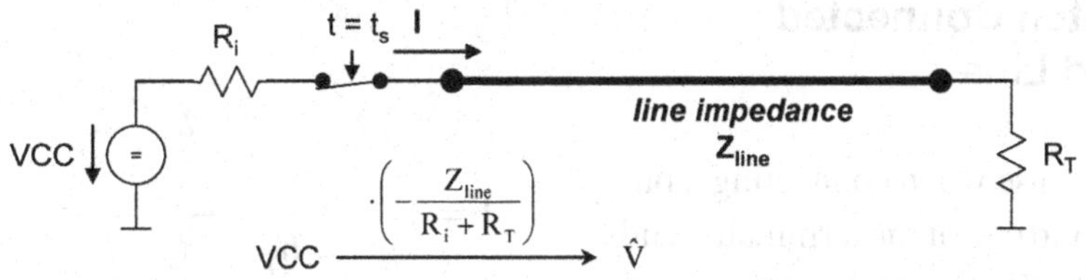

is equal to the line impedance (see example "surge generated by a pulse generator").

- The termination R_T does not influence on the result in the used scenario (see example "surge generated by a pulse generator").

Special Case: Surge Generated by Opening a Switch Connected to a Stub Line

That is the standard case when connecting a bus-driver or transceiver to a stub line of a passive net. Opening a switch represents the dominant to recessive edge in case of CAN and the active (busy) to idle edge in case of FlexRay™.

- **Switching Operation:**

$$\underbrace{\begin{aligned} V_0 &= VCC \\ I &= \frac{VCC}{R_i + R_T} \\ R_0 &= R_i \end{aligned}}_{t<t_s} \xrightarrow{t=t_s} \underbrace{R_1 \to \infty}_{t>t_s} \quad (42)$$

- **Derivation:**

$$\hat{V}(I) = \underbrace{R_1 || Z_{line}}_{\substack{Z_{line} \\ (R_1 \to \infty)}} \cdot \left[\underbrace{\frac{V_1 - V_0}{R_1}}_{\substack{0 \\ (R_1 \to \infty)}} + I \left(\underbrace{\frac{R_0}{R_1}}_{\substack{0 \\ (R_1 \to \infty)}} - 1 \right) \right]$$

$$= Z_{line} \cdot \left[- \underbrace{\frac{VCC}{R_i + R_T}}_{\substack{\text{current driven by} \\ \text{the VCC source} \\ \text{before the switch} \\ \text{was opened}}} \right] \quad (43)$$

- **Result:**

$$\hat{V} = - \underbrace{Z_{line}}_{\substack{\text{load at the} \\ \text{switch}}} \cdot \underbrace{\frac{VCC}{R_i + R_T}}_{\substack{\text{current driven by the VCC source} \\ \text{before the switch was opened}}} \quad (44)$$

The result is surprisingly simple, too, and allows some general interpretations:

- The equivalent circuit for a line in the use case "coupling a surge into a line by opening a switch" is an ohmic resistor. Its value is equal to the line impedance (see previous examples "pulse generator" and "closing a switch").

- The steady state current I influences on the result. The current depends on the termination R_T. It is not influenced by the line impedance Z_{line}.

- The opening of the switch drops the steady state current I down to zero. A current step-shaped wave with the amplitude -I travels along the line.

- The line impedance transfers the current wave into a corresponding voltage wave.

Special Case: Surge by Opening a Switch Connected to a Terminated Line

That is the standard case when connecting a bus-driver or transceiver to one of the terminated ends of a passive bus. Opening a switch represents the dominant to recessive edge in case of CAN and the active (busy) to idle edge in case of FlexRay™.

- **Switching Operation:**

$$\underbrace{V_0 = VCC \frac{R_b}{R_a + R_b} \quad R_0 = R_a \| R_b}_{t < t_s} \xrightarrow{t = t_s} \underbrace{V_1 = 0 \quad R_1 = R_b}_{t > t_s}$$

$$(45)$$

- **Derivation:**

$$\hat{V}(I) = R_1 \| Z_{line} \cdot \left[\frac{V_1 - V_0}{R_1} + I\left(\frac{R_0}{R_1} - 1\right) \right]$$

$$= -Z_{line} \| R_b \cdot \underbrace{\left[\frac{VCC}{R_a + R_b} + I\frac{R_b}{R_a + R_b} \right]}_{\text{current driven by the VCC source before the switch was opened}} \quad (46)$$

- **Current into the Line:**

$$I = VCC \cdot \frac{\dfrac{R_b}{R_a + R_b}}{R_T + R_a \| R_b} \quad (47)$$

- **Result:**

$$\hat{V} = -\underbrace{Z_{line} \| R_b}_{\substack{\text{load at the} \\ \text{switch}}} \cdot \underbrace{\frac{VCC}{R_a + R_b \| R_T}}_{\substack{\text{current driven by the VCC source} \\ \text{before the switch was opened}}} \quad (48)$$

The result seems to be complicated but some general interpretations are possible, anyhow:

- The equivalent circuit for a line is still an ohmic resistor. Its value is the ohmic line impedance (see previous examples).
- The steady state current I influences on the result. The current depends on the termination R_T. It is not influenced by the line impedance Z_{line}.
- The opening of the switch drops the steady state current I down to zero. A current wave -I travels along the line.
- The equivalent line impedance ($Z_{line} \| R_b$) transfers the current wave into a corre-

Figure 14. Surge generated by opening a switch (terminated line)

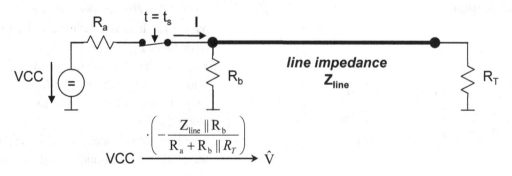

sponding voltage wave. The termination resistor R_b behaves like a line impedance.

Reflection and Refraction of Surges

In the lossless case, any arbitrary voltage wave can propagate on a line without changing its shape. Every voltage wave forces a proportional current wave due to Maxwell's equations. The ohmic line impedance appears as the constant of proportionality. The following derivations are shown by using step shapes; the general availability of the results is not touched by this approach, because each signal shape can be described as sum of various step shapes[8].

If the impedance changes during the propagation (e.g. by connecting a different line and/or by adding a resistor etc.) the ratio of current and voltage will change. To meet the electrical boundary conditions at this joint (well known as Kirchhoff´s laws: continuity of voltage and current), reflections and refractions are going to appear.

Basic scenario that can be used to derivate the most relevant correlations: a voltage step (amplitude $\hat{V}$) travels with velocity v_{line_1} along a 1st line with the impedance Z_{line_1}. When the wave hits upon the joint via the ohmic resistor R_{load} to the

2nd line with the impedance Z_{line_2}, a part of the wave $V_{reflection}$ is reflected, and the rest $V_{refraction}$ is refracted. The reflected part travels with velocity v_{line_1} back along the 1st line. The refracted part travels with velocity v_{line_2} along the 2nd line.

- **Step Wave Traveling Along the 1st Line to the Joint: $\hat{V}$**

- **Load at the End of the 1st Line:**

$$R_T = R_{load} || Z_{line\_2} \qquad (49)$$

(from step wave's point of view)

- **Back into the 1st Line - Reflected Step Wave:**

$$V_{reflection} = \hat{V} \cdot \underbrace{\frac{R_T - Z_{line\_1}}{R_T + Z_{line\_1}}}_{factor\ for\ reflection} \qquad (50)$$

- **Into the 2nd Line - Refracted Step Wave:**

Figure 15. Reflection and refraction at the joint between two lines

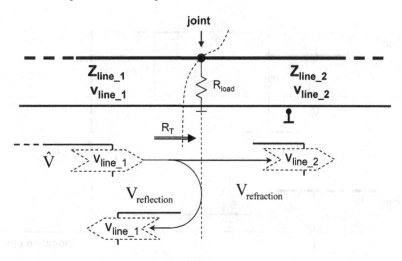

$$V_{\text{refraction}} = \hat{V} \cdot \left(\underbrace{1 + \underbrace{\frac{R_T - Z_{\text{line\_1}}}{R_T + Z_{\text{line\_1}}}}_{\text{factor for reflection}}}_{\text{factor for refraction}} \right) \qquad (51)$$

The way that a joint works can be demonstrated as well by a snapshot of the signal shapes soon after passing the joint. The frozen wave fronts are visible as steps. Due to Kirchhoff´s law the reflected current has to be taken into account by a negative sign.

The scenario and the equations are valid for each combination of the values $\hat{V}$, R_{load}, $Z_{\text{line\_1}}$ and $Z_{\text{line\_2}}$ that is technically implementable. They represent a broad range of applications when dealing with automotive serial communication: using stubs, using branches, connecting different lines (e.g. flat line and twisted pair), using central terminations etc.[9]. They describe the basis features which appear when setting up communication topologies.

The succeeding figure illustrates the overall behavior of reflection and refraction applied to the used scenario. The horizontal axis is built by the normalized termination. The vertical axis is built by the factors which represent the reflection and the refraction portion.

Three main use cases can be distinguished:

1. $R_T \rightarrow 0\ \Omega$ or $R_T \ll Z_{\text{line\_1}}$,
2. (Nearly a) Short circuit at the end of the 1st line.
3. This may happen if squeezing the wires of the line.
4. -100% of the incoming voltage wave is going to be reflected.
5. $R_T \approx Z_{\text{line\_1}}$,
6. Approximately perfect matching termination at the end of the 1st line.
7. That is the standard if connecting two identical cables by an in-line connector designed for this reason.
8. 100% of the incoming voltage wave is going to be refracted. No reflection is going to take place (assuming the connector is designed perfectly).
9. $R_T \rightarrow \infty$ or $R_T \gg Z_{\text{line\_1}}$,
10. (Nearly) Open circuit at the end of the 1st line.

Figure 16. Snapshot of the voltage and the current along the lines soon after the reflection and refraction took place

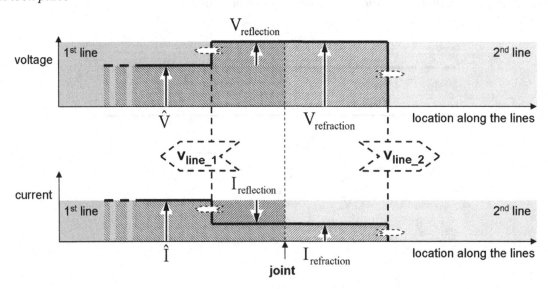

Figure 17. Reflection and refraction (regarding voltage waves)

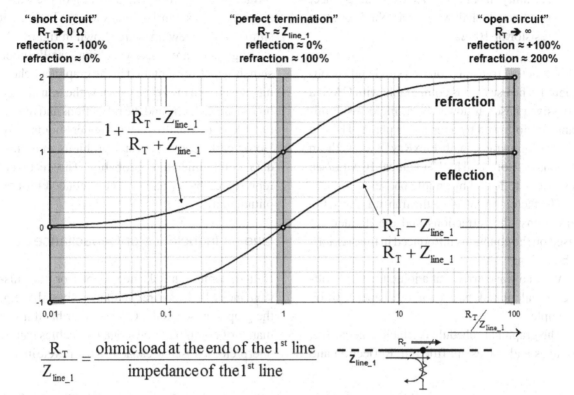

$$\frac{R_T}{Z_{line\_1}} = \frac{ohmic\,load\,at\,the\,end\,of\,the\,1^{st}\,line}{impedance\,of\,the\,1^{st}\,line}$$

11. This can happen if having broken wires or if using un-terminated stubs or branches.

12. 100% of the incoming voltage wave is reflected. A voltage doubling is measured at the joint.

Depending on the termination at the end of a line, the reflection can vary between -100% and +100%. On the other hand the refraction can vary between 0% and 200%.

Graphical Schedule

For somewhat more complex line assemblies, a systematic method for calculating the reflections and refractions lends itself. Fleder (1996) describes a similar more complex and more powerful procedure. The "easy to use" method, the graphical schedule, is introduced with an example.

A transmission link for transmitting data has the following components[10]:

- A direct current voltage source with a serial switch and 50 Ω impedance works as a data bit transmitter. The technically relevant differential operating mode is transferred to a single ended operating mode here.

- Three lossless transmission lines with an impedance of 100 Ω and a delay of 50 ns each work as a passive bus with several high ohmic (and therefore not visible here) receivers.

- Two 100 Ω terminating resistors (beginning and end of the passive bus) work as perfect termination without any mismatch between their ohmic value and the line impedance.

- A faulty in-line connector shall connect two lines. It shall work as an ohmic contact resistor of 100 Ω.

The active transmission of a single bit is considered. The use case is comparable with CAN: a recessive phase - followed by a single 1μs dominant bit - followed by a recessive phase. The use case is comparable with FlexRay™ as well: an idle phase – followed by 10 identical 10 Mbit/sec data bits – followed by an idle phase.

The factors to calculate the transmissions, the reflections, the refractions and the divider are based on the equations introduced in the previous sub-sections.

When a wave arrives at a joint, the final voltage is calculated based on the superposition principle:

The graphical schedule is a table where time, voltages and waves are filled in. Reflection and refraction are visible. The traveling of the waves along the lines can be seen clearly. To keep it simple, only a few lines are shown in figure 19.

Engineers are used watching oscilloscope signals which are included in the graphical schedule as well. On the very first view the transmitted bit is visible clearly and the reflections and refractions lead to small over and under-shoots only. However the contact resistor prohibits a faultless data communication, probably. The receivers represented by V_4 and V_5 get a reduced bit-amplitude only.

Line as Inductance or Capacitance

Data transmission links need cables or transmission lines which require calculations based on e.g. the graphical schedule. On the other hand it is a matter of common knowledge that cables can be interpreted as discrete capacities or discrete induc-

Figure 18. Example for a digital data transmission with its relevant factors for transmission, reflection, refraction and divider

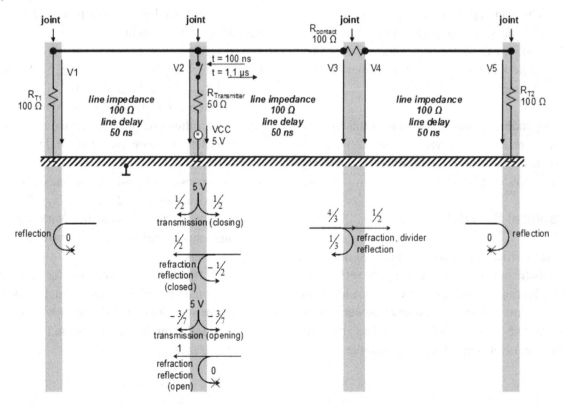

Table 3. Transmission factors and states to characterize the example set-up (part 1)

Verbal	Sketch	Result
Steady state when the switch is open.	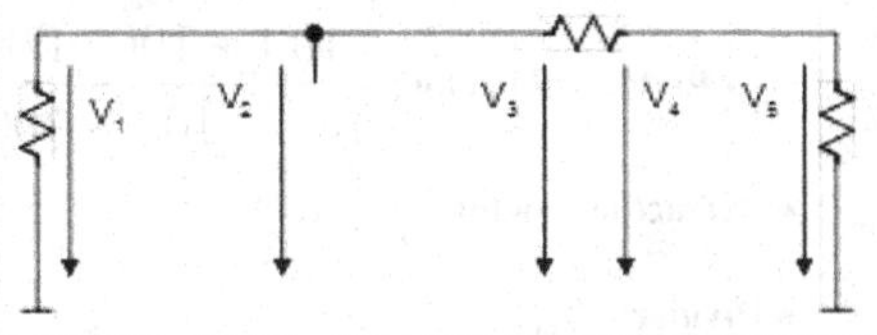	$V_1 = V_2 = V_3 = V_4 = V_5 = 0$ V (52)
Steady state when the switch is closed.	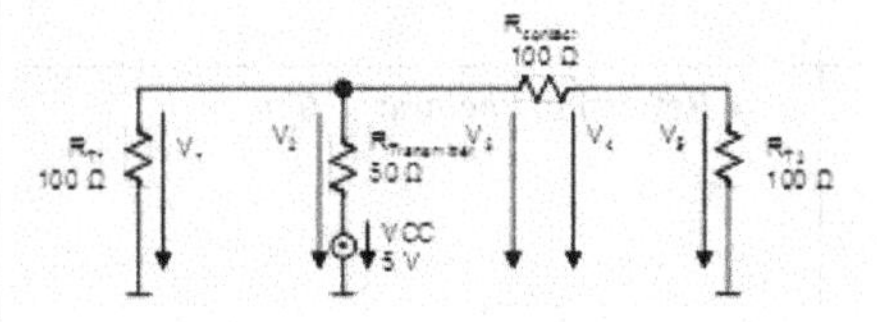	$V_1 = V_2 = V_3 = 5$ V $\dfrac{(100+100)\|\|100}{50+(100+100)\|\|100}$ $= 2{,}86$ V (53) $V_4 = V_5 = \dfrac{V_1}{2} = 1{,}43$ V (54)
1st voltage wave when closing the switch.	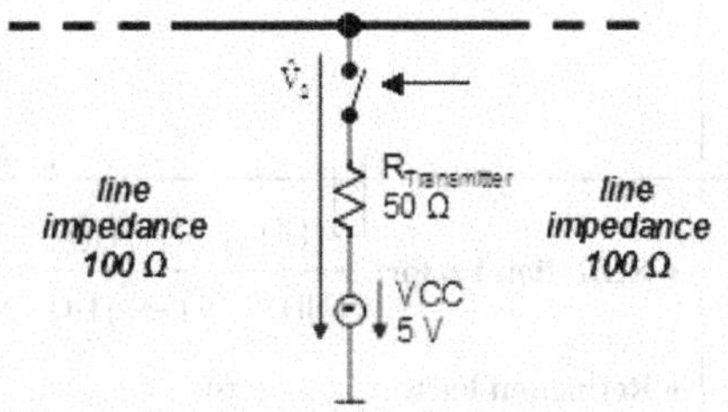	$\hat{V}_2 = 5$ V $\underbrace{\dfrac{100\|\|100}{50+100\|\|100}}_{\text{voltage divider}} = 5$ V $\cdot \dfrac{1}{2}$ (55) • **Transmission Factor:** $\dfrac{1}{2}$ (56)
1st voltage wave when opening the switch.	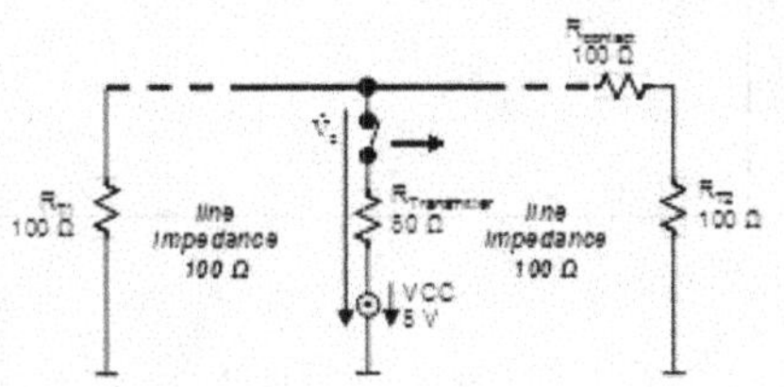	$\hat{V}_2 = -\underbrace{100\|\|100}_{\substack{\text{effective}\\\text{line}\\\text{impedance}}} \cdot \underbrace{\dfrac{5\text{ V}}{50+(100+100)\|\|100}}_{\substack{\text{steady state direct current}\\\text{(switch closed)}}} = -5$ V $\cdot \dfrac{3}{7}$ (57) • **Transmission Factor:** $-\dfrac{3}{7}$ (58)

tances, too. If driving these discrete components with a voltage step source (source impedance R_i) signals shall appear with the parameter "time constant".

- **Capacitive Time Constant:**

$$\tau_{\text{cap}} = C_{\text{line}} \cdot R_i \Big|_{\underbrace{C_{\text{line}} = \frac{\tau_{\text{line}}}{Z_{\text{line}}}}_{\text{line capacitance}}} = \frac{R_i \cdot \tau_{\text{line}}}{Z_{\text{line}}} \qquad (68)$$

- **Inductive Time Constant:**

$$\tau_{\text{ind}} = \frac{L_{\text{line}}}{R_i}\Big|_{\underbrace{L_{\text{line}} = Z_{\text{line}}\, \tau_{\text{line}}}_{\text{line inductance}}} = \frac{Z_{\text{line}}\, \tau_{\text{line}}}{R_i} \qquad (69)$$

The figure below introduces examples characterizing the use cases "line as capacitor" and "line as inductance".

Table 4. Factors to characterize the example set-up (part 2)

Verbal	Sketch	Result
Passing the joint with the contact resistor.	$R_{contact}$ 100 Ω — refraction, divider — line impedance 100 Ω — line impedance 100 Ω — reflection	• **Reflection Factor:** $\dfrac{100 + 100\text{-}100}{100 + 100 + 100} = \dfrac{1}{3}$ (59) • **Refraction Factor:** $\dfrac{4}{3}$ (60) • **Divider:** $\dfrac{1}{2}$ (61)
End of the lines.	R_{T1} 100 Ω — line impedance 100 Ω — R_{T2} 100 Ω — reflection — reflection	• **Reflection Factors:** 0 (62)
At the transmitter (switch open).	line impedance 100 Ω — line impedance 100 Ω — refraction — reflection	• **Reflection Factor:** 1 (63) • **Refraction Factor:** 0 (64)
At the transmitter (switch closed).	line impedance 100 Ω — line impedance 100 Ω — $R_{Transmitter}$ 50 Ω — refraction — reflection	• **Reflection Factor:** $\dfrac{100 \parallel 50\text{-}100}{100 \parallel 50 + 100} = \text{-}\dfrac{1}{2}$ (65) • **Refraction Factor:** $\dfrac{1}{2}$ (66)

Box 1.

$$\underbrace{V\Big|_{t=t_1+0}}_{\substack{\text{voltage just after} \\ \text{the step arrived} \\ \text{at the joint}}} = \underbrace{V\Big|_{t=t_1-0}}_{\substack{\text{voltage just before} \\ \text{the step arrives} \\ \text{at the joint}}} + \sum_{\substack{\text{all} \\ t=t_1}} \Big(\text{incoming wave} + \text{reflected wave}\Big)\Big|_{t=t_1} \tag{67}$$

- **Cable Sample (Lossless):** Line impedance 100 Ω, line delay 1 µs (70)
- **Capacitive Use Case Example:** 1 V step, $R_i = 1$ kΩ (71)
- **Infinite Termination**
- **Inductive Use Case Example:** 1 V step, $R_i = 10$ Ω (72)
- **0 Ω Termination**

The simulation based on surges on lines produces a stair-case shaped exponential function. The simulation with discrete components results in perfect exponential functions. The time constant of each set-up is identical (10 µs). Obviously, the same line can offer capacitive and inductive behavior depending on the used set-up. But – how to distinguish the different use cases? A general answer is hard to give but two statements are available[11]:

Figure 19. Graphical schedule applied to the example transmission link

	time	V_1 termination	line 50 ns delay	V_2 transmitter	line 50 ns delay	V_3 contact	V_4	line 50 ns delay	V_5 termination
steady state	< 100 ns	0 V		0 V		0 V	0 V		0 V
	= 100 ns	0 V		2,5 V		0 V	0 V		0 V
	125 ns		$5V\cdot\frac{1}{2}$		$5V\cdot\frac{1}{2}$				
	150 ns	2,5 V		2,5 V		3,33 V — $5V\cdot\frac{1}{2}\cdot\frac{4}{3}$ →	1,67 V		0 V
	175 ns				$5V\cdot\frac{1}{2}\cdot\frac{1}{3}$			$5V\cdot\frac{1}{2}\cdot\frac{4}{3}\cdot\frac{1}{2}$	
	200 ns	2,5 V	$5V\cdot\frac{1}{2}\cdot\frac{1}{3}\cdot\frac{1}{2}$	2,92 V	$-5V\cdot\frac{1}{2}\cdot\frac{1}{3}\cdot\frac{1}{2}$	3,33 V	1,67 V		1,67 V
	225 ns								
	250 ns	2,92 V		2,92 V		2,78 V — $-5V\cdot\frac{1}{2}\cdot\frac{1}{3}\cdot\frac{1}{4}\cdot\frac{4}{3}$ →	1,39 V	$-5V\cdot\frac{1}{2}\cdot\frac{1}{3}\cdot\frac{1}{4}\cdot\frac{4}{3}\cdot\frac{1}{2}$	1,67 V
	275 ns				$-5V\cdot\frac{1}{2}\cdot\frac{1}{3}\cdot\frac{1}{2}\cdot\frac{1}{3}$				
	...	...		...		...	...		...
	...	...		...		...	...		...
steady state	≲ 1,1 µs	2,86 V		2,86 V		2,86 V	1,43 V		1,43 V
	= 1,1 µs	2,86 V		0,71 V		2,86 V	1,43 V		1,43 V
	1,125 ns		$-5V\cdot\frac{3}{7}$		$-5V\cdot\frac{3}{7}$				
	...			...		...	...		...

Legend (at V3/V4 junction):
* "old" voltage
* + incoming wave(s)
* + outgoing (reflected) wave(s)
* = "new" voltage

Figure 20. Result in the time domain based on the graphical schedule

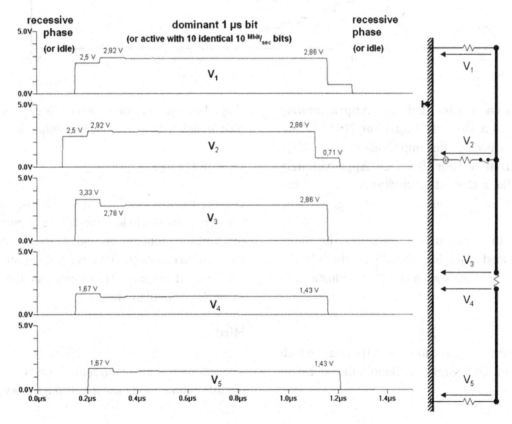

Figure 21. Lossless line used as inductance or capacitor

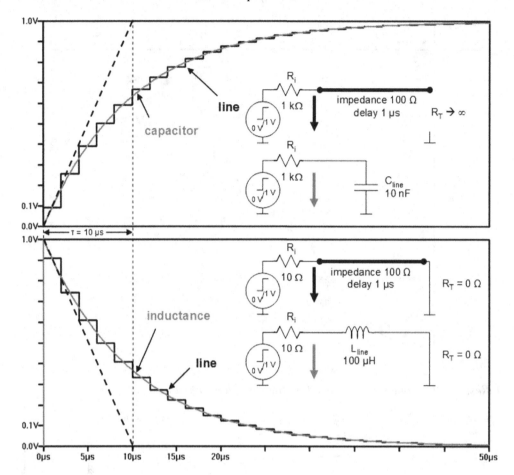

- **Often a Line Behaves Approximately Like a Discrete Capacitor If:** Load impedance >> line impedance (e.g.: >10/1).
- **Often a Line Behaves Approximately Like a Discrete Inductance If:** Load impedance << line impedance (e.g.: <1/10).

The load impedance represents the direct current load of the line. Usually it is built by the impedance of the source and the termination.

Hint

Lossy lines are going to smooth the edges which appear in the lossless simulation; the stair-case shaped exponential function is going to look approximately like an exponential function.

Coupled Lines

Usually the theory of transmission lines covers the behavior of single wires above ground and two wires without any ground. Automotive communication cables (two wires above ground) are not covered directly. They represent the basic example of coupled lines.

Hint

The automotive data communication is based on the differential mode usually. From theory point of view the differential mode behavior can be transferred to a single ended behavior.

UNSHIELDED TWO-WIRE EXAMPLE

An unshielded two-wire cable above ground can be characterized with three line impedance values (e.g. according VDE 0472 part 516):

- The impedance Z_{12} in Ω between the two wires or strands. The value is based on the capacitance between the wires and the inductance between them (in presence of ground).
- The impedance Z_1 in Ω between the one wire or strand and ground. The value is based on the capacitance between wire and ground as well as the inductance between them (in presence of the other wire or strand).
- The impedance Z_2 in Ω between the other wire or strand and ground. The value is based on the capacitance between wire and ground as well as the inductance between them (in presence of the first wire or strand).

In principle cross sections are possible which allow three different values inside a very broad range of the impedances Z_1, Z_2 and Z_{12}. However it is very easy defining a cross section which results in a line impedance Z_{12} much higher than usual

values (e.g. 1 kΩ is possible). On the other hand a twisted pair cable (usually used as two-wire cable in automotive communication systems) represents a special case which allows a simplification. The two wires behave symmetric to ground usually; assuming the manufacturing of twisting and isolation is well done. The impedance Z_1 in Ω is equal to the impedance Z_2 in Ω.

- **Effect of the Twisting:**

$$\overline{Z_{12}} = Z_{12}$$
$$\overline{Z_1} = \overline{Z_2} = Z \qquad (73)$$

- **Impedance of One Wire to Ground:**

$$Z = Z_1 = Z_2 \qquad (74)$$

- **Impedance Between the Wires:**

$$Z_{12} \qquad (75)$$

- **Common Mode Impedance:** $Z_{com} = Z \| Z$ (76)
- **Differential Mode Impedance:** $Z_{dif} = Z_{12} \| (Z+Z)$ (77)

Figure 22. Unshielded twisted pair example

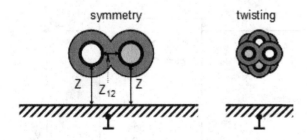
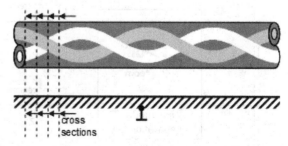

The common mode and the differential mode line impedances and propagation delays can be estimated very easily from oscilloscope measurements in the time domain. In most cases the common mode propagation delay is little less than the differential mode delay.

Approach to Simulate Unshielded Twisted Pair Lines

Hoefer and Nielinger (1985) explain (an example for a broad range of similar sources) how to simulate two parallel wires above ground. The model consists of the super positioning of three transmission lines: two lines to ground with the impedance Z each and a line with the impedance Z_{12} connecting both lines. One important hint is always pointed out: the propagation delay shall be independent from the transmission mode; in other words: common mode propagation delay shall be equal to the differential mode propagation delay or rather all three lines must have the identical propagation delay. Most available two wire cables do not meet theses requirements. However the effect of various speeds can be neglected in many cases. It is recommended to choose the differential modes speed for each of the three lines.

When simulating FlexRay™ systems with this approach the differential mode signal corresponds to the expected signal speed; however the common mode signal could show e.g. ringing shifted in the frequency domain a little.

Alternative Approach to Simulate Unshielded Twisted Pair Lines

The succeeding pages introduce an approach which avoids the explained speed limitation. Additionally the approach allows designing line's common mode behavior completely independent from its differential mode behavior:

- **Differential Mode:** e.g. line impedance Z_{dif}; propagation delay τ_{dif} or signal speed v_{dif}
- **Common Mode:** e.g. line impedance Z_{com}; propagation delay τ_{com} or signal speed v_{com}

The separation of the mixed mode input to the common mode and the differential mode output is done by common known mathematical procedures. The procedure works bidirectional; the correct simulation of e.g. reflections is enabled.

Figure 23. Two-wire-line separated into common mode and differential mode

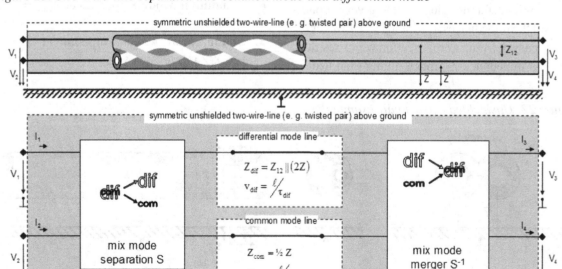

Box 2. Example of a mathematical description as chain matrixes

$$
\begin{pmatrix} V_1 \\ I_1 \\ V_2 \\ I_2 \end{pmatrix} = \underbrace{\begin{bmatrix} \tfrac{1}{2} & 0 & 1 & 0 \\ 0 & 1 & 0 & \tfrac{1}{2} \\ -\tfrac{1}{2} & 0 & 1 & 0 \\ 0 & -1 & 0 & \tfrac{1}{2} \end{bmatrix}}_{\substack{\text{separation} \\ \text{matrix } S}} \cdot \underbrace{\begin{bmatrix} \overbrace{A11_{dm} \quad A12_{dm}}^{\substack{\text{chain matrix} \\ \text{differential mode line}}} & 0 & 0 \\ A21_{dm} \quad A22_{dm} & 0 & 0 \\ 0 \quad 0 & \underbrace{A11_{cm} \quad A12_{cm}}_{\substack{\text{chain matrix} \\ \text{common mode line}}} \\ 0 \quad 0 & A21_{cm} \quad A22_{cm} \end{bmatrix}}_{\substack{\text{block matrix} \\ \text{common mode separated from the differential mode}}} \cdot S^{-1} \cdot \begin{pmatrix} V_3 \\ I_3 \\ V_4 \\ I_4 \end{pmatrix}
\tag{78}
$$

An example visualizes the effects. The line's termination is chosen with a big mismatch to the line impedances. Reflections with positive and negative signs are provoked.

Common mode and differential mode signals appear separated in the time domain so that potentially confusing interferences do not appear.

- At the beginning a triangular common mode pulse is transmitted to the line. Due to the propagation delay it appears after 175 ns at the open end of the lines. The termination mismatches generate reflections each 175 ns.
- The common mode pulses are visible at the common mode line. The differential mode line remains at zero.
- After the common mode reflections disappear (no longer visible in the chosen scal-

ing) a trapezoid (or rectangle) differential mode pulse is transmitted to the line. Due to the propagation delay it appears after 250 ns at the open end of the lines. The termination mismatches generate reflections each 250 ns.

- The differential mode pulses are visible at the differential mode line. The common mode line remains at zero.

The simulation results are satisfying. The expected behavior including all reflections appeared.

Comparison of the Two Approaches

The final example ignores the hint of Hoefer and Nielinger (1985) selecting identical propagation delays to check its relevance. The separation ap-

Figure 24. Separation approach with impedance mismatch

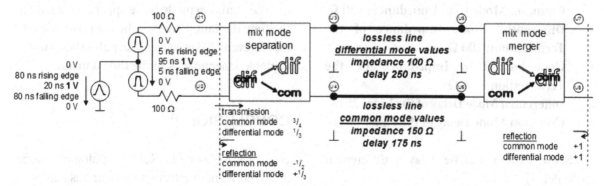

203

Figure 25. Simulation results

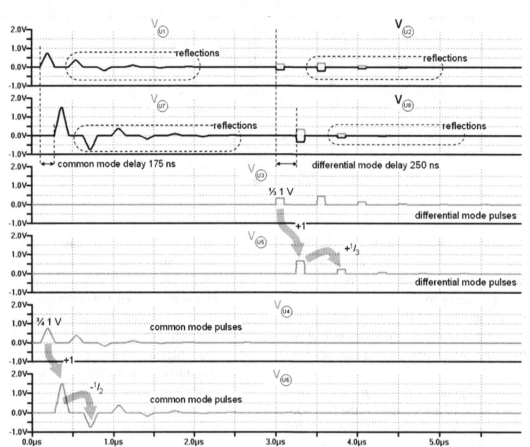

proach is added to enable a direct comparison between both results.

- The set-up is perfectly matched at the end of the line (common mode and differential mode). Any reflection at the end of the line is avoided.
- **Differential Mode Line Impedance:** 100 Ω
- **Common Mode Line Impedance:** 150 Ω
- **Differential Mode Impedance of the Termination:** 100 Ω
- **Common Mode Impedance of the Termination:** 150 Ω
- **Differential Mode Delay of the Line:** 250 ns
- **Common Mode Delay of the Line:** 175 ns

(Skew: Common mode delay ≠ differential mode delay)

- Trapezoid Differential Mode Pulse
- Triangular Common Mode Pulse

When using the approach according to Hoefer and Nielinger (1985) its propagation delay hint is ignored. The three single lines (visible as grey blocks) are designed to support the required common mode and differential mode impedances.

The common mode pulse appears correctly in the time domain. However the data communication relevant differential mode pulse shows unacceptable inaccuracies in the time domain.

CONCLUSION

Neither CAN nor FlexRay™ topologies can be analyzed without interpreting their bus cables as

Figure 26. Set-up to compare the approaches

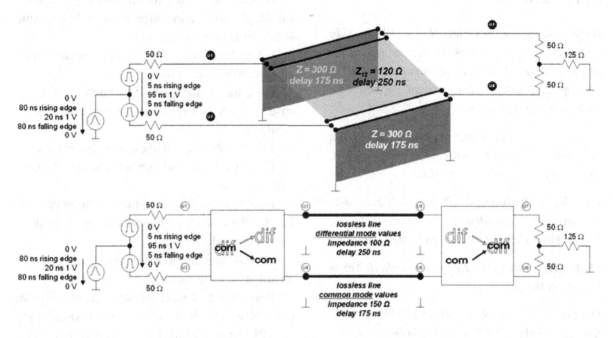

Figure 27. Result of the comparison

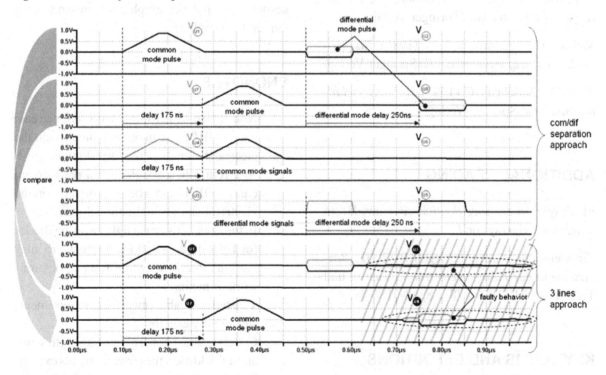

transmission lines. Communication signals are based on steps which travel along cables. Important behaviour models and calculation procedures were introduced in the previous sections. They enable analysing e.g. measuring results gotten in any passive net.

REFERENCES

Bidyut, K. S., & Wheeler, R. L. (1998). *Skin effects models for transmission line structures using generic SPICE circuit simulators*. IEEE 7th Topical Meeting on Electrical Performance on Electronic Packaging.

Bronstein, N. I., & Semendjajew, K. A. (1981). *Taschenbuch der Mathematik*. Thun und Frankfurt, Germany: Verlag Harri Deutsch.

Elco, R. A., & Burke, B. (2003). *SPICE models with frequency dependent conductor and dielectric losses*. High Frequency Design SPICE Models.

Fleder, K. (1996). *The Bergeron method*. Texas Instruments.

Hoefer, E. E. E., & Nielinger, H. (1985). *SPICE*. Springer Verlag. ISBN 3-540-15160-5

Küpfmüller, K. (1973). *Einführung in die theoretische Elektrotechnik*. Springer Verlag.

Meinke, H., & Gundlach, F. W. (1986). *Taschenbuch der Hochfrequenztechnik*. Springer Verlag.

Tietze, U., & Schenk, C. (1993). *Halbleiterschaltungstechnik*. Springer Verlag.

ADDITIONAL READING

FlexRay™. (n.d.). *Specifications*. Retrieved from http://www.flexray.com/

Zimmermann, W., & Schmidgall, R. (2007). *Bussysteme in der Fahrzeugelektronik*. Vieweg und Teubner. doi:10.1007/978-3-8348-9188-4

KEY TERMS AND DEFINITIONS

Active: Status of the communication in case of FlexRay™ when a message or serial data stream is available (see "busy").

Busy: Status of the communication in case of FlexRay™ when a message or serial data stream is available (see "active").

CAN: Abbreviation of controller area network: an automotive serial bus.

Dominant: Status of the communication in case of CAN or LIN when a logical zero bit is available.

ECU: Abbreviation of electronic control unit.

FlexRay™: A made-up word for an automotive serial bus.

Idle: Status of the communication in case of FlexRay™ when not any message or serial data stream is available.

LIN: Abbreviation of local operating network: an automotive serial bus.

Recessive: Status of the communication in case of CAN or LIN when a logical one bit is available.

SPICE: Simulation tool for passive and active electric circuits. Spice represents the simulations kernel typically; the graphical front-ends vary among the suppliers.

ENDNOTES

[1]   Non return to zero: the serial data stream e.g. …00101… can be seen directly on the electrical signals in the time domain

[2]   Non return to zero: the serial data stream e.g. …00101… can be seen directly on the electrical signals in the time domain

[3]   In the optical use case light emitting diodes (or laser diodes) and phototransistors may work as interfaces while fiber optics may work as media.

[4]   Twisted pair cables meet these requirements sufficiently exactly.
      Coaxial cables consisting of a single wires and a shield meet these requirements exactly.

[5]   Transversal electro magnetic fields shall appear only.

[6]   Propagation function: $e^{-\gamma \ell}$

[7]   By the author.

8   According "signal and system theory" with the key-words "sampling theorem" and "convolution".

9   The use case "connecting two lines with completely different cross sections" is not covered.

10   Parasitic components are neglected (e.g. biasing, pin-capacitances etc.)

11   Please take into consideration: Even if one of the rules is applicable, some set-ups are able to show e.g. resonances instead of exponential shape curves.

APPENDIX

Symbols

- ‖: Used when two components are placed parallel to each other (used making equations clearly arranged)
- **V:** Voltage
- **I:** Current
- **R:** Resistor or resistance
- **C:** Capacitor or capacitance
- **L:** Inductivity or inductance
- **,:** The comma is used as decimal comma according ISO. Example: one dot five is written as 1,5

Chapter 7
Optical Wireless Communications in Vehicular Systems

Matthew D. Higgins
University of Warwick, UK

Zaiton Abdul Mutalip
University of Warwick, UK & Universiti Teknikal Malaysia Melaka, Malaysia

Zeina Rihawi
University of Warwick, UK

Roger J. Green
University of Warwick, UK

Mark S. Leeson
University of Warwick, UK

ABSTRACT

This chapter reviews some of the network topologies and technologies within current vehicular systems. This is then followed by a proposal from the authors with initial viability results, into the possibility of implementing optical wireless links to either replace or complement these existing ideas. The initial motivation for this work (Green, 2010) is that there exist multiple pathways within a vehicle such as the engine compartment, within the frame of the chassis, or the internal cockpit that all lend themselves nicely to free space optical propagation. The first specialised study on the viability of optical wireless communications within the vehicles cabin was then published in (Higgins et al, 2012) which provided a further impetus to the concept. It is hoped that through the original results presented here, the reader can gain a basic understanding of the concepts compared to the current technologies, and are then able instigate their own research ideas.

VEHICULAR SUBSYSTEMS AND NETWORK REQUIREMENTS

In general, a vehicle can be segmented into the five subsystems or domains as shown in Figure 1 (Navet et al., 2005, Nolte et al., 2005, Sangiovanni-Vincentelli & Di Natale, 2007). Within the powertrain

DOI: 10.4018/978-1-4666-2976-9.ch007

subsystem, a highly complex and strictly timed set of mechanisms exist that control the engine and transmission. This occurrence of frequent data transmission requires a highly reliable network that also exhibits a high bandwidth to enable such real-time information to be communicated without error. Similarly, real-time, high bandwidth and safety critical networks exist within the chassis domain that is typically thought of encompassing the suspension, steering and brakes. The body and comfort subsystem comprises electronics such as air-conditioning, lighting, wipers, cruise control and locking mechanism for example. The functionality of these components, although required to be reliable, is not considered safety critical, and would generally not require a high bandwidth network for which many elements might be considered as lower in cost. The safety subsystem is formed by a sensor network attached to impact and rollover zones, airbags and seatbelts. The quality of this network is ultimately a factor into the overall success of a vehicles ability to protect the driver and/or passengers from unpredictable and possibly severe events. Whilst the bandwidth may not be an issue (as some of the sensors and actuators will be comparably simple) it must have a high availability. Finally, the infotainment and telematics subsystem deals with the multimedia and mobile communications devices such as audio-visual units, GPS or computer consoles etc. This subsystem is distinctly different to the chassis subsystem as many devices may be considered by vehicle manufactures as optional, or indeed installed and maintained by third parties. This subsystems network is likely to require a very high bandwidth for video and music streaming whilst also being reliable and possibly upgradable given that entertainment and data requirements are possible evolving faster than the time between vehicle purchases.

CURRENT VEHICULAR NETWORK TECHNOLOGIES

Historically, within a given vehicle, is was not uncommon to find up to 70 Electronic Control Units (ECUs) which exchange data around the vehicle (Navet *et al*, 2005). As the functionality within newer vehicles increased, the resultant demand on vehicular networks also increased to the point that an increase in network bandwidth and reliability was desired, but, as was true then as it

is now, to obtain it at lower cost. What transpired is that from an engineering point of view, this problem can be solved by reducing the size and complexity of this "network or networks" whilst maintaining or increasing upon the capabilities of the overall network requirements listed above (Sangiovanni-Vincentelli & Di Natale, 2007, Kopetz & Bauer, 2003). Several excellent wired and wireless solutions, or candidate solutions have been proposed in recent years. A course breakdown of each technology can be seen in Figure 2.

Considering Figure 2 for a moment, a striking, if not obvious observation one could make is that the names of the wired networks are all specific, seemingly specialised to the vehicular market, and only to be understood by vehicular engineers. Contrary to this are the names of the wireless networks, seemingly common to every layperson and synonymous in everyday life. This observation is not uncommon and in fact a virtue to the relative maturities wired vs wireless and industrial uptake. Therefore, considering first the more mature, wired technology, a brief description for each is as follows:

- **Controller Area Network (CAN):** The CAN protocol is a common, robust and inexpensive low-speed serial bus standard used for in-vehicle networks heralding a reasonable level of flexibility whilst maintaining high bandwidth efficiency (Farsi *et al*, 1999). This technology is typically employed within the drivetrain and comfort and convenience subsystems such that the interconnectivity of the engine management system, transmission, ABS/EPS, windows, seats and air conditioning controllers for example can be established with data rates at either 125 Kbit/s or 1 Mbit/s (Tong *et al*, 2007, Othman *et al*, 2006).
- **Time Triggered CAN (TTCAN):** The TTCAN is an extension of the original CAN technology with a deployment emphasis upon safety critical applications

Figure 1. The five embedded vehicular subsystems

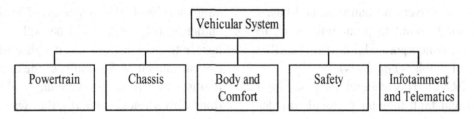

Figure 2. Intra-vehicle networks

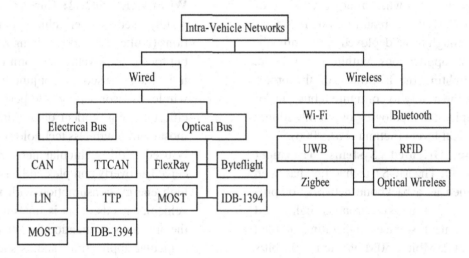

such as the engine management, transmission or the fly-by-wire sensors and actuators. TTCAN employs a bus network topology operating at 1Mbit/s (Leen & Heffernan, 2002a)

- **Local Inerconnect Network (LIN):** The LIN technology is a low speed serial bus operating at 25Kbit/s and is sometimes considered to be an alternative or complement to a low speed CAN (Kopetz & Bauer, 2003) LIN is typically used in simple applications in the body and comfort electrical subsystems connecting the simple sensors and actuators over a single wire. Furthermore, LIN is based upon a master-slave protocol which can take on unconditional, event triggered, random, diagnostic of under defined frame specifications (Navet *et al*, 2005).

- **Time Triggered Protocol (TTP):** A TTP was developed for devices that need a time triggered event solutions. Available as two versions, TTP/A and TTP/C, where TTP/A is based upon a TDMA master-slave topology TTP/C is based upon a TDMA fully distributed topology. In recent years, TTP/C is the most popular variant, set up with a star bus for fault tolerant applications up to 10Mbit/s (Nolte *et al*, 2005, Kopetz & Bauer, 2003).

- **Byteflight:** This 10Mbit/s optical data bus was designed for application that are a combination of either time sensitive or safety critical, for example, airbags. Its flexibility however has meant that is also being deployed in less safety critical electronics within the body and comfort subsystem (Leen & Heffernan, 2002). As the protocol is able to incorporate both time and event driven mechanisms, and based upon single point to point links it is, in part, also incorporated into the FlexRay protocol (Kibler *et al*, 2004).

- **FLEXRAY:** This general purpose, high speed and fault tolerant protocol was developed for time sensitive and safety critical applications. Sometimes it is deployed alongside CAN when a larger data rate of up to 20Mbit/s is required, but it is flexible enough to be deployed across all subsystem applications within the vehicle. Highlighting the flexibility of the protocol is the ability to configure a bus, star of multiple star topologies over both copper of optical layers (Sethna *et al*, 2006).

- **Media Oriented Systems Transport (MOST):** The MOST technology has been designed as a backbone network for providing real time, synchronous high speed infotainment services. Operating at data rates of 25Mbit/s, 50Mbit/s and 150Mbit/s in both copper and optical physical layers, the technology is relatively low cost (Ahmed *et al*, 2007).

- **IDB-1394:** The Intelligent transportation systems Data Bus using IEEE 1394 (IDB-1394) technology was designed as a high performance bus for use in multimedia applications. Currently capable of transferring at rates of up to 400Mbit/s with a possible 1600Mbit/s version on the horizon, it is capable of both isochronous and asynchronous modes. The Isochronous mode is designed to be real time for use with high speed cameras and video streaming (Weihs, 2006).

With the maturity of the wired networks listed above, researchers are now also focusing their attention upon the wireless methods for several reasons. Firstly, as vehicles are still increasing in complexity, with more sensors and actuators present, more wires are necessary. At this most simplistic level, this is perceived by the wider audience as a vehicle will be both heavier and harder to manufacture i.e. it will cost more to operate and build. Secondly, as the end users are themselves involved in more day to day interactions with wireless networks, they want to extend this to the vehicular experience also.

- **Wi-Fi IEEE 802.11:** Currently, Wi-Fi is widely used in inter-vehicle, communications (Nolte *et al*, 2005, Zhang *et al*, 2009) but for the intra-vehicular communications it is mostly used in conjunction with a wireless sensor networks to perform a specific task, e.g. collect data. Although the most mature wireless technology commonly used within vehicular networks is perceived as overly complex and has relatively high energy consumption with regards to vehicular needs. This is not surprising as the original specification for Wi-Fi was not vehicular applications and as such has not gone through the process of modification by many researchers to date.

- **Bluetooth:** The use of Bluetooth in a vehicle is divided into two levels, the user device level and the system level. At the use device level, a network can be made between the users portable multimedia device or mobile electronics such as mobile phones and hands-free accessories for example. This however, to some extent is outside the remit of the vehicle manufacture. At the system level, Bluetooth can be used to function similarly to any other non safety critical wired based network that can accommodate varying degrees of latency (Leen & Heffernan, 2001, Hedderbaut *et al*, 2004).

- **Ultra Wideband (UWB):** Ultra wideband is used in short range applications with low transmission power requirements mainly in the indoor or confined space scenario.

Within the vehicle, UWB can be employed for telematics, multimedia and infotainment applications in addition to vehicular radar systems (Liu *et al*, 2010, Zhang *et al*, 2009). The limitations of UWB come from the challenge of designing an antenna and receiver for the different applications with the knowledge that UWB is particularly susceptible to interference noise generated by other vehicular elements, mostly the engine (Porcino & Hirt, 2003).

- **Zigbee:** The technology behind Zigbee is based upon the IEEE 802.15.4 specifications. Due to its open standard, low power and low cost characteristics, it has become an attractive wireless option compared to Bluetooth or Wi-Fi. Suitable for low data rate and low power applications (Dissanayake *et al*, 2008) it has mostly been deployed so far in vehicle to vehicle and vehicle to infrastructure applications (Tsai *et al*, 2007) with systems based solely on intra-vehicle applications at an early stage in development (Baker, 2005).

- **Radio Freqency ID (RFID):** RFID networks, can be formed by tagging objects within the vehicle that need to be acknowledged, e.g safety harnesses, spare wheels etc. A transponder can be used to communicate via a radio frequency signal to the tag on the system the object to be detected. Several ideas have been proposed in the past but with known limitations with power efficiency and the locations of tags within the vehicle (Tonguz *et al*, 2006).

OPTICAL WIRELESS FUNDAMENTALS

Optical wireless (OW) communication systems that employ an infrared (IR) carrier is best described as a niche technology that is able to combine the high bandwidth opportunities afforded within optical fibre systems, with the advantages of mobility found typically within radio frequency (RF) systems. Compared to an RF system however, OW offers the advantages of high speed short to medium range communications operating within a virtually unlimited and unregulated bandwidth spectrum using typically lower cost components. In addition, OW systems offer the capability of secure deployments with immunity to adjacent communication cell interference due to the inability for the carrier to pass through opaque barriers (Kahn & Barry, 1997).

OW is of course, not limitless in its capabilities or without disadvantages. The inability for the carrier to pass through opaque barriers requires a potentially larger number of base stations to avoid issues in shadowing, the typically used photodetectors are highly susceptible to intense levels of background light from the sun or ambient lighting and finally, a major drawback is the requirement to make an transmitter operated within an acceptable exposure limit (AEL) such that the users sight is not affected (Boucouvalas, 1996).

Pioneering work by Gfeller in the 1970s (Gfeller & Bapst, 1979) began what was, and still is a rapidly expanding segment research in communications systems. Such a review is beyond the scope of this chapter but the reader is directed to several notable reviews such as (Street *et al*, 1997) or (Green *et al*, 2008) if they wish to pursue a more detailed understanding.

In virtually all short range and confined deployments, such as indoors, or as will be seen, in-vehicle, the only practical transmission technique is intensity modulation direct detection (IM/DD). Here, a suitable device such as Laser Diode or LED emits an instantaneous optical power $X(t)$ that propagates through a linear channel with impulse response $h(t)$ until it is incident upon an optical detector at the receiver. Given that typically, the photodetector has a surface area of a size between 10^6 and 10^8 square wavelengths, it is possible to assume optical incoherence and that the received signal $Y(t)$ is simply the integration of the detected optical power over the detectors surface, multiplied by its responsivity, η. If one further assumes, as is typical, that the system is operating in the presence of intense infrared and viable background ration, $Y(t)$ will also comprise a additive white Gaussian noise term $N(t)$ such that:

$$Y(t) = \eta X(t) * h(t) + N(t) \qquad (1)$$

where $*$ denotes convolution. As $X(t)$ is a power quantity, it cannot be negative, such that the average transmitted optical power, P_{av}, is proportional to the time integral of $X(t)$:

$$P_{av} = \lim_{T \to \infty} \frac{1}{2T} \int_{-T}^{T} X(t)\, dt \text{ for } (t) \geq 0 \qquad (2)$$

Assuming $h(t)$ is known, the total optical path gain is given by

$$H(0) = \int_{-\infty}^{\infty} h(t)\, dt \qquad (3)$$

Such that the received power is given by

$$P_{rec} = H(0) P_{av} \qquad (4)$$

where from (4) it is possible to see that the desire should be to maximise P_{rec} through a combination of minimising the path loss and increasing the average transmitted power whilst remaining within the maximum permissible AEL.

Determining, or predicting the impulse response is in itself a great challenge for an optical wireless system designer. The impulse response, $h(t)$ is unique to a given source and receiver configuration within the environment they are deployed within. This means the position and orientation of the transmitter and receiver, with its respective properties such as radiation emission pattern, field of view and active collection area, are all factors of $h(t)$. Coupling these with the dimensions of the environment and any object enclosed, along with their respective reflectivity properties, an essentially infinite number of possible impulse responses exist. Given all these conditions on an optical wireless communication system, one might imagine how it could compete with the established methods described earlier.

OPTICAL WIRELESS VIABILITY

In order to demonstrate the viability, let us consider a sports utility vehicle (SUV) as a test case for system deployment. The SUV cabin structure consisting of the ceiling, floor, seats, steering wheel, fascias and the four passengers can be seen in Figure 3. It was formed through the careful generation and orientation of 560 planar polygons, 296 of them for the fixed interior and 66 of them for each passenger. This model is more complex than the original work found in (Higgins *et al*, 2011, 2012) which was 20cm thinner, 10cm higher and lacking some of the finer details such as 'blind spots' between the seats and door frames. The dimensions and overall topology of the SUV are not specific to a particular marque or model but should provide a familiar resemblance to several popular SUV choices in production.

Traditionally, (Higgins *et al*, 2009, 2011a) when modelling the propagation of infrared (IR) radiation for indoor optical wireless (OW) applications, it is typically assumed that all the surfaces within the environment exhibit a fully diffuse ideal Lambertian reflection profile. For this scenario, such an assumption cannot be made as it is well known that glass, found abundantly within a vehicles interior, exhibits a specular reflection profile. Therefore, in this scenario, it is assumed that the Phong model is more appli-

Figure 3. The internal structure of the sports utility vehicle to be modelled. Adapted from (Higgins et al, 2012)

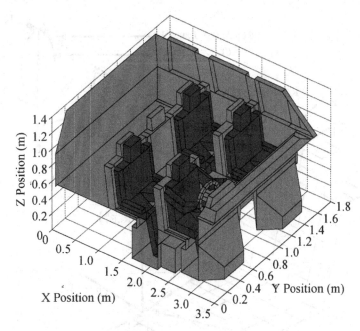

cable, for which, the emitted, or reflected radiation intensity profile $R(\varphi, \phi)$ is given by (Perez *et al*, 2002, Lomba *et al*, 1998):

$$R(\varphi, \theta) = P_S \left[\frac{r_d (n+1) \cos^n (\varphi)}{2\pi} + \frac{(1-r_d)(m+1)\cos^m (\varphi - \theta)}{2\pi} \right]$$

(5)

where, with reference to Figure 4, φ and θ are the angles of observation and incidence relative to the surface normals, respectively. r_d represents the normalised ratio of power reflected diffusely, m is the order of the specular component, n is the order of the diffuse component and P_S is the power of radiation emitted. For clarity, Figure 4 provides some example cases for varying r_d, m and n values. To determine realistic values for application in (5), the open access Columbia-Utrecht Reflectance and Texture (CUReT) database (Dana *et al*, 1999) was employed.

From the databases 61 available materials 'Sample 4 – Rough Plastic' is the most similar to that found on car door interiors, window sills and dashboards. Through the application of an Oren-Nayer fitting process (Oren & Nayer, 1994) it was found that this material exhibited a 0.6 reflectivity with 96% of it being diffuse. 'Samples 7, 18, 19, 42, 44 and 46', which represent the materials of Velvet, Thick Rug, Fine Rug, Corduroy, Linen and Cotton respectively are ideal representatives for the vehicles interior upholstery and passengers cloths. Using the same Oren-Nayer fitting process, the upholsteries reflectivity varied between 0.12 and 0.57 with the lower reflectivities being associated with the heavier upholsteries such as those found on the vehicles floors or boot space. Of these reflectivity values, the power contained within the diffuse component ranged between 94% and 99%, such that, to reduce the computational complexity, all fabrics and plastics will be assumed to be fully diffuse with reflectivities as shown in Table 1. For the skin, a fully diffuse version of CUReT's 'Sample 39 – Human Skin' was assumed with a reflectivity of 0.45. Finally,

Figure 4. Geometrical interpretation of the source, receiver and reflector models. Adapted from (Higgins et al, 2012)

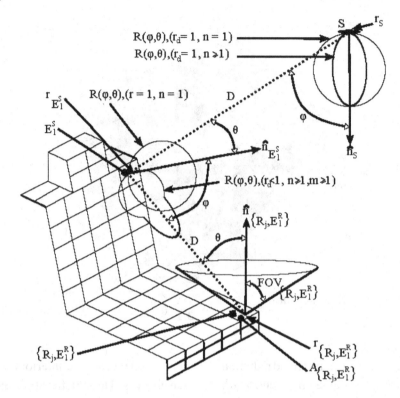

Table 1. Reflectivity properties of the materials used within the vehicle cabin

Material	Γ	r_d	m	n	Material	Γ	r_d	m	n
Glass	0.03	0	280	1	Shoe Leather	0.3	1	1	1
Fabric (Ceiling)	0.5	1	1	1	Fabric (Trousers)	0.4	1	1	1
Fabric (Seats)	0.4	1	1	1	Fabric (Torso Shirt)	0.55	1	1	1
Fabric (Floors/Boot)	0.3	1	1	1	Skin	0.45	1	1	1
Facia Plastic	0.6	1	1	1	-	-	-	-	-

for the reflectivity properties of the glass, the required parameters in Table 1 were derived from the measured results in (Perez *et al*, 2002).

Upon the ceiling of the cabin, a single IR LED source, S, is located with a downwards facing unit length orientation vector $\hat{n}_S$ and position vector $\hat{r}_S = [1.7, 0.82, 1.38]$. The source is assumed to have a radiation profile as (5) with $n = 1, r_d = 1$ and $P_S = 1$ i.e. ideal Lambertian and with unit power. With regards to the receivers,

it is assumed that their exist J identical single element receivers, R_j, such that it is possible simultaneously to describe a system with J receivers or one receiver at J locations. For each receiver, R_j, the location vector $\hat{r}_{Rj}$ and unit length orientation vector $\hat{n}_{Rj}$ is determined by bi-linearly interpolating (Saloman, 2006) each of the polygons in Figure 3 at a resolution of 100 segments per square meter. Furthermore, each receiver is assigned an active collection area

$A_{Rj} = 1\,\text{cm}^2$ and a field of view $FOV_{Rj} = 65°$, defined to be the maximum uniaxial symmetric incident angle of radiation with respect to $\hat{n}_{Rj}$ that will generate a current within the photodiode.

The radiation undergoing propagation between a source S and a receiver, R_j, will undergo either a direct line of sight (LOS) path and/or a path that reflects off k surfaces within the cabin. Therefore, in a manner similar to above, each of the polygons of Figure 3 is bi-linearly interpolated at a resolution of ΔA_k, into L elements E_l, with an associated position vector $\hat{r}_{El}$, unit length orientation vector $\hat{n}_{El}$ and area $A_{El} = 1\,/\,\Delta A_k^2$. Each element will firstly behave as a receiver, E_l^R with a hemispherical FOV, followed by behaving as a source E_l^S, with a radiation emission profile as given by (5) setting the parameters to the respective properties of the surface material, and with P_S equal to the product of the respective reflectivity in Table 1 and the power received when acting as a receiver.

It has previously been shown that for a intensity modulation, direct detection system (IM/DD) channel (Kahn *et al*, 1995), where the movement of the transmitter, receiver or reflectors is slow compared to the bit rate of the system, no multipath fading occurs, and as such can be modelled as a LTI channel with impulse response $h(t; S, R_j)$ given by (Barry *et al*, 1993):

$$h(t; S, R_j) = \sum_{k=0}^{k} h^k(t; S, R_j) \qquad (6)$$

where $h^k(t; S, R_j)$ is the impulse response of the system for radiation undergoing k reflections between S and R_j. The line of sight response is given by:

$$h^0\left(t; S, R_j\right) \approx R(\varphi, \theta)\frac{\cos(\theta)\,A_{Rj}}{D^2} V\left(\frac{\theta}{FOV_{Rj}}\right)\delta\left(t - \frac{D}{c}\right) \qquad (7)$$

where, with reference to Figure 4, $D = \left\| \hat{r}_S - \hat{r}_{Rj} \right\|$ is the distance between the source and the receiver, c is the speed of light, φ and θ are the angles between $\hat{n}_S$ and $\left(\hat{r}_{Rj} - \hat{r}_S\right)$, and between $\hat{n}_{Rj}$ and $\left(\hat{r}_S - \hat{r}_{Rj}\right)$, respectively. $V(x)$ represents the visibility function, where $V(x) = 1$ for $|x| \leq 1$ and $V(x) = 0$ otherwise. For radiation undergoing $k > 1$ reflections, the impulse response is given by:

$$h^k\left(t; S, R_j\right) = \sum_{l=1}^{L} h^{(k-1)}\left(t; S, E_l^R\right) * h^0\left(t; E_l^S, R_j\right) \qquad (8)$$

and the $(k-1)$ impulse response can be found iteratively (Carruthers & Kannan, 2002) from:

$$h^k(t; S, E_l^R) = \sum_{l=1}^{L} h^{(k-1)}\left(t; S, E_l^R\right) * h^0\left(t; E_l^S, E_l^R\right) \qquad (9)$$

where all the $k = 0$ responses in (8) and (9) are found from the careful substitution of the variable in (7). As it is known that the computational time required for evaluating (8) is proportional to k^2 (Carruthers & Kannan, 2002), for the results shown here, k is limited to 3 and for each order, the segmentation resolution of the interpolation is set to $\Delta A_1 = 25$, $\Delta A_2 = 6$ and $\Delta A_3 = 2$. Furthermore, the resultant impulse response in (6) results in a finite sum of scaled delta functions that undergo temporal smoothing by sub-dividing the time into bins of width $\Delta t = 0.1\,\text{ns}$ and summing the total power in each bin (Barry *et al*, 1993).

A ray tracing package capable of evaluating equation (6) was developed as a way of determining where, and with what level of viability, an optical wireless communication link can be used within a vehicular environment. For the results presented here, two areas of the vehicle are given a specific focus, firstly the rear seat region where the passengers physically sit, and secondly the view of those passengers to determine areas where the passengers may physically interact with, namely, the rear or the front seats. The first quantity presented is the total received power for a receiver at a given location placed flush upon the surface, and, following (3) is defined as:

$$H\left(0; S, R_j\right) = \int_{-\infty}^{\infty} h\left(t; S, R_j\right) dt \tag{10}$$

This quantity, commonly known as the DC value of the frequency response, provides an indication as to the how much power is received, such that further analysis can be performed on receiver sensitivity and/or noise analysis for example. However, the temporal information has been lost due to the use of the integral. Not wanting to lose this information, the second quantity considered is the channel bandwidth at each location, found via the DTFT of $h\left(t, S, R_j\right)$. Finally, the RMS delay spread is provided, as given by (Pakravan & Kavehrad, 2001):

$$\Lambda = \sqrt{\frac{\int_{-\infty}^{\infty} \left(t - \Psi\right)^2 h^2\left(t; S, R_j\right) dt}{\int_{-\infty}^{\infty} h^2\left(t; S, R_j\right) dt}} \tag{11}$$

where

$$\Psi = \frac{\int_{-\infty}^{\infty} t h^2\left(t; S, R_j\right) dt}{\int_{-\infty}^{\infty} h^2\left(t; S, R_j\right) dt} \tag{12}$$

VIABILITY RESULTS

To begin, consider the received power, bandwidth and RMS delay spread of the rear passenger seats as shown in Figure 5. Directly under the source, on the middle seat, a maximum received power of 50µW with an associated bandwidth in excess

of 300MHz and negligible RMS delay spread is available. In fact, over the whole of the 'lap' area including the lower arms and tops of the legs, the lowest received power is 22µW. This is ideal for portable devices such as tablet PCs, mobile phones or portable games consoles.

Another area of interest is around the head, shoulder and upper torso area where for example, devices such as wireless IR headphones or hands free telephony equipment may be employed. As shown in Figure 5, the received power ranges between 17µW and 45µW with bandwidth availability in excess of 45MHz and with sub-nanosecond RMS delay spread. These channel characteristics are therefore likely to be adequate for these applications.

Of major concern in the consumer uptake of new technologies is the perceived level of user-friendliness. One test for example is to consider the event that the user drops their portable device into, or stows the devices in, the foot well. In this situation, it is not expected that the devices are meant for high speed communication, but should be able to be polled by the central unit. As can be seen in Figure 5, in areas not directly blocked by the passengers' feet, upwards of 11µW of power is received with bandwidths in excess of 60MHz and 1.5ns maximum RMS delay spreads. These channel characteristics are believed to be sufficient for simple polling operations provided the devices are aligned.

The final area considered from the viewpoint of the passenger is around the window sill and interior door panelling where for example, passenger-vehicle interface devices such as window, AV, air-conditioner or other simple controls may be located. At these locations as per Figure 5 at least 10µW but up to 22µW of power is received with a minimum bandwidth of 190MHz and negligible RMS delay spreads is available. As with all other deployment locations considered so far, the channel characteristics are far in excess of the minimum required.

Figure 5. Rear passenger seat: (a) Received power (in μW). (b) Bandwidth (in MHz). (c) RMS delay spread (in ns)

Figure 6. Rear passenger 'view': (a) Received power (in μW). (b) Bandwidth (in MHz). (c) RMS delay spread (in ns)

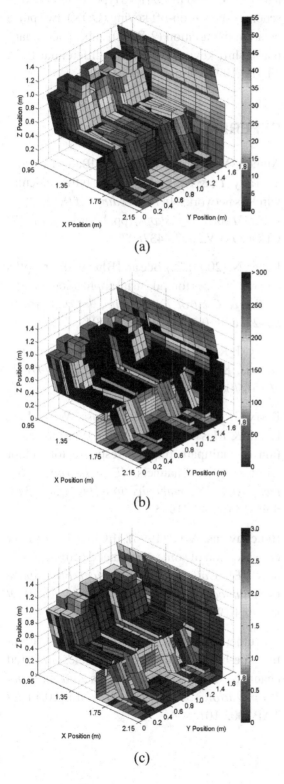

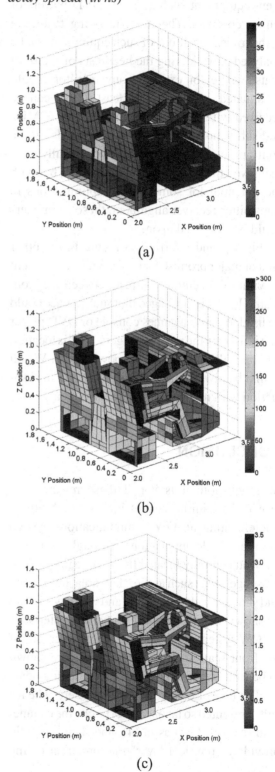

(a)

(a)

(b)

(b)

(c)

(c)

One of the growing areas of vehicular electronics, is in the deployment of AV entertainment equipment such as TV's, DVD players or games consoles. Therefore, referring to Figure 6, the logical positions of such equipment can be analysed. Considering the headrests of the front seats, where for example, any monitors could be situated, a power of at least 21µW can be received with a bandwidth in excess of 62MHz and sub-nanosecond RMS delay spreads. In the configuration shown, this may not be fully optimal given the relatively low bandwidth, but with the potential for some simple optimisations in transmitter-receiver alignment, these parameters could be improved upon.

Finally, and referring to Figure 6, the other area of major promise for OW device deployment is within the central arm-rest between the front seats. Here, it is envisaged that such a space could be used to house a games console or DVD player that is used by the rear passengers. At this location the top is illuminated with a power of 27µW and with a bandwidth in excess of 300MHz in association with negligible RMS delay spread.

CONCLUSION

In conclusion, it is hoped these results have provided an initial insight into the viability of implementing an OW communications system in intra vehicle applications. Through the use of a 1 Watt centrally located IR LED, regions of the passenger seats such as the; 'lap' area; head, neck and shoulders; door interiors and window sills are all shown to have suitable channel characteristics for OW system deployment. Similarly, regions of the front seats such as the headrests are shown to be suitable for OW video screen deployment.

The idea of course does not end here. The results presented so far only incorporate the channel gain (or path loss) as in (4), and therefore only provide a 'physical layer' interpretation of the

problem. The next stage of the work, something at the time of writing is still ongoing, is to compare and contrast the various types of modulation schemes such as on-off keying (OOK), two pulse position modulation (2-PPM), subcarrier binary phase-shift keying (BPSK) or differential PPM (Barry, 1994, Shiu & Kahn, 1999).

REFERENCES

Ahmed, M., Saraydar, C. U., ElBatt, T., Jijun, Y., Talty, T., & Ames, M. (2007). Intra-vehicular wireless networks. In *Proceedings of IEEE GLOBECOM 2007 Workshops,* (pp. 1-9). doi: 10.1109/GLOCOMW.2007.4437827

Baker, N. (2005). ZigBee and Bluetooth strengths and weaknesses for industrial applications. *Computing & Control Engineering Journal, 16*(2), 20–25. doi:10.1049/cce:20050204

Barry, J. R. (1994). *Wireless infrared communications.* Kluwer Academic. doi:10.1007/978-1-4615-2700-8

Barry, J. R., Kahn, J. M., Krause, W. J., Lee, E. A., & Messerschmitt, D. G. (1993). Simulation of multipath impulse response for indoor wireless optical channels. *IEEE Journal on Selected Areas in Communications, 11*(3), 367–379. doi:10.1109/49.219552

Boucouvalas, A. C. (1996). IEC825-1 eye safety classification of some consumer electronic products. *IEE Colloquium on Optical Free Space Communication Links,* 13/1-13/6. doi: 10.1049/ic:19960198

Carruthers, J. B., & Kannan, P. (2002). Iterative site-based modelling for wireless infrared channels. *IEEE Transactions on Antennas and Propagation, 50*(5), 759–765. doi:10.1109/TAP.2002.1011244

Dana, K., Van-Ginneken, B., Nayer, S., & Koenderink, J. (1999). Reflectance and texture of real world surfaces. *ACM Transactions on Graphics*, *18*(1), 1–34. doi:10.1145/300776.300778

Dissanayake, S. D., Karunasekara, P. P. C., Lakmanaarachchi, D. D., Rathnayaka, A., & Samarasinghe, A. T. L. (2008). Zigbee wireless vehicular identification and authentication system. *Proceedings of ICIAFS*, *2008*, 257–260. doi:doi:10.1109/ICIAFS.2008.4783998

Farsi, M., Ratcliff, K., & Barbosa, M. (1999). An overview of controller area network. *Computing & Control Engineering Journal*, *10*(3), 113–120. doi:10.1049/cce:19990304

Gfeller, F. R., & Bapst, U. (1979). Wireless in-house data communication via diffuse infrared radiation. *Proceedings of the IEEE*, *67*(11), 1474–1486. doi:10.1109/PROC.1979.11508

Green, R. J. (2010). Optical wireless with application in automotives. *Proceedings of ICTON*, *2010*, 1–4. doi:doi:10.1109/ICTON.2010.5549164

Green, R. J., Joshi, H., Higgins, M. D., & Leeson, M. S. (2008). Recent developments in indoor optical wireless systems. *IET Communications*, *2*(1), 3–10. doi:10.1049/iet-com:20060475

Heddebaut, M., Deniau, V., & Adouane, K. (2004). In-vehicle WLAN radio-frequency communication characterization. *IEEE Intelligent Transportation Systems*, *5*(2), 114–121. doi:10.1109/TITS.2004.828172

Higgins, M. D., Green, R. J., & Leeson, M. S. (2009). A genetic algorithm method for optical wireless channel control. *Journal of Lightwave Technology*, *27*(6), 720–772. doi:10.1109/JLT.2008.928395

Higgins, M. D., Green, R. J., & Leeson, M. S. (2011). Channel viability of intra-vehicle optical wireless communications. *Proceedings of IEEE GLOBECOM Workshops*, *2011*, 183–817. doi:doi:10.1109/GLOCOMW.2011.6162567

Higgins, M. D., Green, R. J., & Leeson, M. S. (2012). Optical wireless for intravehicle communications: A channel viability analysis. *IEEE Transactions on Vehicular Technology*, *61*(1), 123–129. doi:10.1109/TVT.2011.2176764

Higgins, M. D., Green, R. J., Leeson, M. S., & Hines, E. L. (2011a). Multi-user indoor optical wireless communication system channel control using a genetic algorithm. *IET Communications*, *5*(7), 937–944. doi:10.1049/iet-com.2010.0204

Kahn, J. M., & Barry, J. R. (1997). Wireless infrared communications. *Proceedings of the IEEE*, *85*(2), 265–298. doi:10.1109/5.554222

Kahn, J. M., Krause, W. J., & Carruthers, J. B. (1995). Experimental characterization of non-directed indoor infrared channels. *IEEE Transactions on Communications*, *43*(2,3,4), 1613-1623. doi: 10.1109/26.380210

Kibler, K., Poferl, S., Bock, G., Huber, H. P., & Zeeb, E. (2004). Optical data buses for automotive applications. *Journal of Lightwave Technology*, *22*(9), 2184–2199. doi:10.1109/JLT.2004.833784

Kopetz, H., & Bauer, G. (2003). The time-triggered architecture. *Proceedings of the IEEE*, *91*(1), 112–126. doi:10.1109/JPROC.2002.805821

Leen, G., & Heffernan, D. (2001). Vehicles without wires. *Computing & Control Engineering Journal*, *12*(5), 205–211. doi:10.1049/cce:20010501

Leen, G., & Heffernan, D. (2002). Expanding automotive electronic systems. *Computer*, *35*(1), 88–93. doi:10.1109/2.976923

Leen, G., & Heffernan, D. (2002a). TTCAN: A new time-triggered controller area network. *Microprocessors and Microsystems*, *26*(2), 77–94. doi:10.1016/S0141-9331(01)00148-X

Liu, L., Wang, Y., Zhang, N., & Zhang, Y. (2010). UWB channel measurement and modeling for the intra-vehicle environments. *Proceedings of ICCT*, *2010*, 381–384. doi:doi:10.1109/ICCT.2010.5688815

Lomba, C. R., Valada, R. T., & de Oliveira Duarte, A. M. (1998). Experimental characterisation and modelling of the reflection of infrared signals on indoor surfaces. *Proceedings IEE Optoelectronics, 145*(3), 191–197. doi:10.1049/ip-opt:19982020

Navet, N., Song, Y., Simonot-Lion, F., & Wilwert, C. (2005). Trends in automotive communication systems. *Proceedings of the IEEE, 93*(6), 1204–1223. doi:10.1109/JPROC.2005.849725

Nolte, T., Hansson, H., & Bello, L. L. (2005). Automotive communications-past, current and future. *Proceedings of ETFA, 1*, 985–992. doi:doi:10.1109/ETFA.2005.1612631

Oren, M., & Nayer, S. (1994). Seeing beyond Lambert's law *Proceedings of ECCV*, (pp. 269-280). doi: 10.1007/BFb0028360

Othman, H. F., Aji, Y. R., Fakhreddin, F. T., & Al-Ali, A. R. (2006). Controller area networks: Evolution and applications. *Proceedings of ICTTA, 2*, 3088–3093. doi:doi:10.1109/ICTTA.2006.1684909

Pakravan, M. R., & Kavehrad, M. (2001). Indoor wireless infrared channel chracterisation by measurements. *IEEE Transactions on Vehicular Technology, 50*(5), 1053–1073. doi:10.1109/25.938580

Perez, S. R., Jimenex, R. P., Lopez-Hernandez, F. J., Hernandez, O. B. G., & Alfonso, A. J. A. (2002). Reflection model for calculation of the impulse response on IR wireless indoor channels using a ray-tracing algorithm. *Microwave and Optical Technology Letters, 32*(4), 296–300. doi:10.1002/mop.10159

Porcino, D., & Hirt, W. (2003). Ultra-wideband radio technology: Potential and challenges ahead. *IEEE Communications Magazine, 41*(7), 66–74. doi:10.1109/MCOM.2003.1215641

Salomon, D. (2006). *Curves and surfaces for computer graphics*. Springer-Verlag.

Sangiovanni-Vincentelli, A., & Di Natale, M. (2007). Embedded system design for automotive applications. *Computer, 40*(10), 42–51. doi:10.1109/MC.2007.344

Sethna, F., Stipidis, E., & Ali, F. H. (2006). What lessons can controller area networks learn from FlexRay. *Proceedings of IEEE VPPC, 2006*, 1–4. doi:doi:10.1109/VPPC.2006.364320

Shiu, D. S., & Kahn, J. M. (1999). Differential pulse-position modulation for power-efficient optical communication. *IEEE Transactions on Communications, 47*(8), 1201–1210. doi:10.1109/26.780456

Street, A. M., Stavrinou, P. N., O'Brien, D. C., & Edwards, D. J. (1997). Indoor optical wireless systems: A review. *Optical and Quantum Electronics, 29*, 349–378. doi:10.1023/A:1018530828084

Tong, W., Tong, C., & Liu, Y. (2007). A data engine for controller area network. *Proceedings of Computational Intelligence and Security, 2007*, 1015–1019. doi:doi:10.1109/CIS.2007.137

Tonguz, O. K., Tsai, H. M., Talty, T., Macdonald, A., & Saraydar, C. (2006). RFID technology for intra-car communications: A new paradigm. *Proceedings of IEEE VTC, 2006*, 1–6. doi:doi:10.1109/VTCF.2006.618

Tsai, H. M., Tonguz, O. K., Saraydar, C., Talty, T., Ames, M., & Macdonald, A. (2007). Zigbee-based intra-car wireless sensor networks: A case study. *IEEE Transactions on Wireless Communications, 14*(6), 67–77. doi:10.1109/MWC.2007.4407229

Weihs, M. (2006). Design issues for multimedia streaming gateways. *Proceedings of ICN/ICONS/MCL 2006*. doi: 10.1109/ICNICONSMCL.2006.74

Zhang, J., Orlik, P. V., Sahinoglu, Z., Molisch, A. F., & Kinney, P. (2009). UWB systems for wireless sensor networks. *Proceedings of the IEEE, 97*(2), 313–331. doi:10.1109/JPROC.2008.2008786

Chapter 8
Resilient Optical Transport Networks

Yousef S. Kavian
Shahid Chamran University, Iran

Bin Wang
Wright State University, USA

ABSTRACT

Resilient optical transport networks have received much attention as the backbone for future Internet protocol (IP) networks with enhanced quality of services (QoS) by avoiding loss of data and revenue and providing acceptable services in the presence of failures and attacks. This chapter presents the principles of designing survivable Dense-Wavelength-Division-Multiplexing (DWDM) optical transport networks including failure scenarios, survivability hierarchy, routing and wavelength assignment (RWA), demand matrix models, and implementation approaches. Furthermore, the chapter addresses some current and future research challenges including dealing with multiple simultaneous failures, QoS-based RWA, robustness and future demand uncertainty accommodation, and quality of service issues in the deployment of resilient optical backbones for next generation transport networks.

INTRODUCTION

Nowadays client networks such as Internet Protocol (IP) networks, asynchronous transfer mode (ATM), synchronous optical networking (SONET) and synchronous digital hierarchy (SDH) networks, and so on, employ DWDM optical transport networks to carry data, voice and videos (Murthy and Gurusamy, 2004). The modern optical transport networks are able to provide a huge amount of bandwidth at multiple tetra bits per second (Tb/s) transmission rate by employing fiber optic cables, nodal devices like optical cross-connect switches (OXCs) and optical add-drop multiplexers (OADMs), and the DWDM technology (Kartalopoulos, 2000; Mukherjee, 2000).

The DWDM optical transport networks are being increasingly deployed in the next generation telecommunication networks through the integration of IP over DWDM networks. This may reduce the protocol overhead, the implementation complexity and cost (Rajagolapan, *et al.*, 2000; Ghani, Dixit, & Wang, 2003). Therefore the provisioning of acceptable services in the presence of failures and attacks is a major issue, particularly for future IP services with enhanced quality of service (QoS).

DOI: 10.4018/978-1-4666-2976-9.ch008

The DWDM optical networks are prone to failures such as link and node failures at the optical layer. These failures can potentially lead to a catastrophic loss of data and revenues, producing an unacceptable deterioration in the delivered quality of service (QoS). Therefore, one of the most important optical network design issues is survivability (Zhou, & Subramaniam, 2000), which is the ability of a network to provide continuous services at an acceptable level in the presence of different failure scenarios (Gerstel & Ramasawami, 2000; Zhang, & Mukherjee, 2004). Reasons of designing fault tolerant optical transport networks are summarized as follows:

1. Increasingly, end-users of client networks demand reliable communications and services with assured quality (M´edard & Lumetta, 2002).
2. The unexpected failures of network components such as link and node failures at the optical layer may result in multiple failures at client layers.
3. The aggregated bandwidth on the order of several Tb/s per fiber causes enormous data and revenue loss in the event of network's failures.
4. Fault tolerance and traffic restoration at the optical transport layer have several advantages, such as shorter restoration time, efficient resource utilization, and protocol transparency, over those at the client layers.
5. The failures at the optical layer and data loss would affect the delivered QoS by client networks to end-users.

While survivability in optical networks is highly desirable, redundant resources are necessarily employed to cover the failures, which increases the network cost.

This chapter reviews some major issues of designing fault tolerant Dense-Wavelength-Division-Multiplexing (DWDM) optical transport networks for service provisioning with quality of

service assurance in the presence of failures for the next generation high-speed telecommunication networks. We summarize the main reasons of developing survivable DWDM optical transport networks and the failures scenarios of network components and links. The survivability schemes based on protection and restoration architectures and the traffic demand models are surveyed. The routing and wavelength assignment (RWA) problems which lie at the heart of designing the wavelength-routed DWDM optical transport networks are reviewed. Solution approaches such as integer linear programming (ILP), heuristics, and intelligent genetic algorithm based approaches, are reviewed. The K-shortest path procedure which is needed to design reliable communication networks is presented. The main problems of accommodating future demand uncertainties for designing the robust networks are discussed. Finally the quality of service (QoS) issues that arise in the deployment of resilient DWDM backbones are summarized.

PRINCIPLES

Failure Models and Redundancy

The optical network failure models can be classified into three types, transient, temporary, and permanent failures, which affect the policies of failure recovery in network and the network resource redundancies. Research on survivability in DWDM optical networks has focused on permanent failures employing permanent redundancies such as redundant bandwidth and spare switches for failure recovery. Three main failure models have been considered:

Single Link Failure Model

This is the most popular failure model for DWDM networks (Shenai & Sivalingam, 2005) where the fiber may be cut by workers, fire, earthquake, changing environmental conditions such as temperature, humidity, and so on. In this model the fiber cut in a link of the network is repaired before another fiber cut is assumed to occur in the network (Zhang & Mukherjee, 2004). The link failures are assumed to be independent of each other and the probability of a single link failure is the same for all links, and the failure occurs uniformly in the network. Therefore the quality of protection (QoP) against the failures is the same for all links in the network. To recover from a single link failure, the network topology should be 2-connected, which means that there are two link-disjoint paths between each node pair of the network.

Node Failure Model

The node (OXC, OADM, transmitter, receiver, and amplifier) failure model is also a realistic one that needs to be considered in designing fault tolerant backbone networks (Wang, Cheng & Mukherjee, 2003). While the probability of a node failure is generally much smaller than a link failure due to the built-in redundancy of many network equip-

ments, node failure is still possible and will cause severe service disruption. To recover from a node failure, the network topology should have two node-disjoint paths between each node pair of the network.

Multiple-Link Failure Model

Another failure scenario that should be considered in the next generation high-speed telecommunication networks is multiple near simultaneous link failures where, for example, a new failure would happen before a failure is repaired in the network or more than one link are affected in the network by a disaster.

Survivability Hierarchy

There are different approaches for designing survivable optical mesh networks which are mainly based on protection and restoration architectures. The hierarchy of optical mesh transport network protection and restoration options is illustrated in Figure 1.

Figure 1. The survivability hierarchy in DWDM optical transport networks

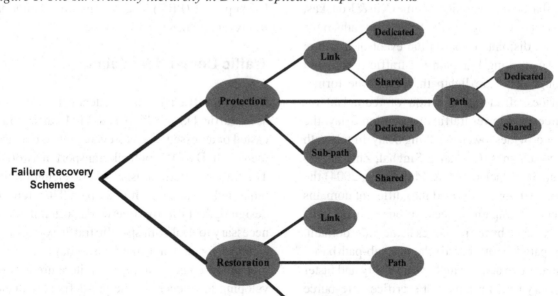

Protection and Restoration Architectures

Two main survivability architectures called protection and restoration provide the basis of the different schemes to the design of fault tolerant optical transport networks. Protection architectures (Ramamuthy & Mukherjee, 1999) establish working and spare lightpaths for arriving requests in advance, whereas restoration architectures (Ramamuthy & Mukherjee, 1999) determine spare lightpaths in the event of a failure based on current network conditions. Whilst the former is static and offers a faster restoration time, the latter is dynamic with greater bandwidth efficiency. Furthermore, the restoration architecture is more flexible against network configuration changes and may be able to recover from multiple link and node failures more easily.

Link, Path, and Sub-Path Survivability

Both protection and restoration can be further divided into link, path, and sub-path survivability schemes. In link-based survivability schemes, traffic is rerouted only around the failed link. In path-based survivability scheme (Xin & Rouskas, 2004) the working and the spare lightpaths must be link disjoint so that in the event of a failure on the working lightpath the traffic is rerouted through its spare lightpath. Whilst the former provides efficient utilization of redundant resources and lower traffic propagation delay, the latter provides lower switching delay. In sub-path based schemes (Ou, Zang, Singhal, Zhu, Sahasrabuddhe, MacDonald, & Mukherjee, 2004) the whole network is divided into different domains where a ligthpath segment in one domain must be protected by the resources in the same domain. Compared to path-based schemes, sub-path based schemes can achieve higher scalability and faster recovery time for a modest sacrifice in resource utilization (Anand, Chauhan, & Qiao, 2002; He, Wen, Li, et al., 2004).

Dedicated and Shared Survivability

Furthermore the survivability schemes can be classified into two major categories, dedicated and shared. In dedicated survivability schemes, backup resources along spare lightpaths are dedicated for that lightpath and cannot be utilized by other spare ligthpaths. However, in shared segment (Xu, Xiong & Qiao, 2003; Ho, Tapolcai & Cinkler, 2004) and shared path (Ho, Mouftah, 2004) schemes, redundant resources can be shared between several spare lightpaths. Dedicated survivability schemes are faster than shared survivability schemes, but shared schemes are more resource efficient. The problem of survivable lightpath provisioning using path sharing in optical mesh networks employing WDM has been investigated in (Ou, Zang, Zang, Sahasrabuddhe, & Mukherjee, 2004). Shared mesh WDM optical networks with partial wavelength conversion capability has been considered in (Ho, & Mouftah, 2004). In addition in (Ho, 2004) the state of the art progress in developing both shared path protection (SPP) and segment shared protection (SSP) including Shared Risk Link Group (SRLG) disjointness and three useful algorithms: Iterative Two Step Approach (ITSA), Potential Backup Cost (PBC), and Maximum Likelihood Relaxation (MLR) were surveyed.

Traffic Demand Modeling

Different traffic patterns of client networks such as IP traffic, DS-n, STS-n, and ATM traffic, forecasted or assessed, are aggregated into a demand matrix in DWDM optical transport networks. The traffic pattern consists of wavelength units requested between each node pair. Therefore to design optical transport networks, often it is not necessary to deal with specific traffic types, only a model for demand matrix is enough. Several models have been employed in literature for developing test benches. The pre-defined demand matrix is the more realistic traffic model for designing transport networks where the incoming traffic including the requesting node pairs and the volume of requested bandwidth is known by network planners during the network set up process before network starts to operate. In dynamic demand matrix, the node pairs and the requesting bandwidth in the demand matrix change during network operation that may have uniform, Poisson or exponential distributions. Random demand matrix may be generated for both the node pairs and the requesting bandwidth. The M/M/1 model is a heavy load traffic model, where every node generates lightpath requests to all other nodes in the network. In the gravity model, demands are generated from a mutual attraction effect proportional to node importance. The demand between a node pair is proportional to node degrees with an inverse distance dependency.

Routing and Wavelength Assignment (RWA) Problems

The routing and wavelength assignment (RWA) problem lies at the heart of designing the wavelength-routed DWDM optical transport networks (Ramaswami, & Sivarajan, 1995) where a variety of services including voice, data, and video, are multiplexed together and routed from origin to destination over a common infrastructure called *lightpaths* or *all- optical communication channels* without any optical-electronic-optical conversion and buffering at the intermediate nodes, known as wavelength routing. Establishment of a lightpath consists of choosing a route and assigning wavelengths. This problem is in general known to be NP-hard.

Static and Dynamic RWA

The RWA problem is called static where the demand matrix is known and static, and a set of lightpaths should be established between demand node pairs while the objective is network resource optimization such as minimizing the number of allocated wavelength channels. The static RWA

problem is employed in protection architectures. The RWA problem is dynamic when the traffic pattern is dynamic and changes during network operation, and the objective is network resource optimization such as minimizing the number of allocated wavelength channels and minimizing blocking ratio (Zang, Jue, & Mukherjee, (2000); Kennington, Olinick, Ortynski, & Spiride, 2003).

Solution Approaches

Heuristics and ILP Approaches

The integer linear programming (ILP) approach has been extensively employed for modeling and solving different survivability problems. The survey chapters (Ramamurthy, Sahasrabuddhe, & Mukherjee, 2000; Kenningtona, Olinick, & Spiride, 2006) and (Kennington, & Olinick, 2004) may be referred to for the body of work in the literature. The ILP approach works well for small problems with linear objective functions and constraints, and for small networks where the number of decision variables is moderate. However, the limitations of ILP models have led to the introduction of a diverse range of heuristic algorithms. In (Zang, Ou, & Mukherjee, 2003) the routing and wavelength assignment (RWA) problem in a network with path protection under duct-layer constraints was addressed. A heuristic algorithm for the design of survivable WDM networks was presented in (Zhang, Zhong, & Mukherjee, 2004). Survivability approaches for optical WDM mesh networks were considered in (Shenai, & Sivalingam, 2005). In (Liu, Tipper, & Siripongwutikorn, 2005) the routing of backup paths and spare capacity allocation in the network to guarantee seamless communication services under failure was addressed. The resource efficient provisioning solutions in WDM mesh networks, which achieved the objective of maximizing resource sharing, was presented in (Yang, Shen, & Ramamurthy, 2005).

The Intelligent Techniques

Among other successful approaches are approaches based on intelligent techniques especially genetic algorithms. A genetic algorithm (GA) is a stochastic global search method based on natural biological evolution and genetics (Goldberg, 1989) that may be employed to solve optimization problems that are not well suited for standard optimization algorithms, including those where the objective function is discontinuous or non differentiable. The complexity of problem, the abilities of genetic algorithms to search solutions for NP-hard problems in large-scale search space (De Jong, & Spears, 1989) and the success of genetic algorithms in designing communication networks (Konak, & Smith, 1999; Arabas, & Kozdrowski, 2001; He, Botham, & O'Shea, 2004; Ahn, & Ramakrishna, 2002) lead to genetic algorithm based models for solving RWA problems in designing the survivable DWDM optical transport networks. Encoding the working and spare lightpaths into chromosomes, creating next generation using crossover, mutation and selection operators, population initializing, developing termination rules and construction the fitness function are the main problems for implementation a GA approach for designing survivable DWDM optical transport networks (Kavian, Ren, Naderi, Leeson, Hines, 2008).

The K-Shortest Paths Problem

The K-shortest path search is often needed to design reliable communication networks (Wei, & Shong, 1999; Nikolopoulos, Pitsillides, & Tipper, 1997). The K-shortest path problem consists of the determination of a set of shortest paths $\Gamma_{od}(\kappa) = \{P_1,...,P_\kappa\}$ between a given pair of nodes (o,d). That is, not only the shortest path to be determined, but also are the second shortest, the third shortest, and so on up to the Kth shortest path. The best known algorithm with a complexity of $O(\kappa(m + n \log n))$ in an undirected graph

with n nodes and m links was proposed in (Lawer, 1972). While the best known complexity bound is $O(\kappa n(m + n \log n))$ in directed graphs (Katoh, Ibaraki, & Mine, 1982). The K-shortest paths algorithm as an extension of the Dijkstra shortest path algorithm can be found in (Mittal, & Mirchandani, 2004).

CURRENT AND FUTURE RESEARCH CHALLENGES

Multiple-Link Failure

Another failure scenario that should be considered in next generation high-speed telecommunication networks is multiple near simultaneous link failures where, for example, a new failure would happen before a failure is repaired in the network or more than one link are affected in the network caused by a disaster (Guo, Li, Cao, Yu, & Wei, 2007; Zhang, Zhu, & Mukherjee, 2004; Guo, Yu, & Li, 2005). To deal with the case of M-simultaneous failures, the network topology should be (M+1)-connected. For example in case of a 2-link failure, the network should be 3-connected that means it needs to establish three link-disjoint paths between each node pair. As shown in Figure 2, during the design of a survivable network, in addition to setting up a working lightpath, (4-6), between each source and destination node pair, two link-disjoint spare lightpaths, (4-5-6), (4-3-2-1-6), need to be established between the node pair, too. The diverse spare ligthpaths are employed as the means to restore the traffic demand when two links are affected in the network.

QoS Based RWA

The routing and wavelength algorithms provide lightpaths that optimizes the network resource utilization; however they may not be suitable to accommodate the QoS requirements (Autenrieth,

Figure 2. A survivable DWDM network with ability of recovering from double near simultaneous link failures

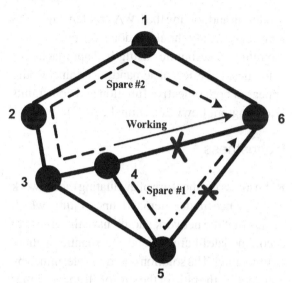

Figure 3. Resource optimization and QoS-based routing and wavelength assignment

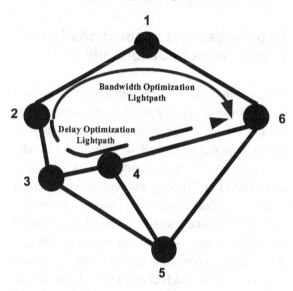

& Kirstadter, 2002; Bejerano, Breitbart, Orda, Rastogi, & Sprintson, 2005; Kavian, Rashvand, Ren, Leeson, Hines, Naderi, 2007). The effect of QoS requirements, for example, propagation delay optimization, on solutions of RWA problems is compared with resource (bandwidth) optimization in the following example. The topology of the network is shown in Figure 3 that includes 6 nodes and 8 links. Table 1 describes the weight of connecting links in fiber kilometers. The demand is a connection request between node pair (2, 6) for 3 units of wavelength bandwidth. The results are tabulated in Table 2. The adjusted delay on each link was assumed to be 1 ms/km. The resource reservation based RWA tried to establish bandwidth-efficient lightpath where the QoS based RWA found a ligthpath with low latency. In resource optimization based RWA the number of allocated wavelengths are 6 wavelengths where the propagation delay of the request is 24ms. In QoS based RWA the number of allocated wavelengths are 9 wavelengths where the propagation delay of the request is 17ms. Consequently, the QoS based RWA tries to establish a lightpath with lower

Table 1. The weight of connecting links in fiber kilometers

Link	Length (km)
(1,2)	12
(1,6)	12
(2,3)	4
(3,4)	3
(3,5)	8
(4,5)	7
(4,6)	10
(5,6)	14

Table 2. Resource optimization and QoS-based RWA

Objective	Lightpath Path/Wavelength	Lightpath Properties (Bandwidth, Delay)
Bandwidth	2-1-6/($\lambda_1, \lambda_2, \lambda_3$)	(6,24)
Propagation Delay	2-3-4-6/($\lambda_1, \lambda_2, \lambda_3$)	(9,17)

latency which may not be as resource efficient compared with QoS oblivious RWA.

Robustness and Accommodation of Future Demand Uncertainty

Rising Internet usage is producing increased traffic uncertainty that must be accommodated in optical core design optimization. The delivery of the required quality of service (QoS) by client networks such as Internet network providers faces uncertain capacity problems where demand volume changes extensively over different periods of network operation. Therefore, existing network capacity is unable to accommodate the demands at all times due to the natural variability of unpredictable loads in periods of high demand. Planning for robust DWDM transport networks capable of meeting future uncertain demands is, therefore, a necessity.

Demand Uncertainty Modeling

Modeling of future demand uncertainty has a big influence on network planning. It is extremely difficult, however, to accurately predict future demand volumes and patterns at the optical layer in uncertain environments. Therefore, different models of demand uncertainty are provided in the literature to design robust transport networks, for example, a set of different scenarios with known probabilities (Leung, & Grover, 2005; Kennington, Lewis, & Olinick, 2003), random variables with known probability distributions (Ukkusuri, Mathew, & Waller, 2007), and sampled scenarios from a normally distributed interval (Atamturk, & Zhang, 2006) that all are based on a set of k discrete future demand scenarios, D_k, and their probability distributions, p_k,;

$$\Im = \{(D_0, p_0), ..., (D_k, p_k)\}.$$

RWA Problem under Demand Uncertainty

Modeling and solving the RWA problem under demand uncertainty should be done to accommodate the future uncertainties using ILP approaches or heuristics. The effect of demand uncertainty would appear in both objective function and constraints (Kennington, Lewis, & Olinick, 2003).

Robustness

Robustness is a measure of evaluating the network performance under demand uncertainty which means that the network should have the ability to accommodate future demand uncertainties without any blocking. The solution could be determined by considering the ratio of the sum of all uncertainties to the sum of the total demands.

Penalty Cost Rules

The penalty cost ratio can be utilized to capture the essential point that unused capacity is expensive for reserving bandwidth in advance by QoS sensitive applications.

Single-Stage and Two-Stage Approaches

Accommodating future demand uncertainty could be achieved by single-stage (SS) or two-stage (TS) schemes. In SS schemes, both certain and uncertain parts of network demands are considered during lightpaths establishment. While in TS schemes, first the network wavelengths are allocated by solving the RWA problems for certain part of network demands then the uncertain future demands are introduced into the RWA problem. The TS schemes are more efficient than the SS schemes in bandwidth tradeoff (Atamturk, & Zhang, 2006).

Differentiated Quality of Service (QoS) Issues

Differentiated quality of service (QoS) refers to the ability of a network to enforce preferential treatment through classification. The QoS of a network can be studied in the context of any combination of properties of a service delivered to the end users including availability, delay, jitter, loss ratio and throughput (Menasce, 2002; Xipeng, Ni, 1999).

Survivability of DWDM networks improves the client network's QoS by tolerance against component failures and avoiding losing data and revenue. The survivability parameters such as restoration time, restoration bandwidth, restoration granularity, and call acceptance ratio, will affect the QoS delivered. The QoS at the optical layer can be differentiated based on differentiated reliability, quality of protection, and quality of recovery (Kaheel, Khattab, Mohamed, & Alnuweiri, 2002). The important role of quality of service (QoS) in deployment of a resilient DWDM backbone for global networks requires critical design-phase planning optimization. In general, QoS requirement extends network resource utilization that is assured through bandwidth tradeoff (Saradhi, Gurusamy, & Zhou, 2004).

CONCLUSION

Survivability is an important and critical issue that must be taken into consideration in the deployment of DWDM optical transport networks as backbones for reliable high-speed next generation telecommunication networks due to their ability to carry large volumes of traffic. This chapter reviewed some major issues of designing fault tolerant DWDM optical transport networks for service provisioning at acceptable level of performance in the presence of failures. Furthermore, the chapter addressed some current and future research challenges including covering multiple simultaneous failures, QoS-based RWA, robustness and future demand uncertainty accommodation, and quality of service issues in the deployment of resilient DWDM backbones for next generation telecommunication networks.

REFERENCES

Ahn, C. W., & Ramakrishna, R. S. (2002). A genetic algorithm for shortest path routing problem and the sizing of populations. *IEEE Transactions on Evolutionary Computation*, 6(6), 566–579. doi:10.1109/TEVC.2002.804323

Anand, V., Chauhan, S., & Qiao, C. (2002). *Subpath protection: A new framework for optical layer survivability and its quantitative evaluation.* Dept. CSE, SUNY Buffalo, Tech. Rep. 2002-01.

Arabas, J., & Kozdrowski, S. (2001). Applying an evolutionary algorithm to telecommunication network design. *IEEE Transactions on Evolutionary Computation*, 5(4), 309–323. doi:10.1109/4235.942526

Atamturk, A., & Zhang, M. (2006). Two-stage robust network flow and design under demand uncertainty. *Operations Research*, 55(4).

Autenrieth, A., & Kirstadter, A. (2002). Engineering end-to-end IP resilience using resilience-differentiated QoS. *IEEE Communications Magazine*, 40(1), 50–57. doi:10.1109/35.978049

Bejerano, Y., Breitbart, Y., Orda, A., Rastogi, R., & Sprintson, A. (2005). Algorithms for computing QoS paths with restoration. *IEEE/ACM Transactions on Networking*, 13(3), 648–661. doi:10.1109/TNET.2005.850217

De Jong, K. A., & Spears, W. M. (1989). Using genetic algorithm to solve NP-complete problems. *Proceeding of Third International Conference on Genetic Algorithms (ICGA).*

Gerstel, O., & Ramasawami, R. (2000). Optical layer survivability- An implementation perspective. *IEEE Journal on Selected Areas in Communications, 18,* 1885–1899. doi:10.1109/49.887910

Ghani, N., Dixit, S., & Wang, T. S. (2003). On IP-WDM integration: A retrospective. *IEEE Communications Magazine,* 42–45. doi:10.1109/MCOM.2003.1232232

Goldberg, D. E. (1989). *Genetic algorithms in search, optimization, and machine learning.* Addison-Wesley.

Guo, L., Li, L., Cao, J., Yu, H., & Wei, X. (2007). On finding feasible solutions with shared backup resources for surviving double-link failures in path-protected WDM mesh networks. *Journal of Lightwave Technology, 25*(25), 287–296. doi:10.1109/JLT.2006.886721

Guo, L., Yu, H., & Li, L. (2005). Protection design for double-link failures in meshed WDM networks. *Acta Eletronica Sinica, 33*(5), 883–888.

He, L., Botham, C. P., & O'Shea, C. D. (2004). An evolutionary design algorithm for ring-based SDH optical core networks. *BT Technology Journal, 22,* 135–144. doi:10.1023/B:BTTJ.0000015503.09815.48

He, R., Wen, H., & Li, L. (2004). Shared sub-path protection algorithm in traffic-grooming WDM mesh networks. *Photonic Network Communications, 8*(3), 239–249. doi:10.1023/B:PNET.0000041236.17592.b4

Ho, P.-H. (2004). State-of-the-art progress in developing survivable routing schemes in mesh WDM networks. *IEEE Communications, 6*(4), 2–16.

Ho, P.-H., & Mouftah, H. T. (2004). A novel survivable routing algorithm for shared segment protection in mesh WDM networks with partial wavelength conversion. *IEEE Journal on Selected Areas in Communications, 22*(8), 1548–1560. doi:10.1109/JSAC.2004.830475

Ho, P.-H., & Mouftah, H. T. (2004). On optimal diverse routing for shared protection in mesh WDM networks. *IEEE Transactions on Reliability, 53*(6), 216–225. doi:10.1109/TR.2004.829141

Ho, P.-H., Tapolcai, J., & Cinkler, T. (2004). Segment shared protection in mesh communications networks with bandwidth guaranteed tunnels. *IEEE/ACM Transactions on Networking, 12*(6), 1105–1118. doi:10.1109/TNET.2004.838592

Kaheel, A., Khattab, T., Mohamed, A., & Alnuweiri, H. (2002). Quality-of-service mechanisms in IP-over-WDM networks. *IEEE Communications Magazine, 40*(12), 38–43. doi:10.1109/MCOM.2002.1106157

Kartalopoulos, S. V. (2000). *Introduction to DWDM technology: Data in a rainbow.* Piscataway, NJ: IEEE Press.

Katoh, K., Ibaraki, T., & Mine, H. (1982). An efficient algorithm for K shortest simple paths. *Networks, 12,* 411–427. doi:10.1002/net.3230120406

Kavian, Y. S., Rashvand, H. F., Ren, W., Leeson, M. S., Hines, E. L., & Naderi, M. (2007). RWA problem for designing DWDM networks delay against capacity optimization. *Electronics Letters, 43*(16), 892–893. doi:10.1049/el:20071219

Kavian, Y. S., Ren, W., Naderi, M., Leeson, M. S., & Hines, E. L. (2008). Survivable wavelength-routed optical network design using genetic algorithms. *European Transaction on Telecommunication, 19*(3), 247–255. doi:10.1002/ett.1263

Kennington, J., & Olinick, E. (2004). *A survey of mathematical programming models for mesh-based survivable networks,* (pp. 1-33). Southern Methodist University, Technical Report 04-EMIS-13.

Kennington, J., Olinick, E., Ortynski, A., & Spiride, G. (2003). Wavelength routing and assignment in a survivable WDM mesh network. *Operations Research, 51*(1), 67–79. doi:10.1287/opre.51.1.67.12805

Kennington, J. L., Lewis, K. R., & Olinick, E. V. (2003). Robust solutions for the WDM routing and provisioning problem: Models and algorithms. *Optical Networks Magazine, 4*(2), 74–84.

Kenningtona, J. L., Olinick, E. V., & Spiride, G. (2006). Basic mathematical programming models for capacity allocation in mesh-based survivable networks. *International Journals of Management Sciences, 35*(6), 1–16.

Konak, A., & Smith, A. E. (1999). A hybrid genetic algorithm approach for backbone design of communication networks. *Proceedings of the Congress on Evolutionary Computation,* Vol. 1, (pp. 1817-1823).

Lawer, E. L. (1972). A procedure for computing the K best solutions to discrete optimization problems and its application to the shortest path problem. *Management Science, 18,* 401–405. doi:10.1287/mnsc.18.7.401

Leung, D., & Grover, W. D. (2005). Capacity planning of survivable mesh-based transport networks under demand uncertainty. *Photonic Network Communications, 10*(2), 123–140. doi:10.1007/s11107-005-2479-z

Liu, Y., Tipper, D., & Siripongwutikorn, P. (2005). Approximating optimal spare capacity allocation by successive survivable routing. *IEEE/ACM Transactions on Networking, 13*(1), 198–211. doi:10.1109/TNET.2004.842220

M'Edard, M., & Lumetta, S. S. (2002). *Network reliability and fault tolerance. Tech. Report., MIT.* Laboratory for Information and Decision Systems.

Menasce, D. A. (2002). QoS issues in web services. *IEEE Internet Computing, 6*(6), 72–75. doi:10.1109/MIC.2002.1067740

Mittal, S., & Mirchandani, P. (2004). *Implementation of K-Shortest path Dijkstra algorithm used in all-optical data communication networks.* SIE 546 Project Report.

Mukherjee, B. (2000). WDM optical communication networks: Progress and challenges. *IEEE Journal on Selected Areas in Communications, 18*(10), 1810–1824. doi:10.1109/49.887904

Murthy, C. S. R., & Gurusamy, M. (2004). *WDM optical networks: Concepts, design, and algorithms.* Prentice Hall.

Nikolopoulos, S. D., Pitsillides, A., & Tipper, D. (1997). Addressing network survivability issues by finding the K-best paths through a trellis graph. *Proceeding of IEEE Infocom,* Kobe.

Ou, C., Zang, H., Singhal, N. K., Zhu, K., Sahasrabuddhe, L. H., MacDonald, R. A., & Mukherjee, B. (2004). Subpath protection for scalability and fast recovery in optical WDM mesh networks. *IEEE Journal on Selected Areas in Communications, 22*(9), 1859–1875. doi:10.1109/JSAC.2004.830280

Ou, C., Zang, J., Zang, H., Sahasrabuddhe, L. H., & Mukherjee, B. (2004). New and improved approaches for shared-path protection in WDM mesh networks. *Journal of Lightwave Technology, 22*(5), 1223–1232. doi:10.1109/JLT.2004.825346

Rajagolapan, B. (2000). IP over optical networks: Architectural aspects. *IEEE Communications Magazine,* 94–102. doi:10.1109/35.868148

Ramamurthy, S., Sahasrabuddhe, L., & Mukherjee, B. (2000). Survivable WDM mesh networks. *Journal of Lightwave Technology, 21*(4), 870–883. doi:10.1109/JLT.2002.806338

Ramamuthy, S., & Mukherjee, B. (1999). Survivable WDM mesh networks: Part I-Protection. *Proceeding of IEEE Infocom Conference,* Vol. 2, (pp. 744-751).

Ramamuthy, S., & Mukherjee, B. (1999). Survivable WDM mesh networks: Part II-Restoration. *Proceeding of IEEE ICC Conference,* (pp. 2023-2030).

Ramaswami, R., & Sivarajan, K. N. (1995). Routing and wavelength assignment in all-optical networks. *IEEE/ACM Transactions on Networking, 3*, 489–500. doi:10.1109/90.469957

Saradhi, C. V., Gurusamy, M., & Zhou, L. (2004). Differentiated QoS for survivable WDM optical networks. *IEEE Communications Magazine, 42*(5), s8–s14. doi:10.1109/MCOM.2004.1299335

Shenai, R., & Sivalingam, K. (2005). Hybrid survivability approaches for optical WDM mesh networks. *Journal of Lightwave Technology, 23*(10), 3046–3055. doi:10.1109/JLT.2005.856273

Ukkusuri, S. V., Mathew, T. V., & Waller, S. T. (2007). Robust transportation network design under demand uncertainty. *Computer-Aided Civil and Infrastructure Engineering, 22*, 6–18. doi:10.1111/j.1467-8667.2006.00465.x

Wang, Y., Cheng, T. H., & Mukherjee, B. (2003). Dynamic routing and wavelength assignment scheme for protection against node failure. *Proceeding of IEEE GLOBECOM'03*, (pp. 2585-2589).

Wei, S., & Shong, C. (1999). A K-best paths algorithm for highly reliable communication networks. *LEICE Transaction on Communication. E (Norwalk, Conn.), 82-B*(4), 586–590.

Xin, Y., & Rouskas, G. N. (2004). A study of path protection in large-scale optical networks. *Photonic Network Communications, 7*(3), 267–278. doi:10.1023/B:PNET.0000026891.50610.48

Xipeng, X., & Ni, L. M. (1999). Internet QoS: A big picture. *IEEE Network, 13*(2), 8–18. doi:10.1109/65.768484

Xu, D., Xiong, Y., & Qiao, C. (2003). Novel algorithms for shared segment protection. *IEEE Journal on Selected Areas in Communications, 21*(8), 1320–1331. doi:10.1109/JSAC.2003.816624

Yang, X., Shen, L., & Ramamurthy, B. (2005). Survivable lightpath provisioning in WDM mesh networks under shared path protection and signal quality constraints. *Journal of Lightwave Technology, 23*(4), 1556–1564. doi:10.1109/JLT.2005.844495

Zang, H., Jue, J., & Mukherjee, B. (2000). A review of routing and wavelength assignment approaches for wavelength-routed optical WDM networks. *SPIE Optical Networks Magazine, 1*(1), 47–60.

Zang, H., Ou, C., & Mukherjee, B. (2003). Path-protection routing and wavelength-assignment (RWA) in WDM mesh networks under duct-layer constraints. *IEEE/ACM Transactions on Networking, 11*, 248–258. doi:10.1109/TNET.2003.810313

Zhang, J., & Mukherjee, B. (2004). Review of fault management in WDM mesh networks: Basic concepts and research challenges. *IEEE Network, 18*, 41–48. doi:10.1109/MNET.2004.1276610

Zhang, J., Zhu, K., & Mukherjee, B. (2004). A comprehensive study on backup reprovisioning to remedy the effect of double-link failures in WDM mesh networks. *Proceeding of IEEE ICC*, Vol. 27, (pp. 1654–1658).

Zhang, J., Zhu, K., Yoo, S. J. B., & Mukherjee, B. (2003). On the study of routing and wavelength assignment approaches for survivable wavelength-routed WDM mesh networks. *Optical Networks Magazine*, 16-27.

Zhang, Z. R., Zhong, W. D., & Mukherjee, B. (2004). A heuristic algorithm for design of survivable WDM networks. *IEEE Communications Letters, 8*, 467–469. doi:10.1109/LCOMM.2004.832772

Zhou, D., & Subramaniam, S. (2000). Survivability in optical networks. *IEEE Network, 14*, 16–23. doi:10.1109/65.885666

Chapter 9
Radio over Fiber Access Networks for Broadband Wireless Communications

Joaquín Beas
Tecnológico de Monterrey, México

Ivan Aldaya
Tecnológico de Monterrey, México

Gerardo Castañón
Tecnológico de Monterrey, México

Gabriel Campuzano
Tecnológico de Monterrey, México

Alejandro Aragón-Zavala
Tecnológico de Monterrey, México

ABSTRACT

In recent years, considerable attention has been devoted to the merging of Radio over Fiber (RoF) technologies with millimeter-wave-band signal distribution. This type of system has great potential to support secure, cost-effective coverage and high-capacity vehicular/mobile/wireless access for the future provisioning of broadband, interactive, and multimedia services. In this chapter, the authors present an overview of an RoF access networks in the context of in-vehicle networks, with special attention to the figures of merit of the system and the basic enabling technologies for downlink/uplink transmission in the RoF land network, which is divided in three main subsystems: Central Station (CS), Optical Distribution Network (ODN) and Base Station (BS). The chapter first reviews the up-conversion techniques from baseband to mm-waves at the CS, and then the different BS configurations. The work finally applies these concepts to the development of an access network proposal for in-vehicle wireless application.

INTRODUCTION

During the past two decades, the personal communications industry has faced an impressive growth in the number of subscribers worldwide, which make use of the different wireless access

DOI: 10.4018/978-1-4666-2976-9.ch009

services and applications offered by service providers. In the early days of mobile radio only voice was demanded, with small amounts of data, which could be sent over the voice oriented channel. This requirement has evolved over the years, as large volumes of information need to be sent from source to destination, even with the additional requirement of on-line connectivity. The objective of mobile broadband wireless access, that aims to provide the above mentioned aspects, has been addressed by the IEEE standard (802.15, Wireless Personal Area Networks) (802.16) and by the ITU evolving standard International Mobile Telecommunication- Advanced (ITU-R). The need for efficient transport systems, which could support such bandwidth requirements, are encouraged to exploit the advantages of both, optical fibers and millimeter-wave (mm-wave) frequencies (Cochrane, 1989). Such systems that use an optical distribution network (ODN) for delivering mm-wave radio signals from a Central Station (CS) to many remote Base Stations (BSs) have long been recognized to allow the increase of user capacity, bandwidth, and mobility (Ogawa H., et al., 1992).

A typical RoF system is shown in Figure 1. There is a CS that contains all the data resources, optical transmitters (TX - lasers), optical receivers (RX – photodetectors), interconnection with the Trunk Network, Internet and switching capacity of the network. In the downlink direction the CS up-converts the electrical signal to optical frequency and uses the ODN to communicate with the BSs, in some cases using Remote Nodes (RN) where the optical signal is split or demultiplexed towards the corresponding BS which converts it back to

electrical domain and radiates it to the Mobile Terminal (MT) end-user in mm-wave bands (Lim, et al., 2010). In the uplink direction, the BS receives the mm-wave signal from the MT, and depending on the configuration of the BS, this signal can be down-converted to baseband before modulating an electro-optical device to transmit the uplink information via the ODN back to the CS.

RoF communication systems have several advantages over conventional wireless systems, which are summarized as follows:

- The distribution of mm-wave frequency signals by the use of optical fibers reduces the attenuation, hence reducing the use of repeaters (amplifiers). This is possible since optical fibers have almost negligible insertion loss for very long distances, something that cannot be achieved in conventional wireless links, where free space loss sets the minimum possible path loss in the link.

- The architecture of an RoF system is often simpler and less expensive since it makes use of the concept of a CS, dealing with signal generation and resource management; and remote BSs can be deployed wherever they are required for wireless access, con-

Figure 1. General RoF system architecture, where central station (CS), optical distribution network (ODN), and base stations (BS) are presented

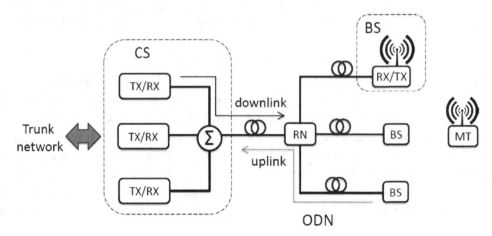

sisting only of an optical-to-electrical converter, RF amplifiers and antennas.

- Low cost expandability can be easily achieved using RoF, as it is virtually modulation format agnostic.

- Higher data rates to accommodate future service demands can be transported through RoF systems (larger bandwidth). This is often a severe constraint in a wireless channel, where the use of sophisticated modulation and coding techniques are required to improve channel capacity, and even though, these data rates are no way comparable to those achieved in optical fiber links.

- Multi-operator and multi-service operation is possible, whereas the same RoF network can be used to distribute traffic from many operators and services, resulting in substantial economic savings.

- Dynamic resource allocation, since functions such as switching, modulation and others are performed at a CS, being able to allocate capacity dynamically.

RoF systems are the subject of many research and development activities, as well as part of commercial products that are currently in use. However, these products are only available for wireless networks operating in the unlicensed bands from 2.4 to 5.8 GHz (Suman Sarkar, Sudhir Dixit, 2007) and they are still unable to supply data rates comparable to wired standards such as gigabit Ethernet and high-definition multimedia interface (HDMI). Thus, the development of high-speed indoor/outdoor wireless access communication systems are expected to operate at higher frequency bands in order to reach higher data rates (over 2 Gbps) (802.15, 2011). Different mm-wave bands have been proposed for high-capacity wireless systems employing RoF. However, the frequency band that has attracted major interest is around 60 GHz, mainly because of two reasons: 1) this frequency coincides with

the oxygen absorption peak, which results in high atmospheric attenuation exceeding 15 dB/Km. High attenuation allows reduction in the cell size and frequency reuse distance in cellular systems, increasing the wireless system capacity. 2) the 60 GHz band offers an unlicensed large frequency range worldwide with the ability to support fairly high data rates: 57 - 64 GHz in North America, 59 - 66 GHz in Japan and Korea, and the 57 - 66 GHz range is under consideration by the European Union (Daniels, R. C., & Heath, W., 2007), which opens the opportunity for worldwide standardization and commercial products.

However, one of the major drawbacks of an RoF system, as wireless frequency increases, is the requirement of mm-wave band optical components at the ODN, large bandwidth photodetectors (PDs) at the BS RXs, and more sophisticated mm-wave up-conversion techniques at the CS. The latter is defined as the process to convert electrical data signal at the CS from baseband to optical domain for transmission through ODN and remote mm-wave generation at the BS. These requirements have led to the development of optical components operating in the mm-wave (Lee, 2007), (Seeds, A.J. & Williams, J., 2006), (Stöhr,et al., 2010), techniques for generation and transmission of mm-wave radio signals over the optical fiber (Yao, 2009) (Jia Z., et al., 2007) and system architectures where functions such as signal routing, processing, handover, and frequency allocation are carried out at CS, rather than at the BS (Pleros, et al., 2009). On the other hand, this type of network demands a large number of BSs to cover a service area. Thus, a stable and cost effective BS simplified to just wireless radiation/reception and electrical/optical conversion should be designed with simple and effective architecture in order to succeed in the market place.

Other challenges for an RoF system operating at high frequencies are: the management of larger number of handovers, because of the reduced cell size, and connectivity issues related to non line-of-sight transmissions between the MT and the BS. Although it might be interesting from an overall availability figure of merit point of view, this chapter does not contemplate the wireless section of the RoF network (Castañón, et al., 2008).

FIGURES OF MERIT IN ROF SYSTEMS

How to assess the RoF system performance is not a trivial issue. It depends on many figures of merit of the different subsystems: CS, ODN, and BS. In some cases, each subsystem can be independently analyzed and the choice of one scheme over another does not affect the other subsystems; however in most cases each part cannot be optimized independently but the whole system should be considered at a time. Because of this, we decided to present figures of merit for the whole system instead of presenting them separately for each subsystem.

Cost and Simplicity

The designed RoF system should satisfy the service requirements at the minimum cost. An integral approach is required, minimizing the overall system cost. Regarding the downlink transmission, it is desirable to avoid high frequency optical modulators and electrical oscillators, as expensive devices are required for achieving low noise mm-wave systems (Olver, 1990) and, in addition, their power consumption increases with the operation frequency for semiconductor made devices (Teshirogi, T., & Yoneyama, T., 2001). On the other hand, in the optical domain, it is important to reduce the number of optical components and to use standard devices as well as to minimize the complexity of the mm-wave generation scheme, since a simpler implementation reduces both, the manufacturing and maintenance costs. The main cost related to the ODN in an RoF system is derived from the fiber installation, more than from the

component themselves. Thus, integrating optical segments into hybrid solutions (fiber sharing) for the next generation of ODNs is a key factor as it enables cost reductions. However, it is important to minimize fiber length while keeping in mind that network topologies with the minimum fiber often offer poor availability performance (Castañón, G., Campuzano, G., & Tonguz, O., 2008). For the BS, as the frequency increases, the number of BSs increases. Therefore, the BS should be kept as simple as possible delegating modulation and control functions to the CS. Simplicity of the BSs drives the reduction of costs associated with site acquisition, site leasing, and energy consumption.

mm-Wave Frequency and Bandwidth

This figure of merit is related to the capacity of the RoF system to generate an mm-wave signal at the desired frequency with the required bandwidth. In this context, each stage has to be designed to fulfill the minimum bandwidth requirements. In some up-conversion techniques, the chosen operation frequency (the central frequency) and the bandwidth are not independent from each other. This is the case, for example, when a single direct modulated laser is used (Hartmann, et al., 2005). The maximum modulation frequency is then limited by the laser's physics and increasing the operation frequency results in a decrease in the available capacity. In contrast, in all-optical up-conversion techniques (which will be introduced in the next subsection), the operation frequency is virtually independent of capacity (Braun, R. P. & Grosskopf, G, 1998), (Insua, et al., 2010). In the ODN, the devices should be chosen to work properly at the operation optical wavelength (preferably according to standardized grids) and with the required bandwidth. This is especially important in the case of the optical filters and optical wavelength multiplexers when they are required. The effect of the fiber dispersion also depends on the bandwidth since as the signal bandwidth increases the effect of the chromatic dispersion is more significant (Kitayama, et al., 2002). The maximum operation mm-wave frequency of the RX at the BS depends on the PD installed (Kato, 1999), while the bandwidth is limited by the electronic components, the band selection filter and the RF amplifier.

RF Output Power and Signal Power

RF output power accounts for the total power available at the output of the PD at the BS (Ito, et al., 2004), which may comprise a single or double modulation sideband and an mm-wave carrier. Generally, the generated mm-wave signal radiated by the BS consists of a single or double modulated sidebands accompanied by an mm-carrier. The information is conveyed in the sidebands while the carrier can be used as a local oscillator (LO) in self-homodyne receiver at MT or as an envelope offset in envelope detection systems. These are the two alternatives proposed to avoid high frequency oscillators in the MT side (Insua, et al., 2010). Even if both the maximum RF output and signal power depend on the whole RoF architecture, the main limiting factor is different whether the ODN is passive or active. In the case of passive ODN, the maximum RF output power is limited by the optical power at the input of the PD, which further depends on the optical output power at the CS and the losses in the ODN. In contrast, in active ODN networks, the main limitation frequently arises from the maximum output power of the PD (Braun, 1998). A common criterion to assess it at a particular frequency is the 1 dB compression point, that is, the power at which the transfer characteristic drops by 1 dB with respect to the ideal linear transfer function. It is worthwhile nothing that the power relation between the carrier and the signal is determined at the CS and remains almost unaltered through the ODN and the BS. All these issues need to be taken into account in the link budget ensuring the power requirement for the air interface is satisfied.

Spectral-Purity and Frequency Accuracy of mm-Wave Carrier

Both spectral-purity and frequency accuracy are figures of merit of the mm-wave carrier quality and are related to the phase stability: the former accounts for short-term phase fluctuations while the latter is related to the long-term stability (Cliche, et al., 2006). For spectral purity or linewidth, single side band phase noise measurement is the best indicator of performance. Phase noise is typically expressed in units of dBc/Hz, representing the noise power relative to the carrier contained in a 1 Hz bandwidth centered at a certain offsets from the carrier.

The frequency accuracy of the generated mm-wave is the degree of conformity to a specified frequency value; it can be expressed as a frequency range or in fractional units (i.e. ppb – parts per billion). Their effects on the signal detected by the MT are highly dependent on how the detection is implemented. If envelope detection is used, phase noise will not degrade the signal as much as if heterodyning is used (Insua, et al., 2010). However, independently on the detection technique, it is important to avoid an excessive frequency deviation, since high frequency excursion tolerance results in larger guard-bands and poorer spectral efficiency.

Modulation Signal Quality

The modulation signal quality can be evaluated using Bit Error Rate (BER) or Error Vector Magnitude (EVM) metrics. The BER is defined as the fractional number of errors in a transmitted sequence, and the EVM is the average of the magnitude difference between the reference and the received constellation at a specific point in the system, it is more significant in complex modulation formats. Generally, these metrics are plotted against the received power in order to be able to calculate the system performance and the sensitivity for a given BER/EVM threshold (Hassun R., et al., 1997).

The main mechanisms that affect the signal quality can be divided in non-linearities, noise and undesired phase modulation or chirp. As a general rule, non-linearities limit the maximum generated, transported or received power, while noise establishes the minimum power boundary. Nonlinear effects in the up-conversion are caused by the nonlinear behavior of the laser if direct modulation is used or by the Mach-Zhender or electro-absorption modulator non-linearities, if external modulation is used. In addition, direct modulation introduces higher noise power and stronger frequency modulation than external modulation. In the ODN, nonlinear effects arise from the fiber (Agrawal, 2001) and, in active ODN, optical amplifier gain compression (Li, Z., et al., 2005), (Kamra, V., & Kumar, V., 2011). Regarding noise, the only noise source within the ODN are the optical amplifiers. In the BS, both the downlink and uplink suffer the non-linearities of the PD and laser, respectively. Actually, one of the key figures of merit of the PDs is the compression power since the higher the photogenerated power, the lower is the required electrical amplification (Ito, et al., 2004).

Scalability

Scalability is defined as the capability of the RoF system to increase coverage, capacity and services within the network to meet usage demand without a significant cost increase.

Aiming at minimizing the system's cost, a large number of low cost BSs should be fed with highly equipped CS and minimal amount of optical components in the ODN (Kitayama, 1998). Therefore, different BSs are fed with the same fiber where different signals are multiplexed. In this way, CS, ODN, and BS design impact significantly on the overall system scalability. Up-conversion techniques using multi wavelength light sources exhibit better capabilities to escalate than others using single wavelength (Nakasyotani, et al., 2006), (Chang, C. H., et al., 2009), (Kim, et al.,

2009). The ODN and BSs need to be designed according to the multiplexing scheme. Several multiplexing possibilities are available as shown in Figure 2, even more if we consider that uplink and downlink do not require the use of the same multiplexing scheme. It is likely that the downlink will use any kind of frequency multiplexing since this simplifies the ODN and makes possible for each channel to be generated independently in the CS. There are three different approaches for frequency multiplexing:

- **Subcarrier Multiplexing (SCM):** Multiplexing is done in the electrical domain, previous to optical up-conversion; in this way signals for different BSs are carried in a single wavelength using frequency division multiplexing as depicted in Figure 2(a). The expense of the up-conversion is shared among different BSs at the cost of higher non-linearities and lower signal power delivered to each BS (Ghafoor, S., & Hanzo, L., 2011).
- **Wavelength Division Multiplexing (WDM):** in contrast to SCM, different signals are transported in different wavelengths as presented in Figure 2(b). This is generally achieved by using a dedicated laser for each channel (Li, et al., 2007); nevertheless, it is possible to use a single source to generate signals at different wavelengths (Toda, H., at al., 2004).
- **Combined SCM and WDM:** Different wavelengths are modulated by subcarrier signals, each for a different BS as shown in Figure 2(c). (Lim, et al., 2003).

In terms of the uplink, Time Division Multiplexing (TDM) has to be added, as well as its combinations (Chang, G., et al., 2009). The uplink multiplexing technique impacts primarily on the BS design. WDM is almost compulsory but within each wavelength SCM or TDM can be implemented. The former requires tunable

Figure 2. Frequency multiplexing techniques: (a) subcarrier multiplexing (SCM), (b) wavelength division multiplexing (WDM), (c) combined SCM and WDM

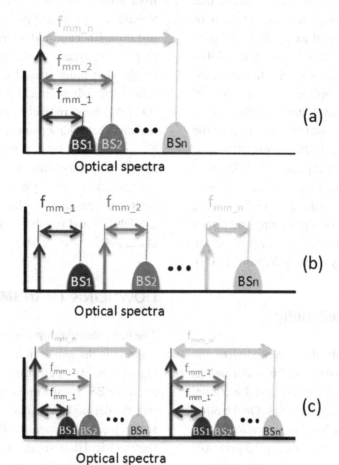

oscillators, at relatively low frequencies, the latter precise time synchronization of the BSs.

Power Consumption

Considerable effort has been dedicated to the development of green, energy-saving communication networks (Baliga, et al., 2008). The power consumption of a network has become an important figure of merit, which can be expressed by the power consumption per user versus the average access rate (Watts/Mbps). Another useful measure of the energy efficiency of a network is the energy consumed per bit of data transferred (Joules per bit).

In an RoF system, a significant reduction in the power consumption needs to be done at each stage (Baliga, et al., 2009). The power efficiency in the up-conversion process should account for both electrical and optical components. Regarding the first, early implementations were based on III-V compound materials due to their better noise and power consumption performance, at the cost of higher price and lower digital integration. In recent years, silicon made complementary metal oxide semiconductor (CMOS) devices have been reported with progressive lower consumption. However, the electronics at mm-wave band are still quite power hungry (Daniels, R. C., & Heath, W., 2007), (Oualkadi, 2011). As for optical components, the power efficiency depends on many factors. It is expected, for example that the consumption increases with the number of components, in this way, if external modulation is used, the consumption is likely to be higher than for directly modulated lasers. In the ODN, the only power consumption is at the optical amplifiers when required (Potenza, 1996). Then, to optimize the network and the selection of the components is important if the consumption is to be kept low. Power consumption of the BSs is of special importance since they are many of them. Some work on how to make green BS is being already done focusing on optimizing the power consumption of the RF amplifier or feeding the BS with reusable energy sources (Schmitt, 2009), (Chen, et al., 2011).

Availability and Reliability

Network reliability indicates the quality of the network to transport traffic, and it is defined as the probability that there exists one functional path between a given pair of entities. On the other hand, availability is the level of operational performance that is met during a measured period of time; it is usually expressed as the percentage of time when the system is operational (Castañón, G., Campuzano, G., & Tonguz, O., 2008).

For an RoF system, we can distinguish two different types of availability and reliability: 1) that related to discrete components in the CS and BS and, 2) that related to the ODN. The difference stems mainly from the Mean Time to Repair (MTTR) and Failure Rate (FR). If a fiber is cut, the MTTR is very long. In this way, if no protection scheme is implemented at the ODN; several BSs can be down for long time. In contrast, devices at the CS and BS are simpler to access and the MTTR is much shorter, however the FR is higher for active components, where redundancy schemes must be considered (Castañón, G., Campuzano, G., & Tonguz, O., 2008). Conventionally, in WDM metropolitan networks for fix Internet access, dual

rings have been employed for protection against fiber failure (Li, M. J., et al., 2005). However, the use of two optical paths increases the cost of installing fiber and adds extra passive components that involve great complexity. For WDM systems, single fiber bidirectional self-healing protection scheme by using Optical Add Drop Multiplexers (OADM) have recently emerged and for metro networks have demonstrated to reduce the required amount of fiber by half and have duplicated the network capacity (Sun, et al., 2007). Furthermore, several architectures of WDM ring access networks for RoF technology have been proposed to improve availability and reliability at a lower cost (Peng, et al., 2007), (Lin, 2005).

DOWNLINK TRANSMISSION

The most challenging stage in the downlink transmission is the generation at the CS of the optical signal that, after detecting by a square-law PD in a remote BS, results in the desired mm-wave signal. We defined this process as the up-conversion technique, which specifically is the conversion of the electrical data signal at the CS from baseband to optical domain for transmission thru ODN and remote mm-wave generation at the BS.

A broad variety of up-conversion techniques have been proposed in the literature, using different approaches to convert a baseband information signal to mm-wave frequencies at the BS (O'Reilly J. & Lane, P., 1994), (Braun, R. P. & Grosskopf, G, 1998), (Jia Z., 2007), (Mohamed, et al., 2008), (Wake, D., Nkansah, A., & Gomes, N, 2010). Two main types of up-conversion techniques can be distinguished: those that perform the up-conversion in two steps and those that do it all-optically. The two-step conversion is as follows: the baseband signal is first electronically up converted to an intermediate or high frequency, and then is further converted into optical frequencies. We refer to such techniques as electro-optical up-conversion, Figure 3(a). Electro-optical up-conversion techniques are classified, in turn, according to whether the conversion to optical domain performs any frequency multiplication. If not, the reason to convert into optical frequencies is merely to allow transmission over the fiber link. In contrast, if frequency multiplication is performed at the electro-optical conversion, the baseband signal has to be electronically up-converted not to the desired mm-wave frequency but to a sub-harmonic of it. The all-optical scheme, Figure 3(b), avoids previous electronic up-conversion.

In addition to the previous classification, the physical implementation should also be considered. It is important to distinguish between techniques that use external modulators (EMs) and techniques that use direct modulation of the light source since the impairments and costs associated to each one of them are completely different and require different design approaches. Table 1 presents the most important up-conversion techniques reported in the literature, classified according to its type and whether external or direct modulation is used. In many cases, new techniques may arise by slightly modifying the proposed ones, for instance, using direct modulation instead of external modulation.

Figure 3. Main types of up-conversion techniques for downlink transmission

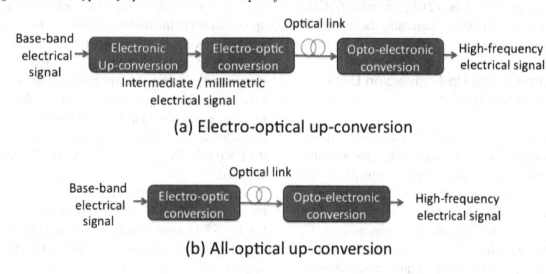

(a) Electro-optical up-conversion

(b) All-optical up-conversion

Table 1. Up-conversion techniques

		Modulation type	
		Direct modulation	External modulation
Generation type	Electro-optical	• Directly modulation laser (Hartmann, et al. 2005) • Directly modulated. OIL-laser (Parekh, et al., 2010) (Ng'oma, et al., 2010)	• Without OFM (Kuri, T., Kitayama, K., & Takahashi, Y., 2003) (Pan, 1988), (Wen H., et al., 2007) • Using OFM (Ng'oma, 2005), (Jia, et al., 2007), (Shih, et al., 2009) and (Qi G., et al., 2005)
	All-optical	• OSBIL (Braun, R. P. & Grosskopf, G, 1998)	• Dual tone laser with external modulation (Wake, et al.,1995), (Zhou, et al., 2008) • Heterodyning of independent lasers with external modulation (Islam A. R. and Bakaul M., 2009) (Insua, et al., 2010) • Supercontinuum source + filter/AWG (Nakasyotani, et al., 2006) (Toda, H., et al., 2004) • FPL + filter/AWG (Lu, et al., 2007)

Electro-Optical Up-Conversion Techniques

Electro-Optical Up-Conversion Using Direct Modulation

A laser is directly modulated with the information signal at the desired mm-wave frequency (fmm), with and without optical injection locking (OIL) as shown in Figure 4. The first is shown in Figure 4 (a), it is very cost effective and simple but it is limited to relatively low frequencies (Hartmann, et al., 2005). Among other improvements, the use of OIL considerably enhances the modulation bandwidth of the injected laser (Lau, et al., 2008), allowing transmissions of 2 Gbps signals at 60 GHz (Parekh, et al., 2010) but increasing the number of optical components as can be seen in Figure 4(b).

Electro-Optical Up-Conversion Using External Modulation

In order to overcome the impairments of the direct modulation, external modulation is the straightforward solution. The simplest implementation consists of a continuous wave (CW) laser followed by an external modulator (EM) that modulates the laser light with an intermediate frequency (IF -fi) or a mm-wave tone (fmm) as shown in Figure 4(c).

The EM used in this technique can be an intensity modulator (a Mach-Zhender Modulator (MZM) (OReilly, et al., 1992), or an Electro Absorption Modulator (EAM) (Kuri, T., Kitayama, K., & Takahashi, Y., 2000)), or a Phase Modulator (PM) (Chien, et al., 2010), whose output is optically filtered. Since the CW-laser is not directly modulated, the main impairments in the transmission of the signal derive from the EM. Nonlinearities in the transmission characteristic result in the generation of higher order harmonics that depending on the architecture can be undesired or required to implement Optical Frequency Multiplication (OFM) (Ng'oma, 2005). The power of these harmonics is controlled by the bias voltage and the modulation index. Without OFM, the reason to convert to optical frequencies is merely to allow the transmission over the fiber link, meaning that the downlink baseband signal was electronically up-converted to the desired mm-wave frequency at the CS (Kuri, T., Kitayama, K., & Takahashi, Y., 2003). In contrast, if OFM is performed at the electro-optical conversion, the baseband signal has to be electronically up-converted not to the desired mm-wave frequency but to a sub-harmonic of it (fi) reducing the complexity of the electronics at the CS (O'Reilly J. & Lane, P., 1994), (Jia Z., et al., 2007).

Considerable effort is being dedicated to reduce the electrical IF and create optical single side band (OSSB) signal, in order to reduce electronics complexity and dispersion effects at the ODN, respectively (Smith G., Novak D., and Ahmed Z., 1997). The higher the order of the OFM, the lower the required IF, but the power efficiency is

Figure 4. Electro-optical up-conversion techniques: (a) a single-mode laser directly modulated (b) single mode laser directly modulated with OIL, and (c) externally modulated single mode laser

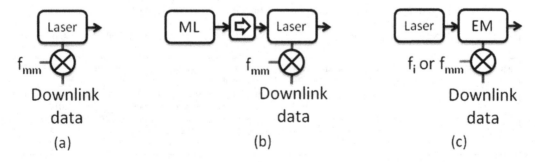

greatly penalized. Because of that, OFM is practically limited to doubling or quadrupling, even if frequency sextupling is also reported (Shih, et al., 2009).

All-Optical Up-Conversion Techniques

All-Optical Up-Conversion Using Direct Modulation

The main advantage of all-optical up-conversion is the reduction of high frequency components, both electronic and optoelectronic. The bandwidth requirements of the electronic/optoelectronic devices is limited by the capacity of the signal to be transmitted and not by the desired mm-wave signal as happens in electro-optical up-conversion. This is the case for Optical Sideband Injection Locking (OSBIL), where the sidebands of a directly modulated Master Laser (ML) with IF signals (fi) are used to correlate two slave lasers

(SLs) at bands separated the desired mm-wave frequency; it sorts out the frequency inaccuracy and phase noise simultaneously (Braun, R. P. & Grosskopf, G, 1998) as shown in Figure5 (a). A ML is sub-harmonically modulated and its output is coupled to the cavities of the two SLs using an optical coupler (OC). Each SL is carefully frequency-tuned to different modulation sidebands by adjusting the bias current or SL temperature. The combined signals of both SLs are coupled with a second OC. Only one of the correlated tones is directly modulated with downlink data (SL 1) and the beating of the two tones results in a modulated signal at the desired mm-wave frequency. OSBIL takes advantage of the modulation characteristics enhancement of OIL and of the phase correlation of two SLs synchronized to different modulation sidebands by the ML (Campuzano G., Aldaya I., and Castañón, 2009). In contrast, the main drawbacks of this technique are the complexity and the high number of optical components compared to directly modulated laser or simple

Figure 5. All-optical up-conversion techniques: (a) optical sideband injection locking (OSBIL), (b) heterodyning of independent single mode lasers, (c) Fabry-Perot laser with external modulation, and (d) super-continuum source with external modulation

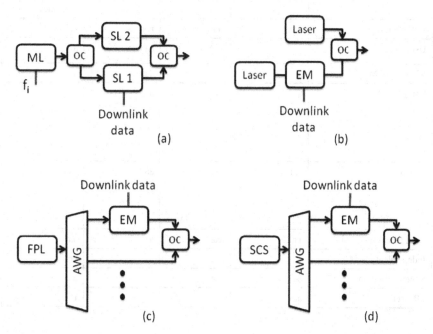

external modulation, which is compensated by the frequency flexibility OSBIL offers.

All-Optical Up-Conversion Using External Modulation

Techniques using all-optical conversion and direct modulation can also be implemented using external modulation, adding the cost of the EM. However, there are some all-optical up-conversion techniques that compulsory require external modulation. In (Insua, et al., 2010) two independent lasers are set up to emit at wavelengths that are spaced at the required mm-wave frequencies. Only one of them is externally modulated as presented in Figure 5(b). It is important to note that direct modulation would result in a chirped signal that will degrade the up-converted mm-wave signal. When a single source is used for generating multiple optical tones external modulation is almost the only alternative. In (Zhou, et al., 2008) a tunable mm-wave tone generation is reported using a dual tone ring laser. The optical tones could be modulated externally to create a modulated signal at mm-wave frequencies. Other alternative is a Super-Continuum Source (SCS), Figure5(c) (Nakasyotani, et al., 2006), (Toda, H., et al., 2004) or a Fabry Perot laser (FPL), Figure5(d), (Lu, et al., 2007) followed by an Array Waveguide Grating (AWG) or filters to select particular wavelengths. In both cases, one of the wavelengths is used as a reference tone while the other is modulated externally. Such techniques are extremely useful when the number of BSs to be fed in the RoF system is high since they can generate multiple mm-waves simultaneously.

Comparison of Downlink Techniques

Table 2 qualitatively compares the different up-conversion techniques in terms of the proposed figures of merit. It is worthwhile noting that the

Table 2. Qualitative comparison table among the different up conversion techniques

			Figure of merit				
			Cost and simplicity	Frequency and BW	Modulation Signal quality	Spectral purity and Frequency Accuracy	Scalability
Electro-optical	Direct Mod.	Without OIL	aaa	**rrr**	**rrr**	**rrr**	**rrr**
		Using OIL	aa	a	aa	a	**rr**
	External Modulation	Without OFM	a	a	aaa	a	**r**
		Using OFM	a	aa	aa	a	a
All-optical	Direct Mod.	OSBIL	**rrr**	aaa	aa	aaa	a
	External Modulation	Dual tone laser	**rr**	aaa	a	aa	**r**
		Independent lasers	a	aaa	aa	**rr**	aa
		Supercontinuum	aa	aaa	a	a	aaa
		FP laser	aa	aaa	aa	**r**	a

Good:a Bad:**r**

performance of a particular technique is very dependent on the operation frequency. In terms of the cost, the cheapest alternative is the directly modulated laser, while, a priori, the most expensive is the OSBIL even if electro-optical up-conversion with external modulation may result more expensive for high frequencies. Regarding frequency and bandwidth, all-optical up-conversion overtakes electro-optical up-conversion. Signal quality is difficult to assess, but directly modulated lasers are supposed to be the worst mainly due to induced frequency chirp. With reference to spectral purity of the mm-wave carrier, those techniques that use correlated tones result to be the best, such as OSBIL, and using different modes from the same SCS or FPL. Finally, the best techniques from the scalability point of view are SCS and FPL, since they are capable of feeding different BSs simultaneously using wavelength multiplexing or subcarrier multiplexing.

UPLINK TRANSMISSION

The key for making an mm-wave RoF system practical for commercial implementation is the development of a low-cost and green BS by installing functionally simple and compact equipment with minimal optical and electronics inventory and with the lowest possible energy consumption.

The main function of the BS in an RoF system is to transport the mm-wave signals to and from the user terminals (MTs) within a coverage area. Since the BSs in the network are connected to the CS over optical fiber. Therefore, it is imperative to provide the BS with an Electric-to-Optic (E/O) converter in the transmission path, and an Optic-to-Electric (O/E) converter in the receiver path.

Several BS configurations using advanced technologies have been proposed to achieve this goal. BS uplink configurations can be divided into two main categories: 1) BS with a laser installed and 2) Laser-free BS. When a laser is installed at the BS, it can be directly modulated with uplink

data or an external modulation scheme can be selected by adding an EM. For laser-free BS, the CS provides an optical signal for uplink transmission. The optical signal from the CS can arrive as a non-modulated or a modulated signal to the BS, the latter requires an erasing process to re-use the optical signal and modulate it with uplink data before it can be retransmitted back to the CS.

For uplink transmission, if the mm-wave wireless signal received from the MT at the BS is down-converted to baseband or low-frequency IF subcarrier, an electrical down-conversion circuit (DCC) is required, it can be composed by an envelope detector or an electrical LO, enabling in this way the use of low-bandwidth E/O converter for direct transmission of the uplink channels. However, the generation of a LO at the BS increases its complexity, therefore several schemes of remote delivery of LO have been proposed (Lim Christina, et al., 2004). On the other hand, if the DCC is not implemented at the BS, then the mm-wave signal from the MT modulates directly the BS TX, and the down-conversion process of uplink data is performed at the CS. This configuration requires that the E/O converter, assigned for uplink transmission, operates at mm-wave frequencies, which increases its cost and complexity (Kuri, et al., 2000). The decision for choosing to down-covert or not the mm-wave signal from the MT at the BS depends on the overall network characteristics and requirements.

In this section we will explore the characteristics of different BS configurations and the key enabling technologies available to support them.

Base Station Using Laser for Uplink

Base Station Using Directly Modulated Laser

The conventional components to provide E/O and O/E capabilities at the BS are the laser and the PD, respectively. Under this scheme, the downlink signal coming from the CS is directed towards the

Table 3. BS schemes

		Modulation type	
		Direct modulation	External modulation
Base Station Configuration	Laser	• Directly modulated laser (Wong, et al., 2009) (Ismail, T. M., & Seeds, A., 2005) • Remotely configurable FPL (Kashima, 2006) • Antenna + MLL laser (Khawaja & Cryan 2009) (Khawaja & Cryan,, 2010)	• Laser + External Modulator (EAM) (Kawanishi, et al., 2001) • Laser + External Modulator (MZM) (Tsuzuki, et al., 2008)
	Laser-Free		• Reusing optical signal from CS +MZM (Lu, et al., 2007), (Chang, Q., Fu, H., & Su, Y., 2008) • Reusing optical signal from CS + EAM (Kuri, et al., 2000), (Chen, et al., 2009) • EAT (Kuri, et al., 2000), (Stohr, et al., 1999), (Kitayama, et al., 2000) • R-EAM (Thakur, et al., 2009) (Lecoche F., et al., 2009) • RSOA (Medeiros, et al., 2007), (Kim, et al., 2009)

PD using a WDM component. On the other hand, the uplink optical signal is generated by directly modulating a laser. The laser installed in the BS can be a vertical-cavity surface-emitting laser (VCSEL) (Wong, et al., 2009) or a Distributed Feedback (DFB) laser (Ismail & Seeds, 2005) as shown in Figure 6(a). The laser can be modulated by the mm-wave signal from MT or by baseband or intermediate down-converted signals depending upon the scheme selected. For mm-wave frequencies higher than 30 GHz a laser direct modulation can be a difficult task because the modulation bandwidth of the laser (Radziunas, et al., 2007), in this case, the implementation

of a down-conversion process from mm-wave frequencies to baseband or IF signals is necessary. Thus, this BS configuration requires two independent optical components (laser and PD), and it needs a DCC for the mm-wave signals from MT. For optical components integration, several approaches to obtain simple and cost effective transceiver architecture including laser and PD had been proposed (Pergola, et al., 2007).

A clear disadvantage of having a fixed laser installed in the BS is that it specifies a fixed operation wavelength, which reduces the network's flexibility. Thus, in order to overcome this limitation a FPL could be installed at the BS, the FPL

Figure 6. BSs with lasers. (a) Fix wavelength laser at the BS, (b) tunable FPL at the BS (UL: uplink, DL: downlink)

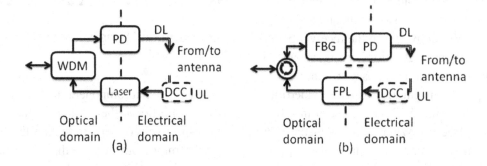

can be remotely configured by the CS. For this configuration, a non-modulated optical signal along with the downlink signal is transmitted from the CS to the BS. A three port optical circulator at the input of the BS directs both signals to a tunable FBG filter, which is used to reflect a portion of the non-modulated signal to the FPL while it allows the downlink signal to be detected by the BS PD as depicted in Figure 6(b). One of the longitudinal modes in the FPL is selectively injection locked by the non-modulated signal, and a single-mode lasing to the appropriate uplink channel is realized (Shen, et al., 2008), (Kashima, 2006). The remotely injection-locked FPL is directly modulated for uplink transmission.

On the other hand, the photonic active integrated-antenna concept is the intimate integration of photonic devices as lasers (Khawaja, B. A. & Cryan, M. J., 2009), (Khawaja, B. A. & Cryan, M. J., 2010) and PDs (Stöhr, A.,et al., 2010) with antennas including resonating matching circuits. It can lead to novel functions with superior performance that exceeds the limits of conventional electronics and optoelectronic integrated circuits.

Wireless hybrid mode-locking is a new concept introduced by (Khawaja, B. A. & Cryan, M. J., 2009) where a mode-locked laser (MLL) is combined with an mm-wave planar antenna such that there can be wirelessly injection locked for baseband data transmission, from the BS to the CS. The mode-locking of the MLL is achieved by injecting a CW external RF signal into the MLL cavity close to its passively mode locked frequency. The MLL repetition rate is then controlled by the applied external RF signal over the locking range. In (Khawaja, B. A. & Cryan, M. J., 2010), the authors explored the possibility of integrating a planar antenna alongside the MLL to produce a highly integrated module at the BS; the remote antenna contains high speed PDs for the downlink and wireless hybrid MLLs for the uplink.

Base station using externally modulated laser

For external modulation scheme, a MZM or EAM needs to be installed in the BS in combination with the laser. This configuration is not common, because it renders to be expensive if the number of BS is high; however it cannot be discarded as another possibility for uplink transmission.

Presently, several approaches integrating the laser and the EM in a single component are being proposed. The EAM monolithically integrated with a DFB is the most practical device, because the high coupling efficiency between the EAM and laser is achievable in the integrated structure. This leads to a high output power of the modulated signal, which provides its advantage over other modulators such as LiNbO MZM, and contributes to reduce the system size and cost (Kawanishi, et al., 2001). On the other hand, wavelength reconfiguration features can be available by using a hybrid integrated module consisting of a wavelength tunable laser and a semiconductor MZM, in this case, the optical coupling of the laser and modulator is accomplished with finely tuned aspheric lenses (Tsuzuki, et al., 2008). Although these component proposals are available for long haul links; depending on the RoF access network requirements, they can also be considered for utilization into a BS.

Laser-Free BS for Uplink

Presently, considerable effort has been done to simplify the BS design combining the use of external modulation and custom-designed optical components, such as looped-back AWG, multiport circulators, and FBG, to achieve functionalities such as OADM and optical carrier reuse (Nirmalathas, et al., 2001). The latter allows removing the laser from the BS which produces significant advantages, as wavelength assignment and source monitoring can be done from the CS if the active sources are only located there. Several laser-free

BS schemes have being proposed, we present the most promising configurations in this section.

Base Station Using an RSOA

Efficient architectures have been proposed incorporating Reflective Semiconductor Optical Amplifiers (RSOAs) located in the BS for the uplink as shown in Figure 7(a). This configuration allows the possibility of dynamic network reconfiguration through the wavelength reassignment to different BS (Medeiros, et al., 2007), (Kim, et al., 2009), (Wake, et al., 2010). The RSOA replaces WDM light sources at the BS, its multi-functionalities, such as colorless operation, re-modulation, amplification and envelope detection, make BSs implementation more compact, less complex and cost effective, improving utilization of wavelength resources in the RoF network with centralized light sources. There are two techniques that can

be used for RSOA implementation, 1) the reuse of the downlink signal for uplink or 2) the utilization of a non-modulated signal coming from the CS. For the downlink signal reuse, an OC is installed in the BS to split the optical power. One optical path carries a portion of the downlink signal (e.g. -3dB) to BS PD, while a second path distributes another portion of the downlink signal to the RSOA which erases the downlink information with low extinction ratio, then modulates the signal with new data (uplink) and reflects it back. Thus, only one wavelength is required for bidirectional transmission between the CS and BS. In this case the RSOA acts as both, i.e. an optical saturator and a modulator (Kim, et al., 2009), (Liu, et al., 2011). On the other hand, to operate the RSOA with a non-modulated signal from the CS, the OC is replaced by a WDM and the CS needs to assign a unique wavelength pair for downlink and uplink (Pato, et al., 2009).

Figure 7. Laser-free BS. (a) RSOA at the BS, (b) External Modulator (EM) at the BS, c) Electro Absorption Transceiver (EAT) at the BS and (d) Reflective Electro Absorption Modulator (REAM) (UL: uplink, DL: downlink).

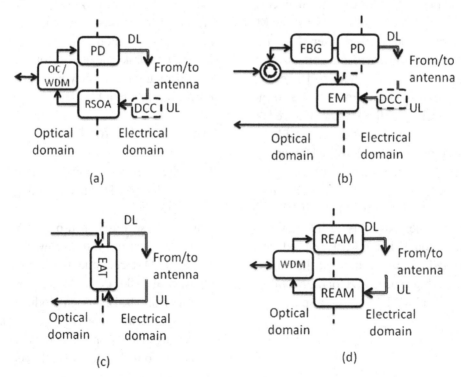

Base Station Using External Modulator

Another configuration used to avoid a laser at the BS consists of an EM for the uplink data. This configuration can be supported by a FPL with an AWG at the CS. Two modes of FPL are filtered out and transmitted to the BS, one optical signal carrying downlink data and the second as a non-modulated optical signal to be used for the uplink. At the BS a three port optical circulator is used to direct the downlink signal to BS PD, the FBG reflects a portion of the non-modulated signal to be used for the uplink light source entering the EM as shown in Figure 7(b). The EM can be a MZM modulated with baseband uplink data (requiring the DCC) (Lu, et al., 2007), (Chang, et al., 2008) or EAM modulated with the mm-wave signal coming from MT (Kuri, et al., 2003), (Chen, et al., 2009). Such architectures are attractive because they enable the BS to share the light source remotely located at the CS.

Base Station Using an EAT

An approach of a single optical component at the BS is the Electro Absorption Transceiver (EAT) depicted in Figure 7(c). This scheme utilizes the EAM as a PD for downlink data and as an optical modulator for the uplink to the CS. Optically, the transceiver receives these two optical signals from the CS, downlink which is absorbed, and uplink which is modulated by the electrical signal to be transmitted from BS to the CS. The important characteristics of this device are not only as a modulator but also as a PD (Kuri, et al., 2000), (Stohr, et al., 1999), (Kitayama, et al., 2000). The EAT has an optical input power to collect both, downlink modulated signal and uplink CW, at the optical output the uplink data is directed back to the CS. If downlink and uplink optical signals need to be transported in a single fiber, a three port circulator is required at the input of the EAT. The EAT module has individual RF input and output ports, each with impedance-matching circuits that enhance detection and modulation efficiencies at 60 GHz, avoiding the requirement of frequency down-conversion at the BS for signals from MT.

Base Station Using an REAM

With very similar functionalities to the EAT, the Reflective-EAM (REAM) consists of an EAM with a reflection layer. Full duplex capability is achieved by the use of two REAMs as demonstrated in (Thakur, et al., 2009), one REAM is configured to operate as a PD for detection of downlink data, while the second is biased to function as modulator, modulating and reflecting a light remotely delivered throughout the network from the CS to the BS. The REAM used in (Thakur, et al., 2009) has 11 GHz RF modulation bandwidth over its working bias range, thus for mm-wave operation in an RoF system, one REAM with this characteristics can be used in combination with a DCC for uplink transmission, and for downlink reception a mm-wave bandwidth PD need to be installed at the BS.

An mm-wave band REAM was designed in (Technologies CIP) for operation at 60 GHz, this device is used and tested for bidirectional optical signal distribution system at 3 Gbps in (Lecoche F., et al., 2009). In this experimentation, the BS requires a WDM component to de-multiplex downlink signal to REAM configured as mm-wave bandwidth PD, and the non-modulated optical signal from CS to an REAM operating in reflective modulator mode as shown in Figure 7(d). In this case, the uplink signal from MT directly modulates the REAM with the mm-wave signal coming from the MT avoiding the use of a DCC at the BS.

Comparison of BS Schemes

Table 4 qualitatively compares the different BS configurations in terms of the proposed figures of merit. According to this table, the most expensive and not simple configuration is the BS using REAM. It is fairly complex because in order to achieve full-duplex communications,

Table 4. Qualitative comparison table amongst the different BS schemes

			Figure of merit				
			Cost and simplicity	Frequency and Bandwidth	Uplink Signal quality	Power Consumption	Scalability
BS Up-link Tech-nol-ogy	Laser at BS	Directly modulated laser	aaa	**rr**	aaa	**rrr**	**rr**
		Remotely configurable FPL	aa	**r**	aa	**rr**	**r**
		Antenna + MLL laser	**rr**	a	**rrr**	**r**	**rrr**
	No-Laser at BS	RSOA	a	**rrr**	**r**	a	aaa
		EM	aa	aaa	aa	aaa	aa
		EAT	**r**	aa	a	aa	aa
		R-EAM	**rrr**	aa	a	aa	aa

Good:a Bad:**r**

the BS requires the installation of two REAMs. On the other hand, devices as EMs and RSOA are still costly, however future RoF deployments will increase their demand making them eligible for mass production processes and thus reduced prices can be obtained. Regarding frequency and bandwidth, the BS configuration using an EM provides better capabilities compared to other configurations, and low cost EAMs are becoming more popular for commercial applications.

The four laser-free configurations have better performance regarding scalability; this is because in a WDM scheme, a BS that can be reconfigured remotely from CS presents great advantages when the network requires upgrading. However, they have a slightly minor performance for the uplink signal quality since the reused optical tone from CS is attenuated by several passive components

(circulators and filters) before it reaches the uplink modulation. In the case of a RSOA, the reused signal can be a portion of the modulated downlink signal, and then the erasing process and uplink re-modulation produce an extra degradation on the signal's quality.

Finally, the better power consumption is for the EM and the EAT single component. On the other hand, since mm-wave LO is required for the directly modulated laser, it is considered as the worst case for BS energy consumption.

NETWORK DESIGN GUIDELINE

Table 5 presents the most important design criteria that must be taken into account in the design of an RoF land access network. These design criteria are

Table 5. Design criteria for the different RoF subsystems

RoF Land Access Network Design Decisions		
CS	**ODN**	**BS**
• Up-conversion Technique • Downlink Multiplexing • Electrical Modulation Format • Equipment redundancy • Energy Supply	• Optical Multiplexing Scheme • Optical Wavelength Range • Network Topology • Fiber Type • Remote Node Configuration • Protection Scheme • Amplification Technology	• BS configuration • Uplink Multiplexing • Number of BSs • Add/Drop Passive optical components at BS • BS capacity • Electrical Modulation Format • Equipment redundancy • Energy supply

divided into three main groups coinciding with the three subsystems that compose the RoF network: CS, ODN, and BSs. The ultimate objective of the network design is to choose the most appropriate techniques and configurations to meet network requirements, attaining a tradeoff between performance and cost. The major challenge in designing an RoF network is considering the system as a whole, since in most cases the implementation of each subsystem (CS, ODN, and BS) affects other subsystems.

The parameter that impacts the design of the network most critically is the mm-wave frequency. On the one hand, the spacing and number of BSs highly depend on the nominal mm-wave frequency, because the wireless propagation characteristics vary significantly as the frequency increases. On the other hand, the required frequency affects the choice of the up-conversion technique, the optical wavelength range, and optical multiplexing scheme, imposing constraints on the ODN and the BS scheme.

The network design, similar to other design processes, is an iterative one and comprises the following steps:

Step 1: A tentative downlink/uplink multiplexing scheme is proposed considering the number of BSs to be supported, the capacity and the operation frequency.

Step 2: A cost effective up-conversion technique that meets frequency and bandwidth requirements, as well as multiplexing scheme is advised.

Step 3: A BS compatible with the proposed multiplexing scheme and the up-conversion technique is chosen.

Step 4: The ODN is designed, giving special attention to the topology, fiber length, reliability, and to RNs. Reliability requirements lead to the selection of a topology (ring, star, bus, etc.), this topology, together with the multiplexing scheme, imposes conditions that RN design must satisfied.

Step 5: A proper power supply for each subsystem is selected. Typically, this does not interfere with the previous steps and, therefore, it is easy to decide.

Step 6: The whole system, from CS to BSs is checked to ensure that the implementation works effectively. It is important to check both uplink and downlink. If any inconsistency is found, the adopted design decision should be fully revised.

RoF ARCHITECTURES FOR VEHICULAR NETWORKS: CASE STUDY

The guidelines presented previously are applied here, into a case study where decisions are made based on comparisons amongst the different techniques described earlier. We first introduce the network requirements for this particular application and then present and justify an implementation proposal.

Network Requirements

The network proposed for the case study is shown in Figure 8. It is an RoF network serving broadband communications to mobile users in a high-speed train. The network configuration consists of a high capacity link connecting the train CS to the trunk network, an ODN communicating the CS and BSs, and an mm-wave wireless link between BSs and MTs. The train is considered to have 10 cars and the number of users per car is assumed to be 100 (Lannoo, et al., 2007). Under worst-case scenario conditions, when cars are at full capacity and all users are connected, the offered capacity should be at least 50 Mbps per user, resulting in a required capacity per car of 5 Gbps. The signal format required is OFDM with mm-wave carrier at 60GHz. OFDM at 60 GHz has been proposed in WPAN standard 802.15.3c for high capacity indoor networks (802.15, 2011).

Figure 8. Network schematic for in-vehicle scenario

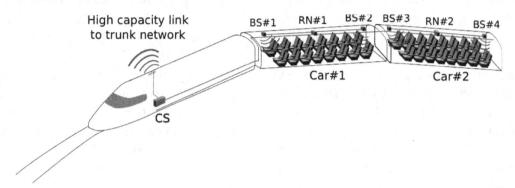

In addition, mm-wave carrier is added to assist the MT in the up-conversion/down-conversion. In order to ensure coverage over the whole car, two BSs per car are assumed. We limit the network design to the RoF land network, which provides the required capacity connection between CS and BSs, assuming the high capacity link connecting the CS to the trunk network is already available (Lannoo, et al., 2007) as well as the mm-wave wireless link. The CS is installed strategically in the train's communication room, while the BSs are located to maximize the coverage on the train cars for mobile users.

Implementation Proposal

Using the guidelines for the RoF access network introduced in the previous section, we present the proposed solution to the case study in Table 6. Because of the use of mm-wave frequency at 60 GHz, an all-optical up-conversion technique is chosen. The CS cost is reduced by using a SCS to simultaneously generate multiple wavelengths (Table 2) as shown in Figure 5(d). This approach reduces the cost by eliminating independent lasers and high frequency LO at CS and BS, respectively. Several experimental demonstration of this scheme are available at 1550nm, thus we selected this wavelength range for our case study.

Based on this up-conversion technique, the optical multiplexing technique that meets our requirements is DWDM having optical signals spaced 60 GHz. In this case, the CS transmits two optical signals to each BS, one carrying the downlink information and the other that acts as a reference, an un-modulated optical tone. Both signals produce the mm-wave signal when they are heterodyned in the PD at the BS.

To provide flexibility and scalability of the BSs and to take advantage of the transmitted refer-

Table 6. Summary of indoor RoF access network design

RoF Access Network Considerations – In-vehicle		
CS Technology	**Distribution Network**	**BS Configuration**
• All-optical with External modulation (EAM) using SCS • DWDM. Unique wavelength per BS • OFDM with mm-Carrier • Redundant SCS and low-bandwidth PDs for uplink as spare. • Grid AC	• DWDM. Two independent paths: downlink and uplink optical signals. • 1550 nm (C band) • Bus • SMF • ADD/DROP RN required • No amplifiers required • No-redundancy	• mm-wave PD and EAM (Reusing Non-modulated optical signal) • DWDM • 20 BS (2 x train car) • 2.5 Gbps per BS • Three port circulator and FBG. • OFDM (base-band) • mm-wave PDs for BS, and full pre-aligned TX. • Solar energy assisted by Grid AC

ence tone, we considered that an appropriate BS configuration is a laser-free BS with a wavelength reuse scheme by the use of an EM (per Table 4) as shown in Figure 7(b). Thus, an optical circulator and a FBG are required at the BS as extra passive optical components.

According to the environment of the RoF access network, we considered to include ODN flexibility by enabling RNs on each train car. Thus, a bus topology where uplink/downlink signals are ADD/DROP to train cars is proposed. Under this scheme and taking into account the network distances, optical amplification is not required for the network proposal.

In terms of network availability and reliability, we are proposing to include redundant equipment for downlink transmission (extra SCS) and modular PD spares for CS and BS. Considering that a single failure on the SCS will disable the full network, the need to have redundancy at this point becomes essential. On the other hand, failures on optical RX are most commonly attributed to the PDs, thus these failures can be contained by available generic replacements, optical PD supporting 1550 C-band, considering no specific requirements per train car are necessary. For the

case of EAMs at the BS, we are considering pre-aligned TX spares, since a new EAM module replacement will require optical alignment, is necessary to have a second BS TX previously aligned to reduce MTTR.

Finally, for power consumption we decided to use AC grid for equipment located at the CS and solar assisted by the AC grid power supplies for the BSs across the train.

The spirit of this design is that all technologies selected were already experimentally demonstrated by several authors as has being referenced during this chapter. Therefore, the RF and optical components mentioned here are available for a real implementation.

The proposed network implementation is shown in Figure 9, where CS, ODN, and BS are identified. Figure 10 presents a detail of the ODN, showing the uplink and downlink fibers and their connections to the RNs.

Downlink Transmission

A SCS at the CS is used to generate multiple optical tones at frequencies spaced 60 GHz. Then, an AWG is used to separate the different

Figure 9. Network implementation for in-vehicle deployment case study

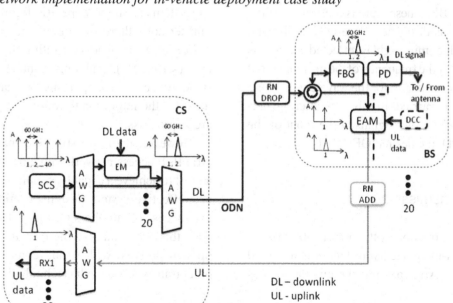

Figure 10. ODN proposed for in-vehicle deployment case study

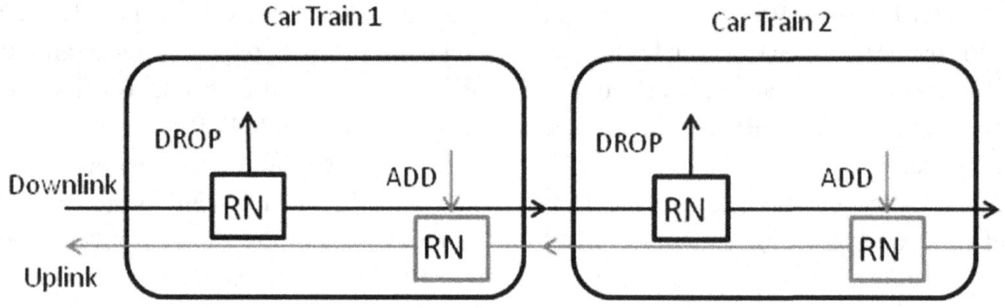

signals. For each BS, two adjacent wavelengths are required, one that is not modulated and another one modulated carrying the downlink information (OFDM signal). Modulation can be performed using a MZM or an EAM; we choose the latter because it does not suffer bias current stability problems. Then, all the wavelengths (modes) are combined using another AWG and are launched into a single fiber. The OADM in each DROP RN extracts the wavelengths required for each train car: there are two BS per train car and each BS requires two wavelengths. Therefore, only the four wavelengths destined to the BSs within the train car are dropped while the others are transmitted with minor attenuation. Once the two wavelengths corresponding to one BS are extracted from the RN, they merge at the three-port circulator at the input of the BS. These signals are routed to the FBG, which is designed to present a stop band (50% of the incoming light is reflected and 50% is transmitted) at the frequency of the un-modulated tone and be transparent for the wavelength where information is carried. In this way, after the heterodyning of the two signals at the input of the mm-wave PD, the desired OFDM signal appears at 60 GHz.

Uplink Transmission

In each BS, a fraction of the power (50%) of the incoming un-modulated wavelength is reflected by the FBG. After passing the circulator, it is

modulated using the EAM by the OFDM baseband uplink signal coming from an envelope detector at the DCC. The uplink signal is sent to the ADD RN that adds the uplink signals to the uplink bus fiber. Then, in the CS, the different uplink signals from each BS are separated by an AWG and detected using the corresponding low-bandwidth PD at an RX.

CONCLUDING REMARKS

In this chapter we have studied several available technologies to support key functionalities at the CS and BS of an RoF access network. Much progress has clearly been made on the development and experimental demonstration of network subsystems to support specific figures of merit. Furthermore, there are commercial applications of RoF access networks operating at low frequencies (<10 GHz), and important efforts in the standardization to support research and development for the migration to mm-wave frequencies are in progress.

However, there are still challenges for specific RoF access network implementations. It is a must to achieve a capable, simple, practical and cost-effective system design to ultimately lead the success of suppliers on this market. To this date, this is still an ongoing research and development process, with strong competition from other transmission media technologies which are still more attractive for broadband deployments. Nevertheless, the anticipation for the use of RoF to support higher data demands is promising for the forthcoming years.

Therefore, there are still abundant opportunities available to provide flexible technologies that adequately assist the overall network requirement and the high-priority figures of merit. Among these new enabling technologies are: cost effective and flexible modular optical systems to provide up-conversion techniques to operate at high frequencies (>60GHz), configurable ODN capable of seamless scalability, self-healing automatic controls across the network for reliable and available services, integrated RF and optical components to provide low-cost, low power consumption and remotely configurable single module BS, and definitely, standards to support design for manufacturing methodologies of new components and network architecture deployments.

REFERENCES

Agrawal, G. P. (2001). *Nonlinear fiber optics (Vol. 3)*. London, UK: Academic Press.

Alliance, W. (2011). *The worldwide UWB platform for wireless multimedia*. Retrieved September 1, 2011, from http://www.wimedia.org/en/about/commonradio.asp?id=abt

Armstrong, J. (2009, February). OFDM for optical communications. *Journal of Lightwave Technology*, 27(3), 189–204. doi:10.1109/JLT.2008.2010061

Baliga, J., Ayre, K., Sorin, R., Hinton, W., & Tucker, R. S. (2008). Energy consumption in access networks. *Conference on Optical Fiber communication/National Fiber Optic Engineers Conference*, (pp. 1-3).

Baliga, J., Ayre, R., Hinton, K., Sorin, W., & Tucker, R. S. (2009, July). Energy consumption in optical IP networks. *Journal of Lightwave Technology*, 27(13), 2391–2403. doi:10.1109/JLT.2008.2010142

Braun, R. P. (1998, September). Tutorial: Fibre radio systems, applications and devices. *24th European Conference on Optical Communication*, (pp. 87–119).

Braun, R. P., & Grosskopf, G. (1998). Optical millimeter-wave systems for broadband mobile communications, devices and techniques. *Proceedings of the International Zurich Seminar on Broadband Communications, Accessing, Transmission, Networking*, (pp. 51-58).

Campuzano, G., Aldaya, I., & Castañón, G. (December de 2009). Performance of digital modulation formats in radio over fiber systems based on the sideband injection locking technique. *ICTON Mediterranean Winter Conference ICTON-MW*, (pp. 1-5).

Castañón, G., Aragón-Zavala, A., Ramirez-Velarde, R., Campuzano, G., & Tonguz, O. (2008). *An integrated availability analysis of RoF networks*. 2nd International Conference on Transparent Optical Networks ICTON.

Castañón, G., Campuzano, G., & Tonguz, O. (2008). High reliability and availability in radio over fiber networks. *OSA Journal of Optical Networking*, 7(6), 603–616. doi:10.1364/JON.7.000603

Chang, C. H., Lu, H. H., Su, H. S., Shih, C. L., & Chen, C. L. (2009, November). A broadband ASE light source-based full-duplex FTTX/ROF transport system. *Optics Express*, 17.

Chang, G., Chowdhury, A., Jia, Z., Chien, H., Huang, M., Yu, J., & Ellinas, G. (2009, September). Key technologies of WDM-PON for future converged optical broadband access networks. *Journal of Optical Communications and Networking*, 1(4), C35–C50. doi:10.1364/JOCN.1.000C35

Chang, Q., Fu, H., & Su, Y. (2008, February). Simultaneous generation and transmission of downstream multiband signals and upstream data in a bidirectional radio-over-fiber system. *IEEE Photonics Technology Letters*, 20(3), 181–183. doi:10.1109/LPT.2007.912992

Chen, L., Yu, J. G., Wen, S., Lu, J., Dong, Z., Huang, M., & Chang, G. K. (2009, July). A novel scheme for seamless integration of ROF with centralized lightwave OFDM-WDM-PON system. *Journal of Lightwave Technology, 27*(14), 2786–2791. doi:10.1109/JLT.2009.2016984

Chen, T., Yang, Y., Zhang, H., Kim, H., & Horneman, K. (2011, October). Network energy saving technologies for green wireless access networks. *IEEE Wireless Communications, 18*, 30–38. doi:10.1109/MWC.2011.6056690

Chien, H., Hsueh, Y., Chowdhury, A., Yu, J., & Chang, G. (2010, February). On frequency-doubled optical millimeter-wave generation technique without carrier suppression for in-building wireless over fiber applications. *IEEE Photonics Technology Letters, 22*(3), 182–184. doi:10.1109/LPT.2009.2037333

Cliche, J., Allard, M., & Tétu, M. (2006). *Ultra-narrow linewidth and high frequency stability laser sources. Optical Amplifiers and their Applications/ Coherent Optical Technologies and Applications.* Technical Digest.

Cochrane, P. (1989, September). The future symbiosis of optical fiber and microwave radio systems. *19th European Microwave Conference,* (pp. 72–86).

Daniels, R. C., & Heath, W. (2007, September). 60 GHz wireless communications: Emerging requirements and design recomendations. *IEEE Vehicular Technology Magazine,* 41–50.

Ghafoor, S., & Hanzo, L. (2011). Sub-carrier-multiplexed duplex 64-QAM radio-over-fiber transmission for distributed antennas. *IEEE Communications Letters,* 1–4.

Gunter, S. (2009). The green base station. *4th International Conference on Telecommunication - Energy Special Conference (TELESCON),* (pp. 1-6).

Hartmann, P., Qian, X., Wonfor, A., Penty, R., & White, I. (October de 2005). *1-20 GHz directly modulated radio over MMF link.* International Topical Meeting on Microwave Photonics.

Hassun, R., Flaherty, M., Matreci, R., & Taylor, M. (1997, August). Efffective evaluation of link quality using error vector magnitude techniques. *Proceedings of Wireless Communications Conference,* (pp. 89–94).

IEEE 802.15. (n.d.). *Task Group 3c (TG3c) Millimeter Wave Alternative PHY.* Recuperado el November de 2011, de http://ieee802.org/15/pub/TG3c.html

IEEE 802.15. (n.d.). *Wireless Personal Area Networks.* Recuperado el November de 2011, de http://www.ieee802.org/15/

IEEE 802.16. (n.d.). *Standard for Wireless Metropolitan Area Networks.* Recuperado el November de 2011, de http://ieee802.org/16/.

Insua, I. G., Plettemeier, D., & Schffer, C. G. (2010, February). Simple remote heterodyne radio over fiber system for Gbps wireless access. *Journal of Lightwave Technology, 28*, 1–1. doi:10.1109/JLT.2010.2042426

Islam, A. R., & Bakaul, M. (2009, December). Simplified millimeter-wave radio-over-fiber system using optical heterodyning of low cost independent light sources and RF homodyning at the receiver. *International Topical Meeting on Microwave Photonics,* (pp. 1–4).

Ismail, T. M., & Seeds, A. (2005). Linearity enhancement of a directly modulated uncooled dfb laser in a multi-channel wireless-over-fibre system. *IEEE MTT-S International Microwave Symposium Digest,* (pp. 7-10).

Ito, H., Kodama, S., Muramoto, Y., Furuta, T., Nagatsuma, T., & Ishibashi, T. (August de 2004). High-speed and high-output InP-InGaAs unitraveling-carrier photodiodes. *IEEE Journal of Selected Topics in Quantum Electronics, 10,* 709–727.

ITU-R. (n.d.). *Recommendation ITU-R M.1645*. Retrieved from http://www.itu.int/ITU-R.

Kamra, V., & Kumar, V. (2011, January). Power penalty in multitone radio over-fibre system employing direct and external modulation with optical amplifiers. *International Journal for Light and Electron Optics, 34*, 44–48. doi:10.1016/j.ijleo.2010.02.001

Kashima, N. (2006, August). Dynamic properties of FP-LD transmitters using side-mode injection locking for LANs and WDM-PONs. *Journal of Lightwave Technology, 24*(8), 3045–3058. doi:10.1109/JLT.2006.878056

Kato, K. (1999). Ultrawide-band/high-frequency photodetectors. *IEEE Transactions on Microwave Theory and Techniques, 47*(7), 1265–1281. doi:10.1109/22.775466

Kawanishi, H., Yamauchi, Y., Mineo, N., Shibuya, Y., Murai, H., Yamada, K., & Wada, H. (2001, September). EAM-integrated DFB laser modules with more than 40-GHz bandwidth. *IEEE Photonics Technology Letters, 13*, 954–956. doi:10.1109/68.942658

Khawaja, B. A., & Cryan, M. J. (2009). A wireless hybrid mode locked laser for low cost millimetre wave radio-over-fiber systems. *International Topical Meeting on Microwave Photonics*, (pp. 1–4).

Khawaja, B. A., & Cryan, M. J. (2010, August). Wireless hybrid mode locked lasers for next generation radio-over-fiber systems. *Journal of Lightwave Technology, 28*, 2268–2276. doi:10.1109/JLT.2010.2050461

Kim, H. S., Pham, T. T., Won, Y. Y., & Han, S. K. (November de 2009). Bidirectional WDM-RoF transmission for wired and wireless signals. *Communications and Photonics Conference and Exhibition*, (pp. 1–2).

Kitayama, K. (1998). Architectural considerations of radio-on-fiber millimeter-wave wireless access systems. *URSI International Symposium on Signals, Systems, and Electronics*, (pp. 378-383).

Kitayama, K., Kuri, T., Onohara, K., Kamisaka, T., & Murashima, K. (2002). Dispersion effects of FBG filter and optical SSB filtering in DWDM millimeter-wave fiber-radio systems. *Journal of Lightwave Technology, 20*(8), 1397–1407. doi:10.1109/JLT.2002.800265

Kitayama, K. I., Stohr, A., Kuri, T., Heinzelmann, R., Jager, D., & Takahashi, Y. (2000, December). An approach to single optical component antenna base stations for broad-band millimeter-wave fiberradio access systems. *IEEE Transactions on Microwave Theory and Techniques, 48*, 2588–2595. doi:10.1109/22.899017

Kuri, T., Kitayama, K., & Takahashi, Y. (2000, April). 60-GHz-band full-duplex radio-on-fiber system using two-RF-port electroabsorption transceiver. *IEEE Photonics Technology Letters, 12*, 419–421. doi:10.1109/68.839038

Kuri, T., Kitayama, K., & Takahashi, Y. (February de 2003). A single light source configuration for full-duplex 60-GHz-band radio-on fiber system. *IEEE Transactions on Microwave Theory and Technique, 51*, 431–439.

Lannoo, B., Colle, D., Pickavet, M., & Demeester, P. (2007, February). Radio-over-fiber-based solution to provide broadband internet access to train passengers. *IEEE Communications Magazine, 45*(2), 56–62. doi:10.1109/MCOM.2007.313395

Lau E. K., Zhao X., Sung H., Parekh D., Chang-Hasnain C., & Wu M. C. (April de 2008). Strong optical injection-locked semiconductor lasers demonstrating > 100-GHz resonance frequencies and 80-GHz intrinsic bandwidths. *Optics Express, 16*(9).

Lecoche, F., Charbonnier, B., Frank, F., Dijk, F. V., Enard, A., & Blache, F. … Moodie, D. (2009, October). 60 GHz bidirectional optical signal distribution system at 3 Gbps for wireless home network. *International Topical Meeting on Microwave Photonics*, (pp. 1-3).

Lee, C. H. (2007). *Microwave photonics*. Taylor & Francis Group.

Li, J., Xu, K., Wu, J., & Lin, J. (2007). A simple configuration for WDM full-duplex radio-over-fiber systems. *33rd European Conference and Ehxibition of Optical Communication (ECOC)*, (pp. 1-2).

Li, M. J., Soulliere, M. J., Tebben, D. J., Neder-lof, L., Vaughn, M. D., & Wagner, R. E. (2005, October). Transparent optical protection ring architectures and applications. *Journal of Lightwave Technology, 23*, 3388–3403. doi:10.1109/JLT.2005.856240

Li, Z., Nirmalathas, A., Bakaul, M., Cheng, L., Wen, Y. J., & Lu, C. (2005). Application of distributed Raman amplifier for the performance improvement of WDM millimeter-wave fiber-radio network. *The 18th Annual Meeting of the IEEE Lasers and Electro-Optics Society*, (pp. 579–580).

Lim, C., Ampalavanapillai, N., Novak, D., Waterhouse, R., & Yoffe, G. (2004, October). Millimeter-wave broad-band fiber-wireless system incorporating baseband data transmission over fiber and remote LO delivery. *Journal of Lightwave Technology, 18*(10), 1355–1363. doi:10.1109/50.887186

Lim, C., Nirmalathas, A., Attygalle, M., Novak, D., & Waterhouse, R. (October de 2003). On the merging of millimeter-wave fiber-radio backbone with 25-GHz WDM ring networks. *Journal of Lightwave Technology, 21*(10), 2203-2210.

Lim, C., Nirmalathas, A., Bakaul, M., Gamage, P., Lee, K., & Yang, Y. (2010, February). Fiber-wireless networks and subsystem technologies. *Journal of Lightwave Technology, 28*(4), 390–405. doi:10.1109/JLT.2009.2031423

Lin, W. P. (2005, September). A robust fiber-radio architecture for wavelength division- multiplexing ring-access networks. *Journal of Lightwave Technology, 23*, 2610–2620. doi:10.1109/JLT.2005.853150

Liu, Z., Sadeghi, M., de Valicourt, G., Brenot, R., & Violas, M. (February de 2011). Experimental validation of a reflective semiconductor optical amplifier model used as a modulator in radio over fiber systems. *IEEE Photonics Technology Letters, 23*, 576– 578.

Lu, H. H., Patra, A., Ho, W. J., Lai, P. C., & Shiu, M. H. (2007, October). A full-duplex radio-over-fiber transport system based on FP laser diode with OBPF and optical circulator with fiber Bragg grating. *IEEE Photonics Technology Letters, 19*, 1652–1654. doi:10.1109/LPT.2007.905077

Medeiros, M., Laurncio, P., Correia, N., Barradas, A., da Silva, H., & Darwazeh, I. ... Monteiro, P. (2007). Radio over fiber access network architecture employing reflective semiconductor optical amplifiers. *ICTON Mediterranean Winter Conference*, (pp. 1–5).

Mohamed, N., Idrus, S., & Mohammad, A. (2008). *Review on system architectures for the millimeter-wave generation techniques for RoF communication link*. IEEE International RF and Microwave Conference.

Nakasyotani, T., Toda, H., Kuri, T., & Kitayama, K. (January de 2006). Wavelength-division-multiplexed millimeter-waveband radio-on-fiber system using a supercontinuum light source. *Journal of Lightwave Technology, 24*, 404–410.

Ng'oma, A. (2005). *Radio-over-fibre technology for broadband wireless communication systems*. Doctoral Thesis.

Ng'oma, A., Fortusini, D., Parekh, D., Yang, W., Sauer, M., & Benjamin, S. (2010, August). Performance of a multi-Gbps 60 GHz radio over fiber system employing a directly modulated optically injection locked VCSEL. *Journal of Lightwave Technology, 28*(16), 2436–2444. doi:10.1109/JLT.2010.2046623

Nirmalathas, A., Novak, D., Lim, C., & Waterhouse, R. B. (October de 2001). Wavelength re-use in the WDM optical interface of a millimeter-wave fiber wireless antenna base-station. *IEEE Transactions Microwave Theory and Techniques, 49*(10), 2006–2012.

O'Reilly, J., & Lane, P. (1994, February). Remote delivery of video services using mm-waves and optics. *Journal of Lightwave Technology, 12*(2), 369–375. doi:10.1109/50.350584

O'Reilly, J. J., Lane, P. M., Heidemann, R., & Hofstetter, R. (1992, November). Optical generation of very narrow linewidth millimetre wave signals. *Electronics Letters, 28*(25), 2309–2311.

Ogawa, H., Polifko, D., & Banba, S. (December de 1992). Millimeter-wave fiber optics systems for personal radio communication. *IEEE Transactions on Microwave Theory and Techniques, 40*, 2285–2293.

Olver, A. (1990). Potential and applications of millimetre-wave radar. *IEE Colloquium on Millimetre-Wave Radar*, (pp. 1-4).

Oualkadi, A. E. (2011, November). *Obtenido de trends and challenges in CMOS design for emerging 60 GHz WPAN applications*. Intechopen. Retrieved from http://www.intechopen.com

Pan, J. J. (1988, May). 21 GHz wideband fiber optic link. *MTTS International Microwave Symposium Digest*, (pp. 977–978).

Parekh, D., & Yang, W. Ng'Oma, A., Fortusini, D., Sauer, M., Benjamin, S., ... Chang-Hasnain, C. (2010). *Multi-Gbps ask and QPSK-modulated 60 GHz RoF link using an optically injection locked VCSEL*. Optical Fiber Communication (OFC), Collocated National Fiber Optic Engineers Conference.

Pato, S., Pedro, J., & Monteiro, P. (2009, July). *Comparative evaluation of fibre-optic architectures for next-generation distributed antenna systems*. 11th International Conference on Transparent Optical Networks.

Peng, P. C., Feng, K. M., Chiou, H. Y., Peng, W. R., Chen, J. J., & Kuo, H. C. (2007). Reliable architecture for high capacity fiber-radio systems. *Optical Fiber Technology, 13*(3), 236–239. doi:10.1016/j.yofte.2007.02.001

Pergola, L., Gindera, R., Jäger, D., & Vahldieck, R. (2007, March). An LTCC-based wireless transceiver for radio-over-fiber applications. *IEEE Transactions on Microwave Theory and Techniques, 55*(3), 579–587. doi:10.1109/TMTT.2006.890527

Pleros, N., Vyrsokinos, K., Tsagkaris, K., & Tselikas, N.. (June de 2009). A 60 GHz radio-over-fiber network architecture for seamless communication with high mobility. *Journal of Lightwave Technology, 27*(12), 1957-1967.

Potenza, M. (1996, August). Optical fiber amplifiers for telecommunication systems. *IEEE Communications Magazine, 34*, 96–102. doi:10.1109/35.533926

Qi, G., Yao, J., Seregelyi, J., Paquet, S., & Blisle, C. (November de 2005). Generation and distribution of a wide-band continuously tunable millimeter-wave signal with an optical external modulation Technique. *IEEE Transactions on Microwave Theory and Techniques, 53*, 3090–3097.

Radziunas, M., Glitzky, A., Bandelow, U., Wolfrum, M., Troppenz, U., Kreissl, J., & Rehbein, W. (2007, January/February). Improving the modulation bandwidth in semiconductor lasers by passive feedback. *IEEE Journal on Selected Topics in Quantum Electronics, 13*, 136–142. doi:10.1109/JSTQE.2006.885332

Schmitt, G. (2009). The green base station. *4th International Conference on Telecommunication - Energy Special Conference (TELESCON)*, (pp. 1-6). Frankfurt, Germany.

Seeds, A. J., & Williams, J. (2006, December). Microwave photonics. *Journal of Lightwave Technology*, *24*(12), 4628–4641. doi:10.1109/JLT.2006.885787

Shen, A., Make, D., Poingt, F., Legouezigou, L., Pommereau, F., & Legouezigou, O. ... Duan, G. (2008). *Polarisation insensitive injection locked Fabry-Perot laser diodes for 2.5Gb/s WDM access applications*. European Conference on Optical Communications (ECOC).

Shih, P., Lin, C., Huang, H., Jiang, W., Chen, J., & Ng'oma, A. ... Chi, S. (2009). 13.75-Gb/s OFDM signal generation for 60-GHz RoF system within 7-GHz license-free band via frequency sextupling. *35th European Conference on Optical Communication*, (pp. 1-2).

Shin, D. S., Kim, W. R., Woo, S. K., & Shim, J. I. (2008, June). Low-detuning operation of electroabsorption modulator as a zero-bias optical transceiver for Picocell radio-over-fiber applications. *IEEE Photonics Technology Letters*, *20*(11), 951–953. doi:10.1109/LPT.2008.922916

Smith, G., Novak, D., & Ahmed, Z. (1997, January). Technique for optical SSB generation to overcome dispersion penalties in fiber-radio systems. *Electronics Letters*, *33*, 74–75. doi:10.1049/el:19970066

Stöhr, A., Babiel, S., Cannard, P. J., Charbonnier, B., van Dijk, F., & Fedderwitz, S. (2010). Millimeter-wave photonic components for broadband wireless systems. *IEEE Transactions on Microwave Theory and Techniques*, *58*(11), 3071–3082. doi:10.1109/TMTT.2010.2077470

Stohr, A., Kitayama, K., & Jager, D. (1999, July). Full-duplex fiberoptic RF subcarrier transmission using a dual-function modulator/photodetector. *IEEE Transactions on Microwave Theory and Techniques*, *47*, 1338–1341. doi:10.1109/22.775476

Suman, S., & Sudhir, D. (2007, November). Hybrid wireless-optical broadband-access network (WOBAN): A review of relevant challenges. *Journal of Lightwave Technology*, *25*(11), 3329–3340. doi:10.1109/JLT.2007.906804

Sun, X., Chan, C. K., Wang, Z., Lin, C., & Chen, L. K. (2007, April). A singlefiber bi-directional WDM self-healing ring network with bidirectional OADM for metro-access applications. *IEEE Journal on Selected Areas in Communications*, *25*, 18–24. doi:10.1109/JSAC-OCN.2007.023305

Technologies, C. (n.d.). *60 GHz reflective electroabsorption modulator*. Obtenido. Retrieved from www.ciphotonics.com

Teshirogi, T., & Yoneyama, T. (2001). *Modern millimeter-wave technologies*. Ohmsha, Ltd.

Thakur, M. P., Quinlan, T., Ahmad Anas, S., Hunter, D. K., Walker, S. D., & Smith, D. W. ... Moodie D. (2009). Triple-format, UWB-WiFi-WiMax, radio-over-fiber co-existence demonstration featuring low-cost 1308/1564 nm VCSELs and a reflective electro absorption transceiver. *Optical Fiber Communication (OFC), Collocated National Fiber Optic Engineers Conference (NFOEC)*, (pp. 1-3).

Thakur, M. P., Quinlan, T., Bock, C., Walker, S. D., Toycan, M., & Dudley, S. (2009, February). 480-Mbps, bi-directional, ultra-wideband radio-over-fiber transmission using a 1308/1564-nm reflective electro-absorption transducer and commercially available VCSELs. *Journal of Lightwave Technology*, *27*(3), 266–272. doi:10.1109/JLT.2008.2005644

Toda, H., Nakasyotani, T., Kurit, T., & Kitayama, K. (2004). WDM mm-wave-band radio-on-fiber system using single supercontinuum light source in cooperation with photonic up-conversion. *IEEE International Topical Meeting on Microwave Photonics*, (pp. 161-164).

Tsuzuki, K., Shibata, Y., Kikuchi, N., Ishikawa, M., Yasui, T., Ishii, H., & Yasaka, H. (2008). 10-Gbit/s, 200 km duobinary SMF transmission using a full C-band tunable DFB laser array co-packaged with InP Mach-Zehnder modulator. *IEEE 21st International Laser Conference*, (pp. 17-18).

Wake, D., Lima, C. R., & Davies, P. A. (1995, September). Optical generation of millimeter-wave signals for fiber-radio systems using a dual-mode DFB semiconductor laser. *Microwave Theory and Techniques*, *43*(9), 2270–2276. doi:10.1109/22.414575

Wake, D., Nkansah, A., & Gomes, N. (2010). Radio over fiber link design for next generation wireless systems. *Journal of Lightwave Technology*, *28*, 2456–2464. doi:10.1109/JLT.2010.2045103

Wake, D., Nkansah, A., Gomes, N., de Valicourt, G., Brenot, R., & Violas, M. (2010, August). A comparison of radio over fiber link types for the support of wideband radio channels. *Journal of Lightwave Technology*, *28*(16), 2416–2422. doi:10.1109/JLT.2010.2046136

Wen, H., Chen, L., He, J., & Wen, S. (2007, October). Simultaneously realizing optical millimeter-wave generation and photonic frequency down-conversion employing optical phase modulator and sidebands separation technique. *Asia Optical Fiber Communication and Optoelectronics Conference*, (pp. 427–429).

Wong, E., Prasanna, A. G., Lim, C., Lee, K. L., & Nirmalathas, A. (2009, April). Simple VCSEL base-station configuration for hybrid fiber-wireless access networks. *IEEE Photonics Technology Letters*, *21*, 534–536. doi:10.1109/LPT.2009.2014393

Yao, J. (2009, February). Microwave photonics. *Journal of Lightwave Technology*, *27*(3), 314–335. doi:10.1109/JLT.2008.2009551

Zhensheng, J., Jianjun, Y., Georgios, E., & Chang, C. G.-K. (2007, November). Key enabling technologies for optical–wireless networks: Optical millimeter-wave generation, wavelength reuse, and architecture. *Journal of Lightwave Technology*, *25*(11), 3452–3471. doi:10.1109/JLT.2007.909201

Zhou, J., Xia, L., Cheng, X., Dong, X., & Shum, P. (February de 2008). Photonic generation of tunable microwave signals by beating a dual wavelength single longitudinal mode fiber ring laser. *Applied Physics, 91*, 99–103.

Chapter 10
WLAN Systems for Communication in Transportation Systems:
Towards the Benefits of a Cooperative Vehicular Approach

Riccardo Scopigno
Istituto Superiore Mario Boella (ISMB), Italy

ABSTRACT

Vehicular Ad-Hoc Networks (VANETs) are wireless networks primarily meant to enforce vehicular safety. The incumbent international VANET solution is based on an adaptation of WLAN to the 5.9 GHz band and to the vehicular environment: it is universally known as IEEE 802.11p. One of the main reasons for the success of IEEE 802.11p lies on the functional requirement of a decentralized solution, that is, one able to work in the absence of infrastructure. While Filed-Operational Tests are being developed world-wide and new VANET applications, not restricted to safety, are being developed, new requisites are emerging. Some limitations of the IEEE 802.11p are coming to light as well: stakeholders must be aware of them to prevent misleading conclusions on reliability and, most importantly, improper solutions for the safety which the protocol is aimed at.

INTRODUCTION

The communication systems adopted by transportation systems can be classified according to their role; for instance, a possible distinction can be made between communications which are

DOI: 10.4018/978-1-4666-2976-9.ch010

restricted on-board and others spanning multiple vehicles. Wireless is basically used for vehicle-to-vehicle communications – when possible, cables are still considered much more reliable. In recent years, wireless has been intensively investigated for transportation. In particular, focusing on the automotive field, cooperation between cars has been proposed with the main goal of enforcing safety; wireless is considered the main enabler of this approach. The rationale behind it is to let vehicles exchange information (such as position, speed, obstacles, etc.) so as to prevent possible accidents, or to allow timely reaction to ones which have already occurred.

For instance, *eCall* is a European initiative whose aim is rapid assistance to drivers involved in a collision anywhere in Europe, as suggested by the European Commission (2011). The project involves deploying a hardware black box, to be installed in vehicles, collecting on-board data - for instance airbag and impact sensor information, GPS coordinates – and forwarding it to local emergency agencies (by dialing 112), in case of an accident. The project is also supported by the European Automobile Manufacturers Association (ACEA) and ERTICO; implementation is expected by 2015. In North America, a similar service is available from OnStar Service, a subsidiary of General Motors. Their new service, called Advanced Automatic Crash Response (AACR), is designed to assist emergency response efforts. The example of eCall is particularly interesting because it demonstrates how much the international community can work jointly and in a concerted way towards a common goal when safety is concerned: at present, 22 EU Member States formally declared their support for a Europe-wide eCall system.

Based on the experience of eCall, one could think that mobile networks can be used for the collection of additional information about other vehicles and roads, contributing to the prevention of accidents. This adheres to the idea of the telematics horizon, a concept which has been explored by several research projects: the extension of a vehicle's field of vision, providing the driver with information spanning hundreds of meters, much beyond the natural visual horizon. In principle, UMTS and WiMax would be two candidates: they both support mobility (even at vehicular speeds) and have migrated to a data-oriented paradigm (rather than a voice one). However there are some non-negligible aspects which hinder their

application. First of all they require an expensive and capillary coverage. This means that, in order to have a reliable service, all the roads would have to be covered by a mobile infrastructure. Furthermore, mobile services are expensive and it is not clear who should pay for them: if a subscription were required, it would probably not be mandatory; on the other hand, if not mandatory, only a few cars would be equipped, so the safety service, which is intrinsically based on the concept of cooperation, would not work. The other alternative would be for the infrastructure to be paid for by governments, but this would hardly be sustainable economically. From a technical point of view, safety services require real-time delivery, roughly within 100ms: given the set-up time of a call, the strict end-to-end delay would imply that each car were always connected – so as to skip the call set-up. The main drawback would be on the scalability of the mobile network: it could hardly manage the number of subscribers of a metropolitan area.

Altogether this led to the conclusion that a wireless network with infrastructure would not be the most appropriate solution for vehicular safety communications. The other possible paradigm is that of *Ad-Hoc Networks*. In the following sections, after a brief introduction to Vehicular Ad-Hoc Networks (VANETs) and Services, the main phenomena affecting VANETs will be covered. Based on this preliminary analysis, the incumbent international solution will then be discussed, highlighting its strengths and weaknesses.

VEHICULAR AD-HOC NETWORKS

The paradigm of ad-hoc networks requires that wireless devices communicate with each other directly: hence, wireless nodes, which are in the respective radio ranges, mutually discover and communicate in peer-to-peer fashion, without involving any central coordinators. While the infrastructure mode (with the so called *access points* - AP) is the most common way in which WiFi is used, the ad-hoc mode is already included in WiFi IEEE 802.11 (2007).

Actually, the ad-hoc mode is easily provided by IEEE 802.11 because the medium access policy is intrinsically distributed (or *decentralized*) and does not require a central coordinator, even when an AP is available. In fact, even in the infrastructure mode, the medium access is ruled by a stochastic approach called Carrier-Sense Multiple Access with Collision Avoidance (CSMA/CA), where the AP has no privilege. As better explained in the following sections, the rationale of CSMA/CA is to let all the nodes sense the medium and transmit after a random time, so as to prevent collisions. Consequently, the AP gains access to the channel exactly in the same way as all the other nodes. The distinctive role of the AP is restricted to *(i)* the management of the network (the so-called Infrastructure Basic Service Set – IBSS), *(ii)* the forwarding of transmissions (in the infrastructure mode all the nodes transmit and receive bridging on the AP) and *(iii)* the prevention of hidden terminals (only for unicast transmissions).

All in all, the CSMA/CA channel access mechanism is intrinsically distributed. This certainly represented a strong advantage for the supporters of IEEE 802.11, when the stakeholders involved started the selection of a possible MAC solution for vehicular communications. The history began in the early 1990s, when the U.S. Department of Transportation (DOT) and the Intelligent Transportation Society of America (ITSA) established a plan for Intelligent Transportation System (ITS) services, recognizing a key potential role for wireless communications. In 1997 the ITSA petitioned the Federal Communications Commission (FCC) for 75 MHz of bandwidth in the 5.9 GHz band: two years later, the spectrum in the 5.85-5.925 GHz range was allocated to DSRC-based ITS radio services. The primary purpose was to enable public safety applications to save lives and improve vehicular traffic; however, other (e.g. entertainment) services were also permitted, in order to

lower the network deployment and maintenance costs and to encourage DSRC development and adoption.

The adoption of a single standard for the physical (PHY) and MAC layers was encouraged and, in 2004, a new IEEE task group (task group *p* of the IEEE 802.11 working group) started to develop an amendment to the IEEE 802.11 standard. The document is known as IEEE 802.11p (IEEE 802.11p, 2010) and ratified in July 2010. It standardizes a solution very close to the Ad-Hoc mode of 802.11a: hence, the name Vehicular Ad-Hoc Network (VANET). At the time of this writing, IEEE has just merged of IEEE 802.11p within IEEE 802.11 (IEEE 802.11, 2012), further emphasizing that it represents just a new flavor of WiFi. Given the global relevance of the topic and of the potential market, international harmonization efforts are being carried out: it is worth mentioning the role played by: FCC (USA), CEPT (Europe), and ITU for frequency regulation; IEEE (USA) and ETSI (Europe) for the MAC and physical layers; SAE (USA), ETSI and CEN (Europe), and ISO for the safety services on top of network/MAC layers.

VANET Services

VANETs should be decentralized and able to work even in the absence of any fixed infrastructure: this is definitely a demanding requirement which weeds out a lot of candidate technologies. However it is not the only requisite: there are additional ones derived from the services, and others which are set by environmental conditions. In this section the analysis will highlight the possible services and their respective requisites on the VANETs.

In recent years, VANET services have been extensively investigated by the research community (Huang, 2010), (Olariu & Weigle, 2009), (Watfa, 2010), (Schoch et al., 2008). A specific research area concerns the analysis of possible applications and services which can best exploit the cooperation enabled by wireless channels, as discussed by

Papadimitratos et al. (2009); most of the services focus on safety: for instance, the enforcement of intersection assistance, as proposed by Dötzer et al. (2005) and collision warning, analyzed also by ElBatt et al. (2006) and shown in Figure 1. However safety applications are not the only ones.

Based on their targets, potential applications for VANETs can be arranged into: safety, transport efficiency, and information/entertainment (infotainment) applications (Hossain et al., 2010). According to others - for instance Schoch et al. (2008) – the classification can be further enriched, making a distinction between active safety and public service. Here a short description is provided based on Schoch's classification.

1.   **Safety:** These services are basically warnings which are broadcasted by vehicles or road-side units (RSUs) to prevent accidents, by warning drivers of hazards. In this category fall, for example, the following possible messages: traffic signal violation

warning, stop sign violation warning, wrong way driver warning, intersection collision warning, pedestrian crossing information, lane change warning, emergency vehicle approaching warning, work zone warning, curve-speed warning, road condition warning, and emergency electronic brake lights (Hartenstein & Labertaux, 2010). Safety can also be enforced by messages supporting the driver, such as intersection assistance, merging assistance, left turn assistance - as discussed by (Schoch et al., 2008) - but also vehicle safety inspection. Finally, informative data has also been proposed in literature for safety, such as: electronic license plate, electronic driver's license (for instance, see Raya et al, 2007), stolen vehicle tracking, post-crash/breakdown warning, pre-crash sensing, event data recording. Safety messages are typically meant to be broadcasted over a single hop, with time-critical delivery, as reliable as possible; the transmission is

Figure 1. An example of safety applications: intersection assistance. The service is meant to prevent collisions at the intersections or, when they occur (as in the example), to limit the effects. Traffic lights and vehicles cooperate to warn other drivers. (Courtesy of ISMB – Source www.ms-aloha.eu)

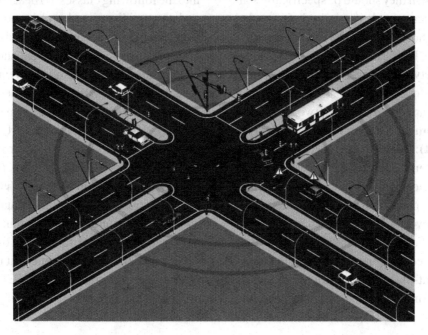

supposed to be either event-triggered (e.g. electronic brake lights) or periodic with a short period (e.g., intersection assistance). Most of them are just transmitted over a single hop.

2. **Transport Efficiency:** Or Traffic Management. Based on experience in some major cities (e.g. the case of Oakland County mentioned in Qingyan Yang et al. (2008), Traffic Operation Centers (TOCs) could play a significant role in traffic management - as suggested by Kroh (2006). In particular, TOCs could be involved in and enable several Traffic Management scenarios such as: collection of crash data, collection of weather data by vehicle probes, collection of origin and destination information, hazardous material cargo tracking, electronic toll payment, and rental car processing. Other services, such as cooperative platooning, and intelligent traffic flow control, could work with or without TOCs. Transport Efficiency applications recall a *unicast* or *anycast* paradigm rather than broadcast. In other words, if messages relevant to traffic management should be collected by a central station, than they should be specifically sent to it (unicast) and, if required, forwarded by some node towards the final destination (anycast). Traffic management is less critical than safety, spans over multiple hops and is likely to require a connection between VANET and the Internet Network.

3. **Infotainment (a.k.a. Enhanced Driver Comfort):** Infotainment services constitute a heterogeneous class of applications which are meant to improve driving comfort, including the identification of points of interests or the improvement of navigation, by means of updates. For instance, parking spot locator, GPS corrections, navigator map download and point-of-interest notification, media broadcasting, internet service provisioning

and info on fueling fall in this class. While these are basically vehicle-to-infrastructure communications, vehicle-to-vehicle communications can also be involved: this is the case of instant messaging (between vehicles) and cooperative positioning.

Obviously, the previous list is far from being exhaustive, and the topic of VANET services is still evolving. For the sake of completeness, different taxonomies of the services have been proposed in literature, depending, for instance, on the nodes involved in the protocol exchanges, according to Hartenstein & Labertau (2010): Vehicle-to-Vehicle (V2V and *vice versa*), Vehicle-to-Infrastructure (V2I), Vehicle-to-Pedestrian (V2P) and, for the niche application of support of Electric Grids, Vehicle-to-Grid (V2G) (Turton & Moura, 2008).

Altogether, the situation is very complex and subtends different communication paradigms: still according to Schoch et al. (2008), five of these can be identified. While the classification is questionable, it may help in understanding the heterogeneous requirements which VANETs need to face. Summing up, communications can fall into the following classes: *(i)* beaconing (periodic updates, unidirectional and broadcasted over a single hop), *(ii)* geobroadcast (data sent over a large area, triggered by an event and warning the approaching vehicles of sudden events or changing road conditions); *(iii)* unicast (connection-oriented services where data is sent to a specific destination - like another vehicle or an RSU. Unicasts can be sent over a single hop or over multiple hops and may be uni- or bi-directional). *(iv)* Advanced information dissemination and *(v)* information dissemination are more conceptual and disputable and are meant to optimize transmissions, by aggregating and prioritizing data over VANET.

For the sake of completeness, it is worth mentioning the importance of VANET scenarios in order to achieve an effective identification of pos-

Figure 2. An example of safety application: if a car stops, its position and state can be announced to the oncoming vehicles, so as to prevent accidents in case of fog. This is, literally, an example where the telematic horizon widens the visual one. (Courtesy of ISMB – Source www.ms-aloha.eu)

sible protocol solutions: also ETSI has addressed the issue and provided an overview in ETSI TR 102 638 (2009).

Conclusions on VANET Requirements

VANET services are, altogether, particularly heterogeneous. Table 1 sums up their characteristics,

depending on the service classification. Again, in this case the classification itself has no ambition to be unerring, as with all the taxonomies; rather, it is meant to deduce directions for the deployment of VANETs.

The following conclusions can be drawn:

- The biggest differences are between safety and non-safety services: since safety

Table 1. Qualitative analysis of service requirements based on their classifications

	Safety	Traffic Management	Enhanced Driver Comfort
Destination	Mainly broadcast Unidirectional	Unicast/Anycast Bidirectional	Unicast/Broadcast Mainly Unidirectional
Forwarding Model	Mainly single hop	Multi-hop forwarding if sparse infrastructure (*e.g.* towards TOC)	Multi-hop forwarding if sparse infrastructure
Generation	Both event-triggered and beaconing	Event-triggered	On demand or continuously sent.
QoS Requirements	QoS demanding, Time-Critical	Best-Effort, Delay-Tolerant	Best-Effort, Delay-Tolerant
Connection	Limited to VANET	Extended to TOC	Possibly over Internet
Georouted[1]	No	Yes	Yes

services are those which primarily motivated VANETs and are the most pressing, they should drive VANET design. Consequently, VANETs should first support timely and reliable delivery of broadcast messages, over single hop, obviously in a distributed way.

- All the other functionalities (multi-hop forwarding, geo-routing, bi-directional communications) are considered requirements with lower priority. Moreover they have less critical implications on medium access and can be carried out by additional blocks.
- VANET architectures, as defined by IEEE and ETSI (in the "Introduction to VANET solutions" section) reflect this perspective: the lower layers alone can fulfill safety services. The other services involve functionalities which are peculiar of upper layers.
- Given the primary relevance of safety services and the state of international standardization, this chapter focuses more on the lower logical layers (physical and MAC) and on the evaluation of their effectiveness, dealing with the service requirements and the physical phenomena which typically affect VANETs.

For the purpose of facilitating the analysis, the following section provides an introduction to the main environmental effects which are expected to affect VANETs.

PHYSICAL PHENOMENA AFFECTING VANETS' PERFORMANCE

The analysis on VANETs is very often restricted to the protocol mechanisms, disregarding physical layer (propagation phenomena) which cannot and should not be neglected. This is due to the typical scientific approach to telecommunications, which parcels out a problem into distinct ones, in

order to make it more manageable; however, this method is sometimes too simplistic. For instance, propagation is usually kept separate from network simulation. As a result, only statistical propagation models are adopted for the analysis of MACs. This way, obstructions cannot be taken into account, and this approximation has recently been demonstrated, by Boban et al. (2011), to be unjustified. What is more, simplifications may lead to wrong conclusions. Therefore, it is important to be fully aware of the simplifications which are applied to any study. This section means to investigate this area, providing a deeper insight into all the physical phenomena which are always dismissed by VANETs simulations.

The first sections investigate the models used for the computations of signal to interference and noise ratio (SINR) and, subsequently, Packet-Error-Rate (PER). For this purpose some preliminary concepts are here shortly recalled.

First of all, given a transmitted power P_t, the received power P_r can be computed as follows:

$$P_r = P_t \cdot G_t\left(\vartheta_t, \varphi_t\right) \cdot G_r\left(\vartheta_r, \varphi_r\right)$$
$$\cdot\left(1 - |\Gamma_t|^2\right) \cdot \left(1 - |\Gamma_r|^2\right) \cdot \left|\vec{a}_t \cdot \vec{a}_r^*\right| \cdot PL_\alpha \cdot \qquad (1)$$

The Equation is known as *Friis' equation* and states that the received power is a function including several terms, not only due to propagation. Usually, in network simulators, such as NS-2 and Omnet++, only the attenuation and fading, which are described by the term PL_α, can be computed. In other words, the received power is computed neglecting the following terms.

$G_t\left(\vartheta_t, \varphi_t\right)$ and $G_r\left(\vartheta_r, \varphi_r\right)$ are the antenna gains, in polar coordinates $\left(\vartheta, \varphi\right)$. Antennas can gain more or less, depending on their emission diagrams: for instance, the emission can be symmetric or asymmetric, thus amplifying forward or backward directions. Additionally, antennas, when mounted on the top of the roof of vehicles, significantly change their diagram. Altogether, the approximation of setting gains isotropically equal to 1 is acceptable in most cases, to understand general performance of VANETs; however the approximation can be arbitrary and even wrong for the evaluation of certain protocols (for instance, some forwarding protocols can be influenced by directional antennas).

Γ_t and Γ_r represent the reflection coefficients for the power received (respectively transmitted) inside the transceiver. They are meaningful only in case of impedance mismatch. So, in most cases, it is acceptable to neglect them.

Finally, the term $\left| \vec{a}_t \cdot \vec{a}_r^* \right|$ accounts for polarization vectors of the transmit and receive antennas, respectively, taken in the appropriate directions. The polarization of an antenna is the polarization of the wave radiated by the antenna and can be *linear* (horizontal, vertical, other), or *elliptical* (right hand, left hand). The received signal is the sum of the direct signal plus a number of reflected signals, and reflections can change the polarization; consequently, the overall polarization of the received signal can change slightly, when compared to the polarization of the transmitted signal. Usually, network simulations assume that the polarization effects are embedded in the fading term (included in PL_α .tern of Equation (1)).

Supposing that the received power is known, and is computed following Friis' equation, the signal to noise and interference ratio (SINR) can be computed. Here it is important to consider both the *noise* term (which is due to environmental physical reasons) and the interference coming from *all* the signals simultaneously transmitted. So, SINR must be computed considering the *sum* of all the interferences.

$$SINR = \frac{P_r}{P_n + \sum_i P_i^{int}} \qquad (2)$$

In the SINR formula, P_n is the power of noise and P_i^{int} the power of the generic *i*-th interfering signal. The cumulative function of the interfering signals is something which cannot be neglected and can strongly influence the realism of results. For instance, this approach was integrated into NS-2 by Mercedes-Benz (Chen et al., 2007), quite recently. Once the SINR is known, the correct reception of the frame can be estimated. This is something which is to be evaluated as well in network simulations. Chen et al. (2007) introduced other improvements, such as differentiated *SINR* acceptance thresholds ($SINR_m$) and signal sensitivity thresholds for the different transfer rates (VANET can exploit different digital modulation and encoding techniques).

The sensitivity of a receiver (or other detection device) is the minimum magnitude of input signal required to produce a specified output signal. It also accounts for the internal noise figure of the receiver. SINR at the input of the receiver is the best figure. Each component in the receiver cascade performs its intended function but also degrades SINR. Friis (1994), introduced the concepts of *Noise Figure* (*F*) to characterize the degradation of SINR by the receiver, as the ratio between the available SINR at the input and at the output of the receiver.

Different types of noise affect receivers: thermal noise (Nyquist, 1928), shot noise and Flicker noise (Hogg & Mumford, 1960). Thermal (a.k.a. Johnson noise) is the main limit on signal sensitivity and is due to thermal agitation of free electrons in the environment (and in the resistors). It is white noise (with flat spectral density) and is proportional to the temperature of the resistor, whose available power is $N_t = kT_oB$, where k is Boltzman's constant ($1,38 * 10^{-23}$ Joule/K), T is the temperature in Kelvin degrees (K) and B the bandwidth. At 290 K, kT_oB is -114 dBm/MHz.

If G is the gain of the receiver (which amplifies both signal and noise), and indexes i and o refer respectively to input and output of the receiver, than the following equations hold:

$$F = \frac{S_i/N_i}{S_o/N_o} = \frac{N_o}{G N_i} = \frac{N_o}{G k T_0 B} = \frac{G k T_0 B + N_R}{G k T_0 B}$$

(3)

where N_R is the noise added by the receiver and it is supposed to be affected mainly by thermal noise (no interferences – best condition). N_R can also be expressed as an equivalent noise temperature of the system. The Equation (3) permits to compute $N_0 = F G k T_0 B$.

Input sensitivity is evaluated by referring the output noise to the receiver's input $N_{0i} = F k T_0 B$.

Thus, environmental noise is considered at the input, and all the internal noise sources of the receiver collapse into figure F. Notably, a rich literature explains how to compose the gain G_i and figure F_i of the distinct blocks composing a receiver cascade into an overall F^2. This means that the input power S_i must exceed a given minimum threshold S_m, so that $S_i/N_{0i} \geq SINR_m$. Then

$$S_m = SINR_m \cdot F k T_0 B.$$

(4)

All in all, a packet can be sensed if the received power exceeds the carrier-sense sensitivity threshold; it can be received if the received power exceeds the *sensitivity* threshold of its transfer rate *and* its *SINR* (which also depends on interference) exceeds the reception threshold (of the transfer rate).

To complete the overview, other phenomena affect the statistics on packet receptions. For instance, due to the effects of digital redundancy codes (the so-called coding rates, as shown in Table 4) and Viterbi decoding, some errors can be corrected. As a results, packet reception can be expressed as a statistical process which is also a function of packet length l. For this reason, based on the Mercedes-Benz's release, Abrate et al. (2011) proposed to define the function *PER* $= f(SINR, l, R)$: the probability of correct reception also considers the role of packet length l and transfer-rate R. The model is depicted in Figure 3: due to its stochastic nature, PER does not fall sharply, but rather decreases smoothly, from 1 (error) to 0 (correct reception), as the SINR grows high.

The simulation detail introduced by Abrate et al. (2011) further improves the realism of simulations; however its effect on result does not seem to be dramatic.

Attenuation

It has been concluded that, considering all the possible interfering signals, SINR can be correctly computed; subsequently, the reception probability

Table 4. How the typical environmental settings are differently affected by the propagation phenomena

	Traffic	Obstruction by Building	Speed	Dominant Phenomena
Urban	+ + + +	+ + + +	-	Heavy Fading, Obstructions (HT), Very High Density
Sub-Urban	+ +	+ +	+ +	Fading, Obstructions, Speed, High Density
Rural	-	+	+ + +	Fading, High Speed
Highway	+ +	+	+ + + +	Fading, Very High mutual speed

Figure 3. PER as a function of packet length l and transfer-rate R, not only of SINR. The paper from Abrate et al. (2011) also proposes an analytical model, whose numerical fitting is here displayed for the case of R = 27Mb/s (64 QAM, coding rate 3/4).

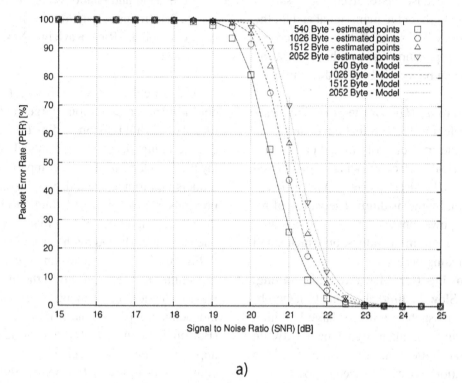

a)

b)

can be evaluated for each given frame-length and transfer-rate. However, the problem is only partially solved, because the received power should be computed, both for the main signal and for the interfering ones. The signal fades over space and time, due to propagation reasons: the propagation phenomena can be mainly broken down into:

- **Attenuation (Or Large-Scale Fading):** This refers to the nominal attenuation due to: *(i)* the spreading of emitted power over a spherical surface growing with squared power, *(ii)* reflections by ground, *(iii)* absorption by the medium. The values of received power obtained by the only attenuation term are mean values, on which small scale fading occurs.

- **Shadowing (Or Medium-Scale Fading, a.k.a. Slow Fading):** It is due to large obstructions such as a hill or large building, obscuring the main signal path between the transmitter and the receiver. These fluctuations are experienced on local-mean powers, *i.e.* they can be highlighted after removing the fast fluctuations due to multipath fading. The amplitude variations due to shadowing are often modeled using a log-normal distribution, which accounts for statistical shadowing (a certain number of obstacles evenly distributed). As an alternative, when the propagation environment is far from being a line-of-sight one (as in the case of a Manhattan-like grid), shadowing and path-loss should be jointly modeled on the specific scenario, as joint processes, as suggested by Sklar (1997) and Jeruchim et al. (2000); this approach has been followed - for urban scenarios - by Giordano et al. (2011), Sommer et al. (2011) and Scopigno & Cozzetti (2010).

- **Fading (Or Small Scale Fading, a.k.a. Multipath or Fast Fading):** Represents variations in the received signal power, of magnitude up to 40 dB, on a spatial scale

of a half-wavelength, due to small scatterers. At 5.9 GHz these are variations on a cm scale and cannot easily be modeled in a deterministic way: for this reason statistical models (Rice, Rayleigh, Nakagami) are used.

The so-called *2-Ray-Ground* (2RG) is a deterministic attenuation (large-scale fading) model which holds for omni-directional antennas; despite being simplistic, the model is useful to lay the foundations of more complex attenuation models and to achieve a deeper insight into the previous classification of fading. 2RG considers the interfering effect of two main rays: a direct one and one reflected by the ground, as depicted in Figure 4(a). For large distances d between the transmitter and receiver, given the hypothesis of ideal reflection for an incidence $\theta_i \to 0$ ($\Gamma \to 0$ independently of the polarization of the antenna (Barton & Leonov, 1997)) and some numerical approximations, the received power P_r can be computed as follows (Rappaport, 1996)[3]:

$$\frac{P_r}{P_t} = G_t G_r \cdot \left(\frac{\lambda}{4\pi d}\right)^2 \cdot \left(2\,\sin\left(\frac{2\pi}{\lambda} \cdot \frac{h_t h_r}{d}\right)\right)^2 \approx \frac{C}{d^4} \tag{5}$$

In the formula, h_t and h_r are respectively the height of the transmitting and the receiving antennas in respect to the ground, G_t and G_r are the respective antenna gains, d the distance between transmitter and receiver and λ the wavelength. At a large distance, when the argument of the sin() function (hyperbolic in d) is small enough, the Equation (5) shows a power 4 decay. The behavior is shown by the continuous line in Figure 4(b).

The following deductions can be drawn from it:

- The model holds after a cut-off distance (at about 120 m in the picture), once the last period of the *sin* () function starts. This means that there is a cut-off distance

Figure 4. Analysis of the two-ray ground model: (a) the scenario on which the interference is computed; (b) the pattern of the received power

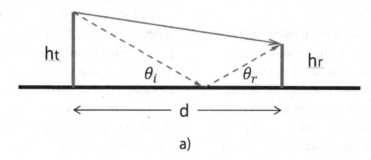

a)

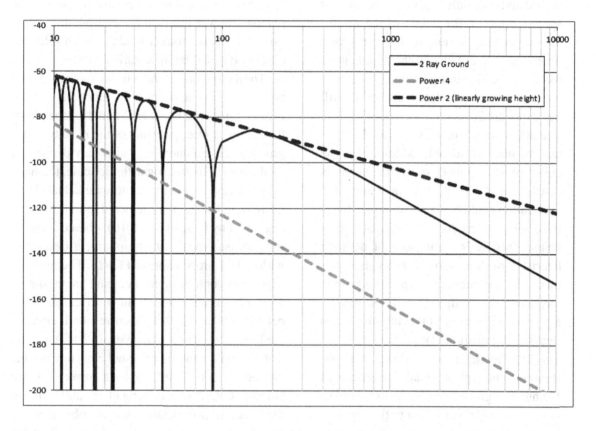

b)

at about $4\pi \cdot (h_t \cdot h_r)/\lambda$ and the 2RG model is meaningful only at longer distances.

- The *free-space* (a.k.a *Friis*) model represents the attenuation due only to the decreasing power density over space: since at distance d the area of the sphere is $4\pi d^2$, the power density and the received power

decreases as power 2 (upper dashed line in the figure). For distances shorter than the cut-off distance of 2RG, the free space model can be used and fluctuations can be superimposed onto it.

- The attenuation of 2RG, after cut-off, follows power 4; this behavior is confirmed by Figure 4 (b), where – using a logarithmic scale – the 2RG graph is parallel to the lower dashed line.

- The power 4 comes from interference: in fact, the envelope of the attenuation is still interpolated by a power 2 decay (upper dashed line), as in the *free-space* model.

- While the received power at a given (fixed) height decreases dramatically when the distance grows, the power is actually preserved and only differently distributed over space. In fact, if the height of the receiving antenna (h_r) is not constant but grows linearly as the distance d, the argument of the sin() function in Equation (3) becomes constant. Hence, the decay, in general, still has exponent 2.

- The attenuation due to absorption by scatters is not considered; additionally, the effect of the ground scattering alone is considered. Consequently, in reality, the attenuation could be a more complex function, with intermediate and variable exponents. This leads to the idea of defining heuristic models derived from measurements, by logarithmic interpolations of the mean values over distances. These models are introduced in the following discussion (leading to Equation (4)).

- The 2RG model also introduces and substantiates the typical signal fluctuations of fading. The ground represents the simplest scatter and, despite this, *(i)* in the first meters the fluctuations are fast, because, in Equation (3) the period is inversely proportional to distance and, initially, comparable to λ. With a purely geometrical approach, one can imagine that *(ii)* each scatter generates multiple rays and, due to multipath, two or more can interfere on a receiver. Even more *(iii)* multiple scatters can be present, especially considering the involved λ. Altogether, any scatters intuitively motivate fading.

Actually, *Free-Space* and *2-Ray-Ground* are two modeling solutions which hold in simplified scenarios. A real propagation environment makes reflections and multipath interferences more complex, leading to different behaviors. Moreover, at the target frequency of 5.9 GHz, the λ is about 5 cm: this means that signal fluctuations due to multipath (and interferences) vary in an unpredictable way, due to the small scale. For this reason, recently a more heuristic approach has been proposed. It can be summarized in two points: *(i)* the average received power is given by interpolating functions; *(ii)* fluctuations around the mean value are described as fading, in a statistical way.

The interpolating function used for the received power is often a decreasing function with an exponent varying between 2 and 4 (respectively the values of free space and of the ideal two-ray-ground); the exponent may also vary depending on distance. Given a logarithmic scale, the most frequent case is that of a continuous, dual slope, decreasing function, as depicted in Figure 4 and suggested by Grau et al. (2010) – based on practical measurements on the platform developed within the European project CVIS. The multiple exponents controlling the decrease are used with a cut-off distance which, in most cases, is set between 100 and 200m and corresponds either to the Fresnel distance (Cheng et al., 2007), *i.e.* the distance where the first Fresnel zone touches the ground (about 200m), or to a lower distance (Masui et al., 2002), (Cheng et al., 2007) due to the effects of pedestrians and other obstacles. By the way, the logarithmic model generalizes the 2RG model, as shown by the comparison between Figure 4(b) and 5: the dual slope can be also identified in the 2RG with the union of the upper dashed segment before cut-off and the continuous curve after cut-off. The varying exponent can be better described by a logarithmic function in a log-log scale, as in the continuous but piece-wise-defined function (6); here, given the logarithmic function, the exponents become multiplying factors (γ_1 and γ_2). The bigger γ, the more severe the

Figure 5. Received power at different distances, according to experimental exponential function (black) and with upper and lower confidence interval (by Nakagami model), with highlighted discontinuities: Nakagami parameter m at 80 and 150m; slope at 100m. Note that, as in Figure 4, the minimum power can have dramatic attenuation spikes: here only confidence intervals are shown.

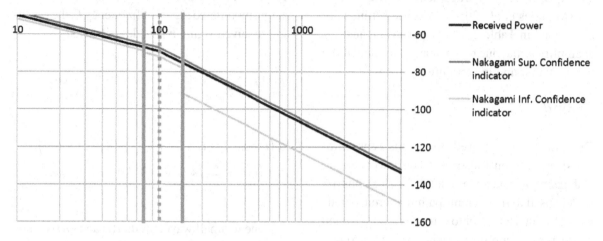

Box 1.

$$
P_{r,\ dBm} =
\begin{cases}
P_{r,\ dBm}(d_0) - 10\gamma_1\ \log_{10}\dfrac{d}{d_0}, & if\ d_0 \le d \le d_1 \\[3mm]
P_{r,\ dBm}(d_0) - 10\gamma_1\ \log_{10}\dfrac{d_1}{d_0} - 10\gamma_2\ \log_{10}\dfrac{d}{d_1}, & if\ d \ge d_1
\end{cases}
\tag{6}
$$

attenuation: if close to 4, then it is similar to two-ray-ground; if close to 2, it is similar to free-space. This is what is modeled by formula (6), with received power ($P_{r,\ dBm}$) computed in dBm (dB of power measured in mW).

More generalized definitions include three or more intervals of the d axis. The model will be called multi-log model.

Actually, the multi-log model is not accurate for very small distances; the reason can be guessed referring the case with exponent 4 - which is associated to the two-ray model. In the two-ray model the power 4 comes from combination of the two rays: however, at short distances, the combination of the two leads to strong oscillations, mainly caused by the constructive and destructive interferences; instead, the multi-log

model foresees infinite received power. This is the reason for the minimum cut-off distance d_0 in the model (between 1 and 10m); at distance d_0 the received power is computed by Friis' model, as shown in (5). d_0 can be used to tune the model according to the experimental data.

$$
P_{r,\ dBm}(d_0) = P_{t,\ dBm} - 10\ \log_{10}\left(\frac{\lambda^2}{(4\pi)^2\ d_0^2}\right).
\tag{7}
$$

Different papers provide distinct settings for the multi-log model (Grau et al., 2010), (Setiawan et al., 2007), (Torrent Moreno et al., 2004). While the differences are ascribed to the environmental settings, two common aspects are found in the available literature: the exponent varies depending on the setting (*e.g.* urban *vs.* highway) and on the distance (to better fit measurements). Exponents are expected to grow when heavy traffic is present (Meireles et al., 2010). Possible settings are mentioned in Table 2, whose purpose is just for exemplification, due to the heavy variance of the parameters found in literature.

Fading

Once the mean received power has been computed, fading must also be taken into account. Nakagami, Rician and Rayleigh distributions are widely used to model multipath scattered signals mutually interfering onto a receiver. The Rician model is more appropriate when a single stronger line-of-sight (LOS) signal survives; Rayleigh's is used to model dense scatters when no LOS is present (Parson, 2001). The Nakagami distribution is a more general model that takes the best of both and can represent them depending on its chosen parameters. For VANETs Rayleigh and Nakagami are the most commonly used models: the former is important because it provides several clues to interpret physical phenomena (including those due to Doppler); the latter is more heuristic and generalizes the former, fitting experimental data better. *Rayleigh's* model was conjectured by Lord Rayleigh in 1889 and was demonstrated by (Jake, 1975); it assumes that a moving node at speed *v* receives a large number (*N*) of multipath

Table 2. How the typical environmental settings are differently affected by the propagation phenomena

Parameter	Highway	Rural	Urban and Suburban
d_1	200m	200m	100m
ϑ_1	1.9-2.0	2.0-2.3	1.9-2.1
ϑ_2	3.8-5.0	3.8-5.0	3.8-5.0

reflected and scattered waves. Each wave comes with an amplitude A_i, a phase φ_i and an angle α_i relative to the direction of *v*, so that the Doppler displacement is $\Delta f_i = (v/\lambda) \cdot \cos \alpha_i$. In the case of a non-modulated signal at frequency f_c, the resulting received signal *r(t)* is described as:

$$r(t) = \sum_{i=1}^{N} A_i \cos\left(2\pi f_c t + \vartheta +_i +2\pi \frac{v}{\lambda} \cos \alpha_i\right) =$$

$$I(t) \cos(2\pi f_c t + \vartheta) + Q(t) \sin(2\pi f_c t + \vartheta)$$
$$= \rho(t) \cos(2\pi f_c t + \vartheta + \psi(t)).$$

$$(8)$$

Due to small wavelength, *I(t)* and *Q(t)* can be considered independent and identically distributed (i.i.d)[4]. Then, thanks to the central limit theorem, *I(t)* and *Q(t)* can be supposed to be (zero-mean) Gaussian random variables, with variance σ^2 (equal to the variance of the i.i.d. variables). When moving to polar coordinates, $\rho(t)$ can be deduced by *I(t)* and *Q(t)* and can be considered uniform in the range [0, 2π]. These hypotheses permit to describe the probability distribution function (*pdf*) of the *r(t)* (that is, of the amplitude of the received signal) in the following way[5]:

$$PDF(\rho) = \frac{\rho}{\sigma^2} e^{-\frac{\rho}{2\sigma^2}}.$$

$$(9)$$

In the Rayleigh model the received power is directly connected to the variance: $P_r = \sigma^2 \ln (4)$. The normalized autocorrelation function of a Rayleigh faded channel is a zero[th]-order Bessel function of the first kind (Clarke, 1968). The Rayleigh model of fading is most applicable when between the transmitter and receiver there is no dominant propagation along a line of sight; additionally it is demonstrated under strong mathematical hypothesis. It is then reasonable to have more heuristic fading models, based on the fitting of experimental measurements. Nakagami distribution provides a probability distribution function (PDF) for the amplitude ρ of the received power, given the mean received power at that distance (obtained by the multi-log model). The PDF is:

$$PDF\left(\rho;m,\omega\right)=\frac{2m^{m}}{\Gamma\left(m\right)\cdot\omega^{m}}\cdot\rho^{2m-1}\cdot e^{-\frac{m}{\omega}\cdot\rho^{2}}$$

(10)

The Nakagami distribution has been demonstrated to result from the sum of multiple independent and identically distributed (i.i.d.) Rayleigh-fading signals (VV.AA, 2011). Nakagami fading occurs for multipath scattering with different clusters of reflected waves. Within any one cluster, the phases of individual reflected waves are random, but the delay times are approximately equal: the envelope of each cumulated cluster signal is Rayleigh distributed. If the envelope is Nakagami distributed, the corresponding instantaneous power is gamma distributed. In the formula (7), $P_r = \omega/2$ is the median power (from attenuation model). The parameter m is in inverse ratio with variance (and represents a shape factor): if $m = 1$, then *Nakagami* distribution becomes a Rayleigh one; if $m > 1$, then Nakagami has lower fluctuations. Once more, in literature several settings have been proposed for Nakagami distribution for VANETs, with parameters which depend on the distance (through the definitions of cut-off distances d_{n1} and d_{n2}) and on the environmental scenario (Rafiq & Patzold, 2008), (Grau et al., 2010), (Cheng et al., 2007). Figure 5 shows how an m depending on distance may impact on the reception, by widening the confidence interval of the received power. The parameter m typically is about 1.5 at short distances (less than 80m), and falls as low as 0.75 at higher distances (which means a heavier spread of fading). Actually, with heavy traffic and denser scatters, in urban areas, values of $m = 0.3$ have been proposed (Cheng et al., 2007).

Concerning the impact of the attenuation and fading modeling, their relevance is quite intuitive.

In fact, a higher received mean power means a higher reception rate - this is obvious. However fading also has a primary effect, with a notable impact on channel capacity, as discussed by Rafiq & Patzold (2008), and on packet reception rate (Cozzetti et al., 2010). All in all, fading is a non-negligible phenomenon.

Occurrence of Hidden Terminals

Also hidden terminals (HT) represent a possible phenomenon hindering wireless communications: they occur when all the nodes do not perceive the same collision domain. The case is depicted in Figure 6: *B* can receive from (or at least sense) both *A* and *C*, but *A* and *C* cannot sense each other. This situation may affect transmissions, whatever the medium access control may be, because *A* and *C* cannot be mutually coordinated and can potentially collide in their transmissions towards *B*. This is particularly true for CSMA/CA, where stations have to sense before transmitting, in order to avoid simultaneous transmissions.

A further step comes from the introduction of the concept of collisions by hidden terminals or hidden collisions. The actual effects of hidden terminals, in fact, can be evaluated only by the statistics of their collisions. In principle, if a protocol – such as MS-Aloha (Scopigno & Cozzetti, 2009) - completely prevents hidden colli-

Figure 6. Representation of the hidden terminal situation. Node B is in the radio range of both A and C, but A and C cannot sense each other.

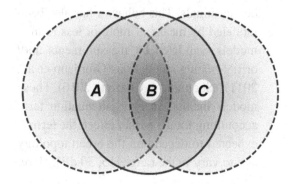

sions, hidden terminals are not an issue; moreover, depending on the traffic load, on the available resources, and on the topology, collisions by hidden terminals can be negligible or prominent. Depending on the distance of the colliding nodes, collisions can be more or less harmful. This leads to some further distinctions: the effects of hidden terminals are expected to depend on the environmental settings.

- On highways, in purely line-of-sight conditions, hidden terminals are met only due to attenuation at long distances. As a consequence, hidden collisions are not expected to be critical, because they cause weak interference and worsen the delivery rate only at longer distances. This was demonstrated by Sjöberg et al. (2011) with simulations set in an ideal scenario with straight highways, in absence of any type of obstructions.
- In urban areas, however, due to buildings obstructing the propagation, hidden terminals may be more dangerous. As shown in Fig. 7(b), in the center of the crossroads, nodes will receive transmissions from the nodes in the legs of the crossroad; conversely, nodes in the legs of the crossroad can receive and transmit only inside the straight road and just some meters around the corner (Giordano et al., 2011), (Sommer et al., 2011), (Meireles et al., 2010), (Giordano et al., 2009), (Mangel et al., 2011). For this reason nodes in the center of the crossroads are exposed to strong interference by hidden terminals. This behavior has also been modeled by means of more or less refined models, validated by measurements and/ or ray-tracing simulations(Giordano et al., 2011), (Scopigno & Cozzetti, 2010). These models include an extra-attenuation term accounting for the corner effect: the term is expected to depend on the urban topology and to vary in the range [15, 30 dB]. Here,

non-line-of-sight propagation is expected to be dominated by reflections and diffusion (not diffractions) because urban apertures are magnitudes greater than the target wavelength (about 5 cm); diffraction would represent a powerful way to turn the corners. This further substantiates the strong extra-attenuations met in literature.

- An even worse case is that of obstructions by vehicles (trucks). This case is depicted in Fig. 7(a). It applies both to urban and highway scenarios. As demonstrated by real measurements by Boban et al. (2011) and Meireles et al. (2010), the obstruction by a single vehicle causes a non-line-of-sight (NLOS) condition which can raise the extra-attenuation as high as 20 dB in static conditions (in a parking lot, with obstruction by a large car); when a truck is involved, values as high as 30 dB can be reached. Similar results are expected under mobility. Notably, in proximity of the obstacle, the attenuation is worse, while it becomes almost negligible at about 400 m: it can be interpreted as an effect of ground wave propagation. When the number of vehicular obstacles increases, the attenuation is expected to get dramatically higher; however there are as yet no models on such phenomena.

In principle, the higher the extra-attenuation induced by the obstructions, the heavier the effects of hidden collisions. For the sake of clarity, in the example of Figure 7(b), supposing that the sensitivity of a node (for a given transmission rate) is -85 dB: then node C could not receive a signal being -85 dB or lower. If the extra-attenuation by the corner is -20 dB and both A and C are received by B with a power of -65 dBm (indeed a high value), then A and C could not sense each other but could cause an interference of -65 dBm on A. This would be a harmful hidden collision. All in all, as discussed in ETSI TR 102 861, the number

Figure 7. Hidden terminals: the interference of A and C on B is strong but they cannot sense each other. Case (a) represents the obstruction by a vehicle (in white); case (b) the obstruction by a building (bottom-left solid square). Radio ranges shown by the shaded shapes.

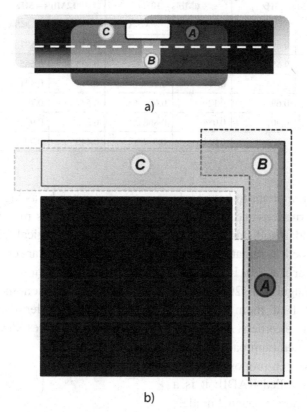

a)

b)

of hidden terminals depends on the number and location of the nodes, the communication environment, and the transmit power level; the number of collisions by hidden terminals also depends on the amount of data traffic of the stations. Hidden collisions dramatically affect CSMA/CA (ETSI TR 102 861), basically because transmission are coordinated by a sensing mechanism (which is impaired by the obstructions).

In order to achieve a deeper insight into the problem, in (Scopigno, 2012) and (Cozzettiet al., 2012) some experiments are reported: they involve the reception rate in a scenario involving 450 VANET nodes in a 750m-wide grid, with obstructions by large blocks (6 x 6 double-lane roads, 150m far the one from the other) under the hypothesis of different uniform data traffic (all the nodes in the VANET generate traffic at the same

rate, with the same packet length and using the same transfer rate). When the reception rate is restricted to the center of the crossroads, the CSMA/CA statistics sharply fall. Being far from congestion, the behavior is ascribed to simultaneous transmissions by hidden nodes: the higher the load (the longer the packet transmitted at a given packet rate), the lower the reception rate (Table 3). The interpretation is further confirmed by the comparison with statistics of MS-Aloha (Scopigno & Cozzetti, 2019) (MSA in Table 3), a connection-oriented protocol which prevents hidden terminals by propagating (appending) information on slot reservations. MSA achieves PDR ten times as high as CSMA (case 6 Mb/s, 10Hz, 900 bytes) just through a better coordination of transmissions and despite its protocol overhead.

Table 3. Results reception rate in the center of crossroads in a Manhattan urban scenario (Scopigno, 2012), (Cozzetti et al., 2012)

		6Mb/s – 5Hz		6Mb/s – 10Hz		12Mb/s – 5Hz		12Mb/s – 10Hz	
		CSMA	MSA	CSMA	MSA	CSMA	MSA	CSMA	MSA
300 byte	0-20m	0.705	0.891	0.423	0.910	0.830	0.897	0.663	0.919
	20-40m	0.536	0.657	0.316	0.693	0.633	0.665	0.498	0.708
600 byte	0-20m	0.465	0.914	0.179	0.869	0.713	0.921	0.448	0.908
	20-40m	0.355	0.694	0.129	0.622	0.537	0.709	0.330	0.686
900 byte	0-20m	0.290	0.907	0.084	0.868	0.595	0.920	0.286	0.903
	20-40m	0.219	0.685	0.061	0.617	0.443	0.707	0.207	0.676

For this reason, it is important to have tools capable of modeling obstructions and, therefore, hidden terminals and hidden collisions. Several models have been proposed in literature (Giordano et al., 2011), (Sommer et al., 2011), (Scopigno & Cozzetti, 2010) (Mangel et al., 2011) (Cozzetti et al., 2012); most of them focus on topologies with sharp square corners. At the time of this writing, the most general approach to urban network simulations has been proposed in (Pilosu et al., 2011) and is called RADII: it is a methodology aimed to generate channel models for any given urban topology. Based on a real 3D urban map and a ray-tracing tool, RADII segments any urban map into areas; for each couple of areas a propagation model is defined, including parameters – such as reachability, mean attenuation and fading spread – which also depend on the position in the respective areas. The approach makes joint physical-and-network layer simulation feasible, preventing possible issues in the scalability of the simulations.

Mobility and Speed

Mobility is a typical characteristic of VANETs. While in mobile networks the users are connected to a base station, in mobile ad-hoc networks (MANETs), including vehicular ones, connections are between mobile nodes. This paradigm involves two main consequences: *(i)* the mutual speed of the nodes can be as high as twice their absolute speed (two nodes may move in opposite directions); *(ii)* the logical topology can vary rapidly. These two effects are considered separately in the following discussion.

The main phenomenon caused by strong mobility is Doppler displacement, which can be computed using the following well known formula:

$$\Delta f = f - f_0 = \left(1 - \frac{v_{sr}}{c}\right) \cdot f_0 - f_0 = \frac{v_{sr}}{c} \cdot f_0 \ ,$$

(11)

Where f_0 is the nominal frequency (5.9 GHz), v_{sr} the speed between receiver and sender (twice the maximum speed, 2*200 km/h in our computation), c is the speed of light. Hence, at a maximum speed of 200 km/h, the displacement (Jiang, 2010) is about 2 kHz, which is below the guard-band of IEEE 802.11p PHY (Jiang, 2010).

So one could conclude that Doppler is not a real issue. However, mobility also causes Doppler spread that is the broadening caused by the scattered components, shifted by a Doppler displacement, depending on their respective mutual speeds. Doppler broadening impacts on fading. This is confirmed by the theory of Rayleigh fading. In fact, the temporal correlation of the envelope of Equation (6) is a Bessel function of order 0 ($R(\tau) = J_0(2\pi\Delta f\tau)$) and, consequently, the power spec-

Figure 8. Examples of reflections leading to Doppler spread: the displacement Δf is computed considering a doubled speed due to mutual mobility. More details in (Mecklenbräuker et. al., 2011).

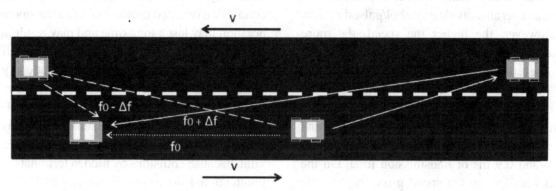

trum (which is the Fourier- transform of the autocorrelation) is of the form:

$$S(f) = \begin{cases} \dfrac{1}{4\pi\,\Delta f \cdot \sqrt{1 - \left(\dfrac{f - f_0}{\Delta f}\right)^2}} & if\,|f - f_0| < \Delta f \\ 0 & otherwise \end{cases}$$

(12)

The resulting power spectrum is displayed in Figure 9. In principle the expected broadening B_D is in the range $[f_0 - \Delta f : f_0 + \Delta t]$. The broadening

Δf can be associated to the so called coherence time T_c of the channel (Sklar, 1997), so that $T_{c1} = 1/(\Delta f)$: the name coherence time comes from the demonstration that the channel becomes uncorrelated over a time $T > T_c$. When T_c is defined more precisely, as the time over which auto-correlation is > 0.5, the relationship between T_c and Δf can be computed as $T_{c2} = 9/(16\pi\Delta f)$, or as the geometric average of T_{c1} and T_{c2} (which is around 300 µs in the case 400 km/h maximum mutual speed and 1 ms at 100 km/h)[6]. In IEEE 802.11p the duration of a symbol is 8.0 µs, so much lower than the coherence time (in most cases between 300 µs and

Figure 9. Doppler spread as defined by the Equation (12)

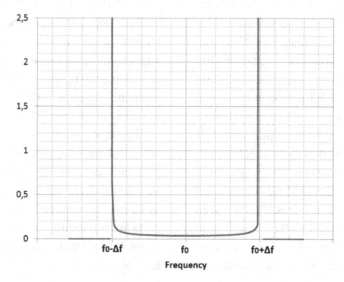

1 ms in VANETs). For this reason the channel is said to be slowly fading, since it stays essentially constant over an individual symbol/pulse duration.

However, the higher the speed, the more rapid the signal variations due to fading - so the more severe the fading. Consider the symbol rate of 802.11p (Table 4), and the transmission of a short frame (64 Bytes = 512 bits): the transmission requires about 1000 symbols at 3 Mb/s and over 110 at 27 Mb/s, implying, respectively, about 8 ms and 0.9 ms of transmission time. On the other hand, when the speed grows higher, the coherence time decreases. So, the case of a short packet at high transfer rate (27 Mb/s) falls within the coherence time for urban speed. However, this is not always true (depending on speed, frame length and the line-rate). Interestingly, the robustness of the coding and the severity of fading cause opposite effects: lower transfer rates require longer transmission times, potentially exceeding the coherence time, but are more robust.

Currently, most of the VANET studies in the literature do not pay attention to coherence time, basically due to the lack of suitable models inside the network simulation platforms.

The other aspect connected to mobility concerns the routes (mobility paths) actually covered by the nodes. Potentially, VANETs can be affected by mobility in several ways: *(i)* medium access control can be influenced by sudden topology changes; *(ii)* upper-layer protocols can be influenced by the topology changes (*e.g.* unicast routing) and *(iii)* by detailed mobility traces (*e.g.* opportunistic delay-tolerant delivery). However, the opposite is also true, that is: *(iv)* VANET protocols can act on mobility through traffic management protocols.

The extensive analysis of these phenomena is beyond the scope of this introduction. However, some brief notes are discussed here. The first aspect (effects of mobility on MAC) is expected to be secondary, because the mechanical speed of vehicles is much lower than the transmission times and, in CSMA/CA (the MAC of incumbent VANET solutions), each frame transmission is

independent of previous ones. For the same reason there may be an impact but only on slotted and connection oriented protocols, because a slot reservation may last a long time and may result in a conflict (unintentional slot re-use) due to mobility patterns (two nodes using the same slot can get closer). This is particularly harmful with urban topologies, which make changes more sudden. For this reason, the reporting mechanism available on MS-Aloha (Scopigno & Cozzetti, 2009) appears vital, because collisions by hidden terminals and unintentional slot re-use can be detected

The aspects connected to upper layer protocols seem to be more conceptual than practical. In fact, the only possible impact is expected on unicast traffic forwarded based on a topology-driven routing protocol (*e.g.* AODV – Ad Hoc On Demand Distance Vector; DSR – Dynamic Source Routing; TORA – Temporally Ordered Routing Algorithm; GSR - Geographic Source Routing). However this is a scenario which seems quite unlikely in the near future, because most VANET traffic is expected to be broadcasted (at least in the first phase) and geo-routed. Incidentally, results from literature (Lochert, 2003) show that topology changes are a critical parameter which may determine the success of a protocol. The same conceptual result holds for delay-tolerant routing as in TOpology-assist Geo-Opportunistic Routing (TO-GO) (Lee et al., 2009), a geographic routing protocol that exploits topology knowledge to select the best target forwarder (accounting for the city topology) and leveraging opportunistic forwarding with the best chance to reach it. The optimization of these protocols requires a thorough understanding of the underlying mobility patterns.

The last phenomenon which has been mentioned about mobility concerns the impact of protocols on mobility. This is currently being studied thanks to joint simulation platforms (such as i-Tetris[7]) where network and mobility simulations are run jointly, so that the interaction of the respective phenomena is mutual. This is a new area of study.

Scalability

Another possible criticality for VANETs is the number of nodes which can be found in an area and compete for channel access. In typical centralized approaches (such as UMTS), the number of subscribers simultaneously connected is restricted by the initial dimensioning by the network providers. If the number of users exceeds the maximum number initially foreseen, then some of them will get blocked and will not be able to communicate. With other approaches – such as CSMA/CA – contention is completely distributed and non-blocking. However, due to collision avoidance, transmissions can be deferred so much that they run out (which is in contradiction with real-time requirement). Additionally, the higher the number of nodes, the higher the probability of simultaneous transmissions which CSMA/CA cannot prevent. This causes a notable worsening in the reliability of broadcast transmissions, even in proximity of the transmitter, as found in literature (ETSI TR 102 861), (Cozzetti & Scopigno, 2011), (Cozzetti et al., 2009), (Bilstrup, 2008), (Sjöberg-Bilstrup, 2009).

One way to counteract the scalability issue of CSMA is to introduce decentralized congestion control methods (DCC - ETSI TS 102 687) for the G5A and G5B bandwidths – although they only partially solve the issue of scalability. For this reason, very recently, ETSI investigated, in a dedicated Specialist Task Force (STF 395), possible slotted and connection-oriented approaches, with the aim of achieving better time and space coordination (ETSI TR 102 861, 102 862). Two main protocols were studied: STDMA and MS-Aloha. They both outperformed CSMA/CA in congested scenarios, while guaranteeing deterministic delay and delivery. Only MS-Aloha (Scopigno & Cozzetti, 2009) (Cozzetti & al., 2009) (Scopigno & Cozzetti, 2010b) (Cozzetti & Scopigno, 2011) was demonstrated to solve the issue of hidden terminals and to be able to coexist with CSMA/CA in a transparent mode, leveraging its pre-emption mechanisms.

Thus, the number of nodes is expected to hamper channel access and stress the coordination capabilities of the MAC, potentially blocking transmissions or deferring them too much. However, the number of nodes may have another effect. Even in the hypothesis of perfect coordination by the MAC, simultaneous transmissions should be permitted to nodes far apart; even more, this spatial multiplexing (not only over time) should be encouraged to manage heavy traffic loads. However, simultaneous transmissions causing low interferences, even below the sensitivity threshold, can worsen the average SINR ratio (of Equation (2)) by some decibels (about 10 dB in (Cozzetti et al., 2010)). This raises the noise beyond the thermal noise, into an average number called background noise (Cozzetti et al., 2010). Background noise can worsen the reception rates at longer distances, where the received signal power is lower.

Finally, some data about high densities of vehicles may be useful: at the time of this writing, some articles found on the web (including *India Today* and *New Geography*) mentioned the cases of Los Angeles, Mumbai and Beijing as the most dramatic congestions. Data refer to about 700-1,000 cars in less than a square km.

Summing up, on one hand, the number of users cannot be foreseen and restricted *a priori*, otherwise the safety would be affected. On the other hand, a high number of nodes should be supported without causing excessive delays or influencing the correct reception in the first meters (say the first 100m). The two aspects, together, set a very challenging problem. In ETSI TR 102 861, a possible solution was identified in a fine spatial multiplexing. In the ideal solution, simultaneous transmissions should be scheduled based on position, so that transmissions occurring at the same time come from nodes far enough apart.

Relevance of the Environment

All in all, several environmental phenomena affect VANETs. However, depending on the actual setting, they may have different relevance. To mention something obvious, on highways mutual speed will be higher, while in rural areas, obstructions by buildings will be negligible and density lower. So two spontaneous questions arise: *(i)* what is the specific characteristic of each setting? *(ii)* What settings are more important to be investigated?

The first question is answered by Table 4; given a distinction into *urban*, *sub-urban*, *rural* and *highway*, for each setting the main phenomena are rated. The number of signs '+' refers to a heavier relevance.

It is then clear that highway and urban feature all the possible effects and, in principle, provide an effective benchmark for the evaluation of VANET solutions. However, this requires adequate models, specifically addressing the respective characteristics of each environment. To complete the analysis, data from (ERSO) may be useful: 55% of all fatal accidents occur in rural areas, due to head-on collisions; there, however, scalability of the MAC protocol is not a major issue. The lowest probability of fatal accidents is in the highway environment (10%). In urban environment, 35% of all fatal accidents occur, often involving vulnerable road users, such as pedestrians and cyclists. This further emphasizes the importance of *(i)* simulation tools for the analysis of urban VANETs performance and *(ii)* the need to enforce safety of pedestrians by a future extension of V2V paradigm by V2P (vehicle-to-pedestrian) communications.

INTRODUCTION TO THE INCUMBENT VANET SOLUTIONS

The functionalities required by VANET solutions are very numerous, ranging from physical and MAC layer to multichannel management, networking, privacy and the definition of safety messages. Additionally, given the specific nature of the VANET networks and of their services, they require solutions which cannot always relay on existing protocols. For instance, referring to the US standard, depending on the type of service, VANETs have two communication options: dedicated messages WSMPs (WAVE short message), which are defined in IEEE 1609.3 for communications which cannot adopt IPv6 protocol (for instance due to the overheads). However, the option of IP is preserved, especially for non-safety service. Other VANET services require custom solutions: for instance, services implying routing and forwarding of messages. Due to the high mobility of nodes, routing can hardly be defined as for fixed networks (e.g. OSPF) or adhere to mobile IP. Conversely, routing should exploit position information (hence the concept of geo-routing studied by ETSI).

An exhaustive discussion of all the functionalities subtended by a VANET logical architecture is beyond our scope. However, an overview of the logical architecture is useful to describe its complexity and, at the same time, its integrated nature.

Figure 10. Architecture of the 1609.x protocol family: logical stack

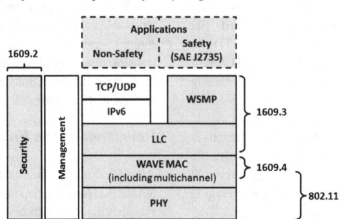

Especially in the US, VANET solutions are referred to as WAVE (Wireless Access in Vehicular Environment) or IEEE 1609.x family, to emphasize their being an integrated protocol suite (or protocol family). Collectively, the IEEE 1609 family describes wireless data exchange, security, and service advertisement between vehicles and roadside devices. These standards provide the basis for the design of applications interfacing with the WAVE environment. The WAVE family integrates with other standards which cover more application-oriented aspects. This is the case of SAE J2735, a standard from the Society of Automotive Engineers which specifies message sets, data frames and data elements for use by applications intended to utilize *"the 5.9 GHz Dedicated Short Range Communications (DSRC)"*. The standard mentions DSRC in its title. For the sake of clarity, the acronym refers both to the existing (legacy) tolling system, which is used in Europe and USA, and to the new WAVE stack. Nowadays DSRC is used as a synonym for WAVE/802.11p.

The same concepts are also found in ETSI work (in Europe), where the protocol architecture foresees cross-layer functions (for management and security), dual network layers (messages can be over IP or directly packetized over the MAC) and custom protocols (e.g. geo-networking, congestion control, application layer safety purposes, etc.). In particular, emphasizing safety purposes of VANETs, ETSI has already defined the Cooperative Awareness Application (CAA), which uses two main protocols: CAM (Cooperative Awareness Messages - ETSI TS 102 637-2) and DENM (Decentralized Environmental Notification Messages - ETSI TS 102 637-3), both working on the so-called control channel.

- The CAM message is sent periodically at a defined repetition rate. Depending on the congestion situation, the repetition rate of the CAM transmission will be adjusted to minimize the potential congestion of the channel.

- The DENM message is sent only after detection of an abnormal situation (*e.g.* danger) which could lead to an accident or potential injury of traffic participants. The DENM is a high priority message which has to be sent over the control channel (CCH) with a higher priority than the CAM messages.

Altogether, the European and American solutions are quite close to one another, but some harmonization effort is still required. After this general introduction, the next sections will focus on the lower layers (PHY and MAC) which are almost completely standardized, very similar in the European and American solutions and, at the same time, the ones which most influence the effectiveness of upper layers. They permit to collect the conclusions which have been drawn by the previous analysis on physical layer phenomena and requirements.

Spectrum Allocated for VANETs

The frequency band assigned to VANETs constitutes a free but licensed spectrum both in the USA and the EU: this means that both the FCC in the USA and CEPT (Conférence Européenne des Administrations des Postes et des Télécommunications) in Europe do not charge a fee for the spectrum usage. This already introduces the novel rationale of VANETs: they are meant to be free but under control, due to their intended goal.

In the US at 5.9 GHz the 75 bandwidth allotted to VANETs is divided into seven 10MHz-wide channels (Figure 11), with the lowest 5 MHz band left free as a guard band. One channel (Channel 178) is used as a control channel (CCH), meaning that it is to be used strictly for safety communications. The remaining 6 channels are service channels (SCH) and are available for both safety and non-safety usage.

Conversely, in Europe, an overall 50 MHz band, still around 5.9 GHz, is allocated for ITS

Figure 11. Spectrum allocation in Europe and the USA

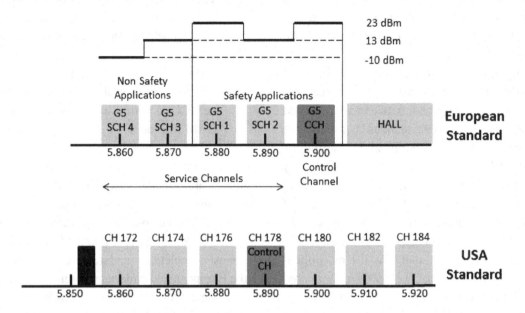

(ETSI ES 202 663 V1.1.0, 2010); hence, in this case, only five channels are available. A distinction between control and service channel holds in Europe, as well: the so-called Control Channel (G5CC) is however assigned the highest frequency; the four others are Service Channels, named G5SC4, G5SC3, G5SC1 and G5SC2 (with central frequencies ranging from 5.86 GHz to 5.89 GHz). An additional channel, Service Channel 5 (G5SC5), is located in the 5.47 – 5.725 GHz band for Radio Local Area Network (RLAN) applications. Finally, the band in the range 5905 to 5925 MHz (the so-called High Availability Low Latency – HALL channel) is left for future use.

With a further classification (Figure 11), channels are grouped into 3 bands, called G5A, G5B and G5C. At the time of this writing, ETSI is specifying the overall channel architecture in a new draft (ETSI TS 102 724), identifying for each of them a specific function (e.g. CCH for safety, including CAM and DENM - only if there is available space, for other services; SCH1 and SCH2 for announcing and offering ITS services for safety and road efficiency, SCH3 and SCH4 for other commercial services…). The same

document also specifies how single-transceiver or multiple-transceiver stations should work (in particular on what channels) depending on whether they are ITS stations for safety, for ITS services or for commercial services. Queuing policies are also specified (for instance, DENM has a higher priority than CAM messages). These details are complementary to the specification of the so-called European Profile ETSI EN 302 663.

Upon deeper analysis, a strong difference emerges between European and American solutions, as far as safety is concerned. In fact, as shown in Figure 11, a precise profile for the maximum transmitted power is foreseen: while G5CC can transmit up to 23 dBm, the neighboring bandwidth can transmit only 13 dBm. This is the reason for the inverted numbering of the channels (between G5SC1 and G5SC2); G5SC2 acts then as guard-band. Additionally, again for the purpose of avoiding interferences G5CC is assigned the highest frequency band.

Summing up, the differences between the American and the European frequency regulations for VANETs range from the control channel location - around 5.89 GHz (channel 178) in the US and around 5.9 GHz (channel 180) in Europe - to the number of available channels. At this stage of standardization, the strong point of the European approach is the enforcement of safety, coming from the design of the multichannel management, while the advantage of the American DSRC seems to be flexibility, coming from the multiple service channels, and potentially facilitating new commercial services. This is further reflected by the way the multichannel is managed (as discussed in the following sections): while the American approach involves single-radio devices switching between CCH and SCH, in Europe two radios will be required to fulfill the requirement of being continuously connected to the CCH. The former solution is cheaper; the latter is more safety-oriented.

IEEE 802.11p: Physical Layer

The physical layer of VANETs is equal in the US and Europe and is a customization of the widely deployed physical layer of WiFi networks (IEEE 802.11a). The key word is Orthogonal Frequency Division Multiplexing: the channel is divided into 52 orthogonal[8] sub-carriers, each modulated separately. Conceptually the OFDM system boosts the available bit rate, leveraging parallel transmissions: thus, very high data rates can be mapped onto low symbol-rate modulation schemes[9]. The main consequence is that the symbols can be relatively long, increasing robustness against inter-symbol interference – as caused by multipath propagation.

However, to further strengthen the physical layer of VANETs, and make transmissions reliable in the harsh vehicular environments, in IEEE 802.11p all time-parameters get doubled, compared to the corresponding ones in 802.11a. This is meant to counteract inter-symbol interference (ISI) (Ström, 2011), as caused by multi-path propagation. As a consequence, the channel width shrinks from 20 to 10 MHz and the data rates are halved compared to 802.11a. IEEE 802.11p supports the same digital modulations and coding rates as IEEE 802.11a, with the final bit-rates halved, due to the doubled duration (8 μs instead of 4 for each OFDM symbol; 1.6 μs instead of 0.8 of guard-time between consecutive symbols; 32 μs instead of 16 for the preamble). Notably, the highest values of RMS (root mean square) delay spread found in literature are in the range [400 ns, 1000 ns], which is below the guard-time (Mecklenbräuker et al., 2011); this confirms the properness of the parameters chosen in IEEE 802.11p. The coherence time, as defined in Equation (10), is much larger than the duration of a symbol or of the preamble, so no dangerous fast fading hinders the frame reception.

The supported digital modulations are: Binary Phase Shift Keying (BPSK), Quadrature Phase Shift Keying (QPSK), 16-point Quadrature Amplitude Modulation (16-QAM), and 64-point Quadrature Amplitude Modulation (64-QAM). Samples of OFDM symbols can be generated by IFFT. Convolution coding increases resilience by coding rates of 1/2, 2/3, or 3/4. As on option, two adjacent channels in 802.11p may be used as one 20 MHz channel. The resulting transfer rates are shown in Table 5.

Table 5. The transfer-rates supported by IEEE 802.11p

Modulation scheme	BPSK	BPSK	QPSK	QPSK	16-QAM	16-QAM	64-QAM	64-QAM
Coding rate	1/2	¾	1/2	3/4	1/2	3/4	2/3	3/4
Data rate [Mbit/s] with 10 MHz ch.	3	4.5	6	9	12	18	24	27
Data rate [Mbit/s] with 20 MHz ch.	6	9	12	18	24	36	48	54
Bits-per-Symbol	0.5	0.75	1	1.5	2	3	4	4.5

The per-channel OFDM sub-channels are 64 in number: 48 are used as data carriers (they are the carriers in the ranges [-26, -22], [-20, -8], [-6, -1], [1,6], [8,20], [22, 26]), 4 are pilots (-21, -7, 7, +21) and the remaining 12 are used as null carriers (in 802.11 they are at the borders of the channel, acting as guard-bands, and correspond to the numbers [-32, -27], [27, 32]). Pilot subcarriers ease the identification of the exact frequency and phase of the signal at the receiver (the nominal frequencies are the same but may differ slightly) and facilitate the estimation of channel coefficients for the equalization of sub-channels (PACE - Pilot-symbol Aided Channel Estimation). The use of pilot sub-channels is a technique which falls into the area of pilot symbol assisted modulation (PSAM). While, in general, PSAM requires that pilot (i.e. well-known) signals are spread over frequency and time (for instance, sent periodically), in 802.11a/p they are continuously sent on dedicated channels, in order to enforce equalization robustness against fading. In addition, initial training sequences complement the equalization (more details follow). Concerning the position of pilots, spacing needs to be carefully evaluated, (Ozdemir, 2007), so that sub-carrier spacing between the pilots in a frequency domain is small enough that the variations of the channel in frequency can be all captured. In 802.11p the spacing is about 2 MHz, (about 8 subchannels) and no sub-channel is more than 7 channels from a pilot (including the edge subcarriers): this satisfies the condition explained by Arlsan & Yucek (2003) for pilot distance[10]. Interposed null carriers have been proposed to enforce CFO estimation (Huang & Letaief, 2006) but they would not help equalization.

Concerning the Doppler effect, the frequency displacement at almost 400km/h mutual speed is about 2kHz, (Jiang, 2010) according to Jake's model. In principle, OFDM is extremely sensitive to carrier frequency offset (CFO), induced by oscillator discrepancies between the transmitter and receiver, and/or Doppler shifts. Frequency offset

can cause interferences and influence orthogonality, leading to ICI. In 802.11p, with 10 MHz channels, the spacing between OFDM sub-channels is 10MHz/64=156.25 kHz. So, the displacements and spread are below the guard-band (null pilots at the edges). Additionally, the subcarrier-spacing is much lower than Doppler spread, which prevents OFDM ICI (Mecklenbräuker et al., 2011), (Faria et al., 2006). The other possible effect of Doppler is the worsening of fading, due to scattering under mobility; however the more severe fading makes ISI harsher, but the doubled symbol duration can counteract this effect. All in all, given the IEEE 802.11p standard, speed does not seem to be a major issue, contrary to what could be imagined.

So far, the design decisions of IEEE 802.11p have been substantiated. A short description of the architecture of a transceiver permits us to mention other mechanisms which are implemented in the devices (Figure 12).

The data feeding a transmitter is first scrambled by a 127-long frame scrambler, to prevent long sequences of equal symbols. Afterwards, data is first encoded by a convolutional encoder to provide error-correction capability: the coding rate can vary between 1/2, 2/3 and 3/4. The following step is performed by an interleaver: every encoded group of bits is mapped so that adjacent bits are spread over non adjacent sub-carriers, and adjacent encoded bits alternatively onto less and more significant bits of the chosen digital constellation (QAM code-set). After pilot insertion 52 QAM values per OFDM symbol are available and modulation onto 52 subcarriers is carried out by applying the Inverse Fast Fourier Transform (IFFT) with 64 points. Before the final conversion, cyclic extensions are implemented: consecutive symbols are extrapolated and mutually matched (in the guard-time interval) to prevent any mismatches in phase and amplitude (spectral regrowth). Finally an IQ modulator converts the signal into analog (digital-to-analog conversion), which is up-modulated to the 5 GHz band, amplified, and transmitted through the antenna.

Figure 12. Transceiver architectures (top: transmitting chain; bottom: receiving chain)

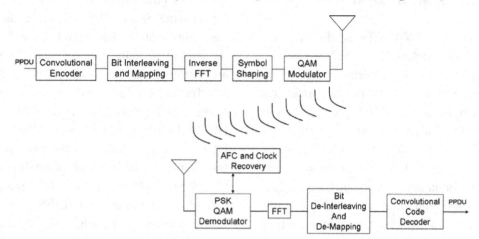

Basically, the receiver performs the same operations the other way around, but involving some additional non-obvious tasks: synchronization (time recovery), carrier recovery and equalization.

- Carrier recovery is carried out by pilots.
- Fading can also differently attenuate the distinct subcarriers, creating problems for OFDM. Equalization deals with such effects by leveraging pilot tones and a well-known initial sequence (preamble) of each frame, which is used as initial value (Ohno et al., 2011). The equalization, based only on the preamble, becomes less effective when the coherence time decreases due to vehicular speed. In fact, in this case, coherence can last less than a packet-time, as already discussed.
- **Synchronization:** The same guard interval and cyclic extension adopted against spectral re-growth also help synchronization (thanks to these, the autocorrelation of the symbol sequence is maximum for a delay equal to the guard interval). Synchronization is required also at frame level and is achieved by a training sequence specifically designed to address this issue.

For the sake of precision, there is an additional logical layer between Physical Medium and MAC layer: it is called Physical Layer Convergence Procedure (PLCP) sublayer and adapts the data units from the data-link layer to the PMD format, and vice-versa. PLCP carries out the following operations:

- Adding information needed by the transmitter and the receiver to properly encode/decode the frame (such as the data rate, the frame check sequence (FCS), and the frame length);
- Multiplexing information by means of the Service field;
- Appending a PHY-specific preamble to start synchronization and indicate the beginning of the PHY frame.

The PLCP is always transmitted at the basic rate allowed by 802.11p - 3 Mbps, BPSK with convolutional coding rate 1/2 - independently of the data rate used to transmit. PLCP Data is included in two fields: preamble (12 symbols) and the signal (1 symbol = 0.5*48 = 24 bits).

IEEE 802.11p: MAC Layer

The MAC layer of VANETs is also defined through the amendment IEEE 802.11p of the MAC layer of WiFi. So, in principle, two sets of MAC mechanisms would be supported and coordinated by the so-called Hybrid Coordination Function (HCF): a centralized (or contention-free) MAC (HCCA – HCF Controlled Channel Access) and a distributed (or contention-based) one (EDCA – Enhanced Distributed Channel Access). Given the requirements of VANETs, only the latter applies. EDCA represents an evolution of the well-known DCF (Distributed Contention Function), aimed at the support of quality of service. The access control of EDCA/DCF is based on Carrier-Sense Multiple Access with Collision Avoidance (CSMA/CA): carrier sense does not allow wireless transmission of a node if another node is transmitting, while collision avoidance is used to improve CSMA performance, by reducing the probability of collision.

Specifically, the access algorithm works as follows: when a station has to transmit, it first senses the medium free (carrier-sense) (see Figure 13). Once the medium is perceived as free it has to remain free for a period of time (say T), before the station can transmit: the time T to be waited includes a fixed component (say T_f) and a variable component (T_v): $T = T_f + T_v$. T_f is the Inter-Frame Space (IFS) and depends only on the priority of the frame and is meant to defer the individual count-down of each station. There are five IFSes; they are, in increasing order: short interframe space (SIFS) which is used for control frames; PCF interframe space (PIFS) used by HCCA; DCF interframe space (DIFS) is used by DCF and is generalized by the arbitration inter-frame space (AIFS), which depends on priority (AIFS[AC0], AIFS[AC1],... AIFS[AC3])); the extended interframe space (EIFS) is involved in re-transmissions of non-acknowledged transmissions. In principle, the longer the IFS, the lower the priority of transmissions; so control frames have priority over data frames and high-priority data frames are assigned an AIFS shorter than the AIFS assigned to a low-priority frame.

After deferral, following a busy medium condition or prior to a new frame transmission attempt, the station selects the back-off counter: it represents the variable waiting time T_v. The time T_v is computed by pinching an integer number over the uniform distribution in the interval [0,CW-1]; the number is then multiplied by a quantity called slot time (13 µs in IEEE 802.11p). The variable waiting time is meant to prevent simultaneous transmissions (collision avoidance) and is further enforced by an exponential mechanism: the pa-

Figure 13. The EDCA access control mechanisms and timings. An example of high priority unicast transmission exploiting RTS/CTS mechanism.

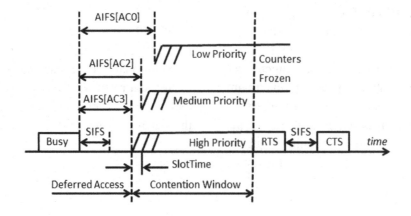

Table 6. EDCA parameters by IEEE 802.11p: they are different from those of IEEE 802.11

Access category	CWmin	CWmax	AIFSN
AC0	15	1023	9
AC1	15	1023	6
AC2	7	15	3
AC3	3	7	2

rameter CW (Contention Window) takes aCWmin as an initial minimum value at the first transmission attempt and is doubled (exponential back-off) at every failed transmission (with an upper limit equal to the maximum size aCWmax). The back-off counter is decremented by one at the end of each idle slot. As soon as the timer reaches zero the station is allowed to transmit. If the channel gets busy during any slot interval, the back-off countdown is frozen; it will be resumed following an idle DIFS (or an EIFS).

While the collision avoidance (back-off) applies both to unicast and broadcast transmissions, the exponential back-off works only for unicast, due to the lack of acknowledgement for non-unicast transmissions. This has a major impact on the effectiveness of collision avoidance of VANET, since most of VANET transmissions are broadcast.

In fact, concerning acknowledgments, IEEE 802.11 networks also employ an immediate positive acknowledgment scheme for unicast frames: upon successful frame reception, the receiver sends back an acknowledgment frame (ACK) to the source. If the ACK is not received, the frame is considered as lost and retransmitted by the source node. In contrast with unicast frames, broadcast and multicast frames are never acknowledged; hence, they cannot be retransmitted.

The *RTS/CTS* exchange is a protocol mechanism of 802.11 aimed at solving hidden node problems: the Request-to-Send (RTS)/Clear-to-Send (CTS) is an optional frame handshaking between the sender and the receiver, preceding the actual frame transmission. Since the receiver has to acknowledge the RTS with its CTS, all the nodes receiving the CTS will not transmit. RTS/CTS can be used (optionally) only for unicast transmissions and involves a time overhead which is hardly compatible with real-time constraints of safety applications.

Overall, three important mechanisms of IEEE 802.11p (exponential back-off for collision prevention, RTS/CTS against hidden terminals, ACKs for reliability) do not apply to broadcast transmissions. This could be a major problem for VANETs.

Coming to the frame formats, the IEEE 802.11 MAC layer uses three types of frames: data, control, and management frames. Data frames carry user data from higher layers. Management frames (*e.g.* beacon) enable stations to establish and maintain communications; they are not forwarded to the upper layers. Control frames (*e.g.* RTS, CTS, ACK) assist in the delivery of data and management frames between stations.

So, to conclude the description, the data frames include the fields shown in Figure 14, which are:

- PLCP header (not shown), including PLCP Preamble and Signal - always sent at the lowest rate (3 Mb/s with 10 MHz channels), and taking 40 μs.
- MAC header; this includes: *(i)* the frame control field, specifying the type and func-

Figure 14. The fields in the 802.11 frame: lengths in bytes are specified; PLCP is omitted

Frame Control	Duration ID	Address1 (source)	Address2 (destination)	Address3 (rx node)	Sequence Control	Address4 (tx node)	Data	FCS
2	2	6	6	6	2	6	0 - 2,312	4

tion of the frame, *(ii)* duration; *(iii)* sequence control, managing fragmentation; *(iv)* QoS field, specifying the priority of the frame; *(v)* 4 48-bit MAC addresses: due to backward compatibility to WiFi, a frame should be able to be delivered through a wireless distribution system (WDS) and this requires 4 MAC addresses – not just source and destination.

- A variable length Frame Body; this consists of the MAC service data unit. The frame body is of variable size; its maximum length is determined by the maximum MSDU size (2304 octets) plus any overhead from security encapsulation.
- Frame Check Sequence, finally, is a 4-byte field which contains a 32-bit cyclic redundancy code (CRC) used for detecting bit errors in the MAC frame. The CRC is computed over all the fields of the MAC header and the Frame Body.

Multichannel Environment

VANETs have multiple channels available, both in the European and American solutions. The multichannel environment in VANETs presents two main issues: *(i)* possible cross-channel interferences and the decisions on the *(ii)* management policy of the multiple channels. In order to discuss the issues of cross channel interferences, some data about the transmitted power need to be recalled. The FCC defines 4 classes of devices (A, B, C and D). Each of them is associated to a maximum transmitted power and to a different frequency profile (see Figure 15). Devices participating in

Table 7. VANET device classes

		Max Power	
Class A		0 dBm	
Class B		10 dBm	
Class C		20 dBm	
Class D		28.8 dBm	
	±4.5 MHz	± 10 MHz	± 15 MHz
Class A	0 dBr	-28 dBr	-40 dBr
Class B	0 dBr	-28 dBr	-40 dBr
Class C	0 dBr	-40 dBr	-50 dBr
Class D	0 dBr	-55 dBr	-65 dBr

Figure 15. Representation of the profile of transmitted power in VANET devices

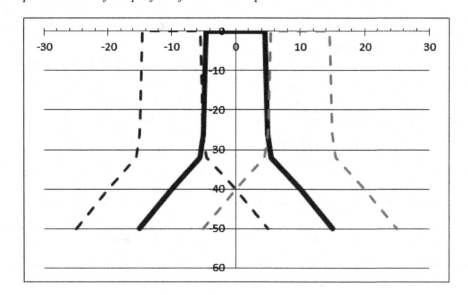

the VANET safety protocol exchanges will normally be in class C.

Cross-channel interference is a topic which has not been yet fully investigated in literature. In order to limit interferences between channels, the standard IEEE 802.11p specifies a transmission spectrum mask which is defined up to 15 MHz far from the center frequency; the mask specification includes then rules for the adjacent channel rejection (ACR) over 10 MHz outside the intended channel. ACR is required to be 12 dB higher for adjacent (and 10 dB higher for non-adjacent) channels, in the respective modulations. The effects of cross-channel interference have been measured (Rai et al., 2007) and simulated (Lasowski et al., 2011) and, these preliminary results confirm that it is an issue which needs further investigation. The masks defined for channel rejections are sufficient to avoid the most harmful interferences between adjacent channels: transmissions on one channel will not defer transmissions on another one (due to collision avoidance and sensing); however these interferences are still expected to contribute to background noise, as already demonstrated by Campolo et al. (2012).

Concerning the management of the multi-channel environment, both the European and the American standards devote a channel to control messages, which are restricted to safety communications. However there is a major difference: in Europe two radios will be required to guarantee a continuous connection to the CCH, while the American approach involves single-radio devices switching between CCH and SCH.

The American approach motivated a standard of the IEEE 1609 family, the IEEE 1609.4. It specifies four channel access switching modes, namely: continuous, alternating, immediate, and extended access. Without entering into detail, all the switching modes subtend *(i)* a common synchronization, with periods of 100 ms and guard-time of 4 ms; *(ii)* a distinction between CCH and SCH sub-periods (whose durations are configurable); *(iii)* different methods to manage

SCH transmissions, with hard or more flexible switching to CCH (for example one can wait for the beginning of the SCH sub-periods or switch to it as soon as the CCH has been sensed).

Finally, for the sake of completeness, the standard IEEE 1609.4 also defines some procedures to let a station behave as *provider* by initiating a new network (Basic Service Set – BSS) on an SCH and announcing its service by specific messages (WAVE Service Advertisement - WSA) over the CCH. This mechanism of providers completes the framework coupling CCH and SCH.

CONCLUSION: ASSESSMENT OF VANET STRENGTHS AND WEAKNESSES

VANETs are mainly intended to improve vehicular safety based on distributed channel access and relaying mainly on broadcast transmissions. A complex protocol suite targets this goal. So far, the physical layer of VANETs seems to meet a nice trade-off between the multiple constraints: it leverages OFDM to support high bit rates; it uses double-length symbols – which IEEE 802.11a does not – to cope with mobility and channel coherence time; it uses multiple channels to increase scalability and support multiple services; it specifies precise transmission masks to prevent cross-channel interference; it includes robust coding mechanisms and multiple possible transfer rates to deal with different requirements (on the delivery rate or maximum transfer-rate).

Overall, given the complex environment and severe requirements, the physical layer seems well thought out and, what is more, it benefits from the knowledge of WiFi systems, which are largely copied by IEEE 802.11p. Conversely, the main open issues seem to concern the upper layer and, most of all, the medium access control. So far, two main issues seem to be potentially critical – both are worsened by a heavier network load.

- Performance (transmission delay and delivery rate).
- Collisions by hidden terminals.

Performance of CSMA/CA networks is known to be critically connected to the network load: under the hypothesis of broadcast transmissions, due to the lack of acknowledgments (hence missing exponential back-off), when the number of transmissions grows, the number of collisions will grow as well. In fact, multiple nodes are expected to simultaneously count their back-off counter down to zero. This will result in simultaneous transmissions, even by nodes which are close to each other, because the transmissions are coordinated over time – not over space.

This situation is expected to be particularly critical with LOS conditions (such as on highways), as shown by simulations in ETSI TR 102 861. Similar results are depicted in Figure 16, where 600 nodes move in a 1 km-wide area in absence of obstruction,; each node transmits 20 800-Byte long packets per second with a 20 dBm transmitted power. In the graph the reception rate of CSMA/CA is compared with the ideal one (that which would be achieved if only one node transmitted at a time); thus, the ideal upper bound considers only attenuation, fading, environmental noise and the dependence of PER on SINR. Altogether, the difference between the two can be interpreted as the effect of MAC and is more than 10% at very short distances[11]. One could object that no other MAC could perform better but, conversely, it was demonstrated by ETSI STF395 that synchronous protocol could be close to the upper bound, at least in the first 100 m: this is shown as well in the Figure 16. These results demonstrate that CSMA/CA is suboptimal in time/space multiplexing and MS-Aloha can coordinate transmissions so that simultaneous transmissions (i.e. slot re-use) takes place only at a certain distances, preventing interferences.

Hidden terminals are another possible issue of CSMA/CA. An RTS/CTS mechanism can, in principle, counteract the problem; however, in practice, it is not viable for three reasons: *(i)* it causes delays which are not compatible with VANET services, *(ii)* it has a detrimental effect on the throughput and *(iii)* it does not work for broadcast transmissions. Simulations in ETSI TR 102 861 demonstrate that hidden collisions in an urban topology can cause a fall of 10% (from 100% to 90%) in close proximity

Figure 16. Comparison between the PDR achieved by CSMA/CA and MS-Aloha. The upper line represents the ideal upper bound (only one sender at a time: losses only due to attenuation and fading). Negligible hidden collisions, which would further enforce MS-Aloha's advantage.

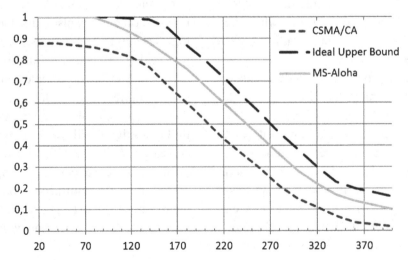

of the transmitter, even in absence of congestion. The percentage is purely exemplificative because the result depends on the topology, on the node density, on the transmission rate and also on the way the figure is computed (in the example it is not restricted to the reception of nodes in the crossroads, which are more exposed to hidden collisions). The result is quite isolated in literature, due to the scarce knowledge of urban propagation models (very recently available): despite this, it is impressive.

Also in the case of hidden terminals, simulations demonstrate that MS-Aloha can resolve this issue (as already shown in Table 3). However, the protocol's practical feasibility and viability still require additional studies. The main open points are: the feasibility of robust synchronization (also in hold-on and under mobility) and the demonstration of mechanisms of coexistence and fall-back to CSMA/CA; both points, so far, have been investigated only theoretically.

The upper layers of VANETs are something which is still evolving, so it is difficult to draw conclusions about. Safety services are more straightforward, easier and robust and, consequently, are at a more advanced stage of development. Conversely, non-safety services require a more complex architecture and an international consensus which is still far from being achieved; this also reflects the main aim of VANETs, which is the enforcement of safety.

All in all, VANET technology is mature for a first deployment, which may be expected by 2015. However some work is still required to further enrich the architecture for a full and turn-key solution. Additionally, it is still worth investigating new and backward compatible solutions, which can overcome the known issues of asynchronous VANET MAC protocols in a future release of the standard.

REFERENCES

Abrate, F., Vesco, A., & Scopigno, R. (2011). An analytical packet error rate model for WAVE receivers. In *Proceedings of the 4th IEEE Symposium on Wireless Vehicular Communications (WiVEC)*, San Francisco, CA, USA.

Akki, A. S., & Haber, F. (1986). A statistical model of mobile-to-mobile land communication channel. *IEEE Transactions on Vehicular Technology, 35*, 2–7. doi:10.1109/T-VT.1986.24062

Arlsan, H., & Yucek, T. (2003), Estimation of frequency selectivity for OFDM based new generation wireless communication systems. *Proceedings World Wireless Congress*, Vol. 1, San Francisco, CA, USA

Barton, D. K., & Leonov, S. A. (1997). *Radar technology encyclopedia*. Boston, MA: Artech House Inc.

Bilstrup, K., Uhlemann, E., Ström, E. G., & Bilstrup, U. (2008). Evaluation of the IEEE 802.11p MAC method for vehicle-to-vehicle communication. In *Proceedings of the 2nd IEEE International Symposium on Wireless Vehicular Communications*, Calgary, Canada.

Boban, M., Vinhoza, T. T. V., Ferreira, M., Barros, J., & Tonguz, O. (2011). Impact of vehicles as obstacles in vehicular ad hoc networks. *IEEE Journal on Selected Areas in Communications, 29*(1), 15–28. doi:10.1109/JSAC.2011.110103

Campolo, C., Cozzetti, H. A., Molinaro, A., & Scopigno, R. (2012). *Overhauling ns-2 PHY/MAC simulations for IEEE 802.11p/WAVE vehicular networks*. IEEE ICC 2012, Workshop on Intelligent Vehicular Networking: V2V/V2I Communications and Applications, Ottawa, Canada

Chen, Q., Schmidt-Eisenlohr, F., Jiang, D., Torrent-Moreno, M., Delgrossi, L., & Hartenstein, H. (2007). Overhaul of IEEE 802.11 modeling and simulation in NS-2. In *Proceedings of the 10th ACM International Symposium on Modeling, Analysis and Simulation of Wireless and Mobile Systems* (MSWiM), Chania, Crete Island, Greece, (pp. 159–168).

Cheng, L., Henty, B. E., Stancil, D. D., Bai, F., & Mudalige, P. (2007). Mobile vehicle-to-vehicle narrow-band channel measurement and characterization of the 5.9 GHz dedicated short range communication (DSRC) frequency band. *IEEE Journal on Selected Areas in Communications, 25*(8), 1501–1516. doi:10.1109/JSAC.2007.071002

Clarke, R. H. (1968). A statistical theory of mobile radio reception. *The Bell System Technical Journal, 47*(6), 957–1000.

Cozzetti, H. A., Campolo, C., Scopigno, R., & Molinaro, A. (2012). Urban VANETs and hidden terminals: Evaluation by RUG, a validated propagation model. In *Proceedings IEEE International Conference on Vehicular Electronics and Safety (ICVES'12)*, Istanbul, Turkey.

Cozzetti, H. A., & Scopigno, R. (2011). Scalability and QoS in slotted VANETs: Forced slot re-use vs pre-emption. In *Proceedings of the 14th International IEEE Conference on Intelligent Transportation Systems (ITSC 2011)*, Washington, DC, USA, Oct. 2011.

Cozzetti, H. A., Scopigno, R., Casone, L., & Barba, G. (2009), Comparative analysis of IEEE 802.11p and MS-Aloha in VANETs scenarios. In *Proceedings of the 2nd IEEE International Workshop on Vehicular Networking* (VON 2009), Biopolis, Singapore.

Cozzetti, H. A., Vesco, A., Abrate, F., & Scopigno, R. (2010). Improving wireless simulation chain: Impact of two corrective models for VANETs. In *Proceedings of the 2nd IEEE Vehicular Networking Conference (VNC 2010)*, Jersey City, New Jersey, USA.

Dötzer, F., Kohlmayer, F., Kosch, T., & Strassberger, M. (2005). Secure communication for intersection assistance. In *Proceedings of the 2nd International Workshop on Intelligent Transportation*, Hamburg, Germany.

ElBatt, T., Goel, S., Holland, G., Krishnan, H., & Parikh, J. (2006). *Cooperative collision warning using dedicated short range wireless communications.* 3rd ACM International Workshop on VANETs, Los Angeles, California, USA.

ERSO - The European Road Safety Observatory. (n.d.). Retrieved from http://erso.swov.nl/

ETSI ES 202 663 V1.2.1. (2012-03). *Intelligent transport systems (ITS); European profile standard for the physical and medium access control layer of intelligent transport systems operating in the 5 GHz frequency band.*

ETSI TR 102 638 V1.1.1. (2009-06). *Intelligent transport systems (ITS); Vehicular communications; Basic set of applications; Definitions.*

ETSI TR 102 861. (2010). *Intelligent Transport Systems (ITS); On the Recommended Parameter Settings for Using STDMA for Cooperative ITS; Access Layer Part.*

ETSI TR 102 862. (2010). *Intelligent transport systems (ITS); Performance evaluation of self-organizing TDMA as medium access control method applied to ITS; Access layer part.*

ETSI TS 102 637-2. (2010). *Intelligent transport systems (ITS); Vehicular communications; Basic set of applications; Part 2: Specification of cooperative awareness basic service.*

ETSI TS 102 637-3. (2010). *Intelligent transport systems (ITS); Vehicular communications; Basic set of applications; Part 3: Specifications of decentralized environmental notification basic service.*

ETSI TS 102 687. (2010). *Intelligent transport systems (ITS); Decentralized congestion control mechanisms for intelligent transport systems operating in the 5 GHz range; Access layer part.*

ETSI TS 102 724. (2010). *Draft: Intelligent transport systems (ITS); Vehicular communications, channel specifications 5 GHz*, v 0.0.10 (2011/10).

European Commission, Information Society and Media. (2011, October). *eCall – saving lives through in-vehicle communication technology*. Retrieved from http://ec.europa.eu/information_society/doc/factsheets/049-ecall-en.pdf

Faria, G., Henriksson, J. A., Stare, E., & Talmola, P. (2006). DVB-H: Digital broadcast services to handheld devices. *Proceedings of the IEEE, 94*(1), 194–209. doi:10.1109/JPROC.2005.861011

Friis, H. T. (1994). Noise figures of radio receivers. *Proceedings of the IRE*, (pp. 419-422).

Giordano, E., Frank, R., Ghosh, A., Pau, G., & Gerla, M. (2009). Two ray or not two ray this is the price to pay. In *Proceedings of the 6th IEEE International Conference on Mobile Ad Hoc and Sensor Systems* (MASS), Macau SAR, China, (pp. 603–608).

Giordano, E., Frank, R., Pau, G., & Gerla, M. (2011). CORNER: A radio propagation model for VANETs in urban scenarios. *Proceedings of the IEEE, 99*(7), 1280–1294. doi:10.1109/JPROC.2011.2138110

Grau, G. P., Pusceddu, D., Rea, S., Brickley, O., Koubek, M., & Pesch, D. (2010). Vehicle-2-vehicle communication channel evaluation using the CVIS platform. In *Proceedings 7th International Symposium on Communication Systems Networks and Digital Signal Processing*, Newcastle upon Tyne, UK, (pp. 449–453).

Hartenstein, H., & Labertaux, K. (2010). *VANET: Vehicular applications and inter-networking technologies*. Wiley 2010.

Hogg, D. C., & Mumford, W. W. (1960). *The effective noise temperature of the sky* (pp. 80–84). The Microwave Journal.

Hossain, E., Chow, G., Leung, V., McLeod, B., Misic, J., Wong, V., & Yang, O. (2010). Vehicular telematics over heterogeneous wireless networks: A survey. *Computer Communications, 33*(7), 775–793. doi:10.1016/j.comcom.2009.12.010

Huang, C. M. (Ed.). (2010). *Telematics communication technologies and vehicular networks: Wireless architectures and applications*. Hershey, PA: IGI Global.

Huang, D., & Letaief, K. B. (2006). Carrier frequency offset estimation for OFDM systems using null subcarriers. *IEEE Transactions on Communications, 55*(4), 813–822. doi:10.1109/TCOMM.2006.874001

IEEE 802.11 Working Group. (2007). *IEEE standard for information technology--Telecommunications and information exchange between systems Local and metropolitan area networks--Specific requirements Part 11: Wireless LAN medium access control (MAC) and physical layer (PHY) specifications.*

IEEE 802.11-Working Group. (2012). *IEEE standard for information technology--Telecommunications and information exchange between systems local and metropolitan area networks--Specific requirements part 11: Wireless LAN medium access control (MAC) and physical layer (PHY) specifications*

IEEE 802.11p Working Group. (2010). *IEEE standard 802.11p, Wireless LAN medium access control (MAC) and physical layer (PHY) specifications: Amendment 6, wireless access in vehicular environments.*

Jakes, W. C. (Ed.). (1975). *Microwave mobile communications*. New York, NY: John Wiley & Sons Inc.

Jeruchim, M. C., Balaban, P., & Shanmugan, K. S. (2000). *Simulation of communication systems* (2nd ed.). New York, NY: Kluwer Academic.

Jiang, T., Chen, H. H., Wu, H. C., & Yi, Y. (2010). Channel modeling and inter-carrier interference analysis for V2V communication systems in frequency-dispersive channels. *The Journal Mobile Networks and Applications, 15*(1), 4–12. doi:10.1007/s11036-009-0177-2

Kroh, R. (2006). *VANETs security requirements.* Deliverable: D1.1, Sevecom project. Retrieved from http://www.sevecom.org/Deliverables/Sevecom_Deliverable_D1.1_v2.0.pdf

Lasowski, R., Scheuermann, C., Gschwandtner, F., & Linnhoff-Popien, C. (2011). Evaluation of adjacent channel interference in single radio vehicular ad-hoc networks. In *Proceedings of Consumer Communications and Networking Conference* (IEEE CCNC 2011), Las Vegas, NV, USA.

Lee, K. C., Lee, U., & Gerla, M. (2009). TO-GO: TOpology-assist geo-opportunistic routing in urban vehicular grids. In *Proceeding of the Sixth International Conference on Wireless On-Demand Network Systems and Services*, (pp. 11-18).

Liberti, J. C., & Rappaport, T. S. (1996). A geometrically based model for line-of-sight multipath radio channels. In *Proceedings of the 46th IEEE Vehicular Technology Conference* (VTC '96), Vol. 2, (pp. 844–848). Atlanta, GA, USA.

Lochert, C., Hartenstein, H., Tian, J., Fussler, H., Hermann, D., & Mauve, M. (2003). A routing strategy for vehicular ad hoc networks in city environments. *IEEE Proceedings of Intelligent Vehicles Symposium*, (pp. 156-161).

Mangel, T., Klemp, O., & Hartenstein, H. (2011). A validated 5.9 GHz Non-Line-of-Sight path-loss and fading model for inter-vehicle communication. *2011 11th International Conference on ITS Telecommunications (ITST)*, (pp. 75-80).

Masui, H., Kobayashi, T., & Akaike, M. (2002). Microwave path-loss modeling in urban line-of-sight environments. *IEEE Journal on Communications, 20*(6), 1151–1155.

Mecklenbräuker, C. F., Molisch, A. F., Karedal, J., Tufvesson, F., Paier, A., & Bernado, L. … Czink, N. (2011). Vehicular channel characterization and its implications for wireless system design and performance. *Proceedings of the IEEE, 99*(7), 1189–1212.

Meireles, R., Boban, M., Steenkiste, P., Tonguz, O., & Barros, J. (2010). Experimental study on the impact of vehicular obstructions in VANETs. In *Proceedings of the 2nd IEEE Vehicular Networking Conference* (VNC 2010), Jersey City, New Jersey, USA.

Nyquist, H. (1928). Thermal agitation of electric charge in conductors. *Physical Review, 32*, 110–113. doi:10.1103/PhysRev.32.110

Ohno, S., Manasseh, E., & Nakamoto, M. (2011). Preamble and pilot symbol design for channel estimation in OFDM systems with null subcarriers. *EURASIP Journal on Wireless Communications and Networking, 2011*, 2. doi:10.1186/1687-1499-2011-2

Olariu, S., & Weigle, M. C. (Eds.). (2009). *Vehicular networks: From theory to practice.* CRC Press. doi:10.1201/9781420085891

Ozdemir, M. K. (2007). Channel estimation for wireless OFDM systems. *IEEE Communications Surveys & Tutorials*, 2nd Quarter.

Papadimitratos, P., La Fortelle, A., Evenssen, K., Brignolo, R., & Cosenza, S. (2009). Vehicular communication systems: Enabling technologies, applications and future outlook on intelligent transportation. *IEEE Communications Magazine, 47*(11), 84–95. doi:10.1109/MCOM.2009.5307471

Parsons, J. D. (2001). *The mobile radio propagation channel.* New York, NY: John Wiley & Sons.

Pilosu, L., Fileppo, F., & Scopigno, R. (2011). RADII: A computationally affordable method to summarize urban ray-tracing data for VANETs. In *Proceedings of the 7th International Conference on Wireless Communications, Networking and Mobile Computing* (IEEE WiCOM 2011), Wuhan, China.

Qingyan, Y., Heng, W., & Jiuchun, G. (2008). Linking freeway and arterial data - Data archiving testing in supporting coordinated freeway and arterial operations. *11th International IEEE Conference on Intelligent Transportation Systems, ITSC*, pp. 259-264. ISBN: 978-1-4244-2111-4

Rafiq, G., & Patzold, M. (2008). The influence of the severity of fading and shadowing on the statistical properties of the capacity of Nakagami-lognormal channels. *IEEE Global Telecommunications Conference, GLOBECOM 2008*, New Orleans, LA, (pp. 1-6).

Rai, V., Bai, F., Kenney, J., & Laberteaux, K. (2007). *Cross-channel interference test results: A report from VSC-A project*. (IEEE 802.11 11-07-2133-00-000p).

Rappaport, T. S. (1996). *Wireless communications, principles and practice*. Prentice Hall.

Raya, M., & Hubaux, J. P. (2007). Securing vehicular ad hoc networks. *Journal of Computer Security, 15*(1), 39–68.

SAE J2735. (2006). *Dedicated short range communications (DSRC) message set dictionary*, V1.0.

Schoch, E., Kargl, F., Weber, M., & Leinmuller, T. (2008). Communication patterns in VANETs. *Communications Magazine, 46*(11), 119-125. ISSN 0163-6804

Scopigno, R. (2012). Physical phenomena affecting VANETs: Open issues in network simulations. In *Proceedings of the International Conference on Transparent Optical Networks (ICTON 2012)*, Coventry, England.

Scopigno, R., & Cozzetti, H. A. (2009). Mobile slotted Aloha for VANETs. In *Proceedings of the IEEE 70th Vehicular Technology Conference* (VTC Fall 2009), Anchorage, Alaska, USA.

Scopigno, R., & Cozzetti, H. A. (2010a), Signal shadowing in simulation of urban vehicular communications. In *Proceedings of the 6th International Wireless Communications and Mobile Computing Conference* (IWCMC), Valencia, Spain.

Scopigno, R., & Cozzetti, H. A. (2010b). Evaluation of time-space efficiency in CSMA/CA and slotted VANETs. In *Proceedings of the IEEE 71st Vehicular Technology Conference* (VTC Fall 2010), Ottawa, Canada.

Setiawan, G., Iskandar, S., Salil, K., & Lan, K. C. (2007). The effect of radio models on vehicular network simulations. *Proceedings of the 14th World Congress on Intelligent Transport Systems*, October 2007.

Sjöberg, K., Uhlemann, E., & Ström, E. G. (2011). How severe is the hidden terminal problem in VANETs when using CSMA and STDMA? In *Proceedings of the 4th IEEE Symposium on Wireless Vehicular Communications* (WiVEC), San Francisco, CA, US.

Sjöberg-Bilstrup, K., Uhlemann, E., Ström, E., & Bilstrup, U. (2009). On the ability of the IEEE 802.11p MAC method and STDMA to support real-time vehicle-to-vehicle communication. *EURASIP Journal on Wireless Communications and Networking, 13*.doi:doi:10.1155/2009/902414

Sklar, B. (1997). Rayleigh fading channels in mobile digital communication systems part I: Characterization. *IEEE Communications Magazine*, 136–146. doi:10.1109/35.620535

Sommer, C., Eckhoff, D., German, R., & Dressler, F. (2011). A computationally inexpensive empirical model of IEEE 802.11p radio shadowing in urban environments. In *Proceedings of the 8th International Conference on Wireless On-Demand Network Systems and Services* (WONS), Bardonecchia, Italy, (pp. 84-90).

Ström, E. G. (2011). On medium access and physical layer standards for cooperative intelligent transport systems in Europe. *Proceedings of the IEEE, 99*(7), 1183–1188. doi:10.1109/JPROC.2011.2136310

Torrent-Moreno, M., Jiang, D., & Hartenstein, H. (2004). Broadcast reception rates and effects of priority access in 802.11-based vehicular ad-hoc networks. *Proceedings of the 1st ACM International Workshop on Vehicular Ad Hoc Networks*, Philadelphia, PA, USA.

Turton, H., & Moura, F. (2008), Vehicle-to-grid systems for sustainable development: An integrated energy analysis. *Technological Forecasting and Social Change, 75*(8), 1091-1108. ISSN: 00401625

Watfa, M. (Ed.). (2010). *Advances in vehicular ad-hoc networks: Developments and challenges*. Hershey, PA: IGI Global. doi:10.4018/978-1-61520-913-2

Zheng, Y. R., & Xiao, C. (2002). Improved models for the generation of multiple uncorrelated Rayleigh fading waveforms. *IEEE Communications Letters, 6*(6), 256–258. doi:10.1109/LCOMM.2002.1010873

ENDNOTES

[1] In VANETs the term Geo-routing refers to the routing towards a destination, exploiting its position. This is linked to the nature of VANETs, where nodes are mobile and

traditional IP routing protocols (such as RIP or OSPF) cannot work.

[2]
$$F = F_1 + \frac{F_2 - 1}{G_1} + \frac{F_3 - 1}{G_1 G_2} + \cdots + \frac{F_N - 1}{G_1 G_2 \cdots G_N}$$

and, notably, the higher F, the worse the performance of the receiver.

[3] The difference in the combining phases by the direct and reflected ray is:
$$\Delta = \frac{2\pi}{\lambda} \left[\sqrt{d^2 + \left(h_t + h_r\right)^2} - \sqrt{d^2 + \left(h_t - h_r\right)^2} \right]$$

which can be approximated as:
$$\Delta \approx \frac{2\pi}{\lambda} d^2 \left[\left(1 + \frac{1}{2} \frac{\left(h_t + h_r\right)^2}{d^2} \right) - \left(1 + \frac{1}{2} \frac{\left(h_t - h_r\right)^2}{d^2} \right) \right] = \frac{4\pi}{\lambda} \cdot \frac{h_t h_r}{d}$$

thanks to the approximation $(\sqrt{1 + \varepsilon} \to 1 + \frac{1}{2}\varepsilon$, if $\varepsilon \to 0)$.

[4] In Jake's model the hypothesis of independence between the two components (phase and quadrature) is fulfilled supposing a uniform distribution along a circle. He demonstrates that some hypotheses (on the topology and phase distribution) can be specified to demonstrate that the quadrature components are independent (null cross-correlation) and the phase is uniformly distributed. In (Liberti & Rappapol, 1996) an elliptic model is discussed to analyze the angles of arrival and departure of the scattered signals. The hypotheses have been recently revisited to enforce the mathematical foundations (Zheng & Xiao, 2002). The convergence is very fast (good approximations when the number of scattered waves is not less than 8).

[5] The demonstration of the Rayleigh distribution is in three steps and leverages the Gaussian distribution of $I(t)$ and $Q(t)$. The steps are: 1) computation of the cumulative distribution function of the received power P;

2) differentiation and computation of the *pdf* of the received power; 3) computation of the *pdf* of the amplitude, based on the *PDF* of the received power. More explicitly:

$$CDF(P) = \left(\frac{1}{\sqrt{2\pi}\sigma}\right)^2 \int_0^{x^2+y^2 \le 2\rho^2} e^{-\frac{x^2+y^2}{2\sigma^2}}$$

$$= \frac{1}{2\pi^2\sigma^2} \int_0^{r^2 \le 2\rho^2} e^{-\frac{r^2}{2\sigma^2}} r\,dr \int_0^{2\pi} d\vartheta = 1 - e^{-\frac{P}{\sigma^2}}$$

. It

can be derived to get the *PDF*, as follows:

$$PDF(P) = \frac{d\,CDF(P)}{dP} = \frac{1}{\sigma^2} e^{-\frac{P}{\sigma^2}} \qquad .$$

Considering the relation between amplitude and power, the PDF of amplitude can be computed as follows:

$$\int_0^\rho PDF(\rho)\,d\rho = \int_0^{P=\frac{1}{2}\rho^2} PDF(P)\,dP$$

, whose

$$= -e^{-\frac{P}{\sigma^2}}\Big|_0^{P=\frac{1}{2}\rho^2} = 1 - e^{-\frac{\rho^2}{2\sigma^2}}$$

derivative is the Rayleigh distribution, for the fundamental theorem of calculus.

[6] For the sake of precision, mobile to mobile communications would require a more detailed discussion and the model of Equation (7) would get more complex, as suggested in (Akki &Haber, 1986), due to a double ring scattering, affecting both the transmitter and the receiver. However, for a general understanding of the phenomena, the information here provided is considered sufficient.

[7] iTETRIS (http://ict-itetris.eu) is a project funded by the European Commission (7th Framework Programme) which is aimed at creating an open, ETSI standard, compliant and flexible platform integrating wireless communications and road traffic simulations. It is based on the integration of NS-3 (http://www.nsnam.org/) and SUMO (http://sumo.sourceforge.net).

[8] The word orthogonal refers to a precise mathematical relationship between the carrier frequencies. The carriers of an OFDM modulator are all synchronous and are arranged in such a way that the sidebands of the individual carriers overlap but, despite this, signals are received without adjacent carrier interference. In OFDM systems the cross-talk between adjacent sub-channels is eliminated so that inter-carrier guard bands are not required.

[9] The concept of using parallel data transmission and frequency division multiplexing was published in the 1960s, with some early development in the 50s. The U.S. patent was filed, and issued in January 1970 (Orthogonal Frequency Division Multiplexing, U.S. Patent No. 3,488,4555. Filed November 14, 1966, issued Jan. 6, 1970).

[10] Arlsan & Yucek (2003) estimate the required spacing among pilots by evaluating how much effective the computation of the channel impulse response (CIR) is, based on channel frequency response (CFR). This operation is performed by an inverse Fourier transformation with size equal to the number of subcarriers. The maximum sampling frequency must be twice the maximum frequency in CFR; the frequency spacing is related to the periodicity (i.e. maximum excess delay τ_{MAX}) of the CFR. Under the hypothesis of an asymmetric extension of CFR (accordingly with the desired odd pilot distribution, with no pilots in 0, and equispaced, symmetric pilot carriers), you get: $\Delta f < 1/\tau_{MAX}$. The higher the delay, the lower spacing is needed to capture all the variations. With excess delay of 400 ns (Mecklenbräuker et al., 2011), $\Delta f < 16*156,25$. Hence the pilot spacing of 14.

[11] DCC policies can only partially mitigate the effects of interferences under congestion, by decreasing the degree of congestion, affecting, in any case, the reception rate. For instance, a lower transmitted power implies a lower reception rate at higher distances; a higher transfer-rate has a similar effect.

Chapter 11
Communication Networks to Connect Moving Vehicles to Transportation Systems to Infrastructure

Kira Kastell
Frankfurt University of Applied Sciences, Germany

ABSTRACT

Communication in transportation systems not only involves the communication inside a vehicle, train, or airplane but it also includes the transfer of data to and from the transportation system or between devices belonging to that system. This will be done using different types of wireless communication. Therefore in this chapter, first, the fundamentals of mobile communication networks are shortly described. Thereafter, possible candidate networks are discussed. Their suitability for a certain transportation system can be evaluated taking into consideration the system's requirements. Among the most prominent are the influence of speed and mobility, data rate and bit error rate constraints, reliability of the system and on-going connections. As in most of the cases, there will be no single best wireless communication network to fulfil all requirements, and in this chapter also hybrid networks are discussed. These are networks consisting of different (wireless) access networks. The devices may use the best suited network for a given situation but also change to another network while continuing the on-going connection or data transfer. Here the design of the handover or relocation plays a critical role as well as localization.

FUNDAMENTALS OF WIRELESS COMMUNICATION

In principle, wireless communication can reach every position intended. But in contrast to fixed line communication more sources of noise and distortion come into play. Fixed line communication mostly suffers from the attenuation of the

communication line and also from its low or band pass characteristic, but the signal only is transmitted along one dedicated path. In wireless communication multipath propagation comes into play.

The source signal is processed to fit the channel and finally transmitted over an antenna into the wireless channel. It is also received by an antenna and then converted to be delivered to the sink in an according format. As the channel is the air, there is no guiding of the electromagnetic wave

DOI: 10.4018/978-1-4666-2976-9.ch011

(including optical waves). The preferred transmission path is the direct path between transmitter and receiver. This only exists if there is a line-of-sight connection between both. If there is an obstacle between transmitter and receiver the wave may penetrate the obstacle suffering from additional attenuation and phase shift. Besides these two paths also reflection, scattering and diffraction play a role. Here the signal does not follow a straight line between transmitter and receiver, but because of the directional characteristic of the antenna, which never focuses on only one single direction but always covers at least an angle of some degrees, is not only transmitted in the direction of the receiver. Nevertheless, these parts of the signal may also reach the receiver taking longer paths. Reflection and diffraction both are based on the same principle, reflection is widely known and occurs if obstacles have the dimension of one wavelength or more. Scattering takes place at smaller obstacles and does not only lead to one reflected signal but the incoming signal is split up and scattered with different attenuation, phase shift and in different directions. Finally diffraction takes place when the signal hits an edge, then it changes its direction, often turning around that edge. Multipath propagation therefore leads to a multitude of received signals where only one signal has been transmitted. They arrive with different time delays, amplitudes and phase shifts. These effects make it more difficult to detect and process the signal at the receiver but they are necessary to be able to transmit the signals everywhere. Otherwise only line of sight communication would be possible, which would require too many transmitters to cover an area completely.

This aspect is even more important for mobile communication (wireless communication with moving devices) needed for transportation systems. But the mobility also contributes to another effect: Doppler shift. This takes into consideration the relative change of position between transmitter and receiver changing the distance and character-

istics of the wireless channel. It results in a shift of the received frequency compared to the transmitted one and has to be taken into consideration in receiver design. In addition the wireless channel also suffers from different sources of distortion, mostly noticeable as attenuation. The conditions of the wireless channel change randomly as moving obstacles are not predictable and rain, smog and other changes of the composition of the air may cause severe signal degradation. Objects covering the antenna have an impact on signal quality as well, e.g. insects, pollen.

A wireless mobile communication network needs to take this into consideration. As for transportation systems we may not be willing to design and setup a dedicated wireless access network here the main principles of system design for existing wireless networks will be explained. The parameter on which the transportation system planer has the most influence is the antenna of the transportation device. Taking into consideration multipath propagation it is preferable to make use of an omnidirectional antenna to be able to receive signals from every direction with (nearly) the same quality. But the antenna may also have a more directional characteristic, if some directions are preferred or should be excluded because of distortions, e.g. signal reception along a track, distortions from a known adjacent transmitter or bright star. Communication to and from a mobile device not only involves the wireless device and its direct counterpart on the other end of the air interface, the network access point (NAP), but also the complete network that is behind the network access point. So the air interface is only a small part of the network. And this, besides the roll out costs, is why communication in transportation systems most likely will make use of existing networks to communicate with other systems.

Before going into details of different already existing or standardized wireless communication systems some basic components of those systems will be explained. In general we can distinguish between complete communication systems having their own backbone network and those only implementing the lower two or three layers of the ISO/OSI reference model and then using another system for the higher layers, converging to all-IP (Internet Protocol) in the future. All wireless mobile systems have a wireless (mobile) device and a network access point. In the following we will use these generic terms rather than alternative expressions used in specific contexts such as mobile station and base station in GSM. The network access points need a control unit and must be connected to a switching unit that also needs to take care of the mobility issues. Depending on the system, several databases are used for mobility handling also incorporating security credentials for authentication, authorization and accounting (AAA). At least one of the switching units needs to work as a gateway to external networks.

Wireless networks have two different modes of operation. Some networks are able to support both of them, some only work with one. Networks are separated into infrastructure-based and ad-hoc. In ad-hoc networks the mobile devices form a network without support of a fixed infrastructure. If an ad-hoc network wants to establish a communication to the world outside it needs at least one mobile device having contact to another network. Typically this is done by using additional infrastructure, thus an infrastructure network. If communication in the transportation system is only needed for exchange between network internal components ad-hoc networks can be used, otherwise at least one gateway to infrastructure, e.g. for accessing the Internet, is needed.

In a wireless infrastructure based access network the mobile device connects to a network access point by a radio link. Each network access point has a certain range within which the mobile device has to stay to be able to connect to this network access point. The area which can be covered by one network access point, i.e. where a certain field strength level can be received, is called a cell. Depending on the technology in use the cell diameter typically varies between 50 m and 35 to 70 km. To cover a larger area with

one technology a multitude of cells and network access points is needed. Besides that, cells are also restricted by the available transmit power, because the attenuation of the physical channel, i.e. air, limits propagation. Another restriction is the equalizer that needs to cope with signal delay caused by the distance between transmitter and receiver and propagation velocity of the signals. The physical resources for mobile communication are mainly frequency bands, i.e. dedicated portions of the frequency spectrum. A certain frequency slot that may further be defined by an additional timeslot pattern or a certain coding is called a channel. This channel cannot not be used in adjacent or nearby cells. As the number of channels is limited this also forces a cellular structure for capacity reasons, because every cell can only handle a certain number of users.

CANDIDATE NETWORKS FOR CONNECTION OF TRANSPORTATION SYSTEMS

There are several mobile communication networks already that may provide sufficient quality for the connection of transportation networks. Which one is best suited strongly depends on the requirements of the transportation system. Among them are reliability and availability, kind and amount of data to be transmitted, frequency and tolerable delay of transmission, needed security level, speed, mobility profile of the devices and cost. In the following some candidate networks are described briefly with regard to the points mentioned above. The networks are GSM, UMTS, LTE, WLAN 802.11, which also has a special amendment for transportation systems, and WiMAX.

Global System for Mobile Communication (GSM)

GSM (Global System for Mobile communication) is a system of the second generation (2G) of mobile communication standardized by ETSI

(European Telecommunications Standard Institute) in 1989 and therefore one of the first digital wireless communication systems, now maintained by 3GPP (3rd Generation Partnership Project). The first GSM network started service in Finland in 1991. Today, all GSM networks together build by far the most successful mobile telecommunication technology in terms of subscribers and coverage. Its enhancement EDGE (Enhanced Data rates for GSM Evolution) has been accepted by ITU (International Telecommunication Union) as part of the IMT-2000 family of third generation (3G) standards (ITU, 2003). The GSMA (GSM Association) of about 800 mobile operators and 200 related companies comments on the success of GSM on its homepage (GSM, 2011) that GSM by now is used in 219 countries by more than three billion users. Thus, in terms of availability GSM is the network with the broadest coverage. In use for 20 years, it also is very reliable as it has undergone live test by billions of users leading to continuous improvement.

GSM has assigned frequencies in the 900 MHz, 1800 MHz, and 1900 MHz bands. The multiple access scheme is a mixture of FDMA (Frequency Division Multiple Access) and TDMA (Time Division Multiple Access). Full frequency division duplex (FDD) is used so that for every link an uplink frequency (lower) and a downlink frequency (higher) are assigned. Every frequency band is subdivided into channels of 200 kHz each (3GPP, 2005) containing 8 time slots. A subset of frequency channels is assigned to every cell. At least one time slot per cell is used to broadcast information about synchronization, cell identity, channel configuration, neighbour cells, etc. In the basic version one time slot is exclusively assigned to one communication link for its whole duration (circuit switched). This has changed when introducing packet switched mode (GPRS, General Packet Radio Service) making channel and time slot assignment more flexible as the time slots are assigned only if needed for communication

at a certain moment. This is especially useful for asymmetric traffic and traffic with communication pauses.

The achievable data rate is low compared to the other candidate networks. GSM started with a maximum data rate of 9.6 kbps on the air interface finally leading to a peak data rate of 384 kbps for EDGE according to the different standardization phases. During on-going connections, handovers from one cell to another are integral parts to cope with the mobility of the user (definition see below). The delay is negligible even for voice communication, but it may play a role during handover for users moving with higher speed (above 120 km/h or 250 km/h).

GSM has some security mechanisms to protect the traffic over the air interface between mobile device and base station. They aim for the same security level as in standard ISDN (Integrated Services Digital Network). Therefore the mobile device is authenticated by the network and the data is ciphered for transmission over the air interface with a 64 bit key.

Universal Mobile Telecommunications Standard (UMTS)

UMTS (Universal Mobile Telecommunications Standard) is maintained by 3GPP. Here roaming to GSM and the interoperability between the two systems plays a role. This extends the coverage of UMTS to all GSM areas as long as the mobile device is capable of GSM and the user accepts that some services are not provided or only with less quality. Therefore the physical structure of UMTS and GSM is very similar. The number of UMTS networks is hard to count as UMTS already has undergone some enhancements leading to network counting in each subdivision. The wider spread standard at the moment may be HSPA (High Speed Packet Access) which by June 2010 had been commercially deployed by over 200 operators in more than 80 countries (UMTS, 2011) serving some hundred million subscribers worldwide. The first commercial UMTS network has been launched by Telenor in Norway December 2001.

The main enhancements in UMTS compared to GSM are the increased data rates of 384 kbps for medium sized cells up to 2 Mbps, theoretically. 384 kbps will be achieved for speeds up to 120 km/h and about 144 kbps for higher speeds. The basic transmission mode is packet switched instead of circuit switched. On the air interface a mixture of FDMA and (W)CDMA ((Wideband) Code Division Multiple Access) is used. Each channel has a bandwidth of 5 MHz and most of the networks are built as single frequency network so that only coding distinguishes between cells and their users. UMTS has two modes of operation FDD with two separate frequencies for up- and downlink as in GSM and additionally TDD (Time Division Duplex) where both directions share one frequency. Therefore certain portions of the time are assigned either to uplink or downlink. This assignment is flexible to cope with different load distribution. The assigned frequency band is at 2 GHz (between 1900 and 2200 MHz), therefore attenuation for UMTS is higher than for GSM as it increases with frequency.

The security mechanisms have also been enhanced compared to GSM: The authentication now is mutual and ciphering is twofold, one key for data and an additional key for integrity protection, both with increased length (128 bit) compared to GSM (64 bit).

Long Term Evolution (LTE)

LTE (Long Term Evolution) (3GPP, 2008) is the newest advancement and successor of UMTS. The first LTE network already went on air in December 2009 servicing Stockholm and Oslo, operated by TeliaSonera. Because of the recent start the coverage is still very small compared to GSM and UMTS but will constantly increase as interoperability to both systems as well as to their

competing systems cdmaOne and CDMA2000 is specified. LTE (Release 8) provides peak data rates of 300 Mbps in the downlink and 75 Mbps in the uplink. The delay is further shortened leading to round-trip times of less than 10 ms. A big step has been the support of scalable carrier bandwidth from 1.4 MHz up to 20 MHz both in FDD and TDD. Also the core or backbone network is transferred to IP-based network architecture. This allows for better interoperability with other networks as no network specific core network is needed.

LTE's access scheme uses a mixture of scalable OFDM(A) (Orthogonal Frequency Division Multiplex (Multiple Access)) and MIMO (Multiple Input Multiple Output) in connection with FDMA. One single cell is able to serve 200 active users compared to 8 in the beginning of GSM. There is no single frequency band for LTE. Instead, bands from 700 MHz to 2.6 GHz are assigned depending on the geographical area. For worldwide roaming the mobile device needs to be able to process all different frequency bands. The security mechanisms are further developed allowing longer key although 128 bit are used right now. The key hierarchy has been extended and control and user plane have been separated. The support for high speeds has been increased allowing all features in good quality up to 350 km/h. The standard even covers speeds up to 500 km/h still very well, depending on the frequency band assigned.

Even LTE already has a successor LTE Advanced which shall provide data rates from 100 Mbps up to 1 Gbps, and is maintained by 3GPP (3GPP, 2011).

Wireless Local Area Networks (WLAN)

The most common standard for wireless local area networks (WLAN) is IEEE 802.11. It consists of a multitude of amendments, e.g. IEEE 802.11a, 802.11b, 802.11g, and 802.11n defining physical layer characteristics. The different amendments offer data rates from 1 to 2 Mbps up to 600 Mbps with different air interfaces and modulation schemes and operate in the license free ISM (Industrial Scientific Medical) bands of 2.4 or 5 GHz. They use either OFDM (Orthogonal Frequency Division Multiplex), FHSS (Frequency Hopping Spread Spectrum) ode DSSS (Direct Sequencing Spread Spectrum). The substandard IEEE 802.11b is commonly referred to as Wi-Fi.

The structure of WLANs is similar to the one of mobile networks containing network access points in cells connected to each other by a local distribution system, mostly Ethernet IEEE 802.3. WLAN initially has not been designed for mobility over more than one cell. To allow for more mobility, roaming and handover as prerequisite for inter cell mobility are addressed in IEEE 802.11f (introducing beacons every 100ms for better mobility support and an authentication server.) and 802.21 (handover). Depending on the protocol in use, which in general is some IP version, mobility may be enabled with different quality, e.g. mobile IP (Johnson, Perkins, Arkko 2004; Perkins, 2002).

In terms of availability WLAN still does not cover a wide area as its name already indicates. The reliability depends on the type of network provider. Private networks may be switched off occasionally while commercial hotspots will continuously stay on air. The devices are very reliable as well if they are set up and maintained correctly. Delay times greatly depend on the number of stations involved beginning with 0.5 ms for only one transmission, around 1 s for average transmission and up to 10 s for fully loaded networks. In WLAN authentication is a prerequisite for registration with the network access point. Thereafter challenge-response using WEP (Wired Equivalent Privacy) or WPA (Wi-Fi Protected Access) is used with one single 64 or 128 bit key for authentication and ciphering. A further enhancement is IEEE 802.11i with an additional authentication server supporting mutual authentication and key agreement. WEP has been

a very weak security mechanism, but WPA and 802.11i enhance the security of WLAN.

Especially for wireless access in vehicular environments (WAVE) a new amendment to IEEE 802.11 has been developed: IEEE 802.11p (IEEE, 2010) dealing with intelligent transportation systems (ITS). The frequency band used is the ITS band from 5.85 to 5.925 GHz which is a licensed band. The typical range is up to 1 km with data rates up to 54 Mbps for distances up to 200 m. For data rates of 1 Mbps it can be as far as 2.5 km whereas Wi-Fi provides 500 m range (Gräfling, Mähönen, Rihijärvi, 2010). The system uses OFDM with bandwidth of 10 MHz and transmit power from 21 to 40 dBm depending on the number of simultaneously used channels. Networks of this type will be built or have to be built especially for use with ITS in vehicular environment, at least the network access points or road side units (RSU). They will be reliable as they are designed to meet the vehicular environment requirements.

The higher layer standard designed to work together with IEEE 802.11p is IEEE 1609 which also deals with architecture, resource management, security, networking, and multi-channel assignment. But the backhaul architecture can consist of WiMAX, IPv6 or cellular networks. IEEE 802.11p can be used for systems based on CALM (Communications, Air-interface, Long and Medium range) architecture, e.g. for toll collection, driver assistance, infotainment and others. Among the topics dealt with are data exchange between vehicles with on-board unit (OBU) and roadside infrastructure or RSU. Another topic is the communication between high-speed vehicles. This leads to short durations of stay in every single cell demanding for short delay times. As the system should provide information about road side accidents it is highly delay sensitive, but delay guarantees are missing in the standard. Research is going on here, e.g. (Böhm, Johnsson, 2009). On the other hand, periodic information exchange is needed for status reports and this is also delay sensitive. To deal with emergency messages as well as with infotainment, different priority categories have been designed: Prioritizing according to FCC (Federal Communications Commission of the US) starts with safety of life applications followed by public safety in the second class and all non priority communication in class three. Classes 2 and 3 will face a higher delay. Up to now the analysis is based on simulations that show delays of some ten ms for priority traffic of class 1. In general a delay of less than 100 ms can be achieved for up to 1000 packets per second. Thereafter delay increases noticeably (Gräfling, Mähönen, Rihijärvi, 2010). The average wait time for channel access is 56 to 264 µs increasing with decreasing priority category (Eichler, 2007).

The security is based on public key infrastructure with a certification authority and link level encryption for all messages. Besides that you are only allowed to broadcast services if approved by a registration authority. Privacy, confidentiality, availability, and integrity have been major concerns when designing the security mechanisms as the system needs information from the vehicles about road, traffic and vehicle condition. But the mobile unit should not be traceable.

Worldwide Interoperability for Microwave Access (WiMAX)

The WiMAX (Worldwide Interoperability for Microwave ACCESS) standard IEEE 802.16 is deployed in 583 networks in 150 countries (WiMAX, 2011), Korea being the first one in 2006. By February 2011, the WiMAX forum cited coverage of over 823 million people. Mobile WiMAX is defined in standard IEEE 802.16e, primarily extending WiMAX with handovers, measurements and power control to cope with mobility. WiMAX covers licensed frequencies from 2 to 66 GHz were 2 to 11 GHz are considered for mobile use. The access scheme is scalable OFDMA. Basic transmission mode is

TDD with a range of about 20.7 km for 3.5 and 7 MHz and 8.4 km for 5 and 10 MHz (Wang et al., 2008). The range can be up to 50 km with general data rates starting from 2 to 6 Mbps over up to 63 Mbps in a 10 MHz downlink channel now extended to 100 Mbps for mobile users and 1 Gbps for fixed users with amendment 802.16m in 2011. Here the supported speed is higher than 120 km/h with scalable bandwidth between 1.25 and 20 MHz achieving a handover delay below 50 ms. The authentication procedure includes key management, mutual authentication (initially not foreseen) and traffic and control encryption with keys of 128 or 256 bits.

NETWORK CHOICE FOR TRANSPORTATION SYSTEM

To choose the best suited network for connecting a given communication system to the outside world at least the following requirements should be taken into consideration:

- Does the system need a continuous connection, a coverage that will be available wherever the mobile device can be or are there dedicated places and times where data exchange takes place? Here the availability of the communication system will play a role.
- How much data is going to be exchanged during which time? This defines the minimum data rate needed and the minimum duration of a single connection.
- Are the data delay-critical for fulfilling the service requirements, e.g. emergency warnings, traffic detection? As some networks need long time to setup a connection or have large delay, especially with handover, the timing constraints play a critical role.
- How fast will the mobile devices move and which mobility pattern will they follow?

The faster the movement, the more (frequent) handovers will occur. This plays a role for reliability and delay constraints.

- What type of security is needed (authorization, authentication, and encryption) and how strong do the mechanisms have to be? As the systems differ in strength of the ciphering key derivation algorithms and may or may not contain mutual authentication or special privacy protection schemes, different services may only be reasonable with a certain system.

The key aspects for each system discussed in the previous sections are summarized in Table 1.

The requirements of the transportation communication systems need to be evaluated first and thereafter a decision about a suited target network can be taken. If none of the networks given here fulfils the requirements one can have a look, if other less common networks may be suited. If not the second step should take into consideration if a combination of two or more networks from the ones mentioned above could provide a solution.

Here different integration or interworking concepts need to be taken into consideration.

The networks could be used in parallel without any interaction and the mobile device connects to the one with the better quality based on its own decision. The more networks the mobile device is allowed and capable to connect to the better is the overall availability. Reliability also is increased as there is redundancy wherever two or more networks cover the same spot. But the reliability achieved does not cover on-going connections when one network or cell drops out of connection and only cells from another network are available. Here networks with at least some interworking capability are needed to maintain the on-going connection. Depending on the entity in the network infrastructure where the interworking takes place or the layer where the interface is integrated, different levels of coupling can be distinguished. The point of integration plays a role in the support of the handover as handovers in general are smoother if the integration is very tight. ETSI (ETSI, 2001) distinguishes between two categories: Loose coupling only affords a

Table 1. Comparison of candidate networks

Network	GSM	UMTS	LTE	WLAN	802.11p	Mobile WiMAX
Reliability / starting date	very high 1991	very high 2001	high 2009	high	not available	high 2006
Availability / Coverage	219 countries very high	> 80 countries high	increasing, low at the moment	depends on location	not available	depends on location
Frequency	900 MHz, 1800 MHz, 1900 MHz	2 GHz	700 MHz to 2.6 GHz	2.4 and 5 GHz	5.85 to 5.925 GHz	2 to 11 GHz
Data rate	384 kbps	2 Mbps	300 Mbps	1 to 600 Mbps	54 Mbps	63 Mbps
Bandwidth	200 kHz	5 MHz	1.4 to 20 MHz	20 to 22 MHz	10 MHz	1.25 to 20 MHz
Delay	< 500 ms	< 200 ms	5 ms	0.5 ms to 10 s	50 to 100 ms	20 ms to 1.2 s
Security key / authentication	64 bit unilateral	2x 128 bit mutual	2x up to 256 bit Mutual	up to 128 bit mutual	enhanced compared to WLAN	up to 256 bit mutual
Theoretical speed limit	500 km/h	500 km/h	500 km/h	not available	200 km/h	125 km/h
Actual speed limit	250 km/h	120 km/h	120 km/h	some 10 km/h	140 km/h	125 km/h

common subscription while in tight coupling the foreign networks appear as an additional radio access network (RAN) which can be handled by a common core network. A more detailed approach (Samarasinghe, Friderikos, Aghvami, 2003) has four categories from open coupling (no coupling, only roaming agreements) over loose coupling (common AAA data base) and tight coupling (only one operator for all networks) up to integration (joint access network). The most detailed description in six levels is given by 3GPP (3GPP, 2011b): scenario 1 has common accounting, scenario 2 adds common (UMTS based) access control and authentication, scenario 3 additionally allows packet transmitted (UMTS) services from WLAN, scenario 4 allows service continuity after handover, scenario 5 does this seamlessly, and scenario 6 enables circuit switched (UMTS) services from WLAN. Other definitions can be found, e.g. in (Tsao, Lin, 2002), (Varma et al., 2003), (Fitzek, Munari, Pastesini, Rossi, Badia, 2003), (Lampropoulos, Passas, Merakos, Kaloxylos, 2005). If no additional entity shall be needed, then loose coupling is the best choice. Here interworking is based on software algorithms only and the only connection is the AAA data base for security credentials.

Loosely coupled networks are called hybrid networks as well. A hybrid network consists of different radio access technologies (RAT) at the first three layers of the ISO/OSI reference model and a hybrid core network as interface between the Internet and the radio access technologies. In contrast to that heterogeneous networks have at least one additional entity compared to the standard network architecture. They have one core layer aggregating all network functionalities, which is seen as one single network by higher layers. As transportation systems are likely to make use of existing networks the approach discussed here in more detail is open or loose coupling and therefore hybrid networks. When considering hybrid networks attention should be drawn especially to

the choice of radio access network, handovers and security mechanisms during and after the change from one radio access network to another. This will be detailed in the following subsections.

Access Network Choice in a Hybrid Network

The choice of the best suited radio access network in a hybrid network depends on several aspects. These aspects define a utility function which serves as the decision measure for the choice. If only one radio access technology is available this will be the best choice. But if the only available radio access technology is provided by more than one operator access costs will play an important role. But costs are not the most important measure as different costs for different networks may be the result of different capabilities and services the radio access networks provide. In general, higher data rates and more bandwidth justify higher prices. Also better availability may increase costs. So, if there are more radio access networks with different radio access technologies available, costs are of lower priority. The decision then should be based on the best suited data rate and quality of service (QoS). The costs may prevent from always choosing the highest data rate. If a lower data rate from another radio access network is sufficient, this will be best suited in terms of the utility function. Other technical features as e.g. bandwidth, call setup time, and transmission delay may also be critical for the transportation system and, if so, should be taken into account when designing the utility function.

But there are additional aspects caused by the nature of the hybrid network: The cells in a hybrid network will differ in size not only because of topological aspects but also because of the different propagation attenuation and transmit power in the single radio access networks. Radio access networks with higher bandwidth tend to have comparatively smaller cells, because they often use higher frequencies and/or low transmit power, e.g. WLAN. If the utility function sets high value on data rate, these networks will be chosen. But if simultaneously the mobile device travels with high speed it will have a short duration of stay in smaller cells and therefore needs to perform many consecutive handovers. As shown in the next section, handovers pose diverse risks on the connection, including call drop, and thus shall be kept to a minimum number. The total throughput of the radio access network may be increased by choosing a network with a smaller data rate but fewer handovers because of larger cells as handovers always contain some time where no data can be transmitted. So the choice of the network is driven by the utility function on one side and the mobility pattern on the other side. Depending on the utility function different parameters will be needed for choosing the best radio access network or target cell.

The combination of complementary radio access networks is the best solution for providing a seamless communication system for vehicular use. It keeps investment costs small as existing infrastructure can be reused. But as the infrastructure and the protocols have not been designed for vehicular use, some modifications have to be made. First of all, for purposes of traffic control and automated tolling, RSUs have to be installed. The investment costs can be reduced if standardized hard- and software is installed instead of proprietary solutions. Therefore, CALM uses P1609 and IEEE 802.11p. At the moment, IEEE 802.11p is regarded as the best suited network for last mile short range communication. Modifications to guarantee delay limits are on their way (Böhm, Johnson, 2009).

As IEEE 802.11p only covers a small coverage area, it is well suited for RSUs providing roadside assistance and traffic status reports. These messages are valid for a certain area only. But there is no WLAN that allows seamless coverage at reasonable costs. So the question is which ac-

cess and distribution networks will be suited for vehicular communication. The answer depends on the use case.

For V2V communication, only ad-hoc network solutions are feasible. On the one hand the topology of the network changes very fast. This would cause a lot of overhead mobility management data in infrastructure mode networks. On the other hand, alerts dealing with road conditions may occur at locations where no RSU is available. V2V communication can prevent traffic jams or accidents without the interaction of a RSU. In this case, the ad-hoc solution must provide the same low latency and the same security level as needed for communication with a RSU. IEEE 802.11p provides a delay below 100 ms and an increased security compared to other 802.11 standards. The use of public key infrastructure (PKI) allows better protection of integrity against attackers. But it contains the identity (ID) of the communication partner and therefore may allow tracking. Lin, Sun, Ho, and Shen (2007) introduce a secure and privacy-preserving protocol for vehicular networks by integrating the techniques of Group Signature and Identity-based Signature (GSIS). But even without this extension the integrity of the data is secured. By this, the most important security threat in traffic flow control is managed.

Besides the communication from one vehicle to another and the communication with RSUs to exchange traffic status reports, other communication needs may arise. For fleet management and traffic control in a larger area, data from several RSUs need to be analysed. Fleet management and public communication may also ask for coverage in areas where no RSU is installed. As RSUs are implemented in hardware they are relatively costly. Therefore they will only cover spots that are regarded as places with dangerous traffic conditions. Among others, these areas may include traffic lights, highway entrances, and railroad crossings. As these areas are neither evenly spaced nor frequently occurring in many areas, additional radio access networks are needed. These networks shall provide seamless service. They will also support the RSUs, e.g. as a connection to the core network where wired connections are too costly. And they may provide additional data, measured by their network components.

If one of the networks mentioned above is suited and which one is best suited strongly depends on the requirements of the transportation system. The reliability of all systems is high as long as no private WLANs are included. These may be switched off by the user without notice. This limits their availability as well. But availability largely depends on the roll-out concept of the provider. Networks such as GSM, UMTS, and LTE are intended for wide area coverage. WLAN and WiMAX are designed to cover smaller areas. WLAN only has a range of some hundred meters while WiMAX may cover several square kilometres. Often, the coverage area is determined by the transmit frequency. And a network can only be used, if the needed frequency band is available at a certain location. From the point of view of seamless coverage GSM, UMTS, and LTE are preferable, but they only provide comparatively small data rates. Depending on the service they will not be best suited compared to WiMAX and IEEE 802.11p, e.g. for real-time video streaming. For traffic control the data rates they provide will be good enough, especially when prioritization is applied. All networks are able to cope with messages sent frequently. GSM, UMTS, and LTE are specified up to 500 km/h, IEEE 802.11p covers speeds up to 200 km/h. IEEE 802.11 and WiMAX are not even specified for vehicular speed; the latter one covers speeds up to 125 km/h (cf. chapter Communication to and from Moving Devices).

The main radio access network differences are delay and handover latency as well as security level. The delay of GSM can be as high as 480 ms. IEEE 802.11 usually has a very small delay, but depending on the load it may be in the range of seconds. This is not acceptable for life-critical

data. The other networks provide a delay below 200 ms. This is fast enough for the use cases described above.

The handover latency has to be dealt with separately. For seamless connectivity the mobile device has to be handed over from one network access point to another. Besides this, a lot of preparing measurements have to be taken. The duration of these measurements varies widely (Kastell, 2009), (Kastell, 2011). Therefore, in some situations UMTS may also not be suitable as it can have a handover latency of several seconds. But the most severe drawback when dealing with life-critical messages is the security. This can be divided in two cases: built-in security mechanisms and security level provided during and after handover.

As both RSU and mobile device may transmit life-critical messages, mutual authentication is a must. Therefore GSM and GPRS, as used in CALM, are not suited for this kind of communication. The other networks provide mutual authentication. If WPA or even IEEE 802.11i is used instead of WEP in WLANs, all networks have a sufficient security level on the air interface. The big difference is in the handover procedure. GSM, UMTS, and LTE perform handovers without integrated authentication. The credentials are forwarded to the next network access point in the core network. Moreover, they allow seamless connectivity between these three networks. The different keys are converted into the suitable one for the new network by publicly known conversion algorithms (Kastell, Jakoby, 2007). This weakens the security of the better protected networks LTE and UMTS. It circumvents the stronger security mechanisms with longer keys and reduces them to the security of GSM. In WiMAX authentication is mandatorily included in the handover completion phase. This makes the WiMAX handover more secure than the one in LTE. It is still weaker than the one in WLAN where the authentication with the new network access point has to take place before the handover execution. Here both systems benefit from the fact that the handover has been designed ex post. It is a combination from call termination followed by an association with the new network access point. This increases the overall handover duration, but the execution itself is not affected very much.

The result of the comparison is that no single network fulfils all requirements. The wide area networks are comparatively insecure and slow. The local area networks cannot provide seamless connectivity. Therefore in Section V we propose a hybrid network consisting of all radio access networks discussed here. This will allow the choice of the best suited network access point for each service. It also provides flexibility to include other radio access networks. Moreover, it therefore also reduces installation cost as every network will be suitable.

Handover

Handover is the transfer of a connection between a mobile device and a third party from a serving network access point to a target network access point without user interaction and without loss or disruption of the connection. Handover procedures allow a user to seamlessly use services while changing from one underlying access technology to another. The rerouted connection is neither lost nor interrupted and in the ideal case the user doesn't even perceive the change of radio access network.

To provide information for handovers in most networks a so-called mobile assisted handover takes place. The mobile device frequently (e.g. every 480 ms in GSM, for more information see (Kastell, Fernandez-Pello, Fernandez, Meyer, Jakoby, 2004; Kastell, 2007; Kastell, 2011) measures the field strength level of the serving network access point and (up to seven) adjacent or neighbouring network access points. It provides the data to the serving network access point as a sorted list. The network access point or the mobile switching centre decide about the need of a handover and the target cell. A handover may take

place because of different reasons. In any case it needs to take place when the field strength of the serving network access point drops below a certain threshold that ensures processability by the receiver. The decision is at least based upon the field strength level. To avoid frequent handovers back and forth along a cell border (ping-pong) most decision algorithms only start the handover procedure if the serving network access point has comparatively lower field strength (e.g. 6 dB) than the target network access point or shows lower values for a certain time. Here the handover needs to take place quite fast, because the mobile device already is close to the border of the serving cell. Other types of handover are less time critical. Among them are handovers because of capacity or load balancing for mobile devices that in terms of field strength may be sufficiently served by more than one network access point. In a hybrid network also handovers from one radio access network to another because of provided services may be desired even if the field strength does not suggest the necessity. Another decision criterion for handovers in hybrid networks is the speed of the mobile. As the mobility pattern together with the cell size affects the frequency of handovers, faster mobile devices may be handed over to larger (umbrella) cells to reduce the number of handovers needed. Especially for high speed above 250 km/h this is recommended, because the measurements may take too long to provide sufficient data for handover decision if the cell size and thus the duration of stay in a single cell is too short.

The handover consists of three phases (ETSI, 2001), (Bargh et al., 2004): handover detection, handover decision, and handover execution. Thereafter some more control messages have to be exchanged for handover completion. In the following, the detection and decision will be subsumed as handover preparation as the decision itself nearly takes no time and therefore will be counted as the end of the detection phase which includes taking the measurements needed for decision. These three phases, preparation, execution, and completion, are more or less prevalent in the different radio access technologies. While handovers are a definitive element of mobile communication networks such as GSM, UMTS, and LTE, the first versions of WLAN and WiMAX did not include handovers. In these networks handovers only enabled local mobility or were initially aimed to replace fixed connections by cheaper air interfaces. These systems often perform a handover as detach from one network access point followed by the setup of a new connection to another network access point. Moreover in none of the systems a change in the wireless interface during the handover is foreseen - except where one communication system is the designated successor of another, e.g. UMTS and GSM. Here a wide range of standardization and research efforts is going on, enabling inter system handovers.

In 2003 3GPP developed an interworking architecture to and from WLAN based on 3GPP subscription data, security mechanisms, service access and accounting. WLAN hereby is an integral component of the local services (for details see (Ahmavaara, Haverinen, Pichna 2003)). Handover between networks with different protocols are investigated in e.g. in (McNair, Akyildiz, Bender, 2000), (Kastell, 2007). The first approach adds a new network element while the latter tries to avoid any changes of existing standards in terms of hardware as well as in changing existing procedures. To see where the problems may arise from in the different handover protocols these protocols are briefly outlined in Figure 1 for GSM/UMTS and in Figure 2 for WLAN and WiMAX. The procedure in UMTS and LTE is quite similar to the one in GSM although the names of commands and entities change.

The comparison of the handover procedures shows that all radio access networks need preparing measurements. These consume a lot time even if they only relate to one message in the given figures. In GSM measurements are taken every 480 ms but some consecutive measurements are needed to make a decision about a handover. The

same is the case with probing and scanning in the other radio access networks that often need an exchange of challenge and response for gathering data. The complete duration of the measurements is between 200 ms in a special UMTS mode to up to 10.8 s as well in UMTS (Kastell et al., 2004; Kastell, 2007). For fast handovers measurements around 500 ms will be tolerable. For WLAN and WiMAX the procedure includes fewer nodes as there is no upper layer network infrastructure involved. If the switching centre is responsible for mobility management resides in the core network as in GSM, UMTS and LTE, more messages need to be exchanged to prepare the handover as more entities and network hierarchy layers play a role. In Figure 1 the most complex procedure is shown as here the transfer of the mobile device from one switching centre to another is shown. This is necessary as one switching centre can only be responsible for a certain number of network access points and connections. The more messages need to be exchanged the more time consuming the decision taking is. This adds up with the measurement time to the overall preparation time.

If we take a closer look at the execution phase one can clearly see that for GSM it is quite fast as the mobile device only receives one message and immediately connects to the target network access point. Notice that 1 always refers to the actual serving network access point while 2 indicates the target network access point. Here the other radio access networks show some more messages. In terms of timing this is a draw back. It is caused by the setup of a new connection including security negotiation. This will be discussed in details in the following section. The time for handover execution plays a critical role as in most radio access networks, besides UMTS, and in all hybrid handovers with change of radio access network the connection is interrupted for the duration of the execution. As human beings are very sensitive to interruptions in voice communication, especially for voice the interruption

Figure 1. Handover procedure for a GSM network

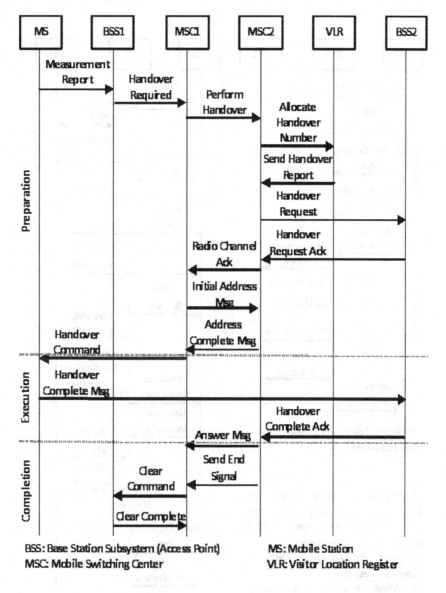

should be kept below 200 ms (ITU, 2003b) for unnoticeable interruption from which we are still far-off (Kastell, 2011). Actually 50 ms maximum interruption are recommended (ITU, 1988), but 200 ms still satisfies user demands.

The handover completion does not contribute to the handover latency but it has to be performed in order to free occupied radio resources and to update data bases keeping track of the mobility issues. Therefore it is needed in traditional handovers but is missing in WLAN where the

connection to the serving network access point is completely terminated before switching to the target network access point.

Security

The security of the data exchange is always weakest during handover. Here a well-established security context between two peers is broken and a new security context to another network access point has to be established. The problem with handover

Figure 2. Handover procedure for a) WLAN IEEE 802.11 and b) WiMAX mobile IEEE 802.16e

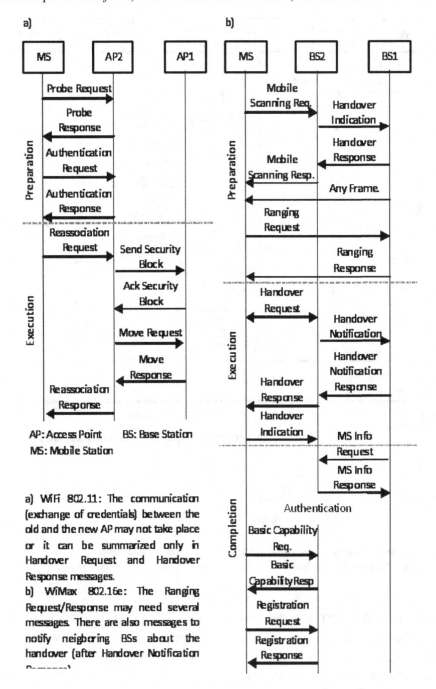

a) WiFi 802.11: The communication (exchange of credentials) between the old and the new AP may not take place or it can be summarized only in Handover Request and Handover Response messages.

b) WiMax 802.16e: The Ranging Request/Response may need several messages. There are also messages to notify neigboring BSs about the handover (after Handover Notification ...

is that it has to take place very fast. That is why there is no authentication included in handovers in GSM and UMTS. The credentials are passed from the serving network access point to the target network access point. In terms of security this has two drawbacks: The transfer of unencrypted credentials over the backbone network may not be secure. Transmitting any keys should strictly be avoided. The transfer method also only works inside one network as different networks use different credentials and algorithms. If we assume that only one network and one operator is involved than we may assume that the operator also owns the backbone network and can influence its security. Then the exchange of credentials may be secure as it has been intended in the standard. But if two operators work together the risk increases. And it further increases if different radio access technologies are involved. For GSM and UMTS the keys and their respective lengths are different. To avoid additional authentication during the handover a key conversion function has been defined. The converted keys are transferred to the target radio access network. Key conversion by a publicly known algorithm poses another security risk. The more modern networks use longer keys and less vulnerable algorithms for authentication and encryption. This has been standardized to enhance the security. It has become necessary because the security protection of GSM has been broken. One can derive the needed keying material in real-time from as few as 4 ms of encrypted traffic (Barkan, Biham, Keller, 2003). If the keys are only converted instead of renegotiated this weakens the keys of better protected radio access networks as well as reverse engineering at least can calculate parts of the keys of the better protected radio access network. Therefore it would be preferable to include authentication and key negotiation into the handover execution. A workaround for handovers that do not allow additional message exchange because of the timing constraints is that immediately after the handover execution a complete authentication is performed. Here the comparatively slow handovers in WLAN and WiMAX show higher security than those of mobile networks. WLAN is the best with respect to the points mentioned above but it has been suffering from very weak security mechanisms in the beginning. By now its protection is increased and it is strictly recommended to use the modern security mechanisms WPA2 or 802.11i instead of WEP.

If the interruption time of the service plays a critical role systems with short handover latency should be preferred. This may be at odds with to the security requirements as security incorporation costs time. If making use of hybrid networks the choice of a hybrid network with an appropriate handover protocol is recommended. Which procedure might be appropriate depends on the requirements. Best choice would be a handover protocol that has low overall latency and contains authentication and key negotiation. Some of these protocols are briefly described in the next section. Here research is still going on.

CONCLUSION

The connection of a transportation system network to the outside world is best done by wireless radio access networks (RAN). These radio access networks provide mobility to the users, e.g. the mobile devices of the transportation system. In this chapter some common terrestrial radio access networks are briefly introduced. For communication with airplanes or vessels satellite radio access networks will also play a role. After the comparison of the terrestrial radio access networks decision criteria for the choice of the best suited network are given, among them are availability, reliability, data rate, bandwidth, delay security mechanisms and speed of the mobile devices as well as costs. Often there will be no single best solution. Then the choice of a hybrid network consisting of different radio access networks may be appropriate. Inside a hybrid network the mobile device can handover between any cells without even noticing a change of the serving radio access network. This enables changes to a better suited radio access network as soon as this becomes available along the path of the mobile device. The hybrid networks pose certain additional challenges to handovers as the preparation may be more time-consuming and integrated authentication is mandatory at least if different providers come into play. Therefore some research efforts are outlined and one solution for a location-based, fast, secure handover in hybrid networks is detailed. The integration of hybrid networks is still in its infancy. Future research needs to focus on faster handovers, requiring better and faster detection of available networks and faster authentication. Traffic load balancing among different networks will be an important challenge. This needs to take into account mobility profiles, user speeds, users' utility functions, and the hybrid network's topology. Localization will play an increasing role, at the same time raising concerns about privacy that may need new security concepts.

REFERENCES

Ahmavaara, K., Haverinen, H., & Pichna, R. (2003). Interworking architecture between 3GPP and WLAN systems. *IEEE Communications Magazine, 41*(11), 74–81. doi:10.1109/MCOM.2003.1244926

Bargh, M. S., Hulsebosch, R. J., Eertink, E. H., Prasad, A., Wang, H., & Schoo, P. (2004). Fast authentication methods for handovers between IEEE 802.11 wireless LANs. *Proceedings from ACM 04: Workshop on Wireless Mobile Applications and Services on WLAN Hotspots,* (pp. 51–60).

Barkan, E., Biham, E., & Keller, N. (2003). Instant ciphertext-only cryptanalysis of GSM encrypted communication. *Advances in Cryptology, 2729.*

Böhm, A., & Jonsson, M. (2009). *Handover in IEEE 802.11p-based delay-sensitive vehicle-to-infrastructure communication.* Technical Report IDE0924, Halmstad University.

Eichler, S. (2007). Performance evaluation of the IEEE 802.11p WAVE communication standard. *Proceedings from VTC 07: 66th IEEE Vehicular Technology Conference,* (pp. 2199-2203).

ETSI. (2001). *ETSI, requirements and architectures for interworking between HIPERLAN/2 and 3rd generation cellular systems.* (Tech. rep. *ETSI TR, 101,* 957.

Fitzek, F., Munari, M., Pastesini, V., Rossi, S., & Badia, L. (2003). Security and authentication concepts for UMTS/WLAN convergence. *Proceedings from VTC 03: 58th IEEE Vehicular Technology Conference,* Vol. 4, (pp. 2343-2347).

Gräfling, S., Mähönen, P., & Riihijärvi, J. (2010). Performance evaluation of IEEE 1609 WAVE and IEEE 802.11p for vehicular communications. *Proceedings from Second International Conference on Ubiquitous and Future Networks,* (pp. 344–348).

GSM. (2011). *GSM World.* Retrieved from www.gsmworld.com

IEEE. (2010). *Part 11: Wireless LAN medium access control (MAC) and physical layer (PHY) specifications amendment 6: Wireless access in vehicular environments.* IEEE 802.11p published standard.

ITU. (1988). *General characteristics of international telephone connections and international telephone circuits.* ITU-TG.114.

ITU. (2003). *Framework and overall objectives of the future development of IMT-2000 and systems beyond IMT-2000.* (Recommendation ITU-R M.1645).

ITU. (2003b). *One-way transmission time.* ITU G.114.

Johnson, D. B., Perkins, C. E., & Arkko, J. (2004). *Mobility support in IPv6.* Retrieved from http://www.ietf.org/rfc/rfc3775.txt.

Kastell, K. (2007). *Sichere, schnelle, ortsbasierte Handover in hybriden Netzen.* Darmstädter Dissertation. Aachen, Germany: Shaker-Verlag.

Kastell, K. (2011). Challenges for handovers in hybrid networks. *Proceedings from IWT 11: International Workshop on Telecommunications.*

Kastell, K., Fernandez-Pello, A., Fernandez, D., Meyer, U., & Jakoby, R. (2004). Performance advantage and use of a location based handover algorithm. *Proceedings from VTC 04: 60th IEEE Vehicular Technology Conference,* Vol. 7, (pp. 5260-5264).

Kastell, K., & Jakoby, R. (2007). A location-based handover for use in secure hybrid networks. *Frequenz, 61,* 162–165. doi:10.1515/FREQ.2007.61.7-8.162

Lampropoulos, G., Passas, N., Merakos, L., & Kaloxylos, A. (2005). Handover management architectures in integrated WLAN/cellular networks. *IEEE Communications Surveys & Tutorials, 7*(4), 30–44. doi:10.1109/COMST.2005.1593278

Mc Nair, J., Akyildiz, I. F., & Bender, M. D. (2000). An inter-system handoff technique for the IMT-2000 system. *Proceedings from Joint Conference of the IEEE Computer and Communications Societies,* (pp. 208-216).

Perkins, C. E. (2002). *IP mobility support for IPv4.* Retrieved from http://tools.ietf.org/html/rfc3344

3rd Generation Partnership Project; Technical Specification Group. (2005). *3GPP TS 05.05 V8.20.0 GSM/EDGE radio access network; radio transmission and reception* (Release 1999), 11/2005.

3rd Generation Partnership Project; Technical Specification Group. (2008). 3GPP TS 36.300 V8.12.0 Evolved universal terrestrial radio Access (E-UTRA) and evolved universal terrestrial radio access Network (E-UTRAN): Overall description (Release 8), 12/2008.

3rd Generation Partnership Project; Technical Specification Group. (2011). 3GPP TS 05.05 V10.5.0 Evolved universal terrestrial radio access (E-UTRA) and evolved universal terrestrial radio access network (E-UTRAN): Overall description stage 2 (Release 10), 10/2011.

3rd Generation Partnership Project; Technical Specification Group. (2011b). 3GPP TR 22.934 V10.0.0 Services and system aspects; Feasibility study on 3GPP system to wireless local area network (WLAN) interworking (Release 10), 03/2011.

Samarasinghe, R., Friderikos, V., & Aghvami, A. H. (2003). *Analysis of intersystem handover: UMTS FDD & WLAN.* London Communications Symposium.

Tsao, S., & Lin, C. (2002). Design and evaluation of UMTS-WLAN interworking strategies. *Proceedings from VTC 02: 56th IEEE Vehicular Technology Conference,* Vol. 2, (pp. 777-781).

UMTS. (2011). *UMTS Forum.* Retrieved from www.umts-forum.org

Varma, V. K., Ramesh, S., Wong, K. D., Barton, M., Hayward, G., & Friedhoffer, J. A. (2003). Mobility management in integrated UMTS/WLAN networks. *Proceedings from ICC 03: IEEE International Conference on Communications,* Vol. 2, (pp. 1048-1053).

Wang, F., Gosh, A., Sankaran, C., Fleming, P. J., Hsieh, F., & Benes, S. J. (2008). Mobile WiMAX systems: Performance and evolution. *IEEE Communications Magazine, 46*(10), 41–49. doi:10.1109/MCOM.2008.4644118

KEY TERMS AND DEFINITIONS

2G: Second generation of mobile communication systems, first digital networks.

3G: Third generation of mobile communication systems.

3GPP: Third generation partnership project, standardization body for UMTS.

AAA: Authentication, Authorization and Accounting, exchange of credentials, getting access to a network and keeping track of billing purposes, mostly combined in one data base.

Ad-Hoc Network: Network built without any additional infrastructure, the (mobile) devices communicate directly with each other.

Air Interface: The air between transmit and receive antenna where wave propagation takes place, substitutes the classical cable.

Authentication: Exchange of credentials to establish a security context, can be unidirectional or mutual.

Availability: Indicates where or with which percentage a certain communication system provides sufficient field strength for communication.

CALM: Communications, Air-interface, Long and Medium range, group of communication systems.

CDMA: Code Division Multiple Access, users are separated by different codes.

CDMA2000: 3G communication standard used in Asia and America, based on CDMA.

Channel: A certain frequency slot with a dedicated bandwidth assigned for a certain communication link, in general cannot be further divided.

Coupling: Indicates the point in the hybrid network where the single networks are connected, from the point of coupling upwards upper layers will see them as one single network.

Credentials: Key or code word with which a device can identify itself and gain access to certain services.

Delay: Time the signal needs for propagation from one point to another, caused by the distance between transmitter and receiver and the propagation velocity of the signal.

DSSS: Direct Sequence Spread Spectrum, spreads the signal over a wider bandwidth to allow for faster transportation, less interference and better signal-to-noise ratio at the receiver, often used in CDMA systems.

EDGE: Enhanced Data rates for GSM Evolution, evolution of the GSM standard, part of the IMT-2000 family of third generation standards.

Encryption: Protection against attackers, mostly done by ciphering with a key.

ETSI: European Telecommunications Standard Institute.

FCC: Federal Communications Commission of the US.

FDD: Frequency Division Duplex, up- and downlink have different frequencies.

FDMA: Frequency Division Multiple Access, users are separated by different frequencies.

FHSS: Frequency Hopping Spread Spectrum, the communication frequency changes regularly following a pre-defined pattern, helps to avoid small band interference.

Frequency Band: Continuous portion of frequency with a dedicated bandwidth.

GSM: Global System for Mobile communication, second generation system for mobile communication.

GSMA: GSM association.

GPRS: General Packet Radio Service, evolution of the GSM standard allowing packet switched data transport.

Handover: Transfer of a connection between a mobile device and a third party from a serving NAP to a target NAP without user interaction and without loss or disruption of the connection.

Handover Completion: Procedure to terminate the connection to the former NAP after handover execution.

Handover Execution: Procedure where the actual handover with change of the NAP takes place.

Handover Preparation: Procedure to prepare the handover including measurements and handover decision, e.g. decision about the necessity and the target cell.

HSPA: High Speed Packet Access, amalgamation of UMTS providing higher data rates.

Hybrid Network: Consists of different radio access technologies at the first three layers of the ISO/OSI reference model and a hybrid core network.

IMT-2000: International Mobile Telecommunications-2000, standard underlying 3G mobile telecommunications.

Infrastructure-based Network: Network consisting of (mobile) devices and additional infrastructure for control, switching and connection to the outside world.

Interworking: If networks with different technology are able to work together, e.g. in processing data or transferring a call.

IP: Internet Protocol.

ISDN: Integrated Services Digital Network, set of communications standards for simultaneous digital transmission of voice, video, and data over the traditional public switched telephone network circuits.

ISM: Industrial, Scientific, and Medical band, unlicensed frequency band for used in the three areas mentioned.

ISO/OSI Reference Model: Model with 7 independent but partially interconnected layers for the description of communication networks.

ITS: Intelligent Transportation System, including information technology.

ITU: International Telecommunication Union.

Localization: Deriving the position of a mobile device.

LTE: Long Term Evolution, evolution of the UMTS standard towards the fourth generation of access networks.

MIMO: Multiple Input Multiple Output, used of two or more antennas at each device resulting in more communication paths between them.

Mobile Communication: Wireless communication with moving devices.

Mobility: Movement of mobile devices determined by speed and probability of change of direction.

Network Access Point (NAP): Fixed component with connection to the core network that uses an antenna to communicate with mobile devices.

OBU: On Board Unit, mobile device in a transportation vehicle.

OFDM(A): Orthogonal Frequency Division Multiplex (Multiple Access), users are separated by different frequencies, all frequency bands are comparatively small and the carriers are orthogonal to each other allowing overlapping of the frequencies bands.

QoS: Quality of Service, parameter to assess the quality of a communication network and its service delivery.

RAN: Radio Access Network.

RAT: Radio Access Technology.

Reliability: Besides availability another measure for the accessibility of a system, indicates the quality of the network.

RSU: Road Side Unit, network access point along a track for communication with the OBU.

Security Mechanisms: Mechanisms to provide some security features as privacy, anonymity, integrity of data, restricted access, etc.

TDD: Time Division Duplex, up- and downlink share one frequency.

TDMA: Time Division Multiple Access, users are separated by different codes.

UMTS: Universal Mobile Telecommunications Standard, standard of the third generation of mobile communication.

Utility function: Measure including different attributes with different weights and interdependencies.

WCDMA: Wideband CDMA.

Wireless Communication: Communication without any cable, the signals are transported over the air.

WAVE: Wireless Access in Vehicular Environments, new amendment to IEEE 802.11 for communication to and between vehicles: IEEE 802.11p.

WEP: Wired Equivalent Privacy, basic encryption in early WLAN.

Wi-Fi: Wireless fidelity, synonym for IEEE 802.11b.

WiMAX: Worldwide Interoperability for Microwave ACCESS, standard IEEE 802.16 for wireless communication.

WLAN: Wireless Local Area Network, although not correct often used as synonym for the standard IEEE 802.11, another WLAN is HiPERLAN.

WPA: Wi-Fi Protected Access, enhanced encryption for WLAN.

Chapter 12
Seamless Communication to Mobile Devices in Vehicular Wireless Networks

Kira Kastell
Frankfurt University of Applied Sciences, Germany

ABSTRACT

Vehicular networks are deployed as hybrid networks, which consist of a cooperation of different radio access networks. Seamless communication in vehicular networks relies on proper network planning and thorough dimensioning of network protocols. Both are assessed and verified by simulation. In transportation networks, the location of the mobile devices and their pattern of movement are very important. Therefore, different mobility models suited to vehicles in transportation networks are introduced. Then, the need for location information is exemplified. Mobility models and location play an important role in the verification of handover protocols. One hybrid handover protocol is given in detail. It provides low handover latency with additional mutual authentication to allow the transfer from one radio access network to another while maintaining the network's built-in security standard. This protocol is easily extensible to include a broad variety of networks.

VEHICULAR WIRELESS NETWORKS

Wireless networks for vehicular communication and communication in transportation systems attract interest from different stakeholders for several reasons. Operators want to establish a new market, governmental authorities want to provide roadside assistance and active traffic management, and transportation authorities look for better fleet management and cost reduction. The different interests of the different stakeholders result in sometimes contradictory constraints. At the same time, high investment costs prevent the simultaneous installation of different redundant wireless networks as well as the setup of a complete new network. Most likely, existing networks will be used wherever possible.

From the different interests described above, very different requirements arise. Among them are reliability and availability, kind and amount

DOI: 10.4018/978-1-4666-2976-9.ch012

of data to be transmitted, frequency and tolerable delay of transmission, needed security level, speed, mobility profile of the devices and cost (cf. chapter Communication Networks for Transportation Systems). Nevertheless, some requirements are mandatory for any vehicular communication. As the mobile devices (MDs) move fast, they will change network access points (NAPs) frequently. Therefore call setup and handover have to take place with a very small delay. Handover is the unnoticeable forwarding of a call from one network access point to a neighbouring network access point due to movement of the mobile device. The call setup needs to be finished before the mobile device leaves the coverage area of the network access point. The handover needs to be even faster, as it has to take place while the mobile device is in the overlapping coverage area of two network access points.

Besides, many of the intended services, such as e.g., roadside assistance, deal with sensitive data, e.g. road condition, traffic light status, position for fleet management and emergency assistance. These data must be protected against malicious attackers. Privacy, confidentiality, availability, and integrity are major concerns when designing the security mechanisms. The system needs information from the vehicles about road, traffic and vehicle condition while at best the mobile device should not be traceable. Thus, the systems shall provide security mechanisms, e.g., encryption and integrity protection. In addition, the access to the network should be restricted to avoid man-in-the-middle attacks. This can be done by authentication. Furthermore, some of the scenarios may require a certain data rate or a certain quality of coverage. Also, the accounting has to be taken into consideration if the network will not consist of closed user groups only. This also requires authentication to attribute the call to the correct bill.

These requirements have to be taken into account in the planning stage of the network already. As the networks are far too large to be set-up in a real-life test bed, simulation plays a major role in mobile communication network planning. The planning is aided by simulation-based planning tools, which are based on radio propagation models, mobility models, traffic data and others. The vehicular environment in transportation systems needs special simulation models to cope with the new requirements, e.g., higher speed, track-bounded movement, data protection. These models can be used to evaluate new protocols that aim for low delay and latency.

Vehicular traffic scenarios have challenges arising from the varying driving speeds, traffic patterns, and driving environments. This multitude of different movement options must be modelled. Vehicular networks will be infrastructure based when including a road side unit (RSU). Nevertheless, for seamless connectivity mobile devices or on board units (OBUs) may also form ad-hoc networks. The combination of these two types of networks has to be reflected in the planning tool. Areas where no infrastructure coverage exists may still be sufficiently covered by ad-hoc networks, depending on the traffic density. In contrast to other mobile networks, these ad-hoc and infrastructure networks provide ample computational and power resources as they are vehicle-based. Especially, this extends the range of the ad-hoc spots and has to be reflected in the ad-hoc propagation models.

Compared to infrastructure networks, additional security and scalability issues arise when wireless communication occurs directly between vehicles, ad-hoc. The hybrid structure of the network which consists of several different radio access networks (RANs), poses new authentication challenges. The security of vehicular networks can be crucial. Life-critical information must not be inserted or modified by a malicious attacker. The system must be able to determine the liability of users while still maintaining their privacy. These problems are difficult to solve because of the large network size, its hybrid structure, the high speed of the vehicles, their relative geographic position, and the randomness of the connectivity between them.

If safety messages are presumed to not include any sensitive private information, then confidentiality or rather encryption is not required. Nevertheless, the exchange of safety messages requires authentication. This may be achieved by digital signatures. Because of the large number of network members and variable connectivity to authentication servers, a Public Key Infrastructure (PKI) is a good way to implement authentication (Ferguson, 2003). With PKI, each mobile device will be given a pair of a public and a private key. Before sending a safety message, it signs it with its private key and includes the Certification Authority (CA) certificate. Because of the use of private keys, a tamper-proof device is needed in each vehicle. This is where the secret information will be stored and the outgoing messages will be signed. The recipient can decrypt the message with the knowledge of the corresponding public key which will be verified by the CA. Therefore it needs to have a connection to verify the certificate if needed. To further lower the risk of compromise by attackers, the device should have its own battery and clock. The clock can be used for time stamping messages and assigning them a certain lifetime.

Spread spectrum techniques offer protection against jamming. However, this requires more bandwidth, which points out the need for protocols to scale well as the available bandwidth shrinks. Also, the protocols must be capable of scaling to a large network size with a high density of vehicles. In this case, the performance of the system is especially important, since reaction times get shorter when following distances between cars become shorter in a dense environment. The protocols must also be able to quickly adapt to a changing network topology.

To fulfil these requirements, IEEE promotes a family of standards for Wireless Access in Vehicular Environments (WAVE), named P1609. These standards define architecture, communications model, management structure, security mechanisms, services, interfaces, and physical access

(IEEE, 2006). These networks are designed for use in intelligent transportation systems (ITSs) in a vehicular environment. Therefore, the primary architectural components are OBU, RSU, and WAVE interface. The aim of these standards is a common basis for a broad range of applications in the transportation environment, including vehicle safety, automated tolling, enhanced navigation, traffic management, and many others. The security is based on PKI with a certification authority and link level encryption for all messages. Also extensions to the physical channel access defined in IEEE 802.11 are provided (IEEE, 2006). IEEE 802.11p (IEEE, 2010) has been especially designed for WAVE. For data rates of 1Mbps it can provide coverage as far as 2.5km (Gräfling, Mähönen, Rihijärvi 2010). This is a very wide coverage range for WLAN which needs to be included in propagation models taking into account the higher available power of the mobile device in the OBU compared to handheld mobile devices.

CALM (Continuous Air interface for Long and Medium distance) is another new standard for automotive communication, currently developed by the International Standardization Organization (ISO TC204/WG16), (Fischer, 2012). The CALM standardized networking terminal will be capable of connecting vehicles and roadside systems continuously and seamlessly. This is accomplished through the use of a wide range of communication media, such as mobile cellular and wireless local area networks, and dedicated short-range communication (DSRC) based on microwave or infrared. This alone makes the network even more hybrid than the use of different mobile communication standards. The simulation environment therefore needs to be able to process different propagation models and communication protocols in parallel to allow for real time simulation of the hybrid network. CALM supports continuous communication by combining GPRS with vehicle-optimized WLAN technology. It supports internet services and ITS and is media independent through DSRC layer 7. IEEE 802.11p and P1609 cooperate with

CALM. In the future, CALM may become the general purpose communication medium that also can be used in safety applications. CALM provides universal access through a number of complementary access networks (ANs).

The character of the vehicular network, its fluctuating ad-hoc composition combined with the use of different infrastructure networks, where available, needs advanced network and protocol planning methods. Both will rely on simulations. Propagation models and protocol test beds already exist but need to be combined in a more powerful simulator. The main challenge for the simulation is the use of proper mobility models. Existing models only cover a part of the possible movement patterns. Therefore they need to be combined and enhanced with mobility models for special scenarios and multi-lane traffic with overtaking procedures or high-speed railway scenarios. Therefore we will now focus on mobility models for vehicular networks. With the help of these models newly designed protocols will be verified.

MOBILITY MODELS FOR VEHICULAR NETWORKS

Mobility models are essential to investigate the performance of protocols and concepts in mobile communication networks. For the assessment of vehicular communication protocols for transportation systems, new mobility models are required. Especially models of train environments and other track-bounded mass transportation systems need to be developed. These models play a critical role in the evaluation of hybrid overlay communication networks for transportation systems. In these networks routing and handover depend on the velocity, preferences and QoS (Quality of Service) requirements of the mobile device, e.g., fast moving devices should not be handed over to very small cells, if possible.

Many mobility models have been proposed. Each model focuses on a different aspect of mobil-

ity such as velocity, acceleration, range of motion, and constrained routes. Discussions on different existing models, mostly designed for vehicular ad hoc networks (VANETs), can be found in (Grilli, 2010), (Sommer, Dressler, 2008), (Lan, Chou 2008), (Wang, Yan 2009), (Fiore, 2006), (Härri, Filali, Bonnet, 2007), (Camp, Boleng, Davies, 2002), (Kastell, 2012). Simulations of VANET scenarios are based on radio propagation models and, therefore, require exact positions of simulated nodes. Only models that describe the behaviour of single vehicles and interactions between them are adequate mobility models for simulated VANET nodes.

When dealing with transportation systems one specialty is the route restriction and the resulting limited degree of freedom of movement compared to a pedestrian. In terrestrial transportation systems, the mobile devices are always track-bounded, e.g., buses on roads, trains on tracks. Moreover, transportation devices such as trains can move with very high speed. The combination of the different speeds and different degrees of freedom of movement is where the standard mobility models are not valid anymore (Kastell, 2012).

CLASSIFICATION OF MOBILITY MODELS

Users can be classified in four groups distinguished by their speed, possible turning angle and acceleration. Each user thereby can change its mobility model when changing the transportation vehicle. To model typical user scenarios the models should be interconnected and applied based on the actual velocity and behaviour.

- **Pedestrian Users:** These users move with low speed but can change their direction and acceleration very abruptly. Therefore they can be described with a random walk model.

- **Cars and Other Motorized Vehicles for Road Traffic in a City Scenario:** Here the degree of freedom is limited to the roads. In an inner city scenario the possibility to turn will be higher and more frequent than in rural areas as the number of intersections is higher in an urban scenario. This also leads to a higher possibility of frequent stops at the intersections. The speed is limited to 50 km/h.

- **Cars and Other Motorized Vehicles for Road Traffic in a Rural Scenario Including Motorways:** Here the speed can be higher. In rural scenarios the limit is about 100 km/h, but we assume 250 km/h as motorways are also included. Therefore the number of intersections is small and the need to stop is decreased compared to the urban scenario.

- **Trains on Tracks, Including High-Speed Trains:** This scenario has the lowest degree of freedom of movement as turning options are very limited. Also lane shifts rarely occur, at least not with high speed. The speed goes up to 500 km/h according to the specifications with already up to 400 km/h in daily operation.

For these different classes existing mobility models will be evaluated. Existing models cover pedestrian and some car movement but they often do not take into consideration multiple lane scenarios as on highways. Besides, a mobility model for high-speed trains will be introduced in addition. This model is used in combination with other mobility models for cars and pedestrians to evaluate the accuracy and usefulness of different localization and handover methods.

In addition to the four mobility classes the simulation environment for radio propagation of the different locations also differs. The well-known propagation models need to be complemented with models for the railway environment. The special propagation characteristics of the railway environment result from different reasons. The most prominent one is the impact of the catenary. It serves as a source of electromagnetic interference in general and of additional noise when the train passes by and sparks throw out. In addition the railway track can only manage a very slight slope. This results in multiple bridges and tunnels along the track. In addition cuttings and forest aisles are very frequent and also have a tunnel like characteristic for the propagating wave. The latter ones cannot be served with leaky cables as well as the real tunnels where the metallic break shoe residue of the trains and the high reliability constraints prohibit the use of leaky cables. This results in small cells with antennas at a very low height (tunnel ceiling). Shadowing effects occur in tunnels where two trains encounter each other. The entrance of a tunnel builds a barrier for most of the waves that will be reflected by the surrounding. Only a small part of the wave enters the tunnels. At the exit, the opposite problem occurs, the wave spreads as soon as the waveguide effect of the tunnel ends. In these situations the overlap areas for handover may be very small (Kastell, Bug, Nazarov, Jakoby 2006).

All mobility models have a random component. There are several basic mobility models, where the movement of a mobile device is generated randomly as in the random waypoint, random direction, or random walk model (Santi, 2012). More specialized models add other features:

- **Models with Temporal Dependency:** In this type of model, the movement of a mobile node is influenced by its movement history, e.g. Gauss-Markov Model (Santi, 2012), Smooth Random Mobility Model (Bettstetter, 2001).

- **Models with Spatial Dependency:** The movement of one mobile device affects other mobile devices and vice versa, e.g. reference point group model (Santi, 2012), set of correlated models.

- **Models with Geographic Restriction:** The movement of a mobile device is restricted by predefined paths such as streets, freeways or obstacles, e.g. pathway mobility model, obstacle mobility model (Bai, Helmy, 2006).

The mobility models also differ in the handling of the limited size of the simulation area. There are four ways to deal with mobile devices reaching the border of the simulation area. The mobile device may bounce back with a 180° change in direction or may be physically reflected at the border. Alternatively, the mobile device will be deleted once it reaches the boundaries of the simulation area and a new mobile device with a new starting point is generated instead. Or the mobile device will leave the simulation area and re-enter at the opposite side with same speed and direction as before.

Description of Basic Mobility Models

Random models are used to simulate the motion of mobile devices because of their simplicity of implementation. The mobile device's speed and direction are chosen randomly and independently from each other. This may cause effects such as sudden stops, sudden acceleration or sharp turn that may not be very realistic unless in a pedestrian environment. In reality the mobile devices usually accelerate incrementally and change direction smoothly.

The random waypoint model is widely used in ad-hoc networks because of its simplicity. The mobile device randomly chooses a destination point from a random speed, a random direction and a constant time interval or a constant travelled distance. The random speed and the random direction are chosen independently from each other and uniformly from a predefined range of speeds and a random angle, respectively. After the mobile device has reached the predefined target, it stops. After a pause time, the procedure starts

from the beginning. The random walk model is very similar to the random waypoint model, but with no pause time.

The smooth random mobility model (Bettstetter, 2001) is an enhanced version of the random direction mobility model to make the movement of the mobile device more realistic. This model can be used either in continuous time or in discrete time. The speed values are chosen from the range of 0km/h and a predefined maximum speed with some preferred speeds with higher probabilities than the others. The speed values therefore are correlated with their previous values. The new target speed is reached by incremental changes from the current speed. The acceleration values follow a uniform distribution. The current speed is accelerated in each time step until it reaches the target speed or the destination. Similar to the speed, the direction is also changed incrementally, but chosen randomly and also adjusted incrementally.

In the Gauss-Markov mobility model the current speed and the current direction are correlated with their previous values, respectively. For each time interval, the mobile device changes its speed as well as its direction based on the value of the previous speed and the previous direction and a random variable to adjust the degree of randomness. Due to the correlation between the new and the current speed as well as the new and the current direction, there are no sharp turns, sudden stops or sudden accelerations in this model.

Fitting Models to Mobility Classes

These models are good to use for pedestrian movement. Pedestrian movement can differ widely depending on the reason for walking. If the pedestrian just strolls around it will be more random than if he/she is moving towards a dedicated target in a certain amount of time. It also depends on the environment. In the latter case the movement pattern of the pedestrian in a city would look like the trace of a car but with lower speed. First of all, the pedestrian behaviour must be analysed. The average speed of a pedestrian varies from 4.716km/h to 4.9752km/h (Scully, Skehill, 2006). It does not frequently change a lot during walking, but pedestrians can change their walking direction immediately without the need of slowing down. However, the probability of the changing the walking direction sharply is very low. Most of the time, the change of direction between any two consecutive steps will be very small and will result in a nearly straight movement. As pedestrians usually perform smooth movements, sharp turns will also inhibit a deviation angle. Therefore smooth models should be preferred.

Vehicular scenarios can also be modelled with the vehicle-borne model (Bratanov, Bonek, 2003), (Dintchev, Perez-Quiles, Bonek, 2004). This model is based on only a few parameters: distance between two intersections (70m to 100m in European cities, 100m to 170m in suburbs, 1km in rural areas), velocity between the intersections (average values of 10km/h, 50km/h and 80km/h respectively, Rayleigh or Rice distributed), and change of direction at the intersection. The relative change of direction at the intersection depends on the type of the intersection. It is possible to prefer right turns to left turns or apply similar constraints. Stops at intersections can also be taken into consideration.

For VANETs several mobility models are proposed, e.g. the stop sign model (SSM), the probabilistic traffic sign model (PTSM), and the traffic light model (TLM) (Mahajan, Potnis, Gopalan, Wang, 2006). In these models the mobile device follows the shortest path along the road to get to its destination.

The stop sign model assumes that a stop sign stands at every intersection. All vehicles must stop at the sign for a specified time. If more than one vehicle arrives at the intersection from the same direction, the vehicles queue. Each vehicle will wait a dedicated time to pass the intersection after the vehicle in front has cleared up. The priority of the vehicles at the intersection is not coordinated among different directions.

The probabilistic traffic sign model is similar to the SSM but instead of stop signs it assumes traffic lights at each intersection. Vehicles stop at red signals and drive through green signals, but the simulation also has no coordination of traffic lights from different directions. When a node reaches an intersection with an empty queue, it stops at the signal with a probability p and crosses the signal with a probability (1 – p). The amount of wait time is randomly chosen between 0 and w seconds. If the vehicle arrives at a non-empty queue it will have to wait for the remaining wait time of the previous node plus one second, which is the start-up delay between queued cars. The nodes cross the signal at intervals of one second when the light is green until there is no car in the queue.

The traffic light model is more detailed compared to SSM and PTSM. The traffic lights at each intersection are coordinated. Single and multiple lanes are simulated. At an intersection with an even number of roads with single-lane traffic when the light turns green the traffic in a single pair of opposing lanes crosses the intersection simultaneously. Vehicles reaching the head of the queue can turn freely to different directions. While one pair of opposing lanes has green signals, the other has red signals. After a predefined time the signals will change from green to red and vice versa. In case of an odd number of roads at the intersection, one of the roads periodically has a green light. In addition, vehicles will accelerate or decelerate gradually with $3m/s^2$ while in SSM and PTSM, vehicles always travel within 5 miles/hour of the street speed limit. When a vehicle enters a new road, it will select the lane which has the least number of vehicles.

MOBILITY MODEL FOR RAILWAY ENVIRONMENT

The mobility pattern of a train differs largely from that of other vehicles. The number of intersections of different tracks is very low when outside a large station. Moreover, if two tracks are in the same location on a map it does not necessarily mean that a change of direction is possible as with roads. In roads, an intersection allows a vehicle to turn in any direction where a road is available, unless prohibited by traffic signs. For tracks, they must be interconnected with a suitable radius to allow a turn in the corresponding direction. This is shown in Figure 1.

The mobility model for high-speed trains can be derived from a random waypoint approach or a vehicle-borne model (Bratanov, Bonek, 2003). Therefore, the direction of movement needs to be restricted as high-speed trains usually move nearly straight with small deviation angles of about 3° to 5°. This is true in 75% of all cases. Plain intersections where two lines meet are about 15%, while larger crossings and inner-city multiple track scenarios cover about 10% (Kastell, Bug, Nazarov, Jakoby, 2006). Moreover, when high-speed trains turn, they will make a very smooth curve and do not suddenly or sharply turn. Therefore the direction is changed incrementally in steps of 5° per time step. One time step is in the range of some seconds. After a turn the movement is limited to straight movement for a certain period of time. The speed will be randomly chosen from 250 to 500km/h, but speed below 330km/h will be preferred. The duration of a single step is in the range of 2 to 3 seconds. In this time the train can run about 200m. In the vehicle-borne model a high-speed track is composed of long high-speed segments and short lower speed segments at crossings and switches (Kastell, 2012).

The speed of the train is adjusted with a constant acceleration of 0.5 m/s² until it reaches the final speed. This final speed depends on the type of the train. Usual values are 120km/h, 250km/h, and 330km/h. Also the maximum speed of 500km/h should be covered as this is the maximum speed for which mobile communication standards are specified. The deceleration for an intended stop is -0.5 m/s². The maximum acceleration is restricted to 0.86 m/s². The maximum deceleration for an emergency stop is -0.79 m/s². These values are derived from movement patterns of the German train ICE3. The acceleration during the travel varies with a probability of 0.15 (Kastell, 2012).

To have a more realistic model, the turning probability is further restricted for a speed above 250km/h. Here no intersections occur and the turning probability is 0. Nevertheless, the deviation and smooth curvatures of the track may still occur, but they are smaller than 45°. Whenever the train approaches an intersection it has to slow down below 250km/h.

LOCALIZATION

Localization becomes more important in mobile communication. On the one hand, for emergency reasons, authorities such as the FCC (Federal Communication Commission) require a localization accuracy of 125m in 67% of the emergency calls (FCC, 1999). On the other hand location-based services become more attractive and location information can enhance QoS-assessment and protocol performance (Kastell, 2007). In transportation networks, location is an important parameter to

Figure 1. Intersection a) without turning possibility, b) with possibility to turn from left downwards.

 Intersection without radius tracks for turns

 Intersection with radius track for a turn from west to south or vice versa

assess the functionality of the network. A variety of localization methods exist, e.g. for GSM and UMTS in total seven localization methods are defined in the standards (3GPP, 2006), (3GPP, 2006b). GSM promotes timing advance (TA)-based, Enhanced Observed Time Difference (E-OTD), Uplink Time Difference of Arrival (U-TDOA), and assisted GPS (A-GPS), in UMTS cell identity (ID), Observed Time Difference Of Arrival (OTDOA) (pure or in combination with Idle Period Downlink (IPDL) or Cumulative Virtual Blanking (CVB)), and assisted GPS (A-GPS) are implemented. The methods differ in their accuracy. The requirements for accuracy depend on the type of service, the actual speed, the network topology, the size of the cell, and the mobility pattern.

Localization Procedures in GSM

In GSM there are three ways of locating a MS (3GPP, 2006): E-OTD, U-TDOA, and A-GPS. In E-OTD the mobile device measures the arrival time of bursts of nearby pairs of network access points. As in most networks there is a lack of synchronization, so that the round-trip time has to be measured. In order to obtain accurate trilateration, at least three distinct pairs of geographically dispersed network access points have to be measured (Kastell, Fernandez-Pello, Fernandez, Meyer, Jakoby, 2004). In contrast to E-OTD, in U-TDOA the network determines the time of arrival (TOA) of a known signal. This signal is sent from the mobile device and received at three or more location measurement units (LMUs) (Kastell, Meyer, Jakoby 2003). Therefore, LMUs need to be added to the network at known locations. With this location information and the TOA, trilateration can be used. The advantage of the method is that for an on-going data transfer with duration greater than 250ms no additional messages have to be sent. With the existing hardware, approximately 100 location estimates per second can be executed. U-TDOA is the best suited localization method for GSM.

Localization Procedures in UMTS

The standard positioning methods in UMTS are the cell ID based method, OTDOA that may be assisted by network configurable idle periods, and A-GPS. The first only provides localization at cell size level. A-GPS needs a mobile device equipped with a GPS receiver. It relies on the signalling between the mobile device's GPS receivers and a continuously operating GPS reference receiver network. This requires clear sky visibility of the same GPS constellation as the assisted mobile devices. The visibility problems described above severely degrade the quality of GPS in dense urban and tunnel environment.

In OTDOA with network configurable idle periods, the position of the mobile device is evaluated from different neighbouring network access points. In addition, the exact geographic position of the transmitter and the reference time delay (RTD) of downlink transmissions is needed. From every pairwise measurement of downlink transmissions a line of constant distance to the network access point is derived as possible location of the mobile device. The intersection of these lines for at least two pairs of network access points gives the location of the mobile device. The accuracy of this method strongly depends on the network topology. Sufficient accuracy can only be achieved with three or more network access points.

Table 1 gives an overview of the achievable accuracy taken from (Ahonen, Laitinen, 2003, Bartlett, Morris, 2001, Borowski, Niemelä, Lempiäinen, 2004, Chiu, Bassiouni, 2000, 3GPP, 2002, Laitinen, Lahteenmaki, Nordstrom, 2001, Lin, Juang, 2005, Porcino, 2001, Siebert, Schinnenburg, Lott, Göbbels, 2004, and Silventoinen, Rantalainen, 1996).

It can be seen that the FCC requirements will not be fulfilled by most of these localization methods. Only E-OTD in rural areas (cell sizes larger than 600m), A-GPS, and OTDOA will meet these requirements. A-GPS cannot be used in railway environment because of the tunnels and forest aisles that decrease visibility (cf. next sub-

Table 1. Accuracy of localization methods

Method	Accuracy in 67%	Accuracy in 95%
TA-based urban TA-based rural	150 m 830 m	185 m 1005 m
E-OTD urban E-OTD rural	141 m **50 – 125 m**	247 m 219 m
U-TDOA urban U-TDOA rural	1005 m 775 m	1690 m 1319 m
A-GPS	**7 m**	10 m
Cell ID UMTS	158 m	436 m
OTDOA OTDOA IPDL OTDOA CVB	**97 m** 30 – 199 m **12 – 24 m**	150 m 50 m

chapter) and also is affected by shadowing from high buildings in urban scenarios. Also, the other methods fulfilling the FCC requirements lack performance as they need data from at least three different network access points at a time. (Silventoinen, Rantalainen, 1996) have shown that in 30% of all scenarios less than three network access points provide a field strength high enough to be evaluated. Consequently, only a combination of several localization methods will lead to sufficient results for location-based handovers.

The location information plays an even more important role in hybrid networks. Here not only the emergency situation and location-based services enhance the quality of the network for the users. Location information is needed to manage the cooperation of the different networks. As these networks will exist in parallel, the network best-suited for the mobile device has to be chosen. Here the location information is needed as in in hybrid networks field strength measurements are not sufficient to decide about the suitability of the network. On the one hand, different networks have different field strength requirements which make it hard to compare measured values in terms of quality. On the other hand, measurements of different air interfaces would require frequent switching between the interfaces or interfaces to be powered up all time. The latter is very power consuming. Third, as hybrid networks may con-

tain very small cells, measurements will probably last too long and thus prevent choosing the best suited network. In addition, using information about the location of the mobile device will help to implement more efficient handover strategies especially in hybrid networks.

The location data can also be used to derive the movement of the mobile device. Knowledge about the speed will further support the decision about the target radio access network, but it will also help to find the best suited target cell if cells of different sizes are available. From the direction of movement also the most appropriate target cell can be determined. This helps with conventional handovers based on signal strength measurements, as the duration of the measurement may be shortened, as well as with hybrid handovers where the choice is more complex because of overlaying networks.

For the hybrid handover protocol discussed hereafter, the absolute position of the mobile device is not relevant. Instead its position relative to the adjacent network access points as well as its direction are needed (Kastell, 2007). The direction can be extrapolated from previous location estimates. Therefore, in a pure line topology, two consecutive measurements are sufficient to calculate the direction. From this knowledge the next cell can be derived, because as soon as the mobile device attaches to the network, the serving cell is known. More accurate localization only has to take place at intersections. Methods to derive the location with the timetable information and RSSI-values (received signal strength indicator) are also available (Kastell, Bug, Nazarov, Jakoby, 2006).

FRAMEWORK FOR A HYBRID NETWORK FOR COMMUNICATION IN TRANSPORTATION SYSTEMS

In hybrid networks, localization can be used minimize handover latency and enhance handover security. In this way the problems for users with high speed can be overcome. The main challenge is the combination of seamless coverage, secure authentication and handover. This should be achieved with minimum investment costs. Therefore, the construction of additional RSUs in places without dangerous traffic conditions is not feasible. The most cost efficient way is the use of already existing wide area networks. GSM is out of the scope because of its unidirectional authentication. Hybrid networks for transportation systems always need mutual authentication because of road condition information that has to be protected from external manipulation. It may be included for private communication but cannot be used for life-critical messages such as in traffic control. UMTS and LTE have security protocols that guarantee a sufficient security level. But this only holds true, if handovers to and from GSM are prohibited or if authentication is included in the handover protocol. The first approach is feasible if GSM shall not be used for any communication from that mobile device. Otherwise a specific network access scheme would be needed.

Here we show a framework supported by a newly designed hybrid handover. The hybrid handover includes authentication without having a higher latency to fulfil the requirements of communication in transportation systems. This is achieved by making use of the location of the mobile device. The location will be needed in traffic control anyways as the messages are strongly related to the actual position. Therefore the location of the mobile device is already known to the system. It also is updated regularly. The location first of all can be used to choose the best suited network access point. Moreover, the movement of the mobile device can be derived from the location data. The speed will further support the decision about the target network access point. The access network has to provide sufficient data rate at the given speed. Also, frequent handovers should be avoided by handing over fast users to larger cells. Further, location information can help to reserve

channel capacity predictively (Chiu, Bassiouni, 2000), to cope with high-speed users based on field strength thresholds (Naghian, 2003), to optimize routing (Ushiki, Fukazawa, 1998), to optimize thresholds for handovers (Wang, Green, Malkawi, 2001), and to enhance the handover decision (Markopoulos, Pissaris, Kyriazakos, Koutsorodi, Sykas, 2004) especially in IP-based networks (Feng, Reeves, 2004). Parameter optimization to guarantee a certain QoS and to prevent call drops and unnecessary handovers are dealt with in (Benson, Thomas, 2002). Extensive research also took place in the European project "CELLO - Cellular Network Optimization Based on Mobile Location" including target cell prediction by GPS, based on a field strength map measured in advance, and data tunnelling in foreign networks (Pesola, Pönkänen, 2003). The location data also need to be managed especially where field strength maps are needed. This may be done by the hybrid information system (HIS) where information gathered by different networks is collected (Siebert, Schinnenburg, Lott, Göbbels, 2004), (Lott, Siebert, Bonjour, von Hugo, Weckerle, 2004). The use of GPS will not be helpful in a railway environment because of the large number of tunnels and forest aisles that have a negative impact on the visibility of the satellites and thus the quality of the localization.

An overview of handover candidates, their advantages and drawbacks is given in (Juang, Lin, Lin, 2005). In the following, an extensible framework for fast, secure, location-based hybrid handovers (Kastell, 2007; Kastell, 2011) will be presented in more detail. The hybrid handover described here can be implemented in every radio access network. It uses the authentication and key negotiation mechanisms already standardized for the radio access network under consideration. Each network will maintain its own authentication and thus always provide the same security level independent of any handovers taking place. Note that the security level in every radio access network of the hybrid network differs. There is no overall security enhancement. But the hybrid handover protects each radio access network against weaknesses of other radio access networks. Also the standardized handover procedure for each radio access network shall be kept. The aim is to minimize changes in the standard, although they cannot be avoided completely. The concept of the hybrid handover is as follows: For speeding up handover preparation, information about the location of the mobile device is used. In (Kastell, 2007; Kastell, Jakoby, 2007) it has been shown that the relative location of a mobile device related to the adjacent cells can be derived with fewer measurements than needed for field strength analysis. Thus, measurements of a subset of adjacent cells will be sufficient for the handover, shortening measurement duration. Moreover, as different air interfaces have to be powered up and switched between for measurement in hybrid networks, fewer measurements help to preserve battery power. And while switching to another air interface no data can be transmitted over the on-going connection. Therefore it also helps to achieve good throughput.

If the location of the mobile device is known, its direction is towards the cell border and it is in the outer region of the cell, handover preparation can start. The threshold distance for the outer region can be set depending on the maximum speed of the mobile device. Therefore it is not necessary that the mobile device already has reached the overlap region of the coverage of serving and target network access point. Figure 2 gives an example of this handover using terms of a GSM network.

The serving network access point will serve as a tunnel to connect the mobile device to the target network access point. Through this tunnel mobile device and target network access point perform a proactive authentication using the credentials of the target radio access network. If the need for a handover is detected thereafter, mobile device and target network access point already have established a trusted relation. Then the handover takes place following the standardized procedure of the each access network in the hybrid network. This includes protocols that are built of protocol parts from RAN1 communicating with NAP1 and mobile device and switches to protocol parts of RAN2 for communication between NAP2 and mobile device.

The new connection will maintain the existing network security if the proactive authentication has been finished before the handover execution begins. Then the mobile device can send its handover execution success message already encrypted with the new keys. But sometimes the time saved by taking into account location information may not be sufficient to allow processing the complete authentication before handover execution needs to start. Then the security could still be maintained if a small change to standard handovers is allowed. Before the first message from the mobile device to the target network access point is sent during handover execution, the protocol needs to include sending the authentication response from the mobile device to the target radio access network to trigger the beginning of the authenticated session. This may be done in an extra message or the handover execution answer may contain the authentication response.

The use of location information as shown in (Kastell, 2007) speeds up any handover. The time saved in most of the cases allows integration of additional authentication with no increase of the overall handover duration. The resulting handover may still be faster as the standard handover. With this hybrid handover all available access networks can be used for vehicular communication. To further improve the quality of the communication we recommend the use of a preference list. In addition, a utility or objective function can be applied. The preferences may be set by the operator, the service provider and/or the user. For example, for traffic control predefined preferences will be set by governmental authorities. For roadside assistance the user may choose a service provider who predefines preferences or may predefine preferences himself without subscribing to a certain service. Moreover, the location information can also be used as an input to a user's preference function

Figure 2. Location-based, fast, secure handover procedure with terms of a GSM network

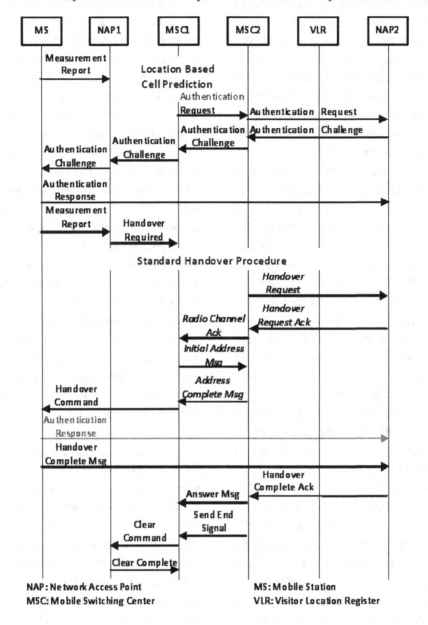

NAP: Network Access Point MS: Mobile Station
MSC: Mobile Switching Center VLR: Visitor Location Register

and thus simplify the choice of the target network in a hybrid network (Kastell, 2007; Kastell, 2011).

Table 2 gives an overview of successful handovers with different hybrid approaches. The data have been derived from simulation with a smooth random mobility model (Bettstetter, 2001) for pedestrians with speeds up to 4km/h, a modified vehicle-borne model (Bratanov, Bonek, 2003) for urban (up to 120km/h) and rural (up to 250km/h)

vehicles and the train mobility model described above for a speed up to 500km/h. The GSM network covered the complete simulation area while UMTS built larger islands or produced a complete coverage. WLAN has been implemented in small islands with up to 200m radius.

The success rates are below 100%. This is because of network load or small overlap areas. The overlap with the chosen target cell may not

Table 2. Handover success rates derived from simulation

Serving network	Target network	Conventional handover	Conventional handover plus subsequent authentication	Hybrid handover	Hybrid handover with late authentication response
GSM	GSM	98.87%	29.45%	92.57%	96.37%
UMTS	UMTS	97.13%	27.67%	90.69%	94.45%
GSM	UMTS	98.99%	Not simulated	98.37%	99.44%
UMTS	GSM	99.03%	Not simulated	99.21%	99.23%
WLAN	GSM	Not specified	Not simulated	77.95%	81.21%
WLAN	UMTS	Not specified	Not simulated	83.57%	87.20%

be large enough to complete the handover successfully. Besides, in this simulation sometimes the target cell is not predicted correctly as the measurement data may be insufficient or inaccurate in some cases.

The conventional handover as specified in the standard does not include any authentication. The two hybrid handovers include authentication as specified in the standards. This means that in GSM only the mobile device is authenticated, in UMTS the authentication is mutual. The late handover response refers to the implementation where the response is sent after the handover complete message.

The conventional handover is a little better than the others. But if we include a complete authentication directly after the handover required message the results change. The handover success rate then drops significantly to 29.45% in intra GSM and 27.67% in intra UMTS handovers. Compared to the handover with authentication the location-based hybrid handover is about three times better in terms of successful completion within a given time (Kastell, 2011). Because of the inacceptable success rate, the standardized handovers up to now do not include authentication.

For all networks it can be seen that the late authentication response significantly increases the success rate in most cases. The rate of successful handovers is higher than 90% for GSM and UMTS and above 75% when WLAN is included. The low success rate in WLAN reflects the small

cell size. In the simulations given here fast mobile devices have not been restricted from accessing small WLAN cells. In these situations a handover failure is very likely. The handover in a WLAN cell should only be allowed for very low speed, i.e., for pedestrians. One way to easily implement this is an automatic change of the model depending on the speed. The handover restriction can be based on the mobility model. This helps to keep the simulation routine fast and simple. If we implement the handover restriction, the number of handovers to WLAN decreases dramatically. This is because in transportation systems users most likely are faster than pedestrians. The use of conventional WLAN only occurs in stations. An exception is IEEE 802.11p for the use of communication with RSUs. To simulate this, the classical scenario for handover evaluation has to be changed. Usually we assume an on-going connection that lasts for some minutes. In this way, we assure that several handovers can take place during the simulation. For the exchange with RSUs there is no on-going connection, but only small messages are exchanged. For this scenario a handover to the IEEE 802.11p RSU is feasible as the connection time only needs to be a few seconds. But still the overlap region of the different cells is very small. Without localization information frequent handovers from one RSU to another will not work.

Another problem may occur: RSUs only serve a small area. Moreover, these areas are rarely overlapping. As RSUs provide life-critical information, the mobile device needs to handover to every RSU available. This is the only way to ensure that the data of the RSU is transmitted to the vehicle. Ad-hoc vehicular networks can solve this problem, but they are not always available. This scenario may increase the number of handovers significantly compared to the simulations made in (Kastell, 2007), where the number of small cells is low compared to the vehicular scenario described in this paragraph. The additional overhead needs to be evaluated. But the framework given here still enhances the connection as otherwise two mobile devices would be needed or no connection would be possible at all at some location. The RSU could also broadcast its messages to the other access network. These can provide the information to mobile devices in a certain geographical area, because the location of the mobile device is known to the network.

CONCLUSION

The implementation of hybrid networks for transportation systems needs to be well prepared. To prove that the adjusted protocols fulfil the requirements of the system regarding latency, delay, reliability, and security, simulations must be run. The simulation environment must be adapted to the special needs of the transportation system. In particular, more precise mobility models have to be used. This is very important as location information do not only play a role for traffic information, but also for the choice of the network and the speeding-up of the handover. In this chapter mobility models for transportation systems have been explained. The importance of localization and its use in hybrid handovers was shown. One hybrid handover was described in more details. This handover fulfils all requirements of transportation networks. It also provides faster and more secure handovers when implemented in a single radio access network while being extensible for the inclusion of further networks.

REFERENCES

Ahonen, S., & Laitinen, H. (2003). Database correlation method for UMTS location. In *Proceedings of the 57th IEEE Vehicular Technology Conference,* Vol. 4, (pp. 2696–2700).

Bai, F., & Helmy, A. (2006). *A survey of mobility models in wireless adhoc networks.* USA: University of Southern California.

Bartlett, D., & Morris, P. (2001). *CVB: A technique to improve OTDOA positioning in 3G networks.* Cambridge Positioning Systems Ltd., White paper.

Benson, M., & Thomas, H. J. (2002). Investigation of UMTS to GSM handover procedure. *Proceedings from VTC 02: 55th IEEE Vehicular Technology Conference*, Vol. 4, (pp. 1829-1833).

Bettstetter, C. (2001). Smooth is better than sharp: A random mobility model for simulation of wireless networks. *Proceedings of the ACM Intl. Workshop Modelling, Analysis and Simulation of Wireless and Mobile Systems*, (pp. 19-27).

Böhm, A., & Johnson, M. (2009). Position-based data traffic Prioritization in safety-critical, real-time vehicle-to-infrastructure communication. In *Proceedings of the International Conference on Communications.*

Borowski, J., Niemelä, J., & Lempiäinen, J. (2004). Performance of cell ID+RTT hybrid positioning method for UMTS radio networks. In *Proceedings of the 5th European Wireless Conference,* (pp. 487 – 492).

Bratanov, P. I., & Bonek, E. (2003). Mobility model of vehicle-borne terminals in urban cellular systems. *IEEE Transactions on Vehicular Technology*, 52(4), 947–952. doi:10.1109/TVT.2003.808795

Camp, T., Boleng, J., & Davies, V. (2002). A survey of mobility models for ad hoc network research. *Proceedings of WCMC 2002: Special Issue on Mobile Ad Hoc Networking: Research, Trends and Applications, 2*(5), 483–502.

Chiu, M., & Bassiouni, M. A. (2000). Predictive schemes, for handoff prioritization in cellular network based on mobile positioning. *IEEE Journal on Selected Areas in Communications*, *18*(3), 510–522. doi:10.1109/49.840208

Dintchev, P. A., Perez-Quikes, B., & Bonek, E. (2004). An improved mobility model for 2G and 3G cellular systems. In *International Conference on 3G Mobile Communication Technologies, 5th 3G 2004*, (pp. 402-406).

Feng, F., & Reeves, D. S. (2004). Explicit proactive handoff with motion prediction for mobile IP. *Proceedings from WCNC 04: 6th IEEE Wireless Communications and Networking Conference*, Vol. 2, (pp. 855-860).

Ferguson, N., & Schneier, B. (2003). *Practical cryptography*. Hoboken, NJ: Wiley.

Fiore, M. (2006). *Mobility models in inter-vehicle communications literature*. Technical report. Retrieved from http://www.tlc-networks.polito.it/oldsite/fiore/papers/mobilityModels.pdf

Fischer, H.-J. (2012). *Standards for C-ITS*. Retrieved from http://its-standards.info/Standards_for_C-ITS.pdf

Gräfling, S., Mähönen, P., & Rihijärvi, J. (2010). Performance evaluation of IEEE 1609 WAVE and IEEE 802.11p for vehicular communications. *Proceedings of the Second International Conference on Ubiquitous and Future Networks*, (pp. 344–348).

Grilli, G. (2010). *Data dissemination in vehicular networks*. PhD thesis University of Rome "Tor Vergata", Rome, Italy.

Härri, J., Filali, F., & Bonnet, C. (2007). *Mobility models for vehicular ad hoc networks: A survey and taxonomy*. Research Report EURECOM. doi:10.1109/SURV.2009.090403

IEEE. (2006). *IEEE trial-use standard for wireless access in vehicular environments (WAVE)—Multichannel operation*. Trial-Use Std. 1609.4.

IEEE. (2010). *Part 11: Wireless LAN medium access control (MAC) and physical layer (PHY) specifications amendment 6: Wireless access in vehicular environments*. IEEE 802.11p published standard.

Juang, R., Lin, H., & Lin, D. (2005). An improved location-based handover algorithm for GSM systems. *Proceedings from WCNC 05: 7th IEEE Wireless Communications and Networking Conference*, Vol. 3, (pp. 1371–1376).

Kastell, K. (2007). Sichere, schnelle, ortsbasierte Handover in hybriden Netzen. *Darmstädter Dissertation*. Aachen, Germany: Shaker-Verlag.

Kastell, K. (2009). Improved fast location-based handover with integrated security features. *Frequenz: Journal of RF/Microwave Engineering. Photonics and Communications*, *63*, 1–2.

Kastell, K. (2011). Challenges for handovers in hybrid networks. *Proceedings from IWT 11: International Workshop on Telecommunications*.

Kastell, K. (2012). Novel mobility models and localization techniques to enhance location-based services in transportation systems. *Proceedings of ICTON 2012*.

Kastell, K., Bug, S., Nazarov, A., & Jakoby, R. (2006). *Improvements in railway communication via GSM-R*. IEEE-Vehicular Technology Conference (VTC) Spring 2006, Melbourne, Australia.

Kastell, K., Fernandez-Pello, A., Fernandez, D., Meyer, U., & Jakoby, R. (2004). Performance advantage and use of a location based handover algorithm. *Proceedings from VTC 04: 60th IEEE Vehicular Technology Conference*, Vol. 7, (pp. 5260-5264).

Kastell, K., & Jakoby, R. (2006). *Fast handover with integrated authentication for hybrid networks.* IEEE-Vehicular Technology Conference (VTC) Fall 2006.

Kastell, K., & Jakoby, R. (2007). A location-based handover for use in secure hybrid networks. *Frequenz, 61,* 162–165. doi:10.1515/FREQ.2007.61.7-8.162

Kastell, K., Meyer, U., & Jakoby, R. (2003). Secure handover procedures. In *Proceedings of the Conference on Cellular and Intelligent Technologies (CIC).*

Laitinen, H., Lahteenmaki, J., & Nordstrom, T. (2001). Database correlation method for GSM location. *In Proceedings of the 53rd IEEE Vehicular Technology Conference,* Vol. 4, (pp. 2504–2508).

Lan, K.-C., & Chou, C.-M. (2008). Realistic mobility models for vehicular ad hoc network (VANET) simulations. *Proceedings of ITST 2008,* Phuket, Thailand, (pp. 362-366).

Lin, D. B., & Juang, R. T. (2005). Mobile location estimation based on differences of signal attenuations for GSM systems. *IEEE Transactions on Vehicular Technology, 54,* 1447–1454. doi:10.1109/TVT.2005.851318

Lin, X., Sun, X., Ho, P.-H., & Shen, X. (2007). GSIS: A secure and privacy-preserving protocol for vehicular communications. *IEEE Transactions on Vehicular Technology, 56*(6), 3442–3456. doi:10.1109/TVT.2007.906878

Lott, M., Siebert, M., Bonjour, S., von Hugo, D., & Weckerle, M. (2004). Interworking of WLAN and 3G systems. *Proceedings from 04. IEE Proceedings. Communications, 151,* 507–513. doi:10.1049/ip-com:20040600

Mahajan, A., Potnis, N., Gopalan, K., & Wang, A. A. (2006). Urban mobility models for VANETs. In *Proceedings of the IEEE Workshop on Next Generation Wireless Networks (WoNGeN).*

Markopoulos, A., Pissaris, P., Kyriazakos, S., Koutsorodi, A., & Sykas, E. D. (2004). Performance analysis of cellular networks by simulating location aided handover algorithms. *Proceedings from EWC 04: 5th European Wireless Conference on Mobile and Wireless Systems beyond 3G,* (pp. 605-611).

Naghian, S. (2003). Hybrid predictive handover in mobile networks. *Proceedings from VTC 03: 58th IEEE Vehicular Technology Conference,* Vol. 3, (pp. 1918-1922).

Pesola, J., & Pönkänen, S. (2003). Location-aided handover in heterogeneous wireless networks. *Proceedings from MLW 03: Mobile Location Workshop,* (pp. 195–205).

Porcino, D. (2001). Location of third generation mobile devices: a comparison between terrestrial and satellite positioning systems. In *Proceedings of the 53rd IEEE Vehicular Technology Conference,* Vol. 4, (pp. 2970–2974).

3rd Generation Partnership Project; Technical Specification Group. (2006). 3GPP GSM 43.059 V7.2.0 radio access network, functional stage 2 description of location services (LCS) in GERAN (Release 7), 04/2006.

3rd Generation Partnership Project; Technical Specification Group. (2006b). *3GPP GSM 45.811 V6.0.0 Radio access network, feasibility study on uplink TDOA in GSM and GPRS* (Release 6), 6/2002.

3rd Generation Partnership Project; Technical Specification Group. (2006b). 3GPP UMTS 25.305 V7.3.0 stage 2 functional specification of user equipment (UE) positioning in UTRAN (Release 7), 12/2006.

Santi, P. (2012). *Mobility models for next generation wireless networks: Ad hoc, vehicular and mesh networks.* Wiley Series on Communications Networking & Distributed Systems. doi:10.1002/9781118344774

Scully, P., & Skehill, R. (2006). Demonstration of aggregate mobility models in Matlab. *IET Irish Signals and Systems Conference*, Dublin Institute of Technology, (pp. 271-276).

Siebert, M., Schinnenburg, M., Lott, M., & Göbbels, S. (2004). Location aided handover support for next generation system integration. *Proceedings from EWC 04: 5th European Wireless Conference on Mobile and Wireless Systems beyond 3G*, (pp. 195-201).

Silventoinen, M. I., & Rantalainen, T. (1996). Mobile station emergency locating in GSM. In *Proceedings of the 1st IEEE International Conference on Personal Wireless Communications*, (pp. 232–238).

Sommer, C., & Dressler, F. (2008). Progressing toward realistic mobility models in VANET simulations. *IEEE Communications Magazine*, 132–137. doi:10.1109/MCOM.2008.4689256

Ushiki, K., & Fukazawa, M. (1998). A new handover method for next generation mobile communication systems. *Proceedings from GTC 98: 41st IEEE Global Telecommunications Conference*, Vol. 5, (pp. 2560-2565).

Wang, J., & Yan, W. (2009) RBM: A role based mobility model for VANET. *Proceedings of the International Conference on Communications and Mobile Computing*, Kunming, China.

Wang, S. S., Green, M., & Malkawi, M. (2001). Adaptive handoff method using mobile location information. *3rd IEEE Emerging Technologies Symposium on Broadband Communications for the Internet Era: Symposium Digest*, (pp. 97-101).

KEY TERMS AND DEFINITIONS

3GPP: Third generation partnership project, standardization body for UMTS.

A-GPS: assisted GPS.

Ad-Hoc Network: Network built without any additional infrastructure, the (mobile) devices communicate directly with each other.

Air Interface: The air between transmit and receive antenna where wave propagation takes place, substitutes the classical cable.

AN: Access Network.

Authentication: Exchange of credentials to establish a security context, can be unidirectional or mutual.

Availability: Indicates where or with which percentage a certain communication system provides sufficient field strength for communication.

CA: Certification Authority.

CALM: Communications, Air-interface, Long and Medium range, group of communication systems.

CDMA: Code Division Multiple Access, users are separated by different codes.

CELLO: Cellular Network Optimization Based on Mobile Location.

Credentials: Key or code word with which a device can identify itself and gain access to certain services.

CVB: Cumulative Virtual Blanking.

Delay: Time the signal needs for propagation from one point to another, caused by the distance between transmitter and receiver and the propagation velocity of the signal.

DSRC: Dedicated Short-Range Communication.

E-OTD: Enhanced Observed Time Difference.

Encryption: Protection against attackers, mostly done by ciphering with a key.

FCC: Federal Communications Commission of the US.

GSM: Global System for Mobile communication, second generation system for mobile communication.

GPRS: General Packet Radio Service, evolution of the GSM standard allowing packet switched data transport.

Handover: Transfer of a connection between a mobile device and a third party from a serving

NAP to a target NAP without user interaction and without loss or disruption of the connection.

HIS: Hybrid Information System, defined by Siebert (2004).

Hybrid Network: Consists of different radio access technologies at the first three layers of the ISO/OSI reference model and a hybrid core network.

ICE: Inter City Express, high-speed train.

ID: Identity.

Infrastructure-Based Network: Network consisting of (mobile) devices and additional infrastructure for control, switching and connection to the outside world.

IP: Internet Protocol.

IPDL: Idle Period Downlink.

ITS: Intelligent Transportation System, including information technology.

LMU: Location Measurement Unit.

Localization: Deriving the position of a mobile device.

LTE: Long Term Evolution, evolution of the UMTS standard towards the fourth generation of access networks.

MD: Mobile device.

Mobile Communication: Wireless communication with moving devices.

Mobility: Movement of mobile devices determined by speed and probability of change of direction.

Network Access Point (NAP): Fixed component with connection to the core network that uses an antenna to communicate with mobile devices.

OBU: On Board Unit, mobile device in a transportation vehicle.

OTDOA: Observed Time Difference Of Arrival.

PKI: Public Key Infrastructure.

PTSM: Probabilistic Traffic Sign Model.

QoS: Quality of Service, parameter to assess the quality of a communication network and its service delivery.

RAN: Radio Access Network.

RSU: Road Side Unit, network access point along a track for communication with the OBU.

RTD: Reference Time Delay.

Security Mechanisms: Mechanisms to provide some security features as privacy, anonymity, integrity of data, restricted access, etc.

SSM: Stop Sign Model.

TA: Timing advance.

TLM: Traffic Light Model.

TOA: Time Of Arrival.

U-TDOA: Uplink Time Difference of Arrival (U-TDOA).

UMTS: Universal Mobile Telecommunications Standard, standard of the third generation of mobile communication.

VANET: Vehicular Ad-hoc NETwork.

Wireless Communication: Communication without any cable, the signals are transported over the air.

WAVE: Wireless Access in Vehicular Environments, new amendment to IEEE 802.11 for communication to and between vehicles: IEEE 802.11p.

WLAN: Wireless Local Area Network, although not correct often used as synonym for the standard IEEE 802.11, another WLAN is HiPERLAN.

Chapter 13
Radiometric Speed Sensor for Vehicles

Vladimir Rastorguev
Moscow Aviation Institute, National Research University, Russia

ABSTRACT

One of the main parameters for providing traffic safety of vehicles is the knowledge of their speed. In this chapter results of research activities on a microwave radiometric sensor for the measurement of the velocity of land vehicles are presented. The work concentrates on a Radiometric Speed Sensor of Correlation Type (RM SSCT). For the analysis of design principles of the RM SSCT, the main parameters of radiometers are defined. The nature and statistical characteristics of radio thermal radiation of a terrestrial surface and objects are considered. For an estimation of the influence of parameters of the antenna system and the linear path of the receiver on parameters of the signal formed at the output of the correlator, a statistical analysis of the radiometric system of correlation type is carried out. Using a statistical model of the RM SSCT, the parameters of the antenna system were optimized as well as the radiometric receivers for various types of objects present on a terrestrial surface. Statistical results of the performance of the RM SSCT and an analysis of the basic characteristics of the SSCT for various types of objects are finally presented.

INTRODUCTION

One of the most important tasks to increase traffic safety of vehicles and automating their movements is a high-precision measurement of their velocities. Research done recently in different countries has shown that the most effective way to solve the problem is the application of radio sensors which

DOI: 10.4018/978-1-4666-2976-9.ch013

allow to estimate not only the speed of vehicles, but also their acceleration. One of main advantages of radio sensors in comparison, for example, with optical or infra-red sensors, is their continuous performance in any weather condition (rain, fog, snow, smoke, dust etc.), i.e. under conditions of limited optical visibility. They even work if there is no visibility at all.

Two basic radio engineering methods for speed measurement are known: utilization of the Doppler Effect and correlation of signals, whereby the correlation method can realize both active and a passive location of vehicles.

The essence of the correlation method (Borkus et al., 1973, Klyucharev et al., 1995) consists in measuring the time delay between signals received by antennas placed along the axis of the object movement (see Figure 1).

These signals have identical form, but are displaced in time. The amount of time delay is equal to the time the moving object needs for a distance equal to half the distance between the receiver antennas. For the measurement of this shift the signal received by the first antenna is delayed and correlated with the signal of the second antenna until a maximum of the mutual correlation function (MCF) is observed. Thus, the amount of the delay corresponding to the maximum MCF unequivocally defines the traveling speed of the object.

For the realization of the correlation method in a speed sensor it is useful to dispose of some channels of the receiver. The necessary number depends on how many components of the velocity vector are to be determined. For the measurement of two vector components V_x and V_y it is necessary to have three channels. Thus, antennas should be placed both along the longitudinal axis OX, and along the perpendicular axis OY.

Depending on the demands, the speed sensor of correlation type (SSCT) can be used either in a stationary mode or on a mobile object. In the first case the SSCT can serve, for example, as speed sensor for various moving vehicles on roads with multi-line traffic. In the second case the application of the SSCT allows to supervise the speed of movement directly onboard a vehicle, for example, a car. In both cases application of the correlation method provides high accuracy in the determination of the speed.

PRINCIPLES OF DESIGNING A RADIOMETRIC SSCT

In this section the radiometric speed sensor of correlation type (RM SSCT) is considered (Klyucharev et al., 1995, Kukushkin et al., 1991).

Research on radiometric measuring instruments which determine the parameters of radio thermal radiation of a terrestrial surface from an airplane was carried out first in the laboratories of radio receiver devices of the Moscow Aviation Institute (MAI) in 1985. The research was supervised by Professor Alexey Petrovich Zhukovsky, a highly respected scientist of the USSR. As a result of the investigations the electrodynamic theory of radio thermal radiation of a terrestrial surface has been established (Zhukovsky, 1992). Taking into account this theory the methods of designing onboard radiometers of various types have been developed. As a result, modern radiometers have been designed for different frequency ranges. In the late 1990s in this laboratory were started research

Figure 1. Principle of operation of a speed sensor

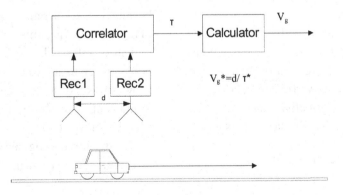

activities on radiometric measuring instruments of movement parameters of vehicles. Therefore, the results in the given chapter are based on a certain experience at the MAI in designing radiometric systems and measuring instruments.

This RM SSCT has a variety of essential advantages in comparison with known Doppler speed sensors. These advantages are:

- Absence of the transmitter in a radiometer allows to simplify the design essentially: lower dimensions and weight of the SSCT as well as less power consumption is achieved.
- The price of a passive SSCT is less than that of a similar Doppler speed sensor, owing to the reduction of the number of blocks and reduction of operational expenses.
- Ability to work in any weather condition (rain, snow, fog, low overcast), and at high smoke content in the atmosphere.
- Absence of radiation allows to provide radio reserve for the performance of control

service of the activities on motorways, and excludes harmful influence on the human body also.

According to the correlation method the measurement of radio thermal radiation is made by two radiometers where the antennas have a distance in space, d_0, as shown in Figure 2. Depending on the application variant of the RM SSCT this sensor can be realized as a sensor where the antenna platform moves with speed V, or the object moves with speed V under the sensor.

If two antennas are placed parallel to each other along the direction in which the object moves (see Figure 1), both will detect practically the same fluctuations of the radio thermal field. The signals at the antennas' exits will show a displacement in time. The time delay is given by the quotient of the distance between the antennas, d_0, and the velocity of the moving object, V:

$$u_1(t) = x(t), \quad u_2(t) = x(t - d_0/V).$$

Figure 2. Geometry of the set-up

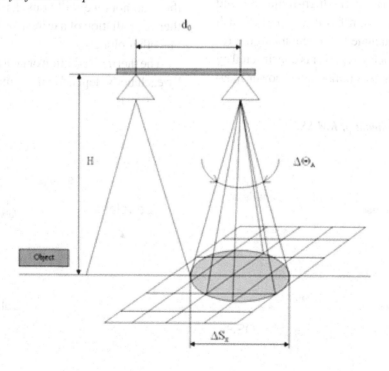

Calculation of the mutual correlation function (MCF) of these signals during a certain interval of supervision, T, allows to estimate a measure of the displacement of these signals $u_1(t)$ and $u_2(t)$ in time τ_0:

$$R(\tau) = 1/T \int_{t-T}^{t} u_1(\tau)\, u_2(\tau - d_0/V)\, d\tau$$

As the distance between antennas, d_0, is known, having defined τ_0, it is possible to calculate a component of the speed of the object: $V = d_0/\tau_0$.

For the calculation of the MCF the signal from the exit of the first radiometer, S_1, is fed to the first input of the correlator through the adjustable device of a time delay. The target signal of the second radiometer, S_2, having a time delay τ, is fed into the second input of the correlator. As a result of the calculation of the MCF the amount of delay, τ_0, corresponding to a maximum MCF of the signals, yields the required speed of the object.

The block scheme of the RM SSCT is depicted in Figure 3.

The measuring instrument can be divided conventionally in two parts: high-frequency and low-frequency. The high-frequency part of each radiometer does include the antenna and the receiver. The radiometric receiver is constructed by the scheme of radiation temperature power mea-surement, i.e. it contains: an entrance broadband high-frequency filter (HFF), a square-law detector and an output low-frequency filter (LFF). The antenna construction should provide the de-manded width of the antenna diagram and should have, whenever possible, minimum size and weight. Finally, the construction of SSCT does not require the application of bulky and expensive gyro-stabilized antenna platforms, as in the case of Doppler speed sensors. The low-frequency part (the signal processor) includes the correlator. It contains the system which continuously tracks the MCF maximum, and the calculator of travel-ing speed.

RADIOMETER PARAMETERS

Results of research on features of radio thermal radiation of a terrestrial surface and various ob-jects, and also methods of the measurement of this radiation are presented in many monographs and articles. Among them are the fundamental monographs of Zhukovsky, 1992, Shutko, 1986, and Sharkov, 2003. Using the results received by these authors we will consider features of radio thermal radiation of a terrestrial surface and vari-ous other objects.

The thermal radiation or heat radiation emitted by each body depends on the material properties

Figure 3. Block scheme of RM SSCT

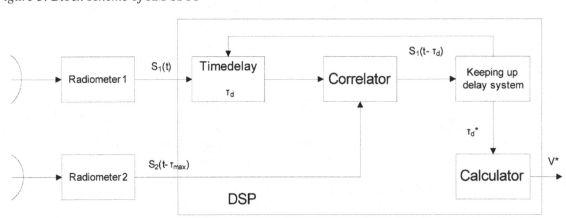

and the thermodynamic temperature. The transmitted thermal energy within a small bandwidth at microwave frequencies is proportional to the temperature of the body.

In contrast to convection (movement of molecules within fluids causing heat and mass transfer) and heat conduction (transport of thermal energy without macroscopic flow of molecules), thermal radiation occurs also in vacuum. Only the possibility of the propagation of electromagnetic waves is required.

This heat radiation or thermal radiation is electromagnetic radiation emitted from each body whose temperature is above absolute zero (-273.15 °C). The radiant flux is distributed over the whole frequency band from gamma rays to long radio waves. All signals are superimposed by white noise. The microwave range emits only a small percentage of the total power. Figure 4 shows an example of the spectral distribution of the radiant power.

The origin of the radiation can be explained by two theoretical complementary approaches (Zhukovsky, 1992, Sharkov, 2003):

- The classic electromagnetic wave theory;
- The quantum theory of radiation.

The energy which is emitted from molecules inside a liquid or solid body is mostly absorbed by their neighbors. There is no energy leaving the surface of a body that comes from deep inside. One can say that radiation emitted by solid or liquid bodies is created within a thin layer under the surface (thickness approximately 1 μm). Therefore, we speak of radiating surfaces instead of radiating bodies.

Solid surfaces and free particles emit various types of electromagnetic waves. These are classified in two categories: spontaneous emission, which takes place without influence from outside and stimulated emission, which is caused by interacting electromagnetic waves. We are considering only the spontaneous thermal radiation (Zhukovsky, 1992).

The black body or ideal body defines the upper limit of the spectral radiance $L_\lambda\left(\lambda, \beta, \varphi, T\right)$ for all bodies. It is a diffuse, in all wavelengths

Figure 4. Hemispherical spectral radiance of black bodies (Shutko, 1986)

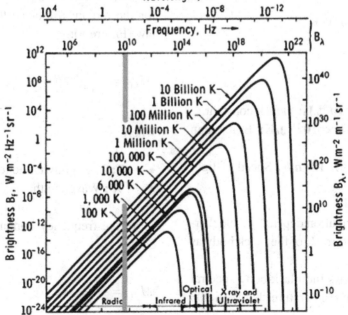

and directions emitting source. Black body spectral radiance $L_{\lambda b}(\lambda, T)$ is independent of the direction; the black body exhibits Lambertian characteristics. It is a universal function only depending on wavelength and the thermodynamic temperature T. There exists a simple relationship between the spectral radiant excitance, $M_{\lambda b}(\lambda, T)$, and the spectral radiance, $L_{\lambda b}(\lambda, T)$ (Sharkov, 2003):

$$M_{\lambda,b}(\lambda, T) = \pi L_{\lambda,b}(\lambda, T) \Omega_0$$

where Ω_0 stands for the unit solid angle 1 sr.

The spectral distribution of radiance, $L_{\lambda b}(\lambda, T)$, of a black body is the basis of many calculations in the field of thermal radiation. The spectral distribution and temperature dependence of the radiance is described by Planck's famous law:

$$\pi L_{\lambda b}(\lambda, T) = M_{\lambda b}(\lambda, T) = \frac{c_1}{\lambda^5 \left[\exp\left(\frac{c_2}{\lambda T} \right) - 1 \right]}$$

The constants c_1 and c_2 are

$$c_1 = 2\pi h c_0^2 = 3.7418 \times 10^{-16} \ W \cdot m^2,$$

$$c_2 = h \frac{c_0}{k} = 14388 \ \mu m \cdot K$$

where:

- $_h$= **6,626068 × 10^{-34} Js:** Planck constant,
- k = **1.3806504 × 10^{-23} J/K:** Boltzmann constant,
- **c$_0$=2.99792458 × 10^8 m/s:** Speed of light in vacuum

The spectral distribution depends on the thermodynamic temperature, T, of the black body as depicted in Figure 4.

This diagram shows the black-body spectral radiance for different temperatures between 100

K and 10 billion K in a double logarithmic representation, exhibiting a typical bell shape. The peak shifts with temperature according to Wien's displacement law:

$$\lambda_{max} T = 2898 \ \mu m \cdot K$$

There is a soft slope of the characteristics on the long wavelength side and a steep slope for the short wavelengths (Shutko, 1986).

For long wavelengths, the term in the exponent of Planck's formula becomes very small. In this case, it is possible to describe Planck's law with the Rayleigh-Jeans law. As depicted in Figure 5, the Rayleigh-Jeans law agrees well with the experimental results for high temperatures or large wavelengths.

The wavelength we consider allows us to simplify Planck's radiation law. If $\frac{hc}{l} \ll kT$, the term in the exponential becomes small and the exponential function is approximated by the first order of a Taylor polynomial as (Sharkov, 2003):

$$e^{\frac{hc_0}{\lambda k T}} \approx 1 + \frac{hc_0}{\lambda k T}$$

This reduces Planck's black-body formula for the radiant excitance to the Rayleigh-Jeans law (Figure 5):

$$M_{\lambda,b} = \frac{2\pi c_0}{\lambda^4} kT,$$

where:

- T: Thermodynamic temperature (Kelvin),
- λ : Wavelength.

In the frequency domain the expression becomes

$$M_{f,b} = \frac{2\pi f^2}{c_0^2} kT$$

Figure 5. Comparison of the laws of Rayleigh-Jeans, Wien, and Planck for T = 300 K

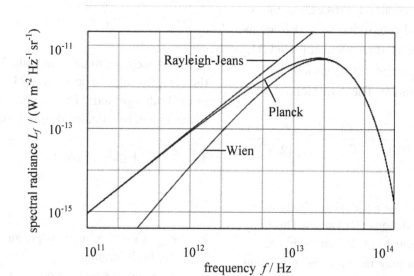

With this equation we can calculate the spectral radiation of a black body for all wavelengths in the microwave range. The radiatiant excitance is fairly constant for a small bandwidth and it is simply proportional to the thermodynamic temperature:

$$M_{\lambda,b} \sim kT$$

It is known (Zhukovsky, 1992, Shutko, 1986) that the radiant excitance, M_λ, of a real body is always smaller than that of a black body, $M_{\lambda,b}$:

$$M_\lambda = \chi M_{\lambda,b},$$

where χ is the emissivity of the body.

It describes the ability of a body to emit radiation and takes on the values $0 \leq \chi \leq 1$, whereby $\chi = 1$ corresponds to a black body. Bodies with constant emissivity are called gray emitters. Bodies whose emissivity depends on temperature and wavelength are called non-gray (real) emitters. In thermodynamic equilibrium a body has to re-emit the absorbed radiation energy, otherwise the temperature would change (Sharkov, 2003).

A radiation temperature (brightness temperature) T_r, of a body may be defined such that at a given temperature T and wavelength λ the spectral radiant excitance, $M_\lambda(T)$, of that body is equal to that of a black body, $M_{\lambda,b}(T_r)$, at temperature T_r:

$$M_\lambda(T) = \chi M_{\lambda,b}(T) = M_{\lambda,b}(T_r)$$

Due to the fact that for microwaves the spectral radiant excitance of an object at a given wavelength is proportional to the thermodynamic temperature, T, it follows that

$$T_r = \chi T \tag{1}$$

where χ, the emissivity, is an integral factor of radiation of an object or a surface.

The emissivity is a material characteristic of a radiating body. This dimensionless quantity is used to determine the specific transmission of real bodies. The factor is based on the emissivity of a black body at the same thermodynamic temperature, T.

The most common type of the emissivity is the directional spectral emissivity χ_f'. It contains the complete information about the dependencies on wavelength, direction and surface temperature. Directional spectral emissivity χ_f' is defined as the ratio of the spectral directional emission M_λ' from a real surface to that from a black body $M_{\lambda b}'$ at the same temperature. The definition is:

$$\chi_f'\left(\lambda,\beta,\varphi,T\right)=\frac{M_\lambda'\left(\lambda,\vartheta,\varphi,T\right)}{M_{\lambda b}'\left(\lambda,\vartheta,T\right)}=\frac{L_\lambda\left(\lambda,\beta,\varphi,T\right)}{L_{\lambda b}\left(\lambda,T\right)}$$

It is common to use an idealized body in the heat transfer for radiation calculations. This body is a gray radiator which is between the black and the real body. The gray radiator has a constant over all wavelength hemispherical spectral emissivity χ and emits diffuse (Sharkov, 2003).

Assuming that we have a non-transmissive surface, we can determine the directional spectral emissivity as:

$$\chi_f'\left(\theta_1,\mu\right)=1-\varrho_f'\left(\theta_1,\varepsilon\right) \tag{2}$$

with $\varrho_f'\left(\theta_1,\varepsilon\right)$ - directional spectral reflectivity

For the investigation of unpolarized radiation, one can split the incident field into two orthogonal components (parallel and vertical). The directional hemispherical spectral reflectivity is then the sum of the square sums of F_{TE} and F_{TM} (Zhukovsky, 1992, Ishimaru, 1991):

$$\varrho_f'\left(\theta_1,\varepsilon\right)=\left|F_{TE}\right|^2+\left|F_{TM}\right|^2 \tag{3}$$

where: $F_{TE, TM}$ – Fresnel factor for vertical and horizontal linear polarization.

The following formulas describe the Fresnel factor for the cases of vertical and parallel polarization separately:

$$F_{TE}=\frac{\varepsilon_2\cos\theta_1-\sqrt{\varepsilon_2-\sin^2\theta_1}}{\varepsilon_2\cos\theta_1+\sqrt{\varepsilon_2-\sin^2\theta_1}}$$

$$F_{TM}=\frac{\cos\theta_1-\sqrt{\varepsilon_2-\sin^2\theta_1}}{\cos\theta_1+\sqrt{\varepsilon_2-\sin^2\theta_1}}$$

A further factor of influence which is changing the complex relative permittivity is the moisture in the body (ground). The complex relative permittivity can be calculated from (Shutko, 1986):

$$\underline{\varepsilon}_{wet}=\underline{\varepsilon}_{dry}+\underline{\varepsilon}_w W-j\lambda\sigma\kappa \tag{4}$$

where:

- $\underline{\varepsilon}_{dry}$: Complex relative permittivity of the dry body/ground,
- $\underline{\varepsilon}_w$: Complex relative permittivity of water,
- W: Relative water content,
- λ: Wavelength,
- σ: Conductivity of the ground,

$$\kappa=\frac{\mu_0 c_0}{2\pi}\approx 60\ \Omega$$

The relative permittivity of water, ε_w, depends on temperature and frequency. The frequency dependence can be described by the following equation:

$$\varepsilon_w=\varepsilon_\infty+\frac{\varepsilon_s-\varepsilon_\infty}{1+j\omega\tau} \tag{5}$$

where

- ε_∞: Relative permittivity of water for infinitely high frequencies,
- ε_s: Static relative permittivity of water for $T=300K$,
- ω: Angular frequency,
- τ: Relaxation time.

Thus, using the analytical expressions (1-5) it is possible to estimate the size of the thermodynamic temperature of a terrestrial surface or object.

The research activities carried out by Zhukovsky, 1992, and Shutko, 1986, have shown, that for extended objects in a viewing sector $\theta \leq 30^0$ the integrated factor of radiation of a terrestrial surface, χ, for the considered wavelength range varies not much, and its size depends first of all on the value of $\acute{e}$. So in the X-band wavelength the value $\acute{e}$ for a terrestrial surface can vary between $2 \leq \acute{e} \leq 8$. The corresponding values of the radiation temperature of the surface are 120 K $\leq T_r \leq$ 170 K.

Certainly, if we measure the thermodynamic temperature of an object which is located on a terrestrial surface from afar, the influence of atmosphere and space radiation must be considered.

In this case, $T_\pounds$ depends on the thermodynamic temperature T of the surface and on noise parameters. These are the atmospheric temperature T_a and the cosmic temperature T_k (Shutko, 1986, Sharkov, 2003), as shown in Box 1.

The disturbance will be integrated over the entire half-space because there is no specific source of atmospheric radiation and cosmic rays we can determine.

BASIC CHARACTERISTICS OF THE RADIOMETRIC SSCT

It is known, that the heart of a radiometer is a broadband receiver of high sensitivity. Receiver key parameters are: a pass-band of a linear path (to the detector) of bandwidth Δf, the noise factor K_N, gain factor G, etc. These parameters should be coordinated with the necessary resolution of the radiometer at minimum measured temperature contrast, ΔT_r, and the time of measurement (integration), t_s. This coordination is established by the following equation (Klyucharev, et al, 1995, Zhukovsky, 1992):

$$\Delta T_r = \frac{T_N}{(\Delta f \cdot t_s + (\Delta G / G)^2)^{1/2}}, \tag{6}$$

where:

- $T_N = T_0 (K_N - 1)$: Noise temperature of the receiver,
- K_N: Noise factor of the receiver,
- $\Delta G / G$: Instability of the receiver's gain.

The value $\Delta G/G$ is possibly very small in stabilized radiometric receivers, which is fundamental for the Dike receiver (modulation type). In this case, for $\Delta G/G = 0$, it is possible to find an optimum value of the necessary bandwidth of the receiver (Rastorguev et al., 2010, 2011):

$$\Delta f = (T_N / \Delta T_r)^2 / t_s \tag{7}$$

The necessary time of measurement t_s, imposes restrictions on the choice of the bandwidth of the receiver on LF (pass-band of LFF) ΔF: $\Delta F = 1.5 / t_s$. Thus, the minimum value for t_s is defined by the maximum value of the measured speed and the spatial distance between antennas, d_0: $t_{s\,min} = d_0 / V_{max}$. For example, for V_{max} = 50 m/s and d_0 = 0.5 m, the bandwidth becomes ΔF = 150 Hz.

Box 1.

$$T_\Sigma = \left[1 - \varrho_f' (\theta_1, \varepsilon)\right] T_{th} + \varrho_f' (\theta_1, \varepsilon) \frac{1}{\pi} \iint\limits_0^\pi T_a(\theta_a, \varphi_a) 6_\pounds^0 (\theta_1, \theta_a, \varphi_a, \varepsilon'') \sin \theta_a d\varphi_a \, d\theta_a$$

$$+ \varrho_f' (\theta_1, \varepsilon) \frac{1}{\pi} \iint\limits_0^\pi \Gamma_a(\theta_a, \varphi_a) T_k(\theta_a, \varphi_a) 6_\pounds^0 (\theta_1, \theta_a, \varphi_a, \varepsilon'') \sin \theta_a d\varphi_a \, d\theta_a$$

Taking for granted that $\Delta G/G = 0$ and taking into account the noise temperature of the radiometer antenna, T_A, equation (6) for the receiver sensitivity will become:

$$\Delta T_{rm} = \alpha(T_A + T_N)\sqrt{\Delta F / \Delta f} \qquad (8)$$

where:

- α: Is a factor depending on the type of the radiometer and
- $T_{rm} = T_A + T_N$: Is the noise temperature of the radiometer.

The parameter of the radiometer which contains the information is the power of the received signal:

$$P_{rm} = k\, T_{rm}\, \Delta f,$$

where k is Boltzmann's constant.

The radiation temperature of an object or a surface measured by the radiometer can be estimated as:

$$T_{rm} = T_A + T_N,$$

where: $T_N = T_0 (K_N - 1)$,

$$T_A = \frac{1}{\Omega_A} \int_0^\pi \int_0^{2\pi} T_r(\Theta,\varphi)\, F_A(\Theta,\varphi) \sin\Theta \ \mathrm{d}\Theta\, \mathrm{d}\varphi$$

- $F_A\,(\Theta, \phi)$: Normalized antenna diagram,
- Ω_A: Normalizing multiplier.

It has been shown (Zhukovsky, 1992) by exact measurements of the radio thermal radiation of various types of objects located on a terrestrial surface that the sensitivity of the radiometer must be: $\Delta T_{rm} \leq 2.5$ K. As a result of these considerations, for a receiver with the noise factor $K_N = 2.5$, the pass-band of HFF of the receiver has to be: $\Delta f \geq 140$ MHz. The given values of noise factor and an input pass-band are easily realized in modern receivers which confirms the possibility of a simple realization of the RM SSCT.

STATISTICAL MODELING OF RADIOMETRIC SSCT

For a statistical modeling of the radiometric SSCT the following factors were considered (Betz et al., 2011):

- Model of the emission of terrestrial surfaces and objects,
- Model of moving objects,
- Model of the influence of the antenna system,
- Model of the noise of the radiometer receiver.

The block scheme of the Matlab/Simulink structure is shown on Figure 6.

Figure 6. Principle of the Matlab/Simulink structure

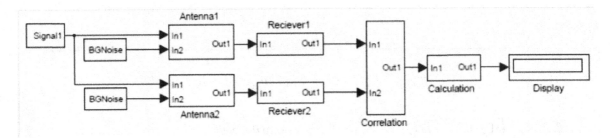

According to this structure we can simulate the properties of an object and the background based on the established formulas. But that does not mean that we have the ability to influence these properties in a future prototype. Therefore we only can vary the system characteristics to find an accurate setting for a possibly high-accuracy speed measurement. These variables are:

- Antenna diagram width,
- High frequency filter characteristic,
- Low pass filter characteristic,
- Distance between antennas.

The Matlab Model of the antenna system is shown in Figure 7.

Once the characteristics of the antenna are generally defined we have to describe accurately the area in which our simulation perceives the thermal signal. We consider a patch antenna which has a square shape and not, as in Figure 8, also a square antenna surface.

The values to be known to build up the model are:

- The height of the antennas above the investigated surface,
- The antenna aperture,
- The sample object,
- The speed of the moving object,
- The distance between two antennas.

The antenna field size is given by the system design (see Figure 2). The antennas receive a continuous background noise during the measurement procedure.

The signal matrix is a two dimensional field like the antenna field. This shape and values contain information about the radio emission body, the temperature and the object size. Figure 8 shows a simulation of an object passing the two dimensional antenna fields. The object evokes a higher emission compared to the background noise surrounded passing the two dimensional antenna fields.

The antenna field represents the main-lobe of the antenna. It is characterized by its parameters and the system geometry. The lobe characteristic is freely selectable like all further parameters in the model. It is simply assumed to have a Gaussian shape in this example (Figure 9). The antenna is used only as a spatial filter. The frequency selection is realized by a high frequency filter incorporated in the receiver.

The two antennas convert electromagnetic waves into electrical signals at the receiver's input (see Figure 10).

These signals are filtered by the UHF bandpass-filter (which will define the desired microwave frequency) in the receiver. Then the signals are passing through the square detector with the low noise amplifier at the output (Figure 11). These two signals are passing through the low frequen-

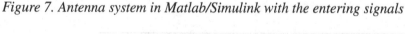

Figure 7. Antenna system in Matlab/Simulink with the entering signals

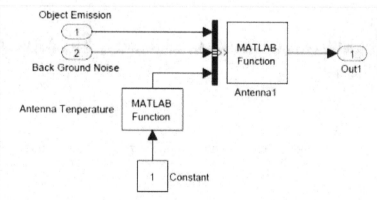

Figure 8. Matlab/Simulink presentation of an object in the antenna field, x-y: local coordinates

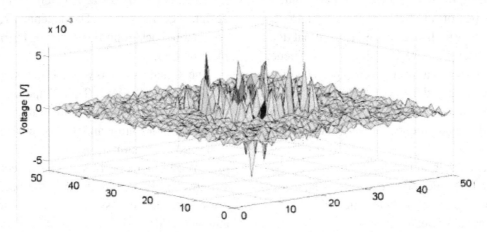

Figure 9. Main-lobe gain in x-y local coordinates

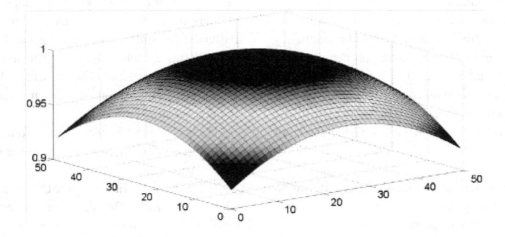

Figure 10. Two antenna signals received from an object

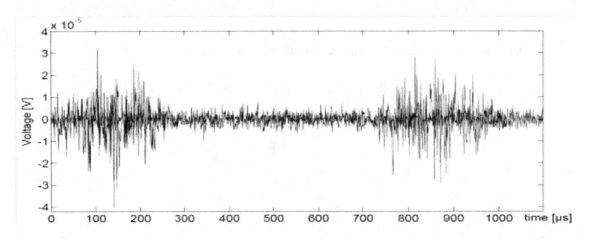

cy filter to smooth the amplitude of the output signals for the correlation operation (Figure 12). The main task of a cross-correlation function is to compare the two signals of the antennas, in order to measure the time difference between both signals. This will be determined by the total receiving time (B) and the maximum in time axis of the correlation function (A). The time between A and B represents the time delay between the first and the second signal (Figure 12). As we know the local spacing between both antennas we can easily calculate the speed of the object passing the antennas.

For example, the statistical modeling of the RM SSCT was done to detect the movement of three objects, namely: a person, a car, a bus. The following parameters were assigned to these objects:

- **Person:** $T_r = 260$ K,
- **Car:** $T_r = 290$ K,
- **Bus:** $T_r = 320$ K.

As a scattering surface asphalt ($T_r = 250$ K $\pm$ 5 K) has been chosen; ambient temperature $T_0 = 270$ K. Possible influence of a rain was not considered.

The height position of the SSCT varied between 3 m and 10 m. The width of the antenna diagram radiometer varied between 3° and 30°. The rating of the receiving antennas changed within 0.25 m and 1.5 m. The noise level of the radiometer was within limits of $\pm$ 1 K.

This modeling has shown that for increasing signal-to-noise ratio in the radiometer receiver the speed estimation error in the RM SSCT will decrease.

Figure 11. Signals after the band-pass filter and gain/square function

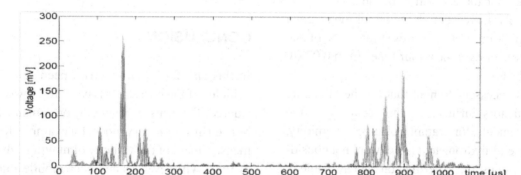

Figure 12. Both signals and the correlation function

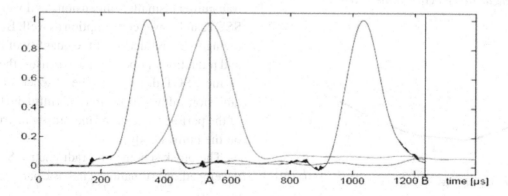

Depending on the height of installation of the RM SSCT and on object type it is also possible to find an optimum of the antennas' distance, d_0, (Figure 13) which provides the minimum error of measurement.

Investigations show that with increasing temperature of the object the contrast between the scattering surface and the object, and therefore, the accuracy of the measurement increases also.

The diagram of Figure 14 shows the dependence of the error of measurement of RM SSCT, σ_v, on the width of the antenna diagram of the radiometer, $\Delta\Theta_A$, for various kinds of moving objects: person (1), car (2), bus (3). The analysis of the figures shows, that a change of this width practically does not change the accuracy of the estimation of the speed of the bus. The accuracy change in this case amounts to about 0.04%V. The effect of smoothing of radio thermal contrast because of the final width of the antenna diagram is largest for the estimation of the speed of the person. An increase in the width of the antenna diagram from 3° to 30° raises the error of the speed sensor by almost four times from 0.07%V to 0.22%V.

It is necessary to note that for the modeling other factors influencing the accuracy of the measurements, for example, change of humidity, presence of hydrometeors, etc. were not considered. Besides, the experimental data available in

Figure 13. Dependence of speed estimation error, σ_v, on the distance between the antennas, d_0

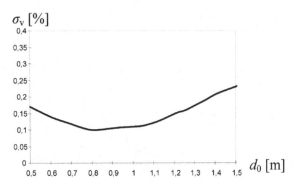

Figure 14. Dependence of speed estimation error on the antenna diagram width

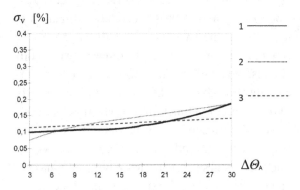

the literature characterizing a radio thermal field are insufficient to draw definitive conclusions. Therefore, the results of this work specify basically on the technical feasibility of the construction of a radiometric SSCT, and define ways of possible perfection of both the measuring instrument and its mathematical algorithm.

CONCLUSION

In this chapter, a radiometric speed sensor for vehicles of the correlator type (SSCT) was considered. This sensor, depending on demands, can be used in a stationary mode, for example, for the measurement of the velocity of moving vehicles on roads with several lanes, or on mobile objects, for example, cars. Application of radiometers on the basis of the SSCT allows to simplify essentially the sensor design (absence of the transmitter), smaller dimensions and weight of SSCT, and power consumption as well. Besides, owing to the reduction of the quantity of blocks and reduction of operational expenses, the price of the SSCT decreases. The absence of radio emission allows to exclude harmful influence of the performance of the measuring instrument on the human body.

The block scheme of a radiometric SSCT is developed, an estimation of its basic parameters is made and it is shown that, using high-frequency devices and digital techniques, a measuring instrument of sufficient simplicity can be realized.

For an estimation of the influence of parameters of the antenna system and of a linear path of receiver RM on parameters of the signal formed at the output of the correlator, a statistical analysis of the radiometric system of correlation type is carried out. For this analysis, the mathematical model of thermal radiation of a terrestrial surface was combined with the phenomenological theory of dispersion of electromagnetic waves.

Results of the statistical modeling of the RM SSCT show that our Matlab/Simulink model confirms the previously applied theoretical knowledge. It allows us to produce thermal radiation with different spectral power, to choose the antenna characteristics, to specify the receiver filter frequencies, thus, to define the receiver's properties as desired. In addition, the system geometry can be selected freely. This includes the height of the antenna structure above the surface, the distance between the antennas and their opening angle. It allows a wide field of applications for all kinds of environments.

Results of the statistical modeling of the radiometric SSCT have allowed to estimate its basic characteristics and have proven that it has the potential to measure speeds with high accuracy.

REFERENCES

Betz, C., Rastorguev, V., & Strobel, O. (2011). Development of a correlation speed measurement for automotive applications. *Proceeding of 19th Annual International Conference on Composites/ Nano Engineering (ICCE - 19), Special Session, Optical and Related Microwave Communication, ICCE-19 Шанхае, Китай, 24-30 июля, 2011*Shanghai, China, July 24-30, (p. 4).

Borkus, M. K., & Cherniy, A. E. (1973). *Correlation's measurements of speed and a deviation corner of flying apparatus* (p. 223). Moscow, Russia: Sovetskoye Radio. (in Russian)

Ishimaru, A. (1991). Wave propagation and scattering in random media and rough surfaces. *Proceedings of the IEEE, 79*(10), 1359–1366. doi:10.1109/5.104210

Klyucharev, M. Y., & Rastorguev, V. V. (1995). Elaboration of automobile radio systems on the radiometer base. In *Proceedings of Third Scientific Exchange Seminar "Radio Technical Systems and Devices of UHF", TUM,* Munich, Germany, (p. 7).

Kukushkin, A. V., & Rastorguev, V. V. (1991). *The device for measurement of object speed.* (Patent of Russian Federation, № 4910299/09/012782).

Rastorguev, V., Betz, C., & Strobel, O. (2011). Optimizations of parameters of the radiometric sensor of movement speed of vehicles. *Proceeding of 13th International Conference on Transparent Optical Networks – ICTON'2011,* Stockholm, Sweden, June 26 – June 30, 2011. ISBN: 978-1-4577-0882-4/11

Rastorguev, V., & Shnajder, V. (2010). Radiometric sensor of movement speed of vehicles. In *Proceeding of 12th International Conference on Transparent Optical Networks – ICTON'2010,* Munich, Germany, June 27 - July 1, 2010. ISBN: 978-1-4244-7797-5

Sharkov, E. A. (2003). *Passive microwave remote sensing of the Earth: Physical foundations* (p. 613). Springer/PRAXIS.

Shutko, A. M. (1986). *The UHF radiometer for a water surface, ground and terrestrial ground.* Moscow, Russia: Nauka. (in Russian)

Zhukovsky, A. P. (1992). *Radiothermal radiation of a terrestrial surface* (p. 76). Moscow, Russia: Publishing MAI. (in Russian)

Chapter 14
Short-Range Ultrasonic Communications in Air

Chuan Li
University of Bristol, UK

David Hutchins
University of Warwick, UK

Roger Green
University of Warwick, UK

ABSTRACT

The idea of this chapter is to give an overview of a relatively new technology – that of using ultrasound to transmit data at short ranges, within a room say. The advances that have made this a useful technology include the ability to utilize a sufficiently wide bandwidth, and the availability of instrumentation that can send and receive ultrasonic signals in air. The chapter describes this instrumentation, and also covers the various aspects of ultrasonic propagation that need to be discussed, such as attenuation, spatial characteristics, and the most suitable forms of modulation. Provided such details are considered carefully, it is demonstrated that ultrasonic systems are a practical possibility for in-room communications.

INTRODUCTION

There are many techniques that can be used for transmitting information from one point to another, and many have been covered within this book. However, there has been relatively little information published on the use of sound waves for transmitting signals in air. This is despite the fact that it is a technique that has been used in under-water communications for some time. There are reasons for this; sound is transmitted much more readily through water at a particular frequency, and has been developed for many years for sonar and other applications. It is thus a mature technique. In general, once below a certain depth, the ocean is a relatively calm environment. In air, propagation is more difficult, in that it is much more susceptible to changes in atmospheric conditions. It is thus not surprising that the communication system described in this Chapter is thus designed for indoor use, using relatively high frequencies

DOI: 10.4018/978-1-4666-2976-9.ch014

to achieve a reasonable data rate. Note, however, that outdoor communication is also possible, but at lower frequencies.

In this Chapter, the principle of using ultrasound for short-range indoor data communication will be investigated, as an alternative to infrared and rf communications. Both are obviously important methods, but in some cases ultrasound can have a possible advantage. The discussion that follows will show that the choice of modulation scheme is important, by optimising the characteristics of the communication channel in terms of available bandwidth. The complications arising from diffraction and attenuation effects will be described, and adaptive equalisation techniques presented.

BACKGROUND TO ULTRASOUND

Sound is generated by the oscillation of particles. For it to propagate, it needs a medium, and that could be solids, gas or liquids. Sound waves are periodic, which can be described by the term frequency, which can be expressed in the following form:

$$f = c/\lambda ,\tag{1}$$

where c donates the speed of sound, and λ is the wavelength.

The audible range of sound for various animal species is different. Humans can hear sound within the range 20 Hz – 20 kHz. For this reason, sound waves with frequencies above 20 kHz are classified as ultrasound, whereas bats are capable of detecting much higher frequencies. In gases, and the majority of liquids, ultrasound propagate as longitudinal waves, where the displacement of particle is parallel to the direction of travel. The propagation velocity of ultrasound varies in different media, as it is controlled by the density and the elasticity of the medium. The velocity of longitudinal waves in gases and liquids is given by

$$c = \sqrt{\frac{K_a}{\rho}}\tag{2}$$

where K_a is the adiabatic bulk modulus, ρ is the mean density of medium. The sound velocity in air is approximately 331 ms^{-1}, although this varies with temperature and humidity. An approximate value for variations in temperature can be calculated using

$$c = c_0 + \gamma d ,\tag{3}$$

where c_0 is the speed of sound in air at atmospheric pressure at 273° K (331.0 ms^{-1}), γ is the temperature coefficient (0.61 for air), and d is the temperature difference from 273° K.

When ultrasonic waves pass though an interface between two materials at an oblique angle, both reflection and refraction take place, provided that the two materials differ in acoustic impedance Z, where $Z = \rho c$ in each medium in question. A large difference in the values of Z in two media A and B results in a large reflection coefficient, which for plane waves at normal incidence can be estimated from

$$R = \left(\frac{Z_B - Z_A}{Z_A + Z_B}\right)^2 .\tag{4}$$

The transmission of ultrasound through air includes effects such as internal friction and thermal conductivity. This results in a finite amount of the oscillation energy being converted to thermal energy; hence, the ultrasonic signal is attenuated as it propagates through the medium. Calculation of this absorption can be complex due to the various energy dissipation mechanisms. Air is especially problematical because of its nature as a variable mixture of gases, making an accurate calculation of ultrasonic absorption difficult (Bass, et al, 1984). However, there have been a number of studies (Bass and Bauer1972; Bass and Shields, 1977;

Lawrence and Simmons 1982) that found there are only two main mechanisms of absorption. The first is known as classical absorption, and is due to internal friction between the molecules of the gas and its thermal conductivity, while the second is known as relaxation loss. This is caused when the translation energy (of the ultrasonic wave) is absorbed into the gas molecules themselves, as internal rotational and vibrational energy states. Hence, the overall absorption loss in air is given by:

$$a = a_{cl} + a_{rot} + a_{vib,O} + a_{vib,N} \qquad (5)$$

where, a_{cl} is the classical absorption, α_{rot} the rotational relaxation loss, $a_{vib,O}$ and $a_{vib,N}$ the vibrational relaxation loss due to oxygen and nitrogen respectively. All of these terms are frequency and temperature dependent. Only the vibrational relaxation loss terms, $a_{vib,O}$ and $a_{vib,N}$, are dependent on humidity (ANSI, 1978) In fact, at frequencies in excess of 300 kHz, this term becomes the dominant mechanism in the absorption of the acoustic wave. Bass et al (1984) evaluated an overall expression for these effects:

$$a_{cr} = 15.895 \times 10^{-11} \cdot \frac{\left(T/T_o\right)^{1/2} \cdot f^2}{\left(P/P_o\right)} \qquad (6)$$

where a_{cr} is the absorption due to classical plus rotational mechanisms (dB/m), T the measured temperature, T_o the reference temperature (293.15 K), P the measured pressure, P_o the reference pressure (101.325 kPa), and f the frequency of the ultrasonic wave. This equation is valid for air temperatures over the range 0 - 40°C, and with pressure below about 2 atmospheres (~200 kPa). It should also be noted that the attenuation is proportional to the square of the frequency. Figure 1 shows the effect of absorption in air on a broadband signal (with an arbitrary initial amplitude of unity) after passing through 20mm of air, at STP. It can be seen that the air has a low-pass filtering effect on the ultrasonic signal, with a –6dB cut-off at approximately 1.4 MHz. This demonstrates that for a practical ultrasonic communication system, the upper frequencies are limited to around 1 MHz, beyond which excessive absorption occurs.

Figure 1. Theoretical amplitude of an infinitely broadband signal attenuated by transmission through 20mm of air at STP

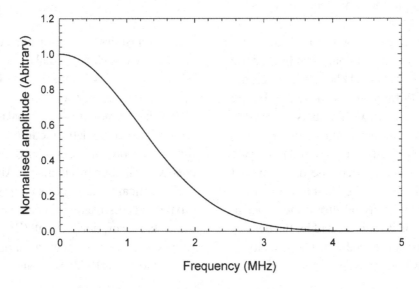

Some knowledge of the diffraction field of sources and receivers is also important. Systems may require transducers with narrow beams, while others may require wider fields of view. The angle of divergence of a transmitted beam is controlled mainly by diffraction, and can generally be determined by the ratio of the size of the transmitting source to the wavelength, λ, of the produced ultrasound (Krautkramer & Krautkramer, 1970). For a transducer with a circular aperture of diameter, D, the divergence angle θ of the beam can be approximated by:

$$\sin \theta \approx 1.22 \frac{\lambda}{D} \qquad (7)$$

Hence, with a fixed wavelength a larger diameter transducer would produce a more directional beam. Likewise, for the same diameter transducer, a shorter wavelength (hence higher frequency) signal would also have the same effect.

For the theoretical prediction of the sound field emanating from a source, a planar piston in an infinite ridged planar baffle solution is commonly used. In this instance, it is assumed that the transducer resembles a uniform vibrating piston held within an infinitely large 'wall', or ridged planar baffle. Numerical simulations often first evaluate an expression for the impulse response of a particular transducer. Once this has been calculated, the resultant variations in the pressure field for a given excitation waveform can then be computed by convolution with the drive waveform. An example is shown in Figure 3. This is a pressure field plot for a 10mm diameter circular piston driven by (a) a 500 kHz, 10 cycle tone burst, and (b) a broadband transient (centre frequency 500 kHz). The two plots show a characteristic shape, consisting of two distinct regions. The first, in close proximity to the source, is known as the nearfield region, where constructive and destructive interference occurs. At further distances, the pressure amplitude eventually decays away to zero;

this region is known as the farfield region. These areas are more clearly shown in Figures 4(a) and 4(b), which shows the on-axis pressure (i.e. $r = 0$) for the 10 mm source with the same driving conditions as before. The differences between the broadband transient, and the more narrow tone burst plot are most evident in the near field region, with greater variations in pressure present in the latter case. Side lobes are also evident in the tone-burst case but not for the broadband transient.

The region between the near and far field is known as the nearfield/farfield boundary. This occurs at the last axial maximum before the pressure decays to zero. This distance (N) is approximated by the following equation:

$$N \approx \frac{a^2}{\lambda} \qquad (8)$$

where a is the radius of the transducer, and λ the wavelength of the emitted ultrasound. Thus, for the above example, $a = 5$mm and $\lambda = 0.68$mm (in air), so $N \approx 36.76$ mm, comparing favourably with Figure 3.

COMPARISON TO EXISTING METHODS

There are many techniques within the RF range, which extends from about 10 kHz to over 300 GHz. RF can go through many types of walls, and this has many uses for areas such as mobile phones, but can cause interference unless precautions are taken. The implementation of a mobile phone network generally requires a license fee for use of the bandwidth. For other systems, such as Wireless LAN (WLAN) or WiFi, casual eavesdropping can be a potential threat to all active users within the network. WiFi has become an alternative for indoor networking, without the need of cable. Bluetooth (Siriam 2002) was born as an open standard that has been adapted by many manufacturers of electronic appliances. It allows an *ad-hoc* approach for various devices to communicate with one another within a nominal 10 meter range. It operates in the 2.4 GHz ISM band (2400 – 2483.5 MHz), and uses a frequency hopping TDD scheme (Crag et al. 2000) for each radio channel, allowing a channel symbol rate of 1 Mbps using GFSK modulation (Lee & Chen 2006).

Wireless infrared communications uses free-space propagation within the near infrared band as a transmission medium. Indoors, it is typically used for communication between one portable device and another, or between a portable device and an access point or base station. As a medium for short-range wireless communications, infrared radiation enjoys several advantages over radio. The primary advantage is a virtually unlimited, unregulated bandwidth. In other words, infrared links generally outperform radio links in terms of data speed. In addition, like visible light, infrared radiation does not pass through walls or other opaque barriers, so that an infrared signal is confined to the room in which it originates. This makes infrared a secure medium, preventing casual eavesdropping. More importantly, it allows neighbouring rooms to use independent infrared links without interference. Finally, unlike radio, infrared links using intensity modulation and direct detection do not suffer from multipath fading, because the detector surface typically spans thousands of wavelengths, providing an inherent spatial diversity (Carruthers and Kahn 1998; Ball 2007). However, infrared communication also has drawbacks. An artificial-light source, like a fluorescent lamp, emits intense infrared noise (Narasimham et al. 1996; Moreia et al. 1996), and hence a photodetector of an infrared receiver is exposed directly to such an environmental noise. The optical noise in typical office environments is mainly due to the daylight, tungsten, and fluorescent lamps. Another limitation is its limited power range due to concerns over eye safety with laser diodes and LEDs (Wilkins 1996).

The technique remains a powerful approach for many communication applications, both indoors and outdoors. The technology is supported by a standardization body, the Infrared Data Association (IrDA) (Williams 2000), which was established in 1995. An IrDA infrared connection is established solely be a directed infrared beam, for bi-directional half-duplex communications. The lowest signalling rate is 9,600 bps, and the highest currently specified is 16 Mbps (Lueftner et al. 2003). Nevertheless, most devices available today operate at a maximum of 4 Mbps. The range of the LOS (Line-of-Sight) connection should be at least 1 meter, but typically, 2 meters can be reached. A low power version is also available with a limited range objective of 20 cm, but at 10 times less power consumption (Akhavan et al. 2002). IrDA and Bluetooth technologies provide complementary implementations for data exchange and voice applications. For some devices, having both Bluetooth and IrDA will provide the best short range wireless solution. For other devices, the choice of adding Bluetooth or IrDA will be based on the types of applications and intended usage.

Research into airborne ultrasonic communications systems has not been widespread to date. This is mainly due to the perceived poor bandwidth of ultrasonic transducers, making ultrasound seemingly inferior in today's increasingly speed-demanding wireless data communication environment. On the other hand, since ultrasonic waves are effectively blocked by walls, it would make eavesdropping outside the room almost impossible. Hence, infrared and ultrasound technology share the same advantage – they allow a secure communication link. But what makes ultrasound even better is that it is, by nature, more immune to interference. This characteristic also applies within the room, as the absorption characteristics of air in terms of ultrasonic propagation allow control of range.

The first generation remote control for TVs used solely ultrasound technology. As shown in Figure 3, the ultrasonic remote control was designed by Dr. Adler from Zenith Electronics Corporation back in 1955, and was named "Zenith Space Command" (Adler & Desmares 1975). The transmitter used no batteries; it was built around aluminium rods that were light in weight and, when struck at one end, emitted distinctive high-frequency sounds. The first such remote control used four rods, each approximately 2½ inches long - one for channel up, one for channel down, one for sound on and off, and one for on and off. They were very carefully cut to lengths that would generate four slightly different frequencies. They were excited by a trigger mechanism – similar to the trigger of a gun – that stretched a spring and then released it so that a small hammer would strike the end of the aluminium rod. It was figured that more than 9 million ultrasonic remote control TVs were sold by the industry during the 25-year reign of Dr. Adler's invention. By the early 1980s, the industry moved to infrared, or IR, remote technology, which is still the most dominant technology in remote controls for home electrics.

In recent years, wireless ultrasound technology has received more interest. For instance, a wireless keyboard using ultrasound has been designed (Li et al 2008), replacing the wire by ultrasonic carriers using capacitive transceivers. Recently, ultrasound has been using for location tracking in hospitals, where time-of-flight from a moving source to a set of fixed receivers can be used (Holm 2005). Other work has looked at simple methods for designing an ultrasonic PC mouse (Zurek et al. 2002) using narrow bandwidth piezoelectric transducers with a centre frequency of 40 kHz in an analogue modulation scheme. An ultrasonic communication system has also been developed where data (for example, characters, images, and voice) could be sent ultrasonically through air or pipes with or without water, using commercially available piezoelectric transducers at a data rate of 100 bits per second (Haynes et al. 2002). For data transmission in air, another study compared digital modulation techniques based on binary phase shift keying and binary frequency shift

keying, 2 common forms of modulation. This demonstrated that signals could be transmitted in air over a 3-m distance at a data rate of 5 kb/s (Keiichi et al. 2007), with BPSK being the better of the 2 types of digital modulation. More recently, researchers are also analysing the ultrasonic communication capabilities of animals such as bats and frogs, so as to further improve the design of current ultrasonic transducers (Paajanen, 2000; Streicher et al. 2003; Streicher et al. 2005).

There have been some studies of acoustic signals for underwater communications, where again RF, infrared and acoustics have all been considered (Otnes & Eqqen 2008). In the acoustic case, phase modulation schemes have been reported (Ochi & Fukuchi 2007), in which effective communication has been demonstrated over quite large ranges, at a data rate of 40 kbps. Piezoelectric transducers were used, of limited bandwidth. This is encouraging, in that it shows that such schemes are possible underwater. However, for use in air, the data rate, bandwidth and other factors will need careful attention, and will affect the performance in a complex way.

It will be seen that ultrasound has some potential benefits to existing technologies for short-range indoor use. These, however, require the choice of a suitable modulation scheme for the bandwidth available, and proper design of instrumentation, with particular attention to the sources and receivers.

INSTRUMENTATION

An essential element in a well-designed ultrasonic communications system is the transducer used for generation and detection in air. It should have enough bandwidth to accommodate modern broadband signalling rates, and a good sensitivity for an effective operational range in air. Piezoelectric based devices were (and still are) by far the most widely used means of detecting and creating ultrasound. This is mainly because piezoelectric transducers are generally inexpensive, rugged, compact, and hence well suited to an industrial environment. They have been successfully developed for use in solids (contact transducers) and liquids (immersion transducers). However, for air-coupled applications a problem arises, due to the high acoustic impedance mismatch of the piezoelectric material, and that of air. This leads to very poor transduction for air-coupled applications. Although there exists various solutions to minimise the large impedance mismatch between the type of device and air, they all come at the expense of either reduced sensitivity or reduced bandwidth. Nevertheless, the latest piezocomposite transducers have demonstrated an improved bandwidth of approximately 1.5 MHz, centred at 1.2 MHz, and with reasonable sensitivity (Haller and Khuri-Yakub, 1994). The drawback of such a device is its complexity in construction, which causes unwanted internal resonances. As an alternative, for air coupled ultrasound, capacitive transducers generally offers much more efficient transduction of ultrasound, better sensitivity, and wider bandwidth without complex add-ons to its simple construction design (Schindel et al., 1993).

The capacitive or electrostatic ultrasonic transducer is primarily used for the transduction of ultrasound into air (or gas). This is mainly due to its inherent low acoustic impedance (see below). The device is similar in form to the condenser microphone (Wong & Embleton, 1995), and is essentially a variable capacitor, consisting of a thin flexible membrane fixed over a ridged conducting backplate. The earliest known reference to such a device (Wente 1917) utilised a thin metallic membrane (or diaphragm) separated from a back electrode by a 25μm air gap. It was demonstrated that this device had an undamped resonant frequency of approximately 17 kHz. Later improvements to the design widened its bandwidth by increasing air damping, with the addition of grooves, or holes, in the backplate (Crandall 1917, Wente 1922a, 1922b). Further work was reported (McLachlan 1934) which

successfully incorporated a solid dielectric layer between the back-plate and front electrode to improve performance. Three years later, a similar device was reported (Sell 1937) with increased centre frequency and bandwidth, and incorporating concentric grooves in the backplate. However, despite of all the remarkable efforts made, their effective bandwidth was still limited within the audio frequency range. It was not until later (Kuhl et al. 1954) that transducers capable of operation in the 50-100 kHz range were produced, a breakthrough of such devices for ultrasonic use. The basic design incorporated a conducting contoured back-plate and a thin (~10μm) metallised polymer film for the membrane, which was considered as the cornerstone for the more modern capacitive transducer designs. Further work shortly ensued (Matsuzawa 1958, 1960), which studied the effects of membrane thickness and backplate surface properties on the characteristics of the transducer.

The construction of the polymer-based capacitive transducer (Kuhl et al 1954) consists of a thin flexible polymer membrane, metallised on one side and fixed with its insulating surface against a ridged contoured conducting backplate, as illustrated in Figure 2.4. During operation, a DC biasing voltage is usually applied between the backplate and the membrane, electrostatically attracting it to the backplate, trapping tiny air pockets. As a detector, an impinging ultrasonic wave causes the membrane to deflect, changing the capacitance of the device. Due to the applied DC bias voltage, this change in capacitance can be observed as a movement of charge on and off the 'plates' of the device, and amplified with a charge amplifier. For a source of ultrasound, the membrane can also be deflected, electrostatically, with the application of pulsed voltages, often superimposed into a DC bias voltage to improve bandwidth. Due to the relatively low acoustic impedance of the membrane/air pocket system, the capacitance transducer is ideally suited for the transduction of ultrasound into air. Various studies have shown that the performance (centre

Figure 2. Angle of divergence (θ) of the sound field emanating from a cylindrical transducer due to diffraction

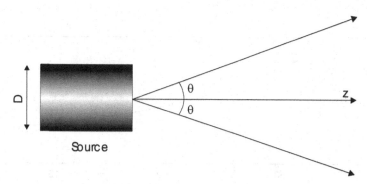

Figure 3. Theoretical sound pressure filed plots (in air) for a 10mm diameter circular piston source driven by (a) a 10-cycle 500 kHz tone-burst, and (b) a broadband transient centred at 500 kHz

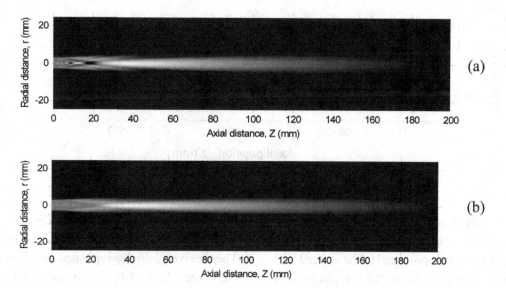

frequency, bandwidth, and sensitivity) of a capacitive transducer is mainly controlled by the surface topology of the backplate (Carr et al. 1992).

Assuming a basic parallel plate capacitor model, with air as the dielectric, the capacitance, C, of the device would be given by the following:

$$C = \frac{A \cdot \varepsilon_o}{x} \qquad (9)$$

where A is the area of one of the plates, ε_o is the relative permittivity of free space, and x the plate separation. Hence, changes in the plate separation, Δx, cause changes in the capacitance of the device, ΔC. This is given by the differentiation of equation (10):

$$\Delta C = -\frac{A \cdot \varepsilon_o}{x^2} \cdot \Delta x \qquad (10)$$

Assuming a constant biasing voltage, V_p, the observed change in charge, ΔQ, on the plates, due to this change in capacitance, ΔC, is given by:

$$\Delta Q = \Delta C \cdot V_p = -\frac{A \cdot \varepsilon_o}{x^2} \cdot \Delta x \cdot V_p \qquad (11)$$

Figure 4. On-axis theoretical pressure field plots in air for a 10mm diameter circular piston source, driven by (a) a 500 kHz 10-cycle tone-burst, and (b) a broadband transient signal with a centre frequency of 500 kHz

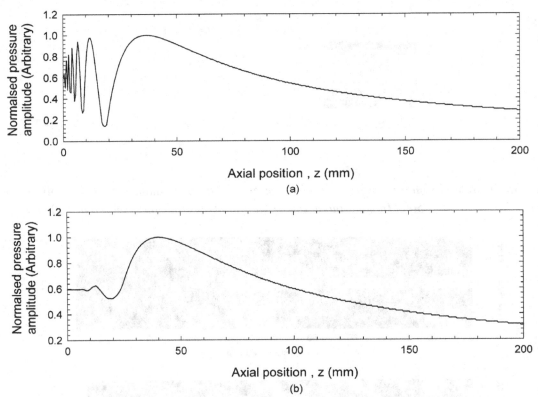

(a)

(b)

Hence, this basic model shows that the observed change in charge (and thus the signal from the transducer) is proportional to the applied bias voltage and the area of one of the plates. It is also inversely proportional to the square of the initial plate separation. However, in practice, the transducers are far from ideal parallel plate capacitors, and the above calculations are only a very basic first principles model of how they operate. Further detail can be found in the publications by Anderson, Sun and others (Anderson et al. 1995; Sun et al. 1997).

Figure 6 shows a typical experimental arrangement for studying ultrasonic communications in air. The transmitter had a membrane thickness of 5 µm, so as to withstand higher excitation voltages without causing damage to the polymer membtane. The source was driven by an arbitrary waveform generator, with a superimposed +100 V dc dias voltage generated by a dc power supply. The receiver, initially separated by 50 mm from the source, had a film thickness of 2.5 µm and was followed by a charge amplifier with a gain of 250 mV/pC. The response was then fed into a digital oscilloscope for signal analysis. Finally, the waveforms were saved on a PC using LabVIEW™ programs (National Instruments, Austin, TX) for offline signal processing. A physical synchronisation link was established between the waveform generator and the oscilloscope. This removed the need for wireless handshaking, which would be needed in a real application.

The bandwidth of the transducers used for ultrasonic communication studies is shown in Figure 7. As can be seen, the spectrum has a full-width, half-maximum (FWHM) value of approximately 1 MHz. Note that signal amplitude is approximately 60 mV peak to peak in a typical measurement, with noise levels at the 1 mV level without signal averaging.

The above factors can be used to predict the expected overall system response of a single communication channel when the above diffraction, attenuation and transducer response effects are taken into account. Figures 8 and 9 show the simulated system response at $z = 200$ mm with $x = 0$ mm and $x = 25$mm respectively for a 1 MHz bandwidth input signal. Consider first the on-axis response, with $x = 0$ (Figure 5.8). Here the spatial field response, due to diffraction effects, is negligible on-axis over the 200 mm propagation distance, with the frequency response staying constant at 1MHz. However, the frequency-dependent attenuation in air causes a large drop in signal with frequency in the manner shown, so that by 700 kHz there is virtually no signal left. Conversely, the transducer response has yet to reach its maximum, with the result that the overall response tends to be a maximum over the approximate range 250-400 kHz. With the receiving point being off-axis by 25 mm, however, the result is very different, as shown in Figure 9. Now diffraction has caused the spatial field response to narrow significantly, and to exhibit the ripples seen earlier. When combined with the effects of attenuation and transducer response, the result is a much more variable response. Hence, the passband ripples introduced by the spatial field response now become an important feature in the overall communication channel system response.

Figure 5. Schematic diagram of the basic random back-plate capacitive transducer

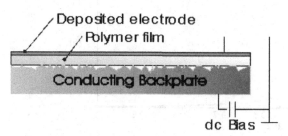

CHOICE OF MODULATION TECHNIQUE

An important aspect of digital data communication is the type of modulation used, in relation to its efficient use of the frequency spectrum, its sensitivity to noise and interference, its simplicity for practical equipment design and its resulting overall economy. The choice of techniques included Quadrature Phase Shift Keying (QPSK) (Saha & Birdsall 1989), Offset Quadrature Phase Shift Keying (OQPSK) (Nelson et al. 2008), Minimum Shift Keying (MSK) (Pasupathy 1979), and 8PSK (Wang & Yongacoglu 1994). All three modulation schemes are able to double the data rate of Binary modulation scheme offers, while preserving similar levels of bit error rate (BER). Their possible use for an ultrasonic communication system will be examined in this Section. Note that quaternary modulation has been successfully adapted to underwater ultrasonic communications (Ochi & Fukuchi 2007), where data was sent at a transmission rate of 80 kb/s over a distance of 350 metres underwater. In the following, a brief overview of these popular quadratic modulation schemes is given. This is followed by simulations and experimental results, from which a choice is made of the most effective modulation scheme for ultrasonic use.

The information capacity of a communication system represents the number of independent symbols that can be carried through the system in a given unit of time. The most basic symbol is the binary digit, also known as bit. Therefore, it is often convenient to express the information capacity of a system in bits per second (bps).

Let us assume that the only effect of the channel is to add thermal noise (Sabatini 1997) to the transmitted signal and that the bandwidth of this noise is very wide relative to the signal bandwidth. The statistics of this noise are assumed Gaussian; the channel is called the additive white Gaussian noise (AWGN) channel. Given these constraints, there exists a maximum rate at which information

Figure 6. Schematic diagram of the apparatus

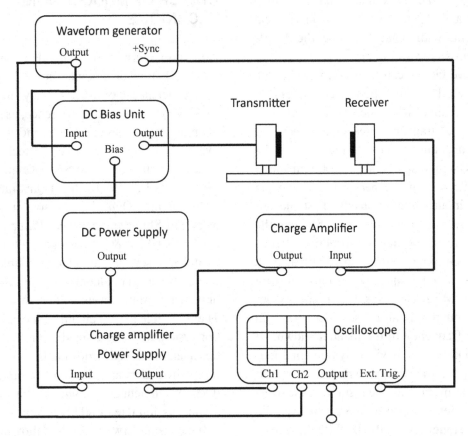

Figure 7. Transducer characteristics, (a) an impulse time response and (b) frequency response

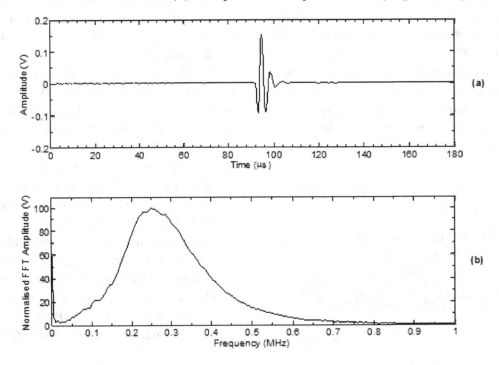

Figure 8. Simulation of magnitude response for attenuation, transducer, field, and channel, for z = 200 mm, and x = 0 mm

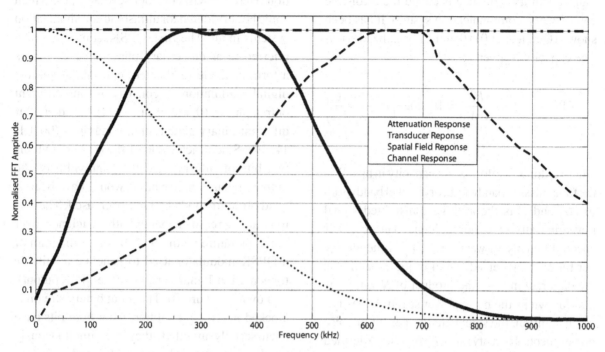

Figure 9. Simulation of magnitude response for attenuation, transducer, field, and channel, for z = 200 mm, and x = 25 mm

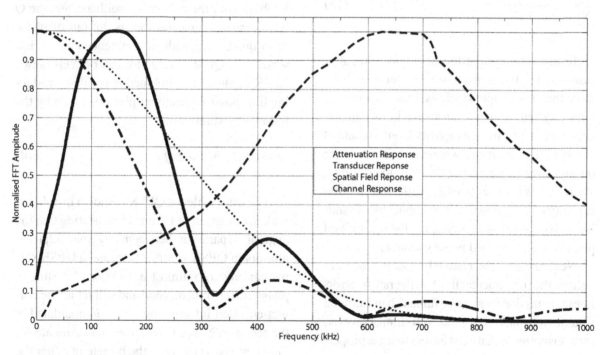

can be transmitted over the link with arbitrarily high reliability. This rate is called the error-free capacity of a communication system. It has been shown (Gallager 2001) that the normalized error-free capacity is given by

$$C = W \cdot \log_2(1 + \frac{P}{N_0 W}) = W \cdot \log_2\left(1 + \frac{E_b}{N_0}\frac{R}{W}\right)$$

$$(12)$$

where C donates information capacity in bps, W is the transmission bandwidth, or channel bandwidth in Hz, and P is the average transmitted signal power in watts, N_0 is the single-sided noise power spectral density in watts/hertz, E_b is the energy per bit of the received signal in joules, and R is the data rate in bps. The product of N_0 and W is also known as the noise power spectral density.

In addition, the signal energy-per-bit to noise power spectral density ratio, E_b/N_0, can be defined in terms of the ratio C/W for an ideal system as: -

$$\frac{E_b}{N_0} = \frac{2^{C/B} - 1}{C/B}.$$

$$(13)$$

In order to achieve the Shannon limit, it is essential for the transmission systems to have more than two output symbols. In other words, a binary system cannot reach such a limit. The dependence of information capacity C on the channel bandwidth W is linear, whereas its dependence on signal-to-noise ratio $P/N_0 W$ is logarithmic. Therefore, it is easier to increase the information capacity of a communication channel by expanding its bandwidth than increasing the transmitted power for a prescribed noise variance.

Nevertheless, the channel capacity theorem defines the fundamental limit on the rate of error-free transmission for a power-limited, band-limited Gaussian channel. To approach this limit, however, the transmitted signal must have statistical properties approximating those of white Gaussian noise.

There are various types of quaternary modulation. Here we will consider several types, which could be considered for ultrasonic communication systems in air. The first, QPSK, sometimes also referred to as quaternary phase shift keying, is a form of angle-modulated, constant-amplitude digital modulation. It consists of four different output phases (0, $\pi/2$, π, and $3\pi/2$), based on four different binary signal input conditions (00, 01, 11, 10). Since each cycle of the carrier contains two bits of information, the bandwidth of the output signal is half what it would have been if quadrature methods had not been used. There is thus an increase in bandwidth efficiency over non-quadrature methods. A block of diagram of a QPSK modulator used to generate a suitable transmission digital waveform (the QPSK output), is shown in Figure 10. The input binary stream is clocked into the bit splitter. After both bits have been serially inputted, they are simultaneously parallel outputted. One bit is directed to the In-phase (I) channel and the other to the Quadrature (Q) channel. The I bit modulates a carrier that is in phase with the reference oscillator, and the Q bit modulates a carrier that is 90^0 out of phase or in quadrature with the reference carrier. Essentially, a QPSK modulator can be seen as two BPSK modulators combined in parallel, producing four possible resultant phasors, given by the following expression:

$$\pm \sin 2\pi f_c t \pm \cos 2\pi f_c t \qquad (14)$$

The truth table for QPSK is shown in Table 1, with the diagram of Figure 11 indicating that the angular separation between any two adjacent phasors in QPSK is 90^0. Therefore, a QPSK signal can undergo almost a $+45^0$ or -45^0 shift in phase during transmission and still retain the correct encoded information when demodulated at the receiver. Nevertheless, the input data are divided into two channels, the bit rate in either the I or Q channel is equal to one-half of the input bit

Figure 10. Schematic diagram of a QPSK modulator

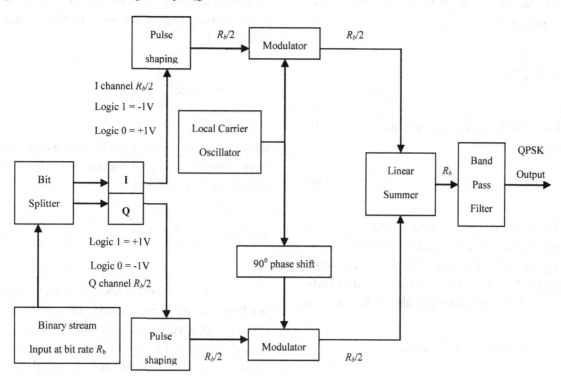

Table 1. Truth table for QPSK modulation

Binary Input		QPSK output Phase
I	Q	
0	0	+45
0	1	+135
1	1	-135
1	0	-45

Figure 11. Constellation diagram for QPSK modulation

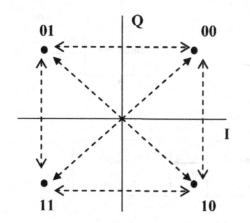

rate R_b. Consequently, the highest fundamental frequency present at the data input to the I or Q balanced modulator is equal to one-fourth of the input data rate. As a result, the output of the I and Q balanced modulators requires a minimum double-side Nyquist bandwidth equal to one-half of the incoming bit rate. Thus, bandwidth compression is realised in QPSK, where the minimum bandwidth is less than the incoming bit rate.

The bandwidth B of a QPSK signal can be expressed as

$$B = (f_c + \frac{f_b}{4}) - (f_c - \frac{f_b}{4}) = \frac{f_b}{2} \tag{15}$$

where f_c is the carrier frequency, and f_b is equivalent to the bit rate R_b in Hz. Mathematically, QPSK signals can be defined as

$$S_{QPSK}(t) = A\cos[\omega_c t + (i-1)\frac{\pi}{2}],$$
$$0 \leq t \leq T_s, \quad i = 1,2,3,4 \tag{16}$$

where the signal amplitude A in volts, can be represented as

$$A = \sqrt{\frac{2E_s}{T_s}}, \tag{17}$$

and angular frequency ω_c in radians, can be defined as

$$\omega_c = 2\pi f_c \tag{18}$$

Here T_s is the symbol duration in seconds, and is equal to twice the bit duration. E_s is the symbol energy in Joules, f_c is the carrier frequency and i is the symbol number. By using trigonometric identities, the above equations can also be rewritten as

$$S_{QPSK} = A\cos[(i-1)\frac{\pi}{2}]\cos(\omega_c t) \\ -A\sin[(i-1)\frac{\pi}{2}]\sin(\omega_c t) \tag{19}$$

The first term represents the in-phase channel, and the second term is the quadrature phase

channel. The average probability of bit error in the additive white Gaussian noise (AWGN) channel is obtained as

$$P_{e,QPSK} = \frac{1}{2}erfc\left(\sqrt{\frac{E_b}{N_0}}\right) \tag{20}$$

where E_b/N_0 is the energy per bit-to-noise spectral density ratio. The complementary error function, denoted $erfc$, is defined as

$$erfc(x) = \frac{2}{\sqrt{\pi}}\int_x^\infty e^{-x^2}dt \tag{21}$$

Coherent detection (Simon & Polydoros 1981) is often used for demodulation, and for this it is essential to have an effective local carrier frequency that is in phase with the original carrier frequency used for transmission. A typical QPSK demodulator is shown in Figure 12. The received QPSK signal, after a bandpass filter to reject any unwanted noise, is passed through two product modulators simultaneously, with the I-channel multiplied by the synchronised carrier and the Q-

Figure 12. Schematic diagram of a QPSK demodulator

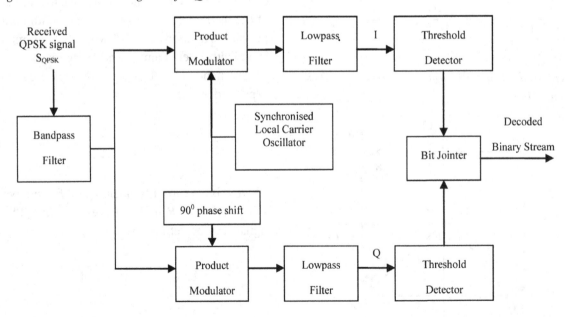

channel multiplied by a 90° phase-shifted carrier. The products are then passed through two lowpass filters to recover the original I-channel and Q-channel baseband signal respectively. Threshold detectors (Fu et al. 2006) are then used to restore the split baseband sequence: if the amplitude of the signal is greater than 0V at the sampling point, logic '1' is decoded; if smaller than 0V, a logic '0' is recovered. Finally, the outputs of the product detectors are fed to the bit joint circuit, where they are converted from parallel I and Q data channels to a signal binary stream.

Mathematically, the demodulation process can be illustrated as follows:

$$S_{tx} = A\cos(\omega_c t) \cdot I(t) - A\sin(\omega_c t) \cdot Q(t) + n(t) \tag{22}$$

where S_{tx} donates the transmitted QPSK signal with noise, $I(t)$ and $Q(t)$ are In-phase channel bit sequence and Quadrature channel bit sequence, respectively. When mix S_{tx} with a local generated In-phase carrier to demodulate $I(t)$, it can be expressed as following in Box 1.

As the symbol rate R_s, is smaller than carrier frequency ω_c, the underlined part can be easily removed using a lowpass filters. Q-channel bit streams is demodulated following the same process as described in Eq. 4.14, with a multiplication of a synchronised local generated carrier that is 90° out of phase.

In QPSK modulation, pulse shaping is normally used to reduce inter-symbol interference (ISI) (Tuncer 2002), which is usually caused by multipath propagation (Katz. & Sruber 1995) and the inherent non-linear frequency response of a channel. Raised cosine (RS) shaping (Alagha & Kabal 1999) of the transmitted bits baseband spectrum is one of the best filter forms to reduce ISI as well as the bandwidth of the signal. It attenuates the starting and end portions of the symbol period. Mathematically, its transfer function can be represented as shown in Box 2.

Here, α is the roll-off factor, which ranges between 0 and 1. When $\alpha = 0$, the raised cosine roll-off filter corresponds to a rectangular filter of minimum bandwidth. As the roll-off factor α increases, the bandwidth B of the filter also increases, according to

Box 1.

$$\begin{aligned} I_d &= I(t) \cdot A\cos(\omega_c t) \cdot \cos(\omega_c t) \\ &\quad -Q(t) \cdot A\sin(\omega_c t) \cdot \cos(\omega_c t) + n(t) \cdot \cos(\omega_c t) \\ &= \frac{1}{2} \cdot I(t) \cdot A + \frac{1}{2}\cos(2\omega_c t) - \frac{1}{2} \cdot Q(t) \cdot A\sin(2\omega_c t) + n(t) \cdot \cos(\omega_c t) \end{aligned} \tag{23}$$

Box 2.

$$H_{RC}(f) \begin{cases} 1 & 0 \le |f| \le \dfrac{(1-\alpha)}{2T_s} \\[3mm] \dfrac{1}{2}\left[1 + \cos\left[\dfrac{\pi(|f| \cdot 2T_s - 1 + \alpha)}{2\alpha}\right]\right] & \dfrac{(1-\alpha)}{2T_s} \le |f| \le \dfrac{(1+\alpha)}{2T_s} \\[3mm] 0 & |f| > \dfrac{(1+\alpha)}{2T_s} \end{cases} \tag{24}$$

$$R_s = \frac{B}{1+\alpha} \qquad (25)$$

where R_s is the symbol rate in symbols per seconds, and B is absolute filter bandwidth.

Offset QPSK is a minor but important variant of QPSK. In OQPSK, the Q channel is shifted by half symbol time (see Figure 13) so that I and Q channel signals do not make a transition at the same time. As a result, phase shifts at a time are limited to be no more than 90° (Figure 14), whereas QPSK can have phase transitions of up to 180°. This yields much lower amplitude fluctuations than non-offset QPSK and is sometimes preferred in practice (Leung & Feher 1993). Although in a linear channel its bit error rate is the same as QPSK, but, in non-linear applications, it often offers a lower BER when operating close to the saturation point of the transmitting amplifier (Chen & Wong 1997).

Minimum Shift Keying (MSK) is a special type of continuous phase-frequency shift keying (Sundberg 1986), where the modulated signal has constant amplitude, and an efficient amplifier operating in class C can be used to minimise power consumption, an important consideration for battery powered devices (Norris & Nieto 2008; Amoroso 1976; Saarnisaari 2007). It is derived as ordinary FSK, with the modulation index set to 0.5, which corresponds to minimum frequency spacing that allows two FSK signals to be coherently orthogonal, and the name minimum shift

keying implies the minimum frequency separation that allows orthogonal detection (Gronemeyer & McBride 1976). In conventional FSK, the digital signal that modulates an FM modulator is a rectangular bipolar Non Return to Zero (NRZ) bit sequence with symbol values $b_i = \{-1,1\}$. A rectangular pulse shape is shown in Figure 15(a). An MSK signal with I-Q components is formed in a similar way as OQPSK, except that modulation signal is shaped by half-sine pulses, as shown in Figure 15(b). This results in a constant-modulus signal, which reduces problems caused by non-linear distortion.

$$g_{RECT}(t) = \begin{cases} U_m, & -\dfrac{T_b}{2} \le t \le \dfrac{T_b}{2} \\ 0 & otherwise. \end{cases} \qquad (26)$$

Figure 14. An offset QPSK constellation

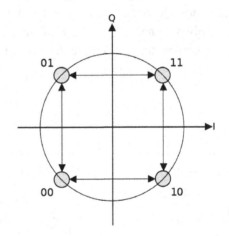

Figure 13. OQPSK encoding

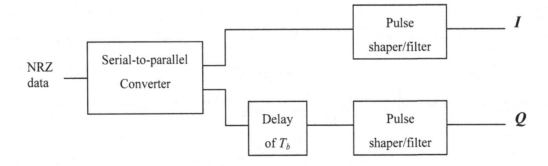

Figure 15. (a) Rectangular pulse shape (b) half-sine pulse shape

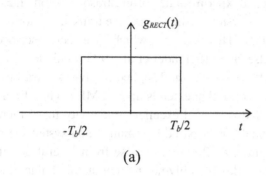

(a)

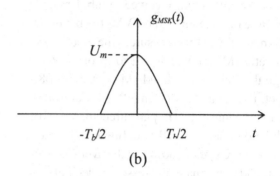

(b)

Mathematically, the I-Q representation of MSK is given by

$$S_{MSK}(t) = \frac{2E_b}{T_b}\begin{bmatrix} b_{In}u_I(t)\cos 2\pi f_c t \\ -b_{Qn}u_Q(t)\sin 2\pi f_c t \end{bmatrix},$$

$$nT_b \le t \le (n+1)T_b \tag{27}$$

where

$$b_I(t) = \cos\frac{\pi t}{2T_b},$$

$$b_Q(t) = \sin\frac{\pi t}{2T_b} \tag{28}$$

are the effective I and Q pulse shapes, and b_{In} and b_{Qn} are the effective I and Q binary data sequences. Note since $b_I(t)$ and $b_Q(t)$ are offset from each other by a time shift of T_b seconds, it might appear that $S_{MSK}(t)$ is indeed in the form of OQPSK with half-sinusoidal pulse shaping.

The minimum shift keying modulation makes the phase change linear and is limited to +-pi/2 over a bit interval T_b. This enables MSK to provide a significant improvement over QPSK. Because of the effect of the linear phase change, the power spectral density has low side lobes that help to control adjacent channel interference. However, the price paid is a wider main lobe than QPSK,

or OQPSK has, and in fact, by a factor of 3/2 if solving the following PSD equations. For QPSK, or OQPSK signal, PSD is given by

$$G_{QPSK}(f) = 4E_b\frac{\sin^2 2\pi fT_b}{(2\pi fT_b)^2}, \tag{29}$$

and for MSK signal, the PSD is given by

$$G_{MSK}(f) = \frac{32E_b}{\pi^2}\frac{\cos^2 2\pi fT_b}{(1-16f^2T_b^2)^2} \tag{30}$$

Note that the error probability of an MSK-based system is the same as that for QPSK.

PERFORMANCE EVALUATION OF MODULATION SCHEMES

There are various factors which are important in a particular modulation scheme, and which help to measure its performance. The Error Vector Magnitude (EVM) is one of such factors (Shafik et al. 2007). This is the magnitude of the vector drawn between the ideal symbol position of the constellation, or hard decision, and the measured symbol position, or soft decision. It is mathematically given by

$$EVM = \frac{E_{RMS}}{S_{MAX}} \times 100\%, \tag{31}$$

where (E_{RMS}) is the RMS error magnitude, and S_{MAX} is the maximum received symbol magnitude. Note that the lower the EVM, the better the performance. Another measure is the modulation error ratio (MER), which is a form of SNR for a digital complex baseband (Mao et al. 2008). In fact, the terms "SNR" and "MER" are often used interchangeably. They are direct measures of modulation quality. MER is defined as the ratio of the average symbol power to the average error power, and is normally expressed in decibels, as follows:

$$MER(dB) = 10\log_{10}\left[\frac{\sum_{j=1}^{N}(I_j^2 + Q_j^2)}{\sum_{j=1}^{N}(\delta I_j^2 + \delta Q_j^2)}\right], \quad (32)$$

where I_j and Q_j are the real (in-phase) and imaginary (quadrature) parts of each sampled ideal target symbol vector, and δl and δQ are the real (in-phase) and imaginary (quadrature) parts of each modulation error vector. In effect, MER is a measure of how fuzzy the symbol points of a constellation are. Ideally, MER should have a value as high as possible.

The above quadrature methods can be investigated experimentally for an ultrasonic communication system, using capacitive transducers for use in air. These devices, which have been described elsewhere (Schindel et al. 1993; Schindel et al. 1995), have been designed to give an ultrasonic response that extends up to 2 MHz. Their broad bandwidth and excellent sensitivity make them ideal for air-based communication systems (Erqun et al. 2003), and arise from careful design of a flexible polymer membrane in conjunction with a rigid, machined back-plate. Note that the output signal amplitude was typically set at 200V peak-to-peak after amplification, and the received signal amplitude was typically around 5 mV rms at 1.2 metres. The experiment was performed in an indoor laboratory, where room temperature was about 25 °C, and where the relative humidity was around 79%. The recorded background noise level was around 600 µV rms, with negligible air turbulence to influence the signal transmission.

The measured overall response of the communication channel (in terms of amplitude and phase) was required for the simulations of the quadrature approach. As shown in Figure 16(a), the magnitude response peaks at 300 kHz, but has a dip at 880 kHz. The 6 dB bandwidth of the

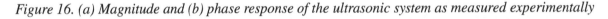

Figure 16. (a) Magnitude and (b) phase response of the ultrasonic system as measured experimentally

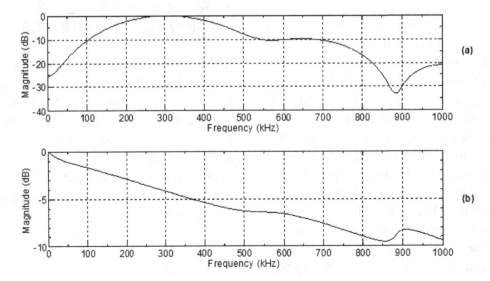

measured channel is about 350 kHz, and the usable frequency range is about 900 kHz. Figure 16(b) shows that the phase response of the channel is roughly linear across the 6 dB bandwidth.

Simulations have been undertaken, using the above experimental response, to determine the expected performance of various RC shaped QPSK signals and the unshaped QPSK signal. It is widely thought that a BER level under 10^{-5} is preferable for a typical wireless data communication system (Hiroshi et al. 2004). The ideal value of E_b/N_o to achieve this was found to be 19 dB. Hence, with an effective bandwidth of 350 kHz, the maximum data rate that could be obtained from the channel is calculated to be 1.3 Mbps. This simulation thus represents the ideal situation, with the correct value of *SNR* being available, and it would be expected that the ultrasonic system would need to approach this value if it was to be useful as a short-range communication system.

A pulse-shaped QPSK signal performs better than one in which the envelope is not shaped in terms of the received signal quality, as shown in Figure 17, when considering the EVM for a given

value of *Eb/No*. This is because the required signal bandwidth to accommodate the pulse can be less than for the unshaped case, thereby reducing the noise bandwidth correspondingly. The comparison is made for a channel simulation in which AWGN was used for the noise model. In this simulation, the *Eb/No* ratio was incremented in 3 dB steps, from 16 to 40 dB. For an example of this improvement, and for a bit-rate of 400 Kbits/s, the required bandwidth to pass the signal was reduced from 400 kHz to 240 kHz, when using a roll-off factor, α, of 0.2. This corresponded to a lower EVM for the pulse shaped approach for values of *Eb/No* up to 28 dB. For higher values of *Eb/No*, and when filtering with higher values of α than 0.2, the EVM became slightly larger. Under higher *Eb/No* values and for the larger values of α, the bandwidth effective occupancy was better and the phase transitions were smoother, leading to improved ISI.

Figure 18 shows the result of applying RC filtering on square pulses with a roll-off factor of 0.2. According to the data, the RC filtered pulses no longer have a constant amplitude, but the phase

Figure 17. Simulation results: comparison of performance with and without pulse shaping

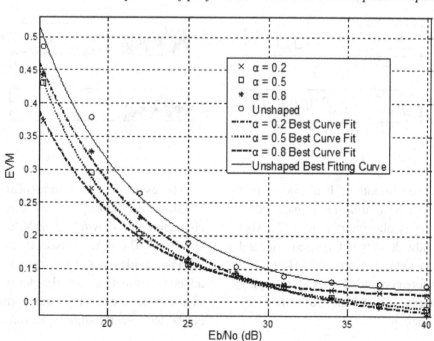

Figure 18. (a) I-channel square pulse and (e) its spectrum; (b) I-channel RC shaped pulse and (f) its spectrum; (c) Q-channel square pulse and (g) its spectrum; (d) Q-channel RC shaped pulse and (h) its spectrum

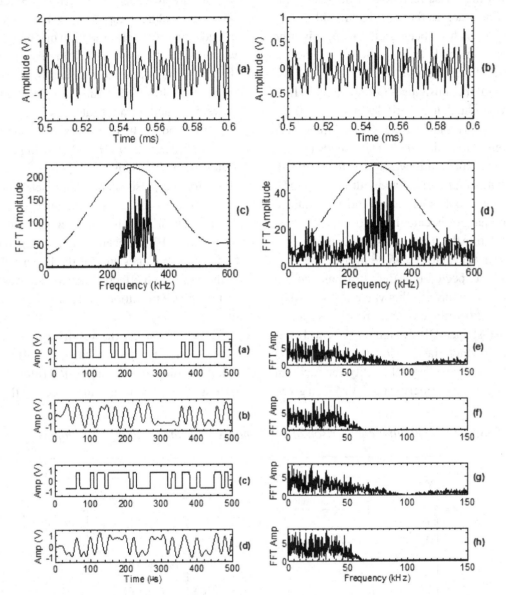

transition has been smoothed. It is evident that the sidelobes have been almost eliminated, thus reducing the inter-symbol interference (ISI). Also the effective bandwidth of the RC filtered baseband signal has been reduced from 200 kHz to 120 kHz. It is therefore concluded that pulse shaping has benefit in terms of both bandwidth efficiency and power efficiency.

The results from simulations of a QPSK scheme are shown in Figure 19, using the experimental characteristics shown earlier. Values of $\alpha = 0.2$ and B = *240 kHz* were used. Results are shown both for an ideal case (no noise), and with noise added to simulate an actual experiment, resulting in a SNR for the received waveform of 10 dB. It can be seen that the signal spectrum is still visible

Figure 19. Simulations of a QPSK ultrasonic communication system after propagation across an air gap of 0.5 m. (a) Transmitted waveform, (b) received waveform with added noise and adjusted for channel response) (c) transmitted spectrum and (d) spectrum of simulated received signal.

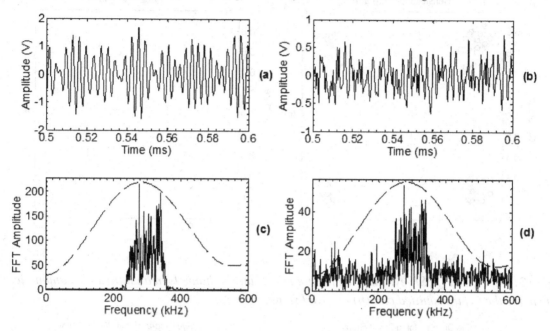

even with the addition of noise. The dashed line indicates the measured channel response at a distance of 1.2 metres.

For the simulations that follow, the baseband signal was chosen to be a 1000-bit binary stream, with a symbol rate R_s of 200 ksps (equivalent to 400 kbps), and the duration of the signal was 2.5 ms. The sampling frequency f_s was chosen to be 10 MHz, *i.e.* 50 samples per symbol. The noise in the channel was assumed to be AWGN, and independent of the signal. Hence, the degrading of received signal quality is mainly due to scattering and absorption.

This is illustrated in more detail by the constellation diagrams, for both transmitted and received signals, in Figure 20. As expected, the received signal demonstrates greater scatter, and a higher EVM value, than originally transmitted. The corresponding eye diagrams for these simulations are shown in Figure 21.

Filtering by the channel response and adding white noise has caused the vertical extent of the central "eye" to be reduced for both channels (an

indication of such factors - a more open "eye" is better). The "eye" is also wider. This means that data can be sampled over more extended time duration, which is also a benefit. Also, on the transmitted signal, distinct traces which cross the horizontal axis are seen, whereas in the simulated received signal these are less distinct – an indication of the presence of noise.

Consider now MSK modulation. Figure 22 shows the I and Q channel NRZ pulses, with an offset of T_b between them. Figure 23 gives the half-sine pulses shape for MSK modulation. The spectrum for the final modulated carrier at 300 kHz are each shown in Figure 24.

The MSK signal has constant envelop and continuous phase. Also, frequencies changes at zero-crossing, and with a minimum phase shifts. It can also be seen that more of the MSK's energy is concentrated toward the band centre, and less in sidelobes, when compared to QPSK or OQPSK. However, for the same data rate of 400 kbps, the main lobe of the MSK spectrum is 1.5 times as wide as that of QPSK and QOPSK.

Figure 20. Simulated QPSK constellation diagrams for (a) the transmitted ultrasonic signal (EVM = 9.1%), and (b) received ultrasonic signal (EVM = 16.4%)

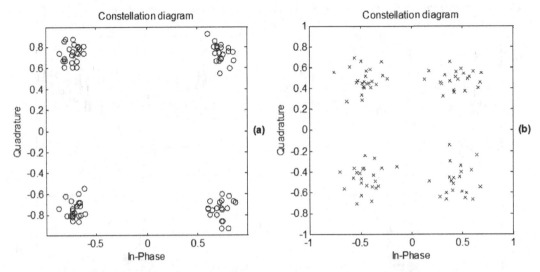

Figure 21. Simulated QPSK eye diagram for both the I and Q channels in an ultrasonic communication system, for the (a) transmitted and (b) received signal

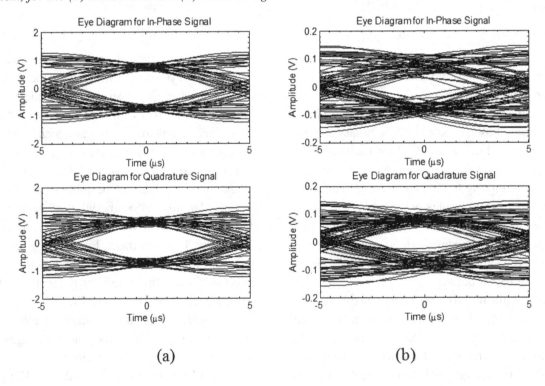

The above simulations indicate that ultrasonic communications based on QPSK signals would be feasible across a distance in excess of 1 m in air. Experiments were thus performed to confirm

that this was the case, and to indicate how the performance was modified by changes in factors such as the roll-off factor (α) of the filter used in the QPSK scheme. The distance between the

Figure 22. I-Q representation of baseband rectangular bipolar pulse

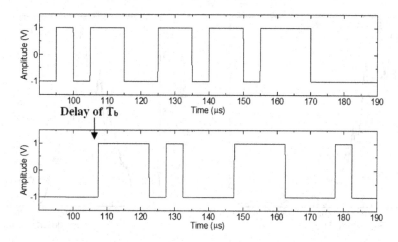

Figure 23. (a) I-Q representation of half-sine shaped pulses in MSK, and (b) final MSK waveforms

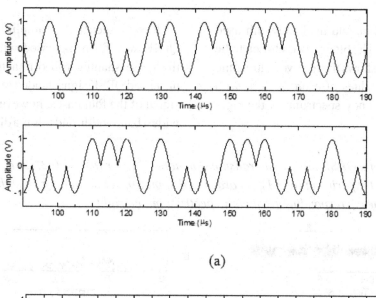

(a)

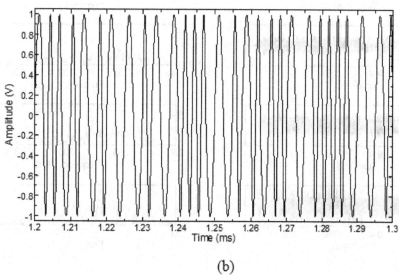

(b)

Figure 24. Simulated spectrum of (a) MSK (b) OQPSK, at 400 kbps

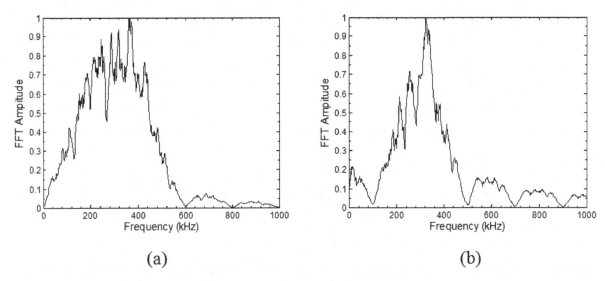

(a)   (b)

transducers was 1.2 m, and the bandwidth used was $120-200$ kHz. The bit rate was chosen to be 200 kbps. Figure 25 shows the received ultrasonic waveform for four values of α on the left, with the equivalent frequency spectrum on the right in each case.

It can be seen that the amplitude of the received QPSK waveform increased with an increase of the pulse shaping roll-off factor. The received unshaped QPSK signal tended to give the strongest signal of the four cases; however, it occupied the widest bandwidth, and this is a disadvantage when

Figure 25. Results of a QPSK ultrasonic transmission across air for (a) $\alpha = 0.2$, (b) $\alpha = 0.5$, (c) $\alpha = 0.8$ and (d) an unshaped experiment ($\alpha = 1$), recorded at distance of 1.2 m. In each case the time waveform is on the left, and the corresponding frequency spectrum on the right.

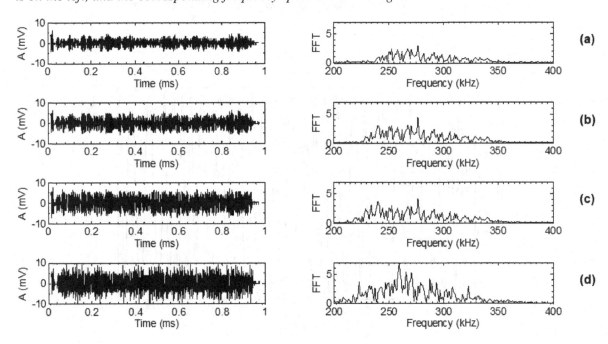

bandwidths are limited in an ultrasonic communication system. It is also clear that in all four spectra, transmitted signals have been 'filtered' by the channel magnitude response, which includes the response of the frequency selective attenuation in air whilst propagating over a relatively long range, such that the higher frequencies are attenuated more than the lower frequencies.

Using the transmitted QPSK as a reference, the experimental performance in terms of MER and EVM defined earlier can be evaluated. Discrete experimental results are presented, together with a best fitting curve to the data in Figure 26, for various values of α. Note that it has been found experimentally that errors start to appear in decoding when the value of EVM is over 0.2 (i.e. 20%); thus, whilst the unshaped response may appear attractive in terms of amplitude, there are other factors to be considered in a real communication system.

This shows that at high E_b/N_o (over 35dB), unshaped QPSK appears to achieve a lower value of EVM, and hence a better performance. However, at low E_b/N_o values (less than 22 dB), a shaped QPSK becomes of more value. It is found in this experiment that an EVM value higher than 0.2 will lead to a severe BER in decoding, which could cause the transmitted information to become unusable. With $\alpha = 0.8$, a reliable communication link could be established when the E_b/N_o is greater than 20 dB. On the other hand, if bandwidth efficiency is the top priority, by setting $\alpha = 0.2$, the channel will not be sufficiently robust unless the E_b/N_o reaches a value of 35 dB. However, with $\alpha = 0.5$, a reasonable compromise between bandwidth occupation and performance can be expected within the range 19dB - 33 dB. It is also evident that the fall in experimental EVM curves is much steeper than those obtained from simulation. This is mainly due to the frequency selective attenuation for ultrasound propagation in air, which will be discussed and modelled in the chapter.

Figures 27-30 show constellation and eye diagrams for four values of α. It can be seen that the linear phase response of the system has kept the QPSK constellation in place, regardless of the unbalance of magnitude response (*i.e.* the centre point of the four constellation groupings remains

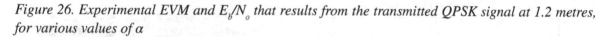

Figure 26. Experimental EVM and E_b/N_o that results from the transmitted QPSK signal at 1.2 metres, for various values of α

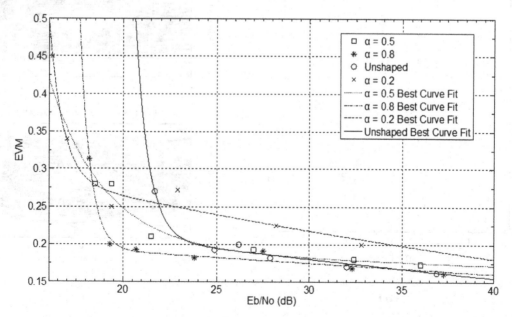

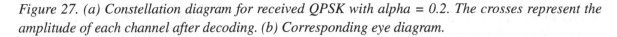

Figure 27. (a) Constellation diagram for received QPSK with alpha = 0.2. The crosses represent the amplitude of each channel after decoding. (b) Corresponding eye diagram.

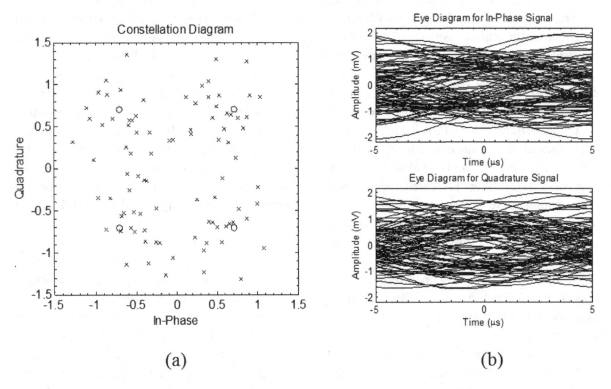

(a)   (b)

Figure 28. (a) constellation diagram and (b) eye diagram for received QPSK with alpha = 0.5

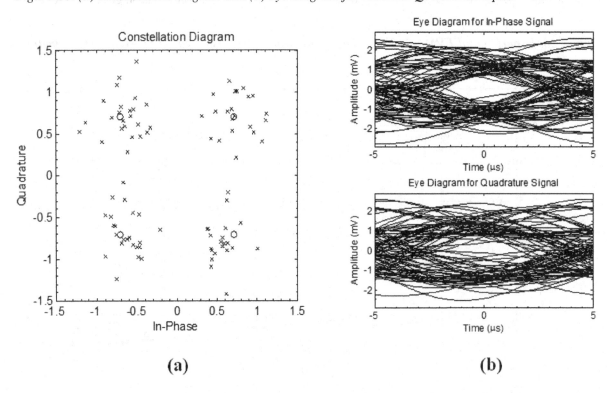

(a)   (b)

Figure 29. (a) constellation diagram and (b) eye diagram for received QPSK with alpha = 0.8.

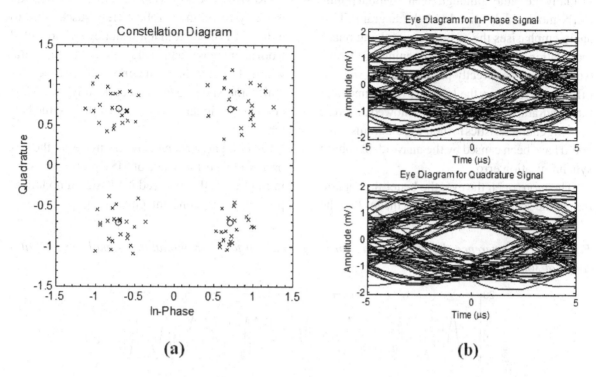

(a)   **(b)**

Figure 30. (a) constellation diagram and (b) eye diagram for received QPSK unshaped

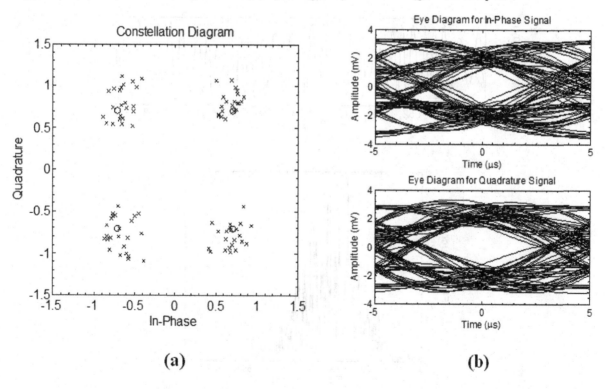

(a)   **(b)**

virtually the same, although the individual points can be more widely spread over the diagram). This again emphasises the advantages of using phase modulation. The level of the opening of the eye was found to be directly proportional to the roll off factor. However, the In-phase (I) channel and the Quadrature (Q) channel have showed different opening characteristics in their eye diagrams. This could have been caused by the unavoidable phase synchronisation error.

The more open the eye, the better the separation in the scatter plot, which also means that the

SNR/MER is better. Hence, signal transmission is likely to be more robust (less susceptible to noise). It is evident that a wider eye has resulted from an increase in the value of α. The horizontal width of the eye diagram represents the time over which the signal can be successfully treated to decode the signal – *i.e.* the wider the eye the better.

A comparison can now be made to the experimental performance of MSK. As can be seen in Figure 31, the received MSK signal no longer preserves the constant envelop property as in

Figure 31. (a) reference (b) received waveform (c) spectrum for MSK modulated signal received at a range of 300 mm

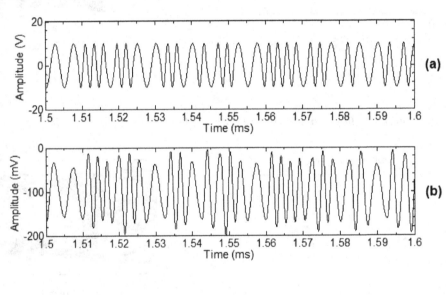

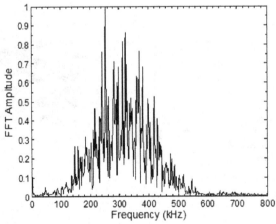

(c)

before transmission. Here, lower frequency tends to have smaller amplitude. But overall, the transition of frequency change occurs at or near zero crossing, and is fairly smooth. The constellation diagrams are shown in Figure 32.

The "minimum phase or frequency shift" has worked well on the capacitive transducer, which has a better reaction on gradually changes rather than sudden changes of phase or frequency. Nevertheless, since MSK has a more energy-concentrated main lob, the band limit of the channel did not have a significant impact on the performance. In addition, the wider opened eye patterns across the *x* axis (longer sampling duration) have proved the benefit of a Frequency Modulation schemes – they are more tolerant on synchronisation errors.

The initial simulations indicate that QPSK modulation would be a good choice for ultrasonic communications in air. Initial experiments identified the frequency response of the ultrasonic system in air in terms of amplitude and phase. This was then used to design the approximate characteristics that would be needed in a QPSK

system for ultrasonic use with the transducers used. Reasonable performance in terms of BER and E_b/N_o was obtained in both simulations and subsequent experiments. The results indicate that a QPSK approach can be used to propagate ultrasonic signals in air over reasonable distances in the 1-2 m range indoors.

The choice of filter seems to have a relatively large effect on performance. This is characterised by the value of α. In most conventional RF communication systems, α tends to be set at a value of around 0.2; this is so that multiple channels can be used over a restricted bandwidth. In the case of ultrasonic communication systems, this was also found to be a useable value. This is because of the limited bandwidth available in practice for operation in air. Thus, whilst the use of no filter might appear attractive in theory, in practice the lack of bandwidth would make this difficult, and a value of $\alpha = 0.2$ is recommended.

The work described within this book chapter was performed over relatively short distances in a controlled laboratory environment. In practice,

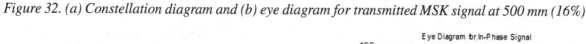

Figure 32. (a) Constellation diagram and (b) eye diagram for transmitted MSK signal at 500 mm (16%)

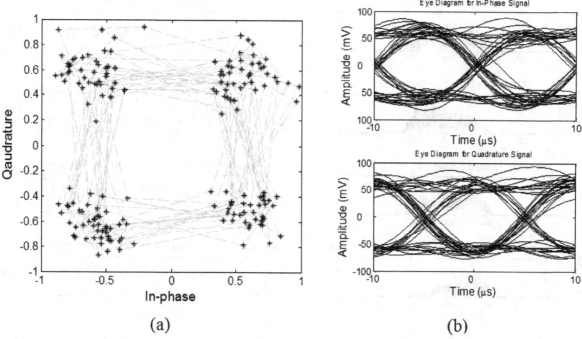

(a) (b)

other factors are likely to influence performance. These include frequency-dependent attenuation in air and the effect of spatial response of the transducer system and multiple reflections. All these factors are discussed in the next section.

RESPONSE OF A REAL SYSTEM WITH QPSK MODULATION

It was seen in the above that frequency-dependant attenuation in air, and the effect of diffraction, will both affect the signal transmitted from an ultrasonic source to a receiver in a communications system. This will thus need to be taken into account. In this illustration, Quadrature Phase Shift Keying (QPSK) has been chosen for the simulation of the expected response, using the known characteristics of the instrumentation and transducers for the model.

Figure 33(a) shows the characteristics of such a modulation scheme, illustrating the spectrum (left) and transmitted waveform that would be used over a signal bandwidth of 240 kHz (chosen to suit the transducers being used in the experiment to follow). Figures 33(b) - (d) then show the predicted effect that the various contributions to the overall channel response are likely to have on such a QPSK signal. The original QPSK signal in Figure 33(a) is firstly shaped by the transducer response, which tends to reduce the amplitude of lower frequencies (Figure 33(b)). Attenuation is then introduced, and this causes a degradation of the upper frequencies (Figure 33(c)). Finally the spatial field response due to diffraction adds further complication to the overall channel response, as can be seen in Figure 33(d). This demonstrates that the effects of geometry and alignment are important in determining the overall frequency response that can be expected from an ultrasonic

Figure 33. Simulation of a filtered QPSK communication channel. (a) Original QPSK waveform (right) and spectrum (left). This is then modified by (b) transducer, (c) transducer + attenuation and (d) transducer + attenuation + spatial field response due to diffraction. The receiver is assumed to be positioned off-axis with z = 200 mm, and x = 25 mm.

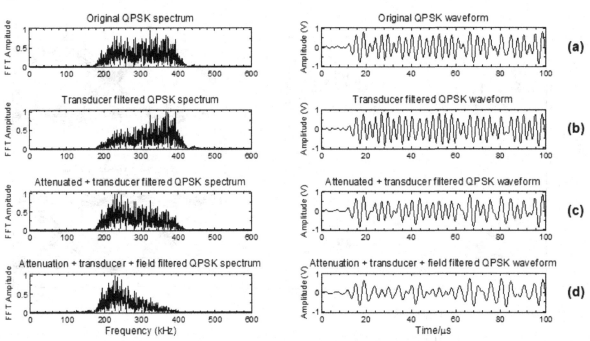

communication system in air, and are in addition to those expected from attenuation and transducer response.

Experiments have been performed to compare with simulations, and to demonstrate that the simulation approach can be used to predict actual experimental results. Consider first the response of the transmitter to an impulsive drive signal, so that in effect the impulse response is being measured. This was the same conditions under which the frequency response of the transducer was measured for input into the simulations above. The results for an axial field position at a range of 200 mm ($z = 200$ mm, $x = 0$ mm) are shown in Figure 34. The frequency response has the expected peak response from 250-400 kHz, and the main bandwidth features of the communication channel have been predicted successfully.

It was now of interest to study the agreement between simulation and experiment for various receiver locations throughout the radiated field of the transmitter, and in particular off-axis. The results are shown in Figure 35, for values of x

ranging from 22.6 mm to 33.9 mm. It can be seen that there is excellent agreement between the predictions of simulation on the left and actual experimental spectra on the right for a QPSK signal with a bandwidth of 240 kHz. In particular, note the position of the minimum in the spectrum, which could be considered as a cut-off frequency for a communication channel. This shifts from approximately 400 kHz down to 280 kHz as the off-axis distance increases, *i.e.* as a greater degree of misalignment takes place.

More detail is shown in Figure 36, where both the waveform and spectrum are presented for a particular geometry. Note the good agreement between experiment and simulation. This agreement is further illustrated in the constellation diagrams of Figure 37. Such diagrams are used to determine the likely performance of a channel, where a better performance would arise from greater separation of the "clusters" when the in phase and quadrature signals of a QPSK modulation signal are plotted against each other. Here, the general nature of the clustering can be seen

Figure 34. Experimentally (a) received pulse, and (b) magnitude response at a point z = 200 mm, x = 0 mm

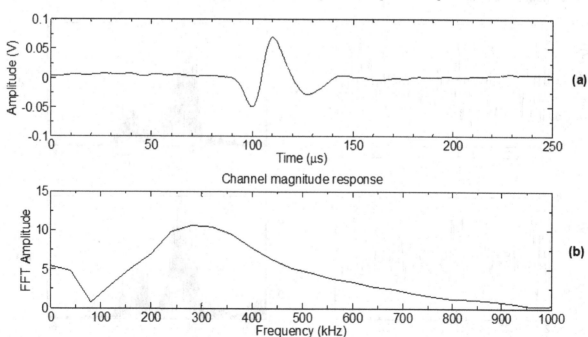

Figure 35. Comparison of QPSK spectra at various off-axis locations at an axial range (z) of 200 mm. The x values of 22.6 mm, 25.4 mm, 28.3 mm, 31.1 mm and 33.9 mm were used for simulating the QPSK spectra on the left for (a) - (e) respectively. The corresponding experimental spectra are shown on the right.

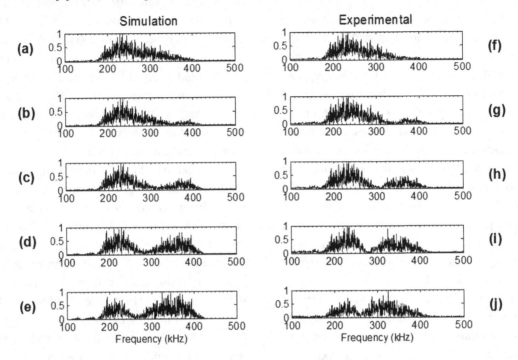

Figure 36. (a) Simulated QPSK signal waveform and its spectrum, in comparison to (b) experimentally recorded waveform and its spectrum, at 200 kbps on a 300 kHz carrier, with a coordination settings of x = 28 mm, z = 200 mm

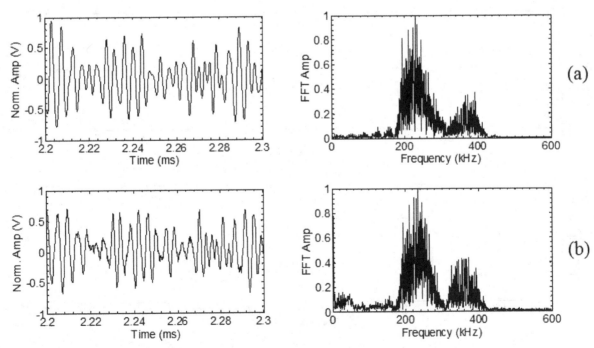

Figure 37. Constellation diagrams for (a) simulation and (b) experiment, at a position x = 28 mm, z = 200 mm

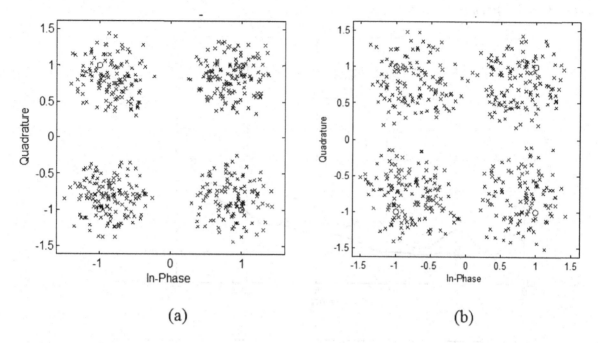

(a)   (b)

to be similar in both cases, underlying the conclusion that the simulations are correctly predicting the performance of the communication channel.

A final variable of interest is the Error Vector Magnitude (EVM), which attempts to illustrate the quality of decoding the QPSK signal after detection at the receiver (to obtain the wanted signal message). Here, the lower the EVM, the better the decoding. Simulation and experiment are compared together at axial distances (z) of 200mm and 400 mm respectively, for various distances (x) off-axis, in Figure 38. It can be seen that when z = 400 mm, the EVM values increase in both simulation and experiment as the off-axis distance increases, across the range x = 0 to 30 mm. This occurs because of the interaction of variations in both radiated field amplitudes and attenuation.

However, for an axial distance z = 200 mm (Figure 38(b)), an improvement of EVM was recorded over the off-axis distance (x) values from 27 to 30 mm. This is because the spatial field response at these coordinates has its cut-off fre-

quency in the centre of the QPSK spectrum (around 300 kHz). This type of channel response has somehow improved the overall frequency balancing of the transmitted signal, in a way that both the lower (180 – 300 kHz) and upper frequency bands (300 – 420 kHz) are being band-passed by the same ripple, or response.

DISCUSSION AND CONCLUSION

The above analysis has demonstrated that it is possible to provide theoretical models that simulate the effects that exist when ultrasound is used in air as a carrier signal for communications. Even though propagation distances are small compared to many other techniques such as r.f. and infrared, the nature of ultrasound leads to complications which must be understood. The first of these is the transducer itself, which invariably will have its own frequency response. Once this is known, then two other propagation effects must be analysed – the effects of diffraction, which lead to

Figure 38. Error Vector Magnitude (EVM) of the decoded QPSK signal for various values of x (i.e. off-axis misalignment). A comparison of simulation and experiment is presented at a range of (a) z = 400 mm and (a) z = 400 mm.

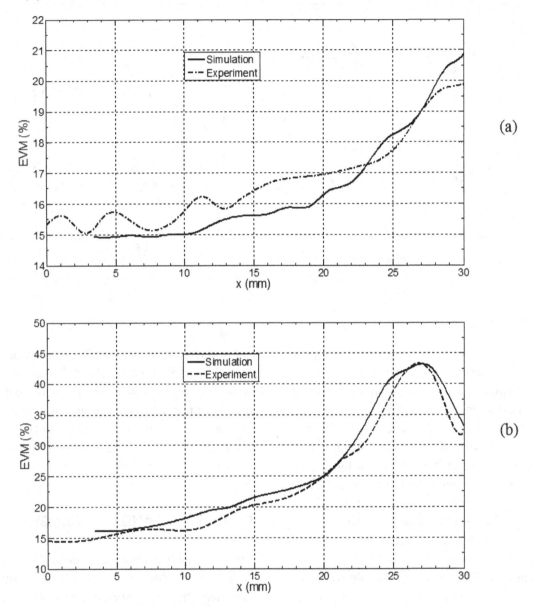

spatial changes in signal amplitude and frequency content, and that of frequency-dependent attenuation. It has been shown that the effects mentioned above can both be modelled and understood. At short ranges, there are large effects due to diffraction, so that off-axis displacement of transmitter and receiver can cause relatively large changes in

frequency response in a single channel. At longer propagation distances, these become less severe, but now attenuation starts to increase and become dominant. Hence, at a particular range, modelling is a useful tool in determining the likely performance of such a system. The results shown here also indicate that decoding is possible, and that predictions of Error Vector Magnitude (EVM) can also be modelled successfully.

It is predicted that short-range ultrasonic communications might have some application in the in-room communication area, in that it could render such a system almost impossible to intercept. In some ways, it is similar to optical free-space communications, but the difference is that attenuation will cause the signal to rapidly reduce in amplitude after a few metres of propagation. Thus, it is highly unlikely that an unwanted interception could take place. In addition, it should be noted that propagation through walls would also be inhibited.

It is thus suggested that ultrasonic communication systems have some promise for secure in-room applications.

REFERENCES

Adler, R., & Desmares, P. (1975). Consumer product applications of ultrasound. *Proceedings of the 1975 IEEE Ultrasonics Symposium*, (p. 553).

Akhavan, K. (2002). High-speed power-efficient indoor wireless infrared communication using code combining – Part I. *IEEE Transactions on Communications*, 50(7), 1098. doi:10.1109/TCOMM.2002.800811

Alagha, N. S., & Kabal, P. (1999). Generalised raised-cosine filters. *IEEE Transactions on Communications*, 47(7), 997. doi:10.1109/26.774849

Amoroso, F. (1976). Pulse and spectrum manipulation in the minimum (frequency) shift keying (MSK) format. *IEEE Transactions on Communications*, 24(3), 381. doi:10.1109/TCOM.1976.1093294

Anderson, M. J. (1995). Broadband electrostatic transducers: modelling and experiments. *The Journal of the Acoustical Society of America*, 97(1), 262. doi:10.1121/1.412310

ANSI. (1978). *Standard S1.26–1978 ASA 23-78, Method for the calculation of the absorption of sound by the atmosphere*. New York, NY: American National Standards Institute.

Ball, D. W. (2007). The electromagnetic spectrum: A history. *Spectroscopy*, 22(3), 14.

Bass, H. E. (1984). Absorption of sound by the atmosphere. *Physical Acoustics*, 17, 145–233.

Bass, H. E., & Bauer, H.-J. (1972). Atmospheric absorption of sound: Analytical expressions. *The Journal of the Acoustical Society of America*, 52(3), 821. doi:10.1121/1.1913183

Bass, H. E., & Shields, F. D. (1977). Absorption of sound in air: High frequency measurements. *The Journal of the Acoustical Society of America*, 62(3), 571. doi:10.1121/1.381576

Carr, H. (1992). Developments in capacitive transducers. *Non-Destructive Testing and Evaluation*, 10(1), 3. doi:10.1080/10589759208952778

Carruthers, J. B., & Kahn, J. M. (1998). Angle diversity for nondirected wireless infrared communication. *IEEE International Conference on Communications*, Vol. 3, (p. 1665).

Chen, H. H., & Wong, S. Y. (1997). Spectral efficiency analysis of new quadrature overlapped modulations over band-limited non-linear channels. *International Journal of Communication Systems*, 10(1), 1. doi:10.1002/(SICI)1099-1131(199701)10:1<1::AID-DAC319>3.0.CO;2-F

Crandall, I. B. (1918). Air-damped vibrating system – Theoretical calibration of the condenser transmitter. *Physical Review*, 11, 449. doi:10.1103/PhysRev.11.449

Erqun, A. S. (2003). Capacitive micromachined ultrasonic transducers: Theory and technology. *Journal of Aerospace Engineering*, 16(2), 76. doi:10.1061/(ASCE)0893-1321(2003)16:2(76)

Fu, W., et al. (2006). Design and performance evaluation of carrier lock detection in digital QPSK receiver. *IEEE International Conference on Communications*, Vol. 7, (p. 2941).

Gallager, R. G. (2001). Claude E. Shannon: A retrospective on his life, work, and impact. *IEEE Transactions on Information Theory, 47*(7), 2681. doi:10.1109/18.959253

Grag, S., et al. (2000). MAC scheduling polices for power optimisation in Bluetooth: A master driven TDD wireless system. *IEEE Vehicular Technology Conference*, Vol. 1, (p. 196).

Gronemeyer, S. A., & McBride, A. L. (1976). MSK and offset QPSK modulation. *IEEE Transactions on Communications, 24*(8), 809. doi:10.1109/TCOM.1976.1093392

Haller, M. I., & Khuri-Yakub, B. T. (1994). Micromachined 1-3 composites for ultrasonic air transducers. *The Review of Scientific Instruments, 65*, 2095. doi:10.1063/1.1145231

Haynes, H. D., et al. (2002). *Ultrasonic communication project phase 1 final report*. National Security Program, Oak Ridge, TN, Office Rep. Y/NSP-252.

Hiroshi, O. (2004). Experiments on acoustic communication with quadrature amplitude modulation in multipath environment. *Japanese Journal of Applied Physics, Part 1: Regular Papers and Short Notes and Review Papers, 43*(5B), 3140. doi:10.1143/JJAP.43.3140

Holm, S. (2005). Airborne ultrasound data communications: The core of an indoor positioning system. *Proceedings IEEE Ultrasonics Symposium*, Vol. 3, (p. 1801).

Katz, E., & Sruber, G. L. (1995). Sequential sequence estimation for trellis-coded modulation on multipath fading ISI channels. *IEEE Transactions on Communications, 43*(12), 2882. doi:10.1109/26.477483

Keiichi, M. (2007). Acoustic communication in air using differential biphase shift keying with influence of impulse response and background noise. *Japanese Journal of Applied Physics, 26*(Part 1), 4541.

Krautkrämer, J., & Krautkrämer, H. (1990). *Ultrasonic testing of materials* (4th ed.). New York, NY: Springer-Verlag.

Kuhl, W. (1954). Condenser transmitters and microphones with solid dielectric for airborne ultrasonics. *Acoustica, 4*(5), 519.

Lawrence, B. D., & Simmons, J. A. (1982). Measurements of atmospheric attenuation at ultrasonic frequencies and the significance for echolocation by bats. *The Journal of the Acoustical Society of America, 71*(3), 585. doi:10.1121/1.387529

Lee, T. C., & Chen, C. C. (2006). A missed-signal GFSK demodulator for Bluetooth. *IEEE Transactions on Circuits and Wystems. II, Express Briefs, 53*(3), 197. doi:10.1109/TCSII.2005.858320

Leung, P. S. K., & Feher, K. (1993). F-QPSK: Superior modulation technique for mobile and personal communications. *IEEE Transactions on Broadcasting, 39*(2), 288. doi:10.1109/11.218001

Li, C. (2008). Short-range ultrasonic digital communication in air. *IEEE Transactions in Ultrasonics, Ferroelectronics, and Frequency Control, 55*, 908. doi:10.1109/TUFFC.2008.726

Lueftner, T. (2003). Edge-position modulation for high-speed wireless infrared communications. *IEE Proceedings. Optoelectronics, 150*(5), 427. doi:10.1049/ip-opt:20030960

Mao, S., et al. (2008). Modelling and measuring of weak frequency selectivity in OFDM modulator. *2008 4th IEEE International Conference on Circuits and Systems for Communications, ICCSC*, (p. 416).

Matsuzawa, K. (1958). Condenser microphones with plastic diaphragms for airborne ultrasonics I. *Journal of the Physical Society of Japan, 13,* 1533. doi:10.1143/JPSJ.13.1533

Matsuzawa, K. (1960). Condenser microphones with plastic diaphragms for airborne ultrasonics II. *Journal of the Physical Society of Japan, 15,* 167. doi:10.1143/JPSJ.15.167

McLachlan, N. W. (1934). *Loudspeakers.* Oxford, UK: Clarendon Press.

Moreia, A. J. C. (1996). Performance of infrared transmission systems under ambient light interference. *IEE Proceedings. Optoelectronics, 143*(6), 339. doi:10.1049/ip-opt:19960696

Narasimhan, R. (1996). Effect of electronic-ballast fluorescent lighting on wireless infrared links. *IEE Proceedings. Optoelectronics, 143*(6), 347. doi:10.1049/ip-opt:19960877

Nelson, T. (2008). Near optimal common detection techniques for shaped offset QPSK and Feher's QPSK. *IEEE Transactions on Communications, 56*(5), 724. doi:10.1109/TCOMM.2008.060155

Norris, J. A., & Nieto, J. W. (2008). Evaluation of a novel constant envelope spread-spectrum modulation technique. *Proceedings IEEE Military Communications Conference MILCOM 2008,* (p. 1).

Ochi, H., & Fukuchi, T. (2007). Development of titled Toroidal beam wideband transducer using quadrature phase shift keying for underwater acoustic communication. *Japanese Journal of Applied Physics, 46*(7), 4961. doi:10.1143/JJAP.46.4961

Otnes, R., & Eqqen, T. H. (2008). Underwater acoustic communications: Long-term test of turbo equalisation in shallow water. *IEEE Journal of Oceanic Engineering, 33*(3), 321. doi:10.1109/JOE.2008.925893

Paajanen, M. (2000). Electromechanically film (EMFi) – A new multipurpose electret material. *Sensors and Actuators, 84,* 95. doi:10.1016/S0924-4247(99)00269-1

Pasupathy, S. (1979). Minimum shift keying: A spectrally efficient modulation. *IEEE Communications Magazine, 17*(4), 14. doi:10.1109/MCOM.1979.1089999

Saarnisaari, H. (2007). A general receiver and constant envelope direct sequence signals. *Proceedings IEEE Military Communications Conference MILCOM 2007,* (p. 1).

Sabatini, A. M. (1997). Stochastic model of the time-of-flight noise in airborne sonar ranging systems. *IEEE Transactions on Ultrasonics, Ferroelectrics, and Frequency Control, 44*(3), 606. doi:10.1109/58.658313

Saha, D., & Birdsall, T. (1989). Quadrature-quadrature phase-shift keying. *IEEE Transactions on Communications, 37*(5), 437. doi:10.1109/26.24595

Sairam, S. (2002). Bluetooth in wireless communication. *IEEE Communications Magazine, 40*(6), 90. doi:10.1109/MCOM.2002.1007414

Schindel, D., et al. (1993). Micromachined capacitance transducers for air-borne ultrasonics. *Proceedings of the Ultrasonics International Conference,* (p. 675).

Schindel, D. W. (1995). The design and characterization of micromachined air-coupled capacitance transducers. *IEEE Transactions on Ultrasonics, Ferroelectrics, and Frequency Control, 42,* 42. doi:10.1109/58.368314

Sell, H. (1937). Eine neue Methode zur Umwandlung mechanischer Schwingungen in elektrische und umgekehrt. *Biomedical Engineering, 18,* 3.

Shafik, R. A., et al. (2007). On the extended relationships among EVM, BER and SNR as performance metrics. *Proceedings of 4ᵗʰ International Conference on Electrical and Computer Engineering, (ICECE 2006)*, (p. 408).

Simon, M. K., & Polydoros, A. (1981). Coherent detection of frequency-hopped quadrature modulations in the presence of jamming – 1. QPSK and QASK modulations. *IEEE Transactions on Communications*, *29*(11), 1644. doi:10.1109/TCOM.1981.1094924

Streicher, A., et al. (2003). Broadband ultrasonic transducer for an artificial bat head. *Proceedings of the IEEE Ultrasonics Symposium*, Vol. 2, (p. 1364).

Streicher, A., et al. (2005). Broadband EMFi ultrasonic transducer for bat research. *Proceedings of the IEEE Ultrasonics Symposium*, Vol. 3, (p. 1629).

Sun, L., et al. (1997). Modelling and optimisation of micromachined air-coupled capacitance transducers. *IEEE Proceedings Ultrasonics Symposium*, Vol. 2, (p. 979).

Sundberg, C. (1986). Continnous phase modulation. *IEEE Communications Magazine*, *24*(4), 25. doi:10.1109/MCOM.1986.1093063

Tuncer, T. E. (2002). ISI-free pulse shaping filters for receivers with or without a matched filter. *Proceedings IEEE International Conference on Acoustics, Speech and Signal Processing*, Vol. 3, (p. 2269).

Wang, J., & Yongacoglu, A. (1994). Performance of trellis coded-8PSK with cochannel interference. *IEEE Transactions on Communications*, *42*(1), 6. doi:10.1109/26.275292

Wente, E. C. (1917). A condenser transmitter as a uniformly sensitive instrument for the absolute measurement of sound intensity. *Physical Review*, *10*, 39. doi:10.1103/PhysRev.10.39

Wente, E. C. (1922). Sensitivity and precision of the electrostatic transmitter for measuring sound intensities. *Physical Review*, *19*, 498. doi:10.1103/PhysRev.19.498

Wente, E. C. (1922). The thermophone. *Physical Review*, *19*, 333. doi:10.1103/PhysRev.19.333

Wikins, G. D. (1996). Eye-safe free-space laser communications. *IEEE Proceedings of the National Aerospacce and Electronics Conference*, Vol. 2, (p. 710).

Williams, S. K. (2000). IrDA: Past, present and future. *IEEE Personal Communications*, *7*(1), 11. doi:10.1109/98.824566

Wong, G. S. K., & Embleton, A. F. W. (1995). *AIP handbook of condenser microphones: Theory, calibration and measurements*. New York, NY: AIP Press. doi:10.1121/1.413754

Zurek, R. A., et al. (2002). *Omnidirectional ultrasonic communication system*. (U.S. Patent 6363139).

Chapter 15
Surround Sensing for Automotive Driver Assistance Systems

Martin Stämpfle
Esslingen University of Applied Sciences, Germany

ABSTRACT

In recent years, driver assistance systems have become a strong trend in automotive engineering. Such systems increase safety and comfort by supporting the driver in critical or stressful traffic situations. A great variety of surround sensors with different fields of view include radar, ultrasonic, laser, and vision systems. These sensors are based on different technologies and measurement principles. They all have their specific advantages and disadvantages and range from low-cost to high-end systems. They also differ in size, mounting position, maintenance, and weather compatibility. Hence, such sensors are used in various configurations to explore the surroundings ahead, sideways, and behind a vehicle. In addition, vehicle dynamics information from speed, steering angle, yaw rate, and acceleration sensors is available. Data fusion algorithms on raw data, feature, or object levels are used to collect all this information and set up vehicle surround models. An important issue in this context is the question of data accuracy and reliability. Situation interpretation of the traffic scene is based on these surround models. Any situation interpretation has to be performed in real-time, independent of the situation complexity. Typically, the prediction horizon is a couple of seconds. Depending on the results of the driving environment analysis critical situations can be identified. In consequence, the driver can be informed or warned. Some driver assistance systems already perform driving tasks like following, lane changing, or parking autonomously. The art of designing new, valuable driver assistance systems includes many factors and aspects and is still an engineering challenge in automotive research.

DOI: 10.4018/978-1-4666-2976-9.ch015

INTRODUCTION

In recent years driver assistance systems have moved into the focus of intensive research activities in the automotive sector. This contribution tries to give an overview on the many topics of sensors and surround models for automotive driver assistance systems. Safety and comfort systems shall disburden the driver from many driving tasks in both critical situations and routine jobs. Safety systems cover all aspects of save driving: They inform, warn, and intervene. Such systems all have in common, that they predict the traffic situation only for a very short period of at most a few seconds. Comfort systems disburden the driver and make driving less stressful. All assistance systems are based on information about the vehicle and the vehicle surroundings. A comprehensive presentation of many topics related to automotive assistance systems can be found in Winner (Winner 2009).

SURROUND SENSORS

Machine Perception

Surround sensors are among the key innovations that have made driver assistance systems possible at all. Most assistance systems need information not only about the vehicle itself but also about the vehicle environment. Surround sensors are no traditional automotive sensors. They have been developed intensively in the last two decades. Many different physical measurement principles including electromagnetic waves, ultrasonic waves, image perception, and laser have been considered and tested. The general key questions in the context of machine perception are:

- Where is something?
- Where is nothing?
- What is the dynamics of the something?
- What kind of thing is detected?

At first glance, these questions look quite general and simple. Especially the second question sounds superfluous. Emergency evading systems need reliable information of sufficient free evading space. This issue is addressed by the second question. Although much research on automotive surround sensors has been carried out giving answers to these questions completely and precisely still is an engineering challenge. Assistance systems requirements can be conflicting. Safety systems need reliable and stable data with little preprocessing whereas comfort systems prefer current data that is prepared in an appropriate way. Model assumptions on the dynamics of a detected object can be generated by classifying the object and by tracking the object over several system cycles. Objects in standstill cannot achieve high velocity in the following time cycle.

Radar Sensors

A radar sensor emits electromagnetic waves via an antenna. Such waves are reflected by an object and are received again by the sensor. Radar echoes are created by electrically conductive material. Most vehicles today are at least partly built of such conductive material. The time delay Δt between sending and receiving directly yields to (twice) the distance d:

$$d = \frac{c \Delta t}{2},$$

where c denotes the speed of light. One big advantage of radar sensors is the ability to directly measure distance as well as velocity. Distance can be computed by the time delay between sending and receiving. Based on the Doppler effect velocity measurements result from a frequency shift:

$$v_{rel} = -\frac{c}{2} \frac{f_D}{f_C}$$

Here, f_C is the frequency of the carrier, and f_D is the frequency shift. Measuring the time delay directly is costly. Therefore, in most cases a frequency modulated continuous wave (FMCW) variant is preferred. Instead of using sending and receiving time the sending and receiving frequency is compared. As a disadvantage of this approach the frequency difference results from time delay as well as from the Doppler effect. The results of distance and speed are not unique but are linearly dependent of each other. This drawback is compensated by using several different FMCW cycles. Automotive radar sensors operate at 24 GHz or 77 GHz. The greater wave length of the 24 GHz technology makes this frequency best suited for mid-range sensors. Most long-range sensors operate at 77 GHz. For a more detailed description see Bosch (Bosch 2004) or Winner (Winner 2009). Figure 1 shows radar sensors for long-range and mid-range applications from Bosch and Continental.

In a first step, radar signal data processing begins with a Fourier transform based spectral analysis. In a second step, new objects are created from detection peaks. Neighboring peaks are clustered together. In a third step, these new objects are matched with already existing objects. Some sensors include additional tracking and filtering

steps. Finally, a list of object data is generated and forwarded to the functional part of the assistance system or put on the sensor bus.

Typical data update rates for radar sensors are around 10 Hz. Some sensors have fluctuating cycle rates depending on the situation complexity and the number of measured peaks. Physical measurements are composed of sending and receiving signals and need not much execution time. The most time-consuming part comes from signal processing within the sensor hardware. The signal processing varies with the amount of measurement data.

The horizontal field of view of a long-range radar typically is about 20 ° with a maximum distance of 200 m. The elevation is small at 5 °. Mid-range sensors cover a range from 1 m to 100 m. In contrast to long-range sensors, mid-range sensors have a broader field of view of about 40°. With multiple beams a horizontal angle resolution is also possible. The ability of lateral object separation is a main quality feature. Some sensors also include mechanical scanning.

Radar sensors have the great advantage of being available in almost any weather condition. An integrated heating prevents the sensor lenses from being covered with snow and ice. However, wet conditions can decrease the radar performance. If water irregularly covers the lenses, the

Figure 1. Automotive long-range and mid-range radar sensor (courtesy of Bosch and Continental)

angular resolution goes down. Moreover, heavy rain decreases the distance range and enables clutter. There is an ongoing discussion about the disadvantages of sensor constraints caused by adverse weather.

The first vehicle with an automotive radar sensor was available in 1998. The sensor was used in an adaptive cruise control system, a system that drastically extends the functionality of a traditional cruise control system. Nowadays, various radar sensors are also used in collision avoidance and collision mitigation systems. Some research projects also discuss the usage of radar sensors in the rear bumpers to monitor the traffic from behind. Moreover, lane aid systems use short-range and mid-range radar sensors at the vehicle sides to supervise the traffic on neighboring lanes.

Ultrasonic Sensors

The main component of automotive ultrasonic sensors typically is a piezoelectric ceramics on a metallic membrane. Such metallic membranes are necessary since in the air acoustic waves require high amplitudes. The distance measurement is based on the time of flight of ultrasonic wave impulses between sending and receiving. The dominant influence on the sensor measurement precision is the temperature of air. Compared with

this factor, most other disturbance parameters can be neglected.

The typical horizontal field of view is around 120 ° for an abject with distance of about 1 m. The vertical field of view has to be designed as small as possible to avoid reflections from the road surface. The typical vertical field of view is around 60 ° which is half the horizontal field of view. For a more detailed description see Bosch (Bosch 2004) or Winner (Winner 2009). Automotive ultrasonic sensors from Bosch and Valeo are shown in Figure 2. Ultrasonic sensors are designed with analog or digital data bus interfaces.

Automotive ultrasonic sensors have a long history of success. The start of automotive series production dates back in the year 1992. Four factors mostly contribute to that success: First, the sensor is quite independent of weather conditions. The only essential limitation results from being covered with snow or ice. Second, the sensor is unrivaled cheap. Third, the sensor is small and light. Ultrasonic sensors can be integrated easily in the shape of front and rear bumpers. Its size allows the mounting at almost any position; the visible part of the sensor can be varnished. And last, the detection quality is, to a large extend, independent of the kind of obstacle. Most hard materials like metal, brick, wood, or plastics have good reflexing properties. Material like light-weight foam plastic absorbs ultrasonic waves. But

Figure 2. Automotive ultrasonic sensors (courtesy of Bosch and Valeo)

such objects are in a distinct minority and thus play little importance in traffic surroundings.

Traditional parking systems use 8 to 12 sensors in the vehicle front and back. More advanced systems also include sensors in the vehicle sides to obtain a continuous 360° field of view in the near range up to 2 m around the vehicle. With such sensor arrays, the length of parking space can be measured by driving along the parking slot. Side view assist systems make use of the detection of other vehicles in the blind spot. Ultrasonic sensors detect vehicles with small relative velocity on neighboring lanes. Furthermore, there is an increasing demand for systems that assist complex parking maneuvers. A 360° sensor belt can estimate the available space in narrow car park environments with walls, pillars, and protrusive bases.

With two ultrasonic sensors the distance to a point-shaped object can be calculated via triangulation. The two sensors and the object form a triangle. Let a be the base distance between the two sensors and let d_1 and d_2 denote the measured distance of sensor 1 and 2. Then, the line through the object orthogonal to the sensor base divides a into parts a_1 and a_2, see Figure 3. Using elementary calculus, three equations with three unknowns can be set up:

$$d_1^2 = d^2 + a_1^2,$$

$$d_2^2 = d^2 + a_2^2,$$

$$a_1 + a_2 = a$$

The first two equations can be resolved with respect to d^2 and combined in one single equation. This gives the symmetric relation

$$d_1^2 - d_2^2 = a_1^2 - a_2^2$$

Finally, the distance d from the sensor base to the object is given by

$$d = \sqrt{d_1^2 - \frac{\left(d_1^2 - d_2^2 + a^2\right)^2}{4a^2}} = \sqrt{d_2^2 - \frac{\left(d_2^2 - d_1^2 + a^2\right)^2}{4a^2}}$$

Both formulas for distance d are valid since the problem is symmetric in the two sensors. This principle of triangulation can be extended to more than two sensors with overlapping fields of view. However, inaccurate measurements produced by different reflection spots then can yield inconsistent results. In theory, a point-shaped object detected by three sensors should generate three circle arcs that intersect in one single point. In practice, in this situation a small triangle with vertices being the pairwise intersections is created.

Another principle of computing the distance to an object with two ultrasonic sensors is called trilateration. In this alternative approach, both sensors receive not only their own direct signal but also the indirect cross signal from the other sensor. Depending on the time difference of the two received signals the distance also can be estimated. This principle can be expanded to more than two sensors as well.

Figure 3. Principle of triangulation to compute an object's distance

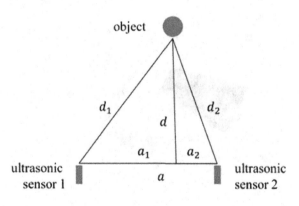

Vision Sensors

In the first generation of automotive vision systems CCD technology has been used. This technology is cheap and integrated in many consumer products. The second generation of automotive vision systems is based on CMOS vision chips. The dynamical range of such sensors is wider in comparison with CCD chips. Typical cycle rates of the imager are around 25 Hz. Up to now most automotive imagers are grayscale chips. The use of color sensors mostly is limited to research projects. Figure 4 shows two typical automotive vision sensors from Bosch and Continental.

There also has been much discussion on whether to favor mono or stereo vision systems. On one hand, apart from additional costs for a second imager, the stereo variant requires much more computational resources compared to the mono variant. Moreover, to get full benefit of the stereo alternative a reasonable camera base width is necessary. This is a design challenge when integrating the hardware into the vehicle. On the other hand, 3D information can be obtained using disparity estimation. This enables to compute depth information of objects.

Some vision sensors already have on-chip integrated preprocessing. Wide-angle lenses often require rectifying the images. This time-consuming signal processing efficiently is accomplished with appropriate hardware. Another issue is the dynamical range of the image. Different lighting conditions require rescaling the brightness and contrast image parameters. Video sensors have to work correctly when driving in the direction of the sun in the morning and evening or when driving through tunnels with low ambient light and bright head lamps.

On a higher level of abstraction more application relevant image processing algorithms are necessary. There is a great variety of methods and filters to detect edges, corners, or areas with constant grayscale value. The images are partitioned or reduced to one bit information patterns. Image filters can be classified into three different groups. Point operators process each pixel value independently of all other pixel values. Local operators adjust the pixel value in dependence of some neighboring values. Using global operators the new value of a pixel depends on the whole old image.

On the top level of image processing, objects are clustered from image features. Vehicles can be represented by 2D or even 3D rectangular boxes. Object detection algorithms include assumptions that hold for real world objects. Vehicles mostly

Figure 4. Automotive vision sensors (courtesy of Bosch and Continental)

show symmetry along the longitudinal vehicle axis. The license plate typically is somewhere in the middle of the rear. The gap between vehicle and road typically is dark because of the vehicle's shadow.

Pedestrians can be detected by other features. A moving pedestrian can be found by searching for leg movements. If arm motion is also detected, this can increase the object's plausibility. The longer a pedestrian can be observed, the more reliable the classification hypothesis becomes. In this context, the image resolution is crucial. Pedestrians cover only a very small part of an image. In consequence, only few pixels contribute to the shape of a human person.

Digital images can be compressed and processed using the discrete cosine transform. For this, an image is partitioned into small blocks. A special compression uses $m = 8$, which means that the image is partitioned into 8×8 pixel blocks. If the pixel size of the image is 320×200, then 40×25 blocks are transformed. Other choices for m, preferable powers of 2, are also possible. For each block, the cosine coefficients

are computed. Let A be an $m \times m$ image block matrix. Then, the discrete cosine transform is defined as

$$B = C^T A C,$$

$$C = \left(\sigma_k \cos \frac{(2j+1)k\pi}{2m} \right)_{j,k=0,...,m-1},$$

$$\sigma_k = \begin{cases} \sqrt{\dfrac{1}{m}} & if \ k = 0 \\ \sqrt{\dfrac{2}{m}} & if \ k \neq 0 \end{cases}$$

Matrix B consists of the m^2 cosine coefficients. Figure 5 shows 320×200 pixel images of a highway scene. The upper left is the original image. The other three images result from a discrete cosine transform using 4, 16, and 36 of 64 possible cosine coefficients. The rates of compression result from the percentage of data after the transform: 6.25%, 25%, and 56.25%. The cosine

Figure 5. Highway scene: Original image (upper left) and compressed images using the discrete cosine transform with 4, 16, and 36 of 64 possible coefficients

transform is like the Sobel or Scharr transform a local operator, whereas other transformations like the Hough transform operate global.

Lidar Sensors

A Lidar sensor is based on laser scanner technology and distance measurement. The sensor emits laser pulses and receives the reflected light. The distance d is calculated from the time delay Δt using the speed of light c:

$$d = \frac{c\Delta t}{2}$$

Due to the exact distance measurement, velocity can be approximated using numerical differentiation and filtering. Some sensors operate with a fixed multibeam arrangement. Other lidar sensors include mechanical moving or rotating parts. The principle of rotating parts in automotive applications is discussed controversial among experts. Figure 6 shows two lidar scanners from Ibeo and Velodyne.

Since the emitted laser light is not visible to humans, sensor safety is a big issue. Automotive lidar sensors comply with the laser class 1 definition. They are designed to be eye-safe in any situation. The wave length is between 850 nm to 1000 nm. The pulse peak power output lies between 10 W and 50 W, and the typical average power output is between 1 mW and 5 mW. This implies a large ratio of pulses and pauses.

Many lidar sensors have large fields of view up to 360° in horizontal direction with high angle resolution. The vertical field of view is quite different. Some sensors scan on several levels. Thus, three-dimensional high resolution data is generated. This implies that the amount of raw data can be immense with lidar sensors. The bandwidth of the CAN bus is by far not sufficient to handle uncompressed lidar raw data. Typical cycle update rates are around 10 Hz. High data rate makes these sensors ideal for demanding perception applications in 2D and 3D applications.

Lidar sensors are designed for obstacle detection and navigation of autonomous ground vehicles. As a big advantage these sensors capture the shape of objects in a natural way. The width is an important object property. Compared with radar sensors, the laser alternative delivers this object feature quite accurately. In sensor configuration comparisons often the combination of radar and vision is discussed versus the usage of a single lidar sensor. Lidar sensors are multi-purpose sensors in many aspects. However, one big disadvantage is the production costs. This fact has inhibited mass production until now. Today, many lidar systems serve as reference systems to obtain ground truth information during the development process.

Figure 6. Automotive lidar scanners (courtesy of Ibeo and Velodyne)

Sensor Configurations

The number of sensors and their geometric arrangement define the total field of view. Figure 7 shows a typical sensor arrangement. According to the function requirements and the principles of redundancy, complementary, and cooperation, the sensor types, numbers, and positions are selected and configured. A standard sensor configuration includes radar for far distances ahead, a vision sensor with a broad field of view in the medium range and some ultrasonic sensors for near distances. The rear region can be covered with a second vision system. From an economic point of view there is always the trade-off between few multi-purpose sensors and a larger number of low-cost single utilization sensors. It is common, to divide the overall field of view into three different ranges.

The short-range field of view typically covers the distance up to a few meters. The distance measurement precision is relatively high. Parking systems need distance information with centimeter precision. All short-range sensors establish a sensor belt around the vehicle. High-end driver assistance systems include a sensor belt that covers the full angle spectrum of 360 °. Hence, front,

side, and rear traffic environment can be observed. Ultrasonic sensors are the most common short-range sensors.

The mid-range varies significantly among different sensor types. A typical distance range begins at 1m and stretches up to 50 m or 100 m. Most mid-range sensors are mounted in front or rear positions to observe preceding and following traffic. In the last years, lane systems also include mid-range sensors at side positions to cover traffic on neighboring lanes. In general, the horizontal field of view of mid-range sensors is smaller than that of short-range sensors. Vision systems are classical mid-range sensors.

The long-range field of view is the domain of radar sensors. Automotive long-range sensors have a detection distance of up to 250 m. Long-range sensors have small horizontal viewing angles and are mounted in front position in most cases. The problem of small horizontal viewing angles is the tracking of detected objects through bended curves. The vertical field of view is small likewise. This could cause problems if the driver initiates hard braking. The vehicle's pitch angle increases and the sensor looks down on the road. As a consequence, the target object might be lost.

Figure 7. Vehicle surround sensing. A variety of sensors with different fields of view cover the vehicle surroundings (courtesy of Bosch)

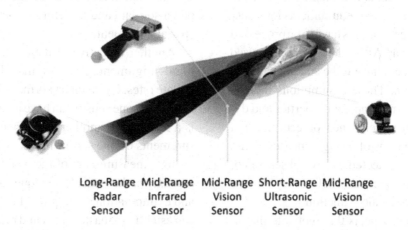

| Long-Range Radar Sensor | Mid-Range Infrared Sensor | Mid-Range Vision Sensor | Short-Range Ultrasonic Sensor | Mid-Range Vision Sensor |

SENSOR DATA FUSION

Data Fusion and Information Fusion

Data fusion and information fusion formerly has been developed for military applications (Waltz et al. 1990). Nowadays, it is a prosperous field of research with many applications in automation engineering, robotics, and automotive engineering (Hall et al. 1992, 1997, 2001). Detailed information on this topic can be found in Bar-Shalom (Bar-Shalom et al. 1988, 1990, 1992, 1993, 2000) or Blackmann-Popoli (Blackmann-Popoli 1999). Basic fusion techniques have been adapted to automotive systems (Becker 2002, Becker et al. 2004, and Niehsen 2005). Many data fusion concepts with special sensor configurations have been discussed (Amditis et al. 2005, Beauvais et al. 2000, and Fang 2002). Moreover, many contributions refer to specific environment objects (Klotz et al. 2004 and Kojima 2002). There are three key quality features of data fusion algorithms:

- Simplicity
- Transparency
- Reliability

These fusion axioms sound short and trivial. But as in many other disciplines, hard work is necessary to find clear and aesthetic solutions. In addition, there are special requirements for safety systems. All data estimations have to be processed in a conservative way. All model assumptions used in the fusion process have to be stated clearly in the documentation. These assumptions include any aspects concerning object properties and dynamics. For safety systems, new objects must not be created on the basis of a single measurement. Objects have to be detected by several sensors or several times by a single sensor before they become relevant for the assistance function.

There are many aspects that motivate the integration of sensor data fusion and information fusion techniques into driver assistance systems.

The term data fusion is used if raw or preprocessed measurement data is fused together, whereas the term information fusion is more general. Information sources like digital maps and car-to-car or car-to-infrastructure communication are no sensors in the traditional meaning. But these sources also can provide data for assistance functions. If several sources are available all data has to be merged in a fusion process. Ideas of the more general information measures can be found in Arndt (Arndt 2001).

There is a principal dilemma between data reliability and data processing which cannot be resolved. In general, sensor fusion input data should consist of unchanged physical measurements or at least slightly preprocessed data based on physical reasoning. Without any further model assumptions sensor data fusion just can loop through data without any further data processing. This especially excludes clustering, association, or filtering, and is a very strong constraint. With any additional assumption concerning the model, fused data reliability decreases.

One objective of data fusion is the transformation into a uniform coordinates system. Sensor data is shifted and scaled from different coordinates into one single system with a common origin. The unique origin can be the center of gravity or the midpoint of the rear axle of the vehicle. This transformation addresses the spatial components of sensor data and therefore sometimes is called spatial alignment.

Another objective of data fusion is the temporal alignment. Sensors may have different update rates. Cycle intervals may not be constant. Moreover, one sensor can have an update rate that is a multiple of another one. Then, several measurements of one sensor have to be merged with a single measurement of the other sensor. In this context, mathematical techniques of interpolation and extrapolation are applied. Fusion algorithms can establish a master clock and refer all data with respect to that clock.

Data Fusion Design

Data fusion algorithms are designed according to their performance requirements. Comfort functions are less critical and benefit from any new piece of sensor data immediately. False data is corrected in the next fusion cycle and has no severe consequences. For safety functions, the situation is completely different. Model assumptions are as conservative as possible. Data has to be confirmed over several fusion cycles to achieve high plausibility before it is passed on to assistance systems.

Common fields of view and common sensor attributes imply data redundancy. In case of redundancy, fused data is of higher precision compared to the accuracy of individual sensor data. Moreover, fused data is more robust to sensor failures and malfunction. A further advantage is that detected objects can achieve high plausibility faster. This is especially important for safety functions. Of course, redundancy means more hardware which increases the overall system costs.

A configuration with different fields of view or sensor attributes is based on complementarity. In this case, the total set of all available attributes is larger compared to the individual sensor attributes. As an example, the speed measurement of a radar sensor can be combined with the width measurement of a vision sensor. The total field of view is composed of disjoint or only partly overlapping fields of view. Higher plausibility can be achieved with complementary sensors based on different physical measurement principles.

The principle of cooperation also can be integrated into data fusion methods. Cooperation means, that additional information can be obtained by merging the data of two or more individual sensors. Without this merging process additional information would not be available. Using two vision sensors, stereo vision with depth information is possible. Such depth data is not available if the two vision sensors work independent of each other.

The sensor data fusion process can neglect data history using an instantaneous model. This model describes the vehicle environment only at the current point in time, regardless of any information from the past. No history is included in the model. As an advantage only few data has to be stored. Also, no matching or association with old data is necessary. However, a big disadvantage is the lacking possibility of correcting measurement errors or faults.

A history-based model describes the vehicle environment for a preset time interval. The left interval point is some point of time in the past and the right interval point is the current point in time. This alternative approach has several advantages: First, decisions can be made that are based on previous, historic data sets. Second, it is not necessary to predict sensor data to intermediate points in time. Third, if the sensor data fusion operates in large time cycles, sensor data from several intermediate cycles can be included into the model.

Data Fusion Levels

Sensor data fusion can be implemented on different levels. Fusion on a signal level means that raw data is fused directly. As an example, the electric output signals of radar sensors are mixed together. Data fusion on the signal level can be realized both directly in hardware and in software. The fusion of digital images on the pixel level also serves as an example of raw data fusion. Since the amount of data is huge on a signal level, this kind of fusion is quite expensive. Digital communication on the signal level between different controllers requires special LAN or MOST networks.

Data fusion on an object level needs preprocessed object data. Objects are described by lists of features or attributes. Data on the feature level is already compressed. Thus, one big advantage of object fusion is the mild storage and processing requirements. Sensor and fusion data on the object level can be interchanged by the traditional CAN serial bus system. Most current assistance systems in the market operate on the object level.

Data Fusion on the Object Level

A well-established way to represent surroundings data is the usage of objects. All relevant environment elements like vehicles, pedestrians, buildings, and roads are represented by objects. Objects are created, maintained in lists, associated, merged, tracked, and deleted. Sensor data fusion on the object level requires that all sensors also deliver object lists. This implies individual preprocessing in each sensor. Objects consist of several characteristic attributes. These attributes depend on the object classification. A standard object may have the following attributes:

- Relative position
- Relative velocity
- Relative acceleration
- Size information like width, height, or length
- Shape information like square, rectangle, or polygon
- Classification information like passenger vehicle, van, truck, or cyclist
- Age and plausibility or existence probability
- Quality measures like variances or confidence intervals for all attributes above

The list of object properties should include an attribute that measures the plausibility or existence probability. Plausibility is a measure for data reliability and truth. The plausibility attribute is maintained as long as an object exists. New objects have low plausibility. The longer an object is tracked the higher its plausibility gets. Safety functions have strong existence probability requirements. Here again, the dilemma between fast reaction on new objects on one hand and the necessity of having high reliable objects on the other hand is apparent.

One main fusion objective is the generation and maintenance of a common object list. Such a list is built up from several individual sensor object lists. If different assistance functions make use of the common list, the fusion attributes have to be calculated such that all individual function requirements are fulfilled. This is a severe challenge for the fusion algorithms if both safety and comfort systems share the same object list.

The elements of a common object list sometimes are also called fusion tracks. Associating new objects to existing fusion objects is called matching. There is a great variety of algorithms that are addressed to this subtask. One on hand, if a sensor object cannot be associated with an existing common object, a new common object is created. On the other hand, if several sensor objects are close to an existing object, both can contribute to its data update.

Matching or data association requires distance measures. A simple and commonly used measure is the Euclidean distance between the reference point of the sensor object and the reference point of the common object. Most applications use two-dimensional coordinates. More sophisticated approaches define data vectors including additional values like size or classification information and then take the distance between such vectors.

Measurement data always is uncertain. Some fusion algorithms make use of variances or confidence intervals. The Mahalanobis distance is based on the sum of track covariance matrices and can be formulated as a quadratic form. In two dimensions, objects with the same Mahalanobis distance lie on an ellipse. Different signals like the -coordinate and the -coordinate of an object's position can be weighted differently. This feature helps to incorporate signals of different precision obtained with different physical measurement principles.

Matching is closely related with the usage of gates. Objects are encircled with gates that define the boundary of possible associations. Gates can be defined using different distances. Circles, ellipses, or squares are examples in two dimensions. Objects can only be matched if the position of one object lies within the gate of the other. Thus, gating is a mechanism that enables to create new common objects because existing objects cannot be associated with them on one hand, and delete old common objects because sensor objects do not support them any longer on the other hand.

Using the nearest neighbor algorithm, each sensor object is matched with the nearest common object. Practical implementations of this variant mostly also include gates. If a sensor object is far away from the nearest common object it becomes a common object itself. Of course, this approach includes cases where several sensors detect the same real object and thus contribute to the data update. Details can be found in Bar-Shalom (Bar-Shalom et al. 1988).

In contrast to the nearest neighbor approach the Munkres algorithm in some sense generates a globally optimal association. The objective is to minimize the sum of all distances between associated sensor and common objects. The computational complexity of the Munkres algorithm is higher than that of the nearest neighbor variant. Especially in complex traffic situations where many objects are in the surroundings, this optimization approach is time-consuming and may exceed the time limit given by the data fusion update rate.

Having matched new sensor objects with existing common objects the next step addresses the data fusion itself. Many fusion algorithms are based on the calculation of weighted means. Weights are created in dependence of the distance. If covariance information influences the weights, precise data is weighted higher in comparison with uncertain data. In general, the weighted means approach works with any number of associated objects. Among other methods are the covariance intersection and the information matrix fusion.

For the idea of filter techniques in the context of data association and data fusion see Anderson (Anderson et al. 1979) or Kalman (Kalman et al. 1960 and 1961). The use of Kalman filters is very popular. Moreover, probabilistic data association filters are extensions of the classical Kalman filter. Post-processing of common fusion objects also comprises the subtask to smoothen the resulting tracks. Sliding mean algorithms can compensate sensor measurement oscillations.

Tracking

The filter-based object tracking is the classical approach. It is well-established since many decades and is used as a workhorse for a great variety of problems concerning data filtering, tracking, and fusing. One main advantage of object tracking is the small amount of data that has to be stored and processed. Moreover, many different sensors provide preprocessed object data. Thus, fast and multi-purpose object tracking is very common.

The probabilistic grid tracking is an alternative approach using occupancy grids with sparse matrices. This approach has been developed in the context of mobile robots. Occupancy grid mapping refers to a family of algorithms in probabilistic robotics which address the problem of generating grids and maps from noisy and uncertain sensor measurement data. The basic idea of the occupancy grid is to generate a map of the surroundings as a field of binary random variables each representing the presence of an obstacle at that location.

Using deterministic tracking, the trajectory calculation is based on vehicles' dynamics models. These models describe the acceleration ranges of vehicles. The set of all possible vehicle trajectories can be computed on the basis of these models. With information about the driver's intention such a set can be delimited. Additional information about the road geometry and the space accessible by vehicles also can reduce the number of physically possible trajectories. Some collision mitigation systems are based on deterministic tracking. As a disadvantage, the computational effort is high.

Performance Evaluation

Finally, the question arises how the quality of sensor data fusion algorithms can be measured. To find an answer, ground truth must be available (Chang 1996 and Holz 2003). The method of performance evaluation uses performance indices as quality measures (Stämpfle et al. 2005). This method calculates performance indices on several evaluation levels. Since the number of indices decreases from low to high levels, the indices can be arranged in a pyramidal structure, see Figure 8. Performance indices can depend on the specific algorithm, the scene, the ground truth track, the fusion track, and the evaluation criterion.

At the bottom the pyramid splits into two blocks. The object block is built of indices that depend on an object matching. The scene block contains indices that are based on the whole scene. Towards the top of the pyramid, the two blocks merge, and the set of indices gets smaller. The top of the performance pyramid consists of a single scalar value. This number represents the overall fusion quality. While a large number of indices on lower levels describe many aspects and details of the evaluation process, few indices on higher levels give a more overall answer to the performance issue. Figure 9 shows the comparison process between fusion data and ground truth.

SURROUND MODELS AND SITUATION ANALYSIS

Road and Lane Models

Roads consist of lanes. Typically, there are one or more lanes in each direction. Many models describing the course of roads or lanes include the curvature. The curvature κ describes the bending of a curve and is the reciprocal value of the radius of the circle of curvature ρ:

$$|\kappa| = \frac{1}{\rho} .$$

In literature, the curvature sometimes is defined as an unsigned positive real number and sometimes as a signed value. It is a basic principle of road construction that the curvature κ changes proportional to the curve length s:

$$\kappa(s) = c_0 + c_1 s .$$

For each road segment, the parameters c_0 and c_1 are constant. The parameter c_0 can be interpreted as the curvature for $s = 0$, and the parameter c_1 is the derivative of the curvature at $s = 0$. Since the curvature is roughly proportional to the

Figure 8. Performance index pyramid to evaluate data fusion algorithms

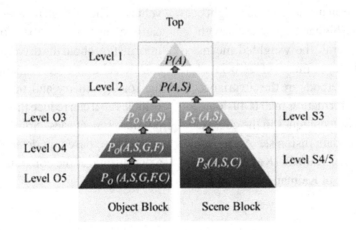

Figure 9. Process of performance evaluation of automotive sensor data fusion

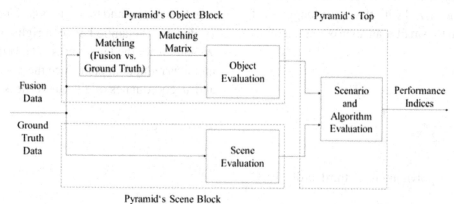

lateral acceleration, this also means that with constant speed the lateral acceleration is proportional to the curve length. This approach implies safe and comfortable driving. Curves of such a form are cald clothoids. In Cartesian coordinates these curves are given by

$$x(t) = \alpha\sqrt{\pi} \int_0^t \cos\frac{\pi\tau^2}{2}\, d\tau,$$

$$y(t) = \alpha\sqrt{\pi} \int_0^t \sin\frac{\pi\tau^2}{2}\, d\tau,$$

where α denotes the clothoid parameter and t some independent variable, e.g. time. Using the definition of curvature and arc length the following relation results:

$$\kappa = c_0 + \frac{\sqrt{\pi}}{\alpha}t,$$

$$s = \alpha\sqrt{\pi}t \Rightarrow \kappa = c_0 + \frac{1}{\alpha^2}s,$$

$$c_1 = \frac{1}{\alpha^2}$$

The differential geometric description can be transformed into a Cartesian one: The integral of the curvature

$$\varphi(s) = \varphi_0 + \int_0^s \kappa(\tau)\, d\tau = \varphi_0 + c_0 s + \frac{1}{2}c_1 s^2$$

describes the lane angle. A second integration yields Cartesian coordinates:

$$x(s) = x_0 + \int_0^s \cos\varphi(\tau)\, d\tau,$$

$$y(s) = y_0 + \int_0^s \sin\varphi(\tau)\, d\tau$$

Inserting the expression for $\varphi(s)$ into the Cartesian form gives

$$x(s) = x_0 + \int_0^s \cos\left(\varphi_0 + c_0\tau + \frac{1}{2}c_1\tau^2\right) d\tau,$$

$$y(s) = y_0 + \int_0^s \sin\left(\varphi_0 + c_0\tau + \frac{1}{2}c_1\tau^2\right) d\tau$$

This is the Cartesian representation for the curve, parameterized with the curve length s . If $\varphi_0 = 0$, then for small s an approximation reads

$$x\left(s\right) \approx x_0 + s,$$

$$y\left(s\right) = \ y_0 + \frac{1}{2}c_0 s^2 + \frac{1}{6}c_1 s^3,$$

which gives a polynomial of third order in y-direction.

Object Models

In the context of surround models, objects are static or dynamic parts of the traffic environment that have spatial distension. Typical objects are buildings and other road users. They are described using sets of properties like position, size, velocity, and classification. Since data is collected by sensors, also properties referring to object plausibility or object tracking age are used. In addition, advanced object model descriptions include probability measures and variances for all characteristic values.

The position of objects typically is described by a reference point RP . The size of parallelogram-shaped objects can be defined in various ways. The bounding box variant uses the four numbers

$$s_{xl} \geq 0, \ s_{xr} \geq 0, \ s_{yl} \geq 0, \ s_{yr} \geq 0$$

to describe the x -coordinates and the y -coordinates of the left and right vertex relative to the reference point. Alternatively, the angle-leg variant comprises the four numbers

$$\varphi \in \left[0, \frac{\pi}{2}\right], \ \alpha \in \left(0, \pi\right), \ R \geq 0, \ L \geq 0$$

Variable φ describes the angle between the negative y -axis and the right object leg. Variable α denotes the angle between right and left leg. R and L are the leg lengths. The vertex variant includes redundant data since the x -coordinates and y -coordinates of all four vertices are stored:

$$p_1, p_2, p_3, p_4$$

with

$$p_1 - p_2 \parallel p_3 - p_4$$

and

$$p_1 - p_4 \parallel p_2 - p_3$$

and

$$p_{1x} = 0, \quad p_{2y} + p_{4y} = 0$$

The vertices are numbered in mathematically positive orientation. In case of rectangular objects, additional constraints hold. For the bounding box variant

$$\left(s_{xl}, s_{yl}\right) \perp \left(s_{xr}, s_{yr}\right)$$

For the angle-leg variant, the angle α between right and left leg is 90 degrees:

$$\alpha = \frac{\pi}{2}$$

In the same way, the vertex variant has the additional condition that the two legs are orthogonal:

$$p_1 - p_2 \perp p_1 - p_4$$

All three variants are depicted in Figure 10.

Figure 10. Three variants of the description of rectangular objects

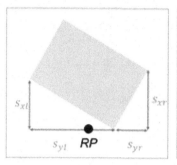 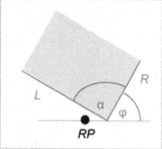 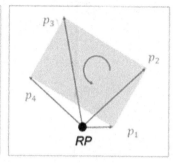

Vehicle Models

One basic principle of vehicle dynamics is the Kamm circle. Due to limited tire friction, a vehicle can either accelerate or decelerate with maximum longitudinal force or change direction with maximum lateral force. The Kamm circle or ellipse is a rough, first-order approximation of the possible distribution of vehicle accelerations:

$$a_x^2 + a_y^2 = g^2$$

or

$$a_x^2 + (ca_y)^2 = (\mu g)^2$$

Here, a_x denotes the longitudinal and a_y denotes the lateral acceleration, g is the constant of gravity, c is the proportional factor between x-direction and y-direction, and μ is the road friction coefficient. Of course, these equations are a first, rough approximation for the decomposition of longitudinal and lateral accelerations and decelerations.

Driver assistance systems that operate not on the stabilization level often make use of the bicycle model. It is assumed that the two wheels of each axis can be compressed into a single one. A simple model includes the steering angle δ, the global position x and y, the yaw angle E, and

the vehicle slip angle β, see Figure 11. Using the velocity v, the global position coordinates are

$$\dot{x} = v \, \cos(\psi + \beta),$$

$$\dot{y} = v \, \sin(\psi + \beta)$$

Let l_r be the length between center of gravity and rear axle, and let l_f be the length between center of gravity and front axle, then, the first derivative of the yaw angle is

$$\dot{\psi} = \frac{v \cos \beta}{l_r + l_f} \tan \delta,$$

$$\beta = \tan^{-1}\left(\frac{l_r \tan \delta}{l_r + l_f}\right)$$

Thus, with given functions $v(t)$ and $\delta(t)$ the global position x and y and the orientation ψ can be computed. A comprehensive overview of vehicle dynamics and control can be found in Rajamani (Rajamani 2006). Simultaneous simulation methods are based on the triangulation of the parameter space (Stämpfle et al. 1998, 1999, 2001). With such techniques sets of infinite many initial-value problems can be solved with one single integration process.

Figure 11. Vehicle bicycle model

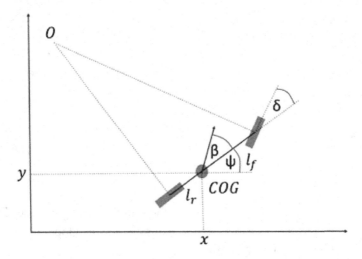

Situation Interpretation

All preprocessed information about the vehicle and its surroundings is collected and stored in surround models. These models include models of static objects like buildings, bridges, or tunnels. Another class of models describes other road users like passenger cars, trucks, cyclists, or pedestrians. Furthermore, road models consist of lanes, lane markings, parking space, and crossings, and also contribute to the surround models. Finally, surround models also comprise descriptions of the dynamics of all moving objects including the ego vehicle.

Depending on the classification of objects, assumptions on the dynamical range can be made. Heavy-duty trucks are more or less one-dimensional road users. They need much time to change lanes. This is one of the reasons why collision avoidance systems for trucks are less challenging as for sportive cars. Motorbikes or sportive cars are assumed to have a wide range of lateral acceleration. Cyclists and pedestrians also have their specific dynamical range. In general, the dynamical range is limited to the laws of physics. Due to limited tire forces a vehicle cannot perform maximal deceleration and cornering at the same time.

Surround models essentially describe the current situation. The real world is mapped to reduced models consisting of information for the current point in time. Like a snapshot, surround models represent the traffic environment at a certain instant. Data fusion algorithms temporally align all sensor data and thus provide consistent model information. Typically, surround models are recomputed at a fixed update rate.

Situation Prediction

Once such models are established, they can be used for situation prediction with respect to the next few seconds. Such predictions are extremely helpful to interpret and evaluate the traffic situation appropriately. Situation analysis needs assumptions on the dynamical range of moving objects. This means, that any algorithm can predict the situation correctly only with a certain probability.

Traffic situations can be interpreted as regular. This is the normal case and means that there is no evidence that the current situation is or can become critical in the next few seconds. All road users behave reasonable and move with sufficient safety distance between each other. Also, the lane assignment of all vehicles is without any inconsistency. No collision potential is derived from the situation prediction.

Traffic situations are classified as critical, if a dangerous situation might evolve with a significant probability. Due to surround model prediction, safety distances are likely to decrease to critical values. Another indicator is high relative dynamics. If road users move with high relative speed and short distance, this may lead to critical situations. In case of an intended lane change maneuver, the situation can be evaluated as critical, if traffic approaches on neighboring lanes from behind.

Traffic situation also can be interpreted as fatal. Based on model assumptions about the dynamical range of all road users, a collision may become unavoidable. This means that no matter how all traffic participants react, a collision will take place within the next few seconds. The surround models provide the conclusion that in any circumstance, an accident will happen in near future. This kind of situation analysis is the basis for any assistance system that addresses collision mitigation.

DRIVER ASSISTANCE SYSTEMS

Classification and Complexity

One possible classification of driver assistance systems divides these systems into safety and comfort systems. An alternative classification is based on the length of the prediction interval. Assistance systems on the stabilization level have the shortest prediction horizon. This horizon typically is about one second. Stabilization systems operate with very short update rates of only few milliseconds. The general objective is to stabilize the vehicle and thus prevent it from skidding. Most systems on the stabilization level are safety systems. On this level, vehicle sensors contribute most information.

Driver assistance systems on the maneuver level have a prediction horizon of a couple of seconds. They include cruising, lane changing, and parking. On this level, both safety systems as well as comfort systems are in the market. In contrast to stabilization systems, maneuvering systems offer different degrees of assistance ranging from simple information to full autonomous operation. One challenge of maneuver level systems is the fusion of vehicle and surround sensor data.

Assistance systems on the navigation level help to find appropriate routes to the destination. Navigation systems optimize the route with respect to time, types of roads, and stopovers. The result is either static and computed only once at the beginning of the tour or dynamic and recalculated permanently including information of traffic channels. Most systems on the navigation level are comfort systems. On this level, surround sensors and map data are the main sources of information.

Driver assistance systems include many algorithms on different levels of signal processing packed with lots of mathematics. This trend will be strengthening even more in next generation assistance systems. Basic mathematics including algebra and analysis is offered in literature in a great variety. A modern student presentation can be found in Koch (Koch et al. 2010). An easy to read introduction into numerical analysis can be found in Atkinson (Atkinson et al. 2004) or Burden (Burden 2011). A classical, comprehensive formula overview is given in Bronstein (Bronstein et al. 1981).

System Architecture

Driver assistance systems are embedded systems in the automotive context. They are designed as internal controllers integrated in other components like the hydro aggregate or as external controllers with separate hardware. This variety also implies that signal processing can be concentrated on one single chip or be segmented on different hardware components connected by bus systems. Figure 12 shows four different variants of possible system architectures for driver assistance systems.

Variant 1 is the simplest one and is called a 1-1-system. Each assistance function has its own sensor. These sensors are optimized to fulfill

Figure 12. Different variants of the architecture of driver assistance systems

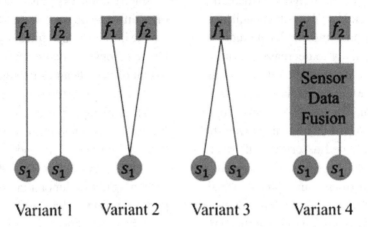

specific function requirements. Since they are function-optimized they hardly can be used as multi-purpose hardware. Sensor data is preprocessed within the sensor hardware to fit best the function's requirements. The first adaptive cruise control systems were designed in this way.

System architecture variant 2 is a 1-n-system and includes a multi-purpose sensor. Such a sensor has signal output that can be forwarded to more than one function. A typical use case is a key function and several value-added functions. The customer benefits from more than one function using just one single sensor hardware. A vision sensor can deliver data to both a traffic sign recognition function and an advanced cruise control system.

Variant 3 shows a complex function. The architecture is called m-1-system. The realization of that function makes it necessary to include more than one sensor. One reason to use multiple sensors is that the field of view can be extended. Another reason is the high reliability requirement of a safety function. Two sensors cover the same field of view and are redundant; data output of the two sensors is compared. Parking systems with several ultrasonic sensors are of this architecture variant.

System architecture variant 4 is denoted m-n-system. It has an open architecture. Several sensors contribute to the surround model. Such a model is passed to the assistance functions. Variant 4 allows high flexibility and benefits most from all sensors. A sensor data fusion can combine all specific sensor features. The distance of an object can be measured by a radar sensor, whereas the width of the object is obtained by image processing in a vision sensor.

Communication Bus Systems

Driver assistance systems interact with other automotive systems like the brake controller or the drivetrain controller. Information is interchanged via bus systems. A comprehensive overview can be found in Zimmermann (Zimmermann et al. 2007). Modern passenger cars in the market already embody several bus systems. Since modern vehicles consist of up to 100 separate controllers the requirements on bandwidth and failure treatment are quite inhomogeneous.

Standard convenience systems like sunroofs, electrically operated window regulators, or restraint systems communicate on cheap serial bus systems with low bandwidth like LIN. Such a bus system can handle data rates up to 100 Kbit/s. The data transfer is asynchronous. The LIN bus system operates in a fail-safe mode. In the event of failure, the device responds in a way that will cause a minimum of harm.

Safety systems, including all vehicle dynamics components like drivetrain, gears, transmission, and brakes together with all vehicle state sensors share a separate bus. The classical bus system for this group of components is the CAN. This bus operates asynchronously with a maximum of 1 Mbit/s. Comfort systems are connected together on a second CAN. If necessary, gateways are added to enable the link between several CAN busses.

The next generation bus system is FlexRay. In contrast to the state-of-the-art CAN, the serial FlexRay is designed as a synchronous bus with data rates up to 10 Mbit/s. The development of FlexRay has been initialized by the requirements expected for drive-by-wire systems. These requirements include a fail-operational behavior in case of an occurring error. In 2006 the first model with a FlexRay system has been introduced into market.

Comfort systems that need higher bandwidth communicate on a MOST bus system. This communication bus is capable to transfer sensor raw data or digital images. Data rates with a maximum of 100 Mbit/s are possible. The MOST protocol comprises no special features in case of failure. The data transfer is asynchronous.

Cruise Systems

The well-established cruise control keeps a preset velocity as long as the driver does not press the brake pedal. This functionality is independent of the surrounding traffic. If the driver intervenes by braking, the cruise control system is switched off. Hence, the typical use case is driving on highways with sparse traffic. In platooning situations with oscillating velocity the driver has to brake and subsequently switch on the system again and again. The benefit and thus the acceptance of such systems then go down.

The next generation are adaptive cruise control systems. These systems are already in the market since more than 10 years and nowadays available even for compact passenger cars. The technological key step from the ordinary cruise control to the adaptive variant is the integration of vehicle surround sensor information. The distance to the preceding vehicle is measured. Thus, the system can hold a constant time gap to the car in front. Such systems harmonize best with automatic transmission. An advanced cruise control system from Daimler is shown in Figure 13.

With additional short-range sensors adaptive cruise control systems can be extended to full speed range systems. Low speed following requires precise distance data to generate smooth acceleration and deceleration signals. Moreover, it has to be defined whether or within which time delay the vehicle automatically can start again after standstill. Such a function feature addresses legal aspects. Until now, the driver is always responsible for the action of driving independent of the amount of technical support. Full speed range systems also have extended requirements to the sensor detection range. A mid-range sensor with a large horizontal viewing angle is mandatory. Preceding vehicles have to be tracked stable on curved roads and highway entries and exits. Vehicles on neighboring lanes can become target objects when cutting in.

The application of adaptive cruise systems is a complex engineering challenge. Two aspects are discussed in the following. Firstly, a cruise system is parameterized to obey the safety distance regulations by law. This can provoke situations where following vehicles overtake and cut in sharply again. As a result, braking is initiated by the system to reconstitute the safety distance. This effect of falling back might cause acceptance problems. Secondly, a critical situation can occur, if the system follows a vehicle across an intersection when the traffic lights switch from green to red. The preceding vehicle passes without danger whereas the vehicle with the assistance

Figure 13. Mercedes-Benz Distronic Plus (courtesy of Daimler)

system would automatically follow and cross the intersection in a critical moment if the driver does not intervene.

Lane Systems

Lane systems expand the spatial complexity from 1D to 2D. For systems like cruise control, only vehicles on the own lane are relevant. This is different for lane systems. Two or more neighboring lanes have to be observed simultaneously. This aspect creates a new level of complexity for the surround models and the situation interpretation algorithms. There are systems that just inform and warn, whereas more sophisticated systems actively support the driver when changing lanes.

Many functions detect and inform about vehicles that approach on neighboring lanes from behind. In addition, they warn if a driver's intention to change lanes is detected. Most systems also cover the blind spot. They are called side view assist or lane change aid. Many systems use radar sensors to cover a large field of view that ranges up to 100 m. Thus, vehicles that are approaching fast are also detected. A side view system from Audi is depicted in Figure 14. It is common to

place small warning lights in the near of the left and right rear mirror visible to the driver.

Another class of functions address lane keeping. These functions are based on information of lane markings. Lidar sensors or vision sensors are able to detect these markings. Passive systems give an alert to the driver if the vehicle is about to tend towards the markings. Active systems autonomously intervene and give additional forces to the steering actuator. In this case, forces of the driver and forces of the assistance system are superimposed.

A loose guidance restricts additional steering forces to situations where the vehicle already touches the lane markings. In such cases, the system intervention is quite intense, since the vehicle's orientation relative to the lane has to be corrected quickly. Such a function design gives maximal freedom to the driver with minimal phases of intervention and uses a rectangular shaped characteristic diagram for the steering moment. Systems with a tight guidance use a symmetric bi-linear characteristic diagram for the steering moment. The steering angle correction is proportional to the distraction from the middle of the lane. This implies that the lane keeping function is active permanently. Even in

Figure 14. Audi side assist (courtesy of Audi)

situations where the vehicle's lateral position is not far away from the ideal center line, a moderate correction is performed. A comfortable guidance uses a parabola-like characteristic diagram for the steering moment. Small perturbations from the lane's center are handled with low corrections. The degree of intervention increases disproportionally if the vehicle tends to the lane markings. Comfortable guidance, on one hand, scarcely overrides the driver's input in uncritical situations, and, on the other hand, prevents the vehicle from leaving the lane.

Parking Systems

There is a vast range of different parking systems in the market. They differ in the types and numbers of sensors they use. Also, the degree of assistance is different. Simple and cheap systems just inform the driver by displaying information about the size and type of a parking slot. More sophisticated systems evaluate the sensor information and give advice to the driver. High-end systems control the steering, the brakes and the drive-train and autonomously perform a parking maneuver. More complex parking strategies also can be computed by adapting cellular automata techniques used for mobile robots (Stämpfle 1996).

Information systems just give information to the driver. Typically, ultrasonic sensors detect objects in the near of the vehicle. Such systems are available in mass production cars since many years. Newer systems use multi-purpose short-rage radar sensors with a larger field of view compared with ultrasonic sensors. Some systems include vision sensors in the rear that generate images and display them to the driver. Such systems also measure the size of parking slots and forward this information to the driver. Figure 15 shows a Bosch parking system.

Guiding systems already process and evaluate information. If vision images are available, guiding curves can be integrated into the images to give the driver an impression of the best parking trajectory. These trajectories consist of straight lines and parts of circles. With this approach the trajectory can be followed step by step since steering is possible in standstill. This class of systems already requires precise information of the vehicle position and orientation. Thus, sensors measuring the vehicle state are integrated into these systems. Guiding systems do no intervene.

Semi-autonomous systems assist the driver during the parking maneuver by controlling the lateral direction of motion. This requires an electro-mechanical servo steering actor or a conventional steering system with an additional steering motor. Semi-autonomous systems disburden the driver from setting the steering wheel angle correctly. The driver can concentrate on controlling the speed while observing the parking space with full attention.

Figure 15. Bosch parking system (courtesy of Bosch)

Autonomous systems are still a challenge in research today. The problem in this context is the great variety of parking slots. Also, many different obstacles have to be taken into account. Bushes in the wind, running children, small pillars, high road curbs, or irregular slot geometry are all big challenges for full autonomous systems. Moreover, there is an increasing demand of systems that assist navigating and maneuvering in large parking garages.

Collision Mitigation and Avoidance Systems

About 60% of all rear-end crashes and nearly one third of all frontal crashes could be avoided if drivers would react only half a second earlier. In nearly 50% of all collision accidents the driver does not apply the brakes at all and in further 20% the driver brakes too unassertively. There is a great variety of collision systems. Many of them are already in market, others are more complex and still a technical challenge (Lages 2001). Some highly developed collision avoidance systems address longitudinal traffic (Stämpfle et al. 2008). Newer types also include lateral motion. The design of collision mitigation and avoidance systems always includes the tradeoff between reliable information on the vehicle surroundings on one side and the early warning or intervention on the other.

The information phase is the first phase of a collision mitigation system. The traffic situation tends to become critical. A collision is quite probable in the next few seconds. The driver is informed about the situation. Such information can be a text message in the conventional display or, more advanced, in a head-up display. Most systems display the information visually. No sound or acoustic signal is given.

During the warning phase, the situation escalates. No action has been taken by the driver based on the information in the first phase. The warning can be given in a visible, an audible, or a haptic way. The safety belt can be tightened or a warning jerk by shortly applying the brakes can be performed. In addition, an acoustic alert is given to make the driver aware of the critical situation. Some systems also prepare the brake assist and the brake circuits.

The intervention phase is the next level of escalation. The situation has evolved quite dangerous. Only parts of a second remain before a collision almost surely happens. The assistance system intervenes by actuating the brakes or the steering. Since at this time there is still a remaining risk of misinterpreting the situation the braking force is only half the maximum in most systems. The begin of intervention depends on the assumptions of the dynamical abilities of all involved vehicles.

The last phase is called mitigation phase. Considering physical limits of vehicle dynamics a collision now is unavoidable. No matter how the driver and all other objects behave there is no possible alternative to avoid a collision. Typically, such a situation is confirmed only milliseconds before the crash. In this phase full autonomous braking decreases the relative collision speed. In addition, emergency calls can be made automatically by the system.

Another type of collision avoidance systems addresses emergency stops, especially on highways. If the driver suffers a sudden loss of health, like a heart attack, the system completely overtakes control and initiates a collision-free stopping maneuver. Such an autonomous maneuver also includes one or more necessary lane changes. On one hand, the traffic complexity on several lanes can be quite challenging. High velocities on highways require long-range rear sensors. On the other hand, the scenario is embedded into a specific situation and lasts only for a short period in time. Figure 16 shows the BMW emergency stop assist.

The analysis of the current traffic situation comprises a prediction of what will happen in the next few seconds. This prediction is based on model assumptions of the surrounding objects. Deterministic and probabilistic approaches are in discussion. With increasing prediction length the deterministic variant has to cope with an exploding number of combinatorial possibilities. In case of the probabilistic variant, due to the increasing uncertainty the probability of a specific scenario tends to zero. Both approaches require a trade-off between function reliability and function benefit.

Advanced collision avoidance or mitigation systems intervene by braking, steering, or a combination of both. All alternatives are computed and evaluated. This requires a situation analysis in 2D. In this context, the following key questions arise:

- When is which maneuver possible?
- When is the last point in time at which a collision can be avoided?

Figure 16. BMW emergency stop assist (courtesy of BMW)

Firstly, for a simple braking maneuver, the necessary braking distance d_{xB} with initial velocity v_0 and longitudinal deceleration a_x can be formulated:

$$d_{xB} = -\frac{1}{2a_x} v_0^2$$

This formula shows a quadratic dependence of d_{xB} on the velocity. Secondly, evading is discussed. Let s be the lateral evading shift. Assuming constant velocity in longitudinal direction and constant lateral acceleration a_y, the evading distance is given by the formula

$$d_{xE} = \sqrt{\frac{2s}{a_y}}\, v_0$$

Now, the dependence of d_{xE} on the velocity is linear. Likewise, assuming constant velocity in the direction of the vehicle, the corresponding trajectory is a part of a circle. In this case,

$$d_{xE2} = \sqrt{s\left(\frac{2v_0^2}{a_y} - s\right)}$$

Other approximations include sigmoid or clothoid curves and are more complicated. Figure 17 shows the diagram with the four possible cases. The limit braking distance d_{xB} is a straight line. The limit evading distance d_{xE} is a parabola. The two lines intersect at velocity v_{lim}. Below the value v_{lim}, the braking distance is smaller than the evading distance. This means, that the initiation of a braking maneuver is possible at a later point in time. Above the value v_{lim}, the evading distance is smaller than the braking distance. Evading needs less longitudinal space and thus is capable to completely avoid an imminent collision, whereas braking can only mitigate the collision. All velocities in this context are relative velocities.

Schimmelpfennig (Schimmelpfennig et al. 1985) investigated the maximum lateral acceleration that a normal driver with average skills is willing to accept. The empirically found correlation based on studies is

$$a_y(v) = 0.155(3.6v)\, e^{-\left(\frac{3.6v}{70}\right)^{1.85}}$$

Figure 18 shows this dependence between velocity and accepted lateral acceleration. It is remarkable that the acceleration values are quite

Figure 17. Braking or evading? Four different cases depending of velocity and distance

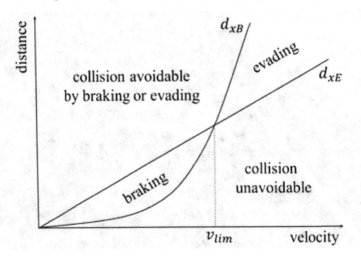

Figure 18. Maximum lateral acceleration accepted by a normal driver according to the Schimmelpfen-nig formula

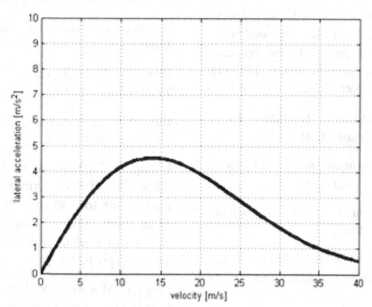

small in general. They do not exceed half of the gravity constant. Especially for higher velocities, the tolerated acceleration is very small and tends to zero. As a consequence, the normal driver would accept an evading maneuver initiated by an assistance system with moderate lateral acceleration quite early. Another result of these studies is that the human perception of accelerations in longitudinal and lateral direction is quite different. Hard braking maneuvers are associated with lower acceleration values than hard evading maneuvers.

There is a principal difference between the vehicle dynamics of passenger cars and heavy-duty trucks in terms of collision avoidance: Trucks need much time and space to change lanes. Thus, their degree of freedom concerning collision avoidance is much smaller in comparison with passenger vehicles. The alternative of evading in a critical situation practically has not to be taken into account. As a consequence, emergency braking can be initiated much earlier. Systems for

trucks that are already in the market apply full braking quite early and thus are able to drastically reduce the collision velocity.

CONCLUSION

Surround sensing for automotive driver assistance systems is a very broad field of research. Contributions come from many different disciplines. Hardware and software components on various levels have to be compliant with each other. The overall system complexity often is underestimated. The mathematical share concealed in many algorithms definitely will increase in future systems. Sensors may become smaller and cheaper. In a few years, color and stereo vision may enter series production. Intersection assistance or full collision avoidance will remain research topics for the next time. But all progress contributes to the ultimate goal of safe, comfortable, and autonomous driving.

REFERENCES

Amditis, A., Polychronopoulos, C., Floudas, N., & Adreone, L. (2005). Fusion of infrared vision and radar for estimating the lateral dynamics of obstacles. *Information Fusion*, *6*, 129–141. doi:10.1016/j.inffus.2004.06.002

Anderson, B., & Moore, J. B. (Eds.). (1979). *Optimal filtering*. Prentice Hall.

Arndt, C. (2001). *Information measures*. Springer. doi:10.1007/978-3-642-56669-1

Atkinson, K., & Han, W. (2004). *Elementary numerical analysis*. Wiley.

Bar-Shalom, Y. (Ed.). (1990). *Multitarget-multi-sensor tracking: Advanced applications (Vol. 1)*. Artech House.

Bar-Shalom, Y. (Ed.). (1992). *Multitarget-multi-sensor tracking: Advanced applications (Vol. 2)*. Artech House.

Bar-Shalom, Y. (Ed.). (2000). *Multitarget-multi-sensor tracking: Advanced applications (Vol. 3)*. Artech House.

Bar-Shalom, Y., & Fortmann, T. E. (1988). *Tracking and data association*. Academic Press.

Bar-Shalom, Y., & Li, X. R. (1993). *Estimation and tracking*. Artech House.

Beauvais, M., & Lakshmanan, S. (2000). CLARK: A heterogeneous sensor fusion method for finding lanes and obstacles. *Image and Vision Computing*, *18*, 397. doi:10.1016/S0262-8856(99)00035-9

Becker, J. (2002). *Fusion der Daten der objekter-kennenden Sensoren eines autonomen Straßen-fahrzeugs*. Fortschritt-Berichte VDI.

Becker, J. C., & Stämpfle, M. (2004). Fusion and filtering of multidimensional objects for driver assistance systems. *2004 IEEE Intelligent Vehicles Symposium*, Parma, (pp. 613-618).

Blackmann, S., & Popoli, R. (1999). *Modern tracking systems*. Artech House.

Bosch. (2004). *Sicherheits- und Komfortsysteme*. Vieweg.

Bronstein, N. I., & Semendjajew, K. A. (1981). *Taschenbuch der Mathematik*. Harri Deutsch.

Burden, R. L., & Faires, J. D. (2011). *Numerical analysis*. Brooks/Cole.

Chang, K. C., Saha, R. K., Bar-Shalom, Y., & Alford, M. (1996). Performance evaluation of multisensor track-to-track fusion. *Proceedings of the 1996 IEEE International Conference on Multisensor Fusion and Integration for Intelligent Systems*, (pp. 627-632).

Fang, Y., Masaki, I., & Horn, B. (2002). Depth-based target segmentation for intelligent vehicles: Fusion of radar and binocular stereo. *IEEE Transactions on Intelligent Transportation Systems*, *3*(3), 196–202. doi:10.1109/TITS.2002.802926

Hall, D. L. (1992). *Mathematical techniques in multisensor data fusion*. Artech House.

Hall, D. L., & Llinas, J. (1997). An introduction to multisensory data fusion. *Proceedings of the IEEE*, *85*(1), 6–23. doi:10.1109/5.554205

Hall, D. L., & Llinas, J. (Eds.). (2001). *Handbook of multisensor data fusion*. CRC Press. doi:10.1201/9781420038545

Holz, D. (2003). *Automatisierte Bewertung von Fusionsverfahren zur Fahrzeugumfelderfassung*. Master Thesis, FH Karlsruhe und Robert Bosch GmbH.

Kalman, R. E. (1960). A new approach to linear filtering and prediction problems. *Transactions ASME . Journal of Basic Engineering*, *82*, 35–45. doi:10.1115/1.3662552

Kalman, R. E., & Bucy, R. S. (1961). New results in linear filtering and prediction theory. *Transactions ASME . Journal of Basic Engineering, 83,* 95–108. doi:10.1115/1.3658902

Klotz, A., Sparbert, J., & Hötzer, D. (2004). Lane data fusion for driver assistance systems. *Proceedings of the International Conference on Fusion 2004.*

Koch, J., & Stämpfle, M. (2010). *Mathematik für das Ingenieurstudium.* Hanser. doi:10.3139/9783446425507

Kojima, Y., Yamada, K., & Ninomiya, Y. (2002). *Three-dimensional road structure estimation by fusion of a digital road map and an image.* (SAE 2002, 2002-01-0758).

Lages, U. S. (2001). *Untersuchungen zur aktiven Unfallvermeidung von Kraftfahrzeugen.* Fortschritt-Berichte VDI.

Niehsen, W., Garnitz, M., Weilkes, M., & Stämpfle, M. (2005). Informationsfusion für Fahrerassistenzsysteme . In Maurer, M., & Stiller, C. (Eds.), *Fahrerassistenzsysteme mit maschineller Wahrnehmung.* Springer. doi:10.1007/3-540-27137-6_3

Rajamani, R. (2006). *Vehicle dynamics and control.* Springer.

Schimmelpfennig, K. H., & Nackenhorst, U. (1985). Bedeutung der Querbeschleunigung in der Verkehrsunfallrekonstruktion - Sicherheitsgrenze des Normalfahrers. *Verkehrsunfall und Fahrzeugtechnik, 4,* 94–96.

Stämpfle, M. (1996). Cellular automata and optimal path planning . *International Journal of Bifurcation and Chaos in Applied Sciences and Engineering, 6*(3), 603–610. doi:10.1142/S0218127496000291

Stämpfle, M. (1998). *Numerische Approximation des Flusses eines dynamischen Systems mit adaptiven Triangulierungsverfahren.* Shaker.

Stämpfle, M. (1999). Dynamical systems flow computation by adaptive triangulation methods. *Computing and Visualization in Science, 2*(1), 15–24. doi:10.1007/s007910050023

Stämpfle, M., & Branz, W. (2008). *Kollisionsvermeidung im Längsverkehr - Die Vision vom unfallfreien Fahren rückt näher.* München, Germany: Tagung Aktive Sicherheit durch Fahrerassistenz.

Stämpfle, M., Holz, D., & Becker, J. C. (2005). Performance evaluation of automotive sensor data fusion. *8th International Conference on Intelligent Transportation Systems,* Wien, (pp. 590-595).

Stämpfle, M., Hunt, K. J., & Kalkkuhl, J. (2001). Efficient simulation of parameter-dependent vehicle dynamics. *International Journal for Numerical Methods in Engineering, 52,* 1273–1299. doi:10.1002/nme.254

Waltz, E., & Llinas, J. (Eds.). (1990). *Multisensor data fusion.* Artech House.

Winner, H., Hakuli, S., & Wolf, G. (Eds.). (2009). *Handbuch Fahrerassistenzsysteme.* Vieweg Teubner. doi:10.1007/978-3-8348-9977-4

Zimmermann, W., & Schmidgall, R. (2007). *Bussysteme in der Fahrzeugelektronik.* Vieweg Teubner. doi:10.1007/978-3-8348-9188-4

KEY TERMS AND DEFINITIONS

ACC: Abbreviation of adaptive cruise control.

ADAS: Abbreviation of advanced driver assistance system. A high-end automotive driver assistance system that supports the driver.

Automotive System: An electronics or mechatronics system that is integrated in a vehicle.

CAN: Abbreviation of controller area network. CAN is an automotive serial bus system.

CAS: Abbreviation of collision avoidance system.

CC: Abbreviation of cruise control.

CCD: Abbreviation of charge-coupled device. This is a vision chip technology.

CMOS: Abbreviation of complementary metal oxide semiconductor. This is a vision chip technology.

CMS: Abbreviation of collision mitigation system.

Comfort System: A driver assistance system that addresses automotive comfort or convenience aspects.

Convenience System: Same as comfort system.

DAS: Abbreviation of driver assistance system. An automotive system that supports the driver.

DCT: Abbreviation of discrete cosine transform.

DFT: Abbreviation of discrete Fourier transform.

ECU: Abbreviation of electronic control unit.

FlexRay: FlexRay is an automotive serial bus system.

FMCW: Abbreviation of frequency modulated continuous wave. This principle is used in many radar sensors.

Imager: Video chip that captures images.

Laser: Abbreviation of light amplification by stimulated emission of radiation.

LAN: Abbreviation of local area network.

LCA: Abbreviation of lane change aid.

Lidar: Abbreviation of light detection and ranging.

LIN: Abbreviation of local interconnect network.

Mono System: A vision system including one single imager chip.

MOST: Abbreviation of media oriented systems transport.

PP: Abbreviation of parking pilot.

PSS: Abbreviation of predictive safety systems.

Radar: Abbreviation of radio detection and ranging.

Safety System: A driver assistance system that addresses automotive safety aspects.

SDF: Abbreviation of sensor data fusion.

Sensor: A device or system that measures and processes one or several signals.

Sensor Data Fusion: Algorithms to fuse data from more than one sensor. Such sensors can include vehicle and surround sensors.

Stereo System: A vision system including two imager chips.

Surround Sensor: A sensor that measures the vehicle's surroundings.

Ultrasonic Sensor: A sensor based on ultrasonic waves.

Vehicle Sensor: A sensor that measures the dynamic state of a vehicle.

Video Sensor: Same as vision sensor.

Vision Sensor: A sensor based on an image sensor.

Compilation of References

(Product Code: J1211: 20090407). Retrieved from http://standards.sae.org/j1211_200904/

3 rd Generation Partnership Project; Technical Specification Group. (2005). *3GPP TS 05.05 V8.20.0 GSM/EDGE radio access network; radio transmission and reception* (Release 1999), 11/2005.

3 rd Generation Partnership Project; Technical Specification Group. (2006). 3GPP GSM 43.059 V7.2.0 radio access network, functional stage 2 description of location services (LCS) in GERAN (Release 7), 04/2006.

3 rd Generation Partnership Project; Technical Specification Group. (2006b). *3GPP GSM 45.811 V6.0.0 Radio access network, feasibility study on uplink TDOA in GSM and GPRS* (Release 6), 6/2002.

3 rd Generation Partnership Project; Technical Specification Group. (2006b). 3GPP UMTS 25.305 V7.3.0 stage 2 functional specification of user equipment (UE) positioning in UTRAN (Release 7), 12/2006.

3 rd Generation Partnership Project; Technical Specification Group. (2008). 3GPP TS 36.300 V8.12.0 Evolved universal terrestrial radio Access (E-UTRA) and evolved universal terrestrial radio access Network (E-UTRAN): Overall description (Release 8), 12/2008.

3 rd Generation Partnership Project; Technical Specification Group. (2011). 3GPP TS 05.05 V10.5.0 Evolved universal terrestrial radio access (E-UTRA) and evolved universal terrestrial radio access network (E-UTRAN): Overall description stage 2 (Release 10), 10/2011.

3 rd Generation Partnership Project; Technical Specification Group. (2011b). 3GPP TR 22.934 V10.0.0 Services and system aspects; Feasibility study on 3GPP system to wireless local area network (WLAN) interworking (Release 10), 03/2011.

Abrate, F., Vesco, A., & Scopigno, R. (2011). An analytical packet error rate model for WAVE receivers. In *Proceedings of the 4th IEEE Symposium on Wireless Vehicular Communications* (WiVEC), San Francisco, CA, USA.

Adamek, T. (2010). Statistik 1. Retrieved from http://www.uni-stuttgart.de/bio/adamek/numerik/statistik1.pdf

Adams, W. G., & Day, R. E. (1876). The action of light in selenium. *Proceedings of the Royal Society of London, 25*, 113. doi:10.1098/rspl.1876.0024

Adler, R., & Desmares, P. (1975). Consumer product applications of ultrasound. *Proceedings of the 1975 IEEE Ultrasonics Symposium*, (p. 553).

Agrawal, G. P. (2001). *Nonlinear fiber optics (Vol. 3)*. London, UK: Academic Press.

Ahmavaara, K., Haverinen, H., & Pichna, R. (2003). Interworking architecture between 3GPP and WLAN systems. *IEEE Communications Magazine, 41*(11), 74–81. doi:10.1109/MCOM.2003.1244926

Ahmed, M., Saraydar, C. U., ElBatt, T., Jijun, Y., Talty, T., & Ames, M. (2007). Intra-vehicular wireless networks. In *Proceedings of IEEE GLOBECOM 2007 Workshops*, (pp. 1-9). doi: 10.1109/GLOCOMW.2007.4437827

Ahn, C. W., & Ramakrishna, R. S. (2002). A genetic algorithm for shortest path routing problem and the sizing of populations. *IEEE Transactions on Evolutionary Computation, 6*(6), 566–579. doi:10.1109/TEVC.2002.804323

Ahonen, S., & Laitinen, H. (2003). Database correlation method for UMTS location. In *Proceedings of the 57th IEEE Vehicular Technology Conference,* Vol. 4, (pp. 2696–2700).

Akhavan, K. (2002). High-speed power-efficient indoor wireless infrared communication using code combining – Part I. *IEEE Transactions on Communications, 50*(7), 1098. doi:10.1109/TCOMM.2002.800811

Akki, A. S., & Haber, F. (1986). A statistical model of mobile-to-mobile land communication channel . *IEEE Transactions on Vehicular Technology, 35*, 2–7. doi:10.1109/T-VT.1986.24062

Alagha, N. S., & Kabal, P. (1999). Generalised raised-cosine filters. *IEEE Transactions on Communications, 47*(7), 997. doi:10.1109/26.774849

Alliance, W. (2011). *The worldwide UWB platform for wireless multimedia.* Retrieved September 1, 2011, from http://www.wimedia.org/en/about/commonradio.asp?id=abt

Amditis, A., Polychronopoulos, C., Floudas, N., & Adreone, L. (2005). Fusion of infrared vision and radar for estimating the lateral dynamics of obstacles. *Information Fusion, 6*, 129–141. doi:10.1016/j.inffus.2004.06.002

Amoroso, F. (1976). Pulse and spectrum manipulation in the minimum (frequency) shift keying (MSK) format. *IEEE Transactions on Communications, 24*(3), 381. doi:10.1109/TCOM.1976.1093294

Anand, V., Chauhan, S., & Qiao, C. (2002). *Sub-path protection: A new framework for optical layer survivability and its quantitative evaluation.* Dept. CSE, SUNY Buffalo, Tech. Rep. 2002-01.

Ananenkov, A. (2008). *Characteristics of radar images in radio vision systems of the automobile. ICTON MW 2008.* Morocco: Marrakech.

Anderson, B., & Moore, J. B. (Eds.). (1979). *Optimal filtering.* Prentice Hall.

Anderson, M. J. (1995). Broadband electrostatic transducers: modelling and experiments. *The Journal of the Acoustical Society of America, 97*(1), 262. doi:10.1121/1.412310

ANSI. (1978). *Standard S1.26–1978 ASA 23-78, Method for the calculation of the absorption of sound by the atmosphere.* New York, NY: American National Standards Institute.

Arabas, J., & Kozdrowski, S. (2001). Applying an evolutionary algorithm to telecommunication network design. *IEEE Transactions on Evolutionary Computation, 5*(4), 309–323. doi:10.1109/4235.942526

Arlsan, H., & Yucek, T. (2003), Estimation of frequency selectivityfor OFDM based new generation wireless communication systems. *Proceedings World Wireless Congress,* Vol. 1, San Francisco, CA, USA

Armstrong, J. (2009, February). OFDM for optical communications. *Journal of Lightwave Technology, 27*(3), 189–204. doi:10.1109/JLT.2008.2010061

Arndt, C. (2001). *Information measures.* Springer. doi:10.1007/978-3-642-56669-1

Arrue, J., Zubia, J., Durana, G., & Mateo, J. (2001). Parameters affecting bending losses in graded-index polymer optical fibers. *IEEE Journal on Selected Topics in Quantum Electronics, 7*(5), 836–844. doi:10.1109/2944.979345

Atamturk, A., & Zhang, M. (2006). Two-stage robust network flow and design under demand uncertainty. *Operations Research, 55*(4).

Atkinson, K., & Han, W. (2004). *Elementary numerical analysis.* Wiley.

Autenrieth, A., & Kirstadter, A. (2002). Engineering end-to-end IP resilience using resilience-differentiated QoS. *IEEE Communications Magazine, 40*(1), 50–57. doi:10.1109/35.978049

Bai, F., & Helmy, A. (2006). *A survey of mobility models in wireless adhoc networks.* USA: University of Southern California.

Baker, N. (2005). ZigBee and Bluetooth strengths and weaknesses for industrial applications. *Computing & Control Engineering Journal, 16*(2), 20–25. doi:10.1049/cce:20050204

Baliga, J., Ayre, K., Sorin, R., Hinton, W., & Tucker, R. S. (2008). Energy consumption in access networks. *Conference on Optical Fiber communication/National Fiber Optic Engineers Conference,* (pp. 1-3).

Baliga, J., Ayre, R., Hinton, K., Sorin, W., & Tucker, R. S. (2009, July). Energy consumption in optical IP networks. *Journal of Lightwave Technology, 27*(13), 2391–2403. doi:10.1109/JLT.2008.2010142

Ball, D. W. (2007). The electromagnetic spectrum: A history. *Spectroscopy, 22*(3), 14.

Bargh, M. S., Hulsebosch, R. J., Eertink, E. H., Prasad, A., Wang, H., & Schoo, P. (2004). Fast authentication methods for handovers between IEEE 802.11 wireless LANs. *Proceedings from ACM 04: Workshop on Wireless Mobile Applications and Services on WLAN Hotspots,* (pp. 51–60).

Barkan, E., Biham, E., & Keller, N. (2003). Instant ciphertext-only cryptanalysis of GSM encrypted communication. *Advances in Cryptology, 2729.*

Barry, J. R. (1994). *Wireless infrared communications.* Kluwer Academic. doi:10.1007/978-1-4615-2700-8

Barry, J. R., Kahn, J. M., Krause, W. J., Lee, E. A., & Messerschmitt, D. G. (1993). Simulation of multipath impulse response for indoor wireless optical channels. *IEEE Journal on Selected Areas in Communications, 11*(3), 367–379. doi:10.1109/49.219552

Bar-Shalom, Y. (Ed.). (1990). *Multitarget-multisensor tracking: Advanced applications (Vol. 1).* Artech House.

Bar-Shalom, Y. (Ed.). (1992). *Multitarget-multisensor tracking: Advanced applications (Vol. 2).* Artech House.

Bar-Shalom, Y. (Ed.). (2000). *Multitarget-multisensor tracking: Advanced applications (Vol. 3).* Artech House.

Bar-Shalom, Y., & Fortmann, T. E. (1988). *Tracking and data association.* Academic Press.

Bar-Shalom, Y., & Li, X. R. (1993). *Estimation and tracking.* Artech House.

Bartlett, D., & Morris, P. (2001). *CVB: A technique to improve OTDOA positioning in 3G networks.* Cambridge Positioning Systems Ltd., White paper.

Barton, D. K., & Leonov, S. A. (1997). *Radar technology encyclopedia.* Boston, MA: Artech House Inc.

Bass, H. E. (1984). Absorption of sound by the atmosphere. *Physical Acoustics, 17,* 145–233.

Bass, H. E., & Bauer, H.-J. (1972). Atmospheric absorption of sound: Analytical expressions. *The Journal of the Acoustical Society of America, 52*(3), 821. doi:10.1121/1.1913183

Bass, H. E., & Shields, F. D. (1977). Absorption of sound in air: High frequency measurements. *The Journal of the Acoustical Society of America, 62*(3), 571. doi:10.1121/1.381576

Beauvais, M., & Lakshmanan, S. (2000). CLARK: A heterogeneous sensor fusion method for finding lanes and obstacles. *Image and Vision Computing, 18,* 397. doi:10.1016/S0262-8856(99)00035-9

Becker, J. C., & Stämpfle, M. (2004). Fusion and filtering of multidimensional objects for driver assistance systems. *2004 IEEE Intelligent Vehicles Symposium,* Parma, (pp. 613-618).

Becker, J. (2002). *Fusion der Daten der objekterkennenden Sensoren eines autonomen Straßenfahrzeugs.* Fortschritt-Berichte VDI.

Bejerano, Y., Breitbart, Y., Orda, A., Rastogi, R., & Sprintson, A. (2005). Algorithms for computing QoS paths with restoration. *IEEE/ACM Transactions on Networking, 13*(3), 648–661. doi:10.1109/TNET.2005.850217

Benson, M., & Thomas, H. J. (2002). Investigation of UMTS to GSM handover procedure. *Proceedings from VTC 02: 55th IEEE Vehicular Technology Conference,* Vol. 4, (pp. 1829-1833).

Bettstetter, C. (2001). Smooth is better than sharp: A random mobility model for simulation of wireless networks. *Proceedings of the ACM Intl. Workshop Modelling, Analysis and Simulation of Wireless and Mobile Systems,* (pp. 19-27).

Betz, C., Rastorguev, V., & Strobel, O. (2011). Development of a correlation speed measurement for automotive applications. *Proceeding of 19th Annual International Conference on Composites/Nano Engineering (ICCE - 19), Special Session, Optical and Related Microwave Communication, ICCE-19 Шанхае, Китай, 24-30 июля, 2011* Shanghai, China, July 24-30, (p. 4).

Bidyut, K. S., & Wheeler, R. L. (1998). *Skin effects models for transmission line structures using generic SPICE circuit simulators.* IEEE 7th Topical Meeting on Electrical Performance on Electronic Packaging.

Bilstrup, K., Uhlemann, E., Ström, E. G., & Bilstrup, U. (2008). Evaluation of the IEEE 802.11p MAC method for vehicle-to-vehicle communication. In *Proceedings of the 2nd IEEE International Symposium on Wireless Vehicular Communications,* Calgary, Canada.

Blackmann, S., & Popoli, R. (1999). *Modern tracking systems.* Artech House.

Boban, M., Vinhoza, T. T. V., Ferreira, M., Barros, J., & Tonguz, O. (2011). Impact of vehicles as obstacles in vehicular ad hoc networks. *IEEE Journal on Selected Areas in Communications*, 29(1), 15–28. doi:10.1109/JSAC.2011.110103

Böhm, A., & Johnson, M. (2009). Position-based data traffic Prioritization in safety-critical, real-time vehicle-to-infrastructure communication. In *Proceedings of the International Conference on Communications*.

Böhm, A., & Jonsson, M. (2009). *Handover in IEEE 802.11p-based delay-sensitive vehicle-to-infrastructure communication*. Technical Report IDE0924, Halmstad University.

Borkus, M. K., & Cherniy, A. E. (1973). *Correlation's measurements of speed and a deviation corner of flying apparatus* (p. 223). Moscow, Russia: Sovetskoye Radio. (in Russian)

Börner, M. (1967). *Mehrstufiges Übertragungssystem für Pulscodemodulation dargestellte Nachrichten.* (Patent DE1254513).

Borowski, J., Niemelä, J., & Lempiäinen, J. (2004). Performance of cell ID+RTT hybrid positioning method for UMTS radio networks. In *Proceedings of the 5th European Wireless Conference*, (pp. 487 – 492).

Bosch. (2004). *Sicherheits- und Komfortsysteme.* Vieweg.

Boucouvalas, A. C. (1996). IEC825-1 eye safety classification of some consumer electronic products. *IEE Colloquium on Optical Free Space Communication Links*, 13/1-13/6. doi: 10.1049/ic:19960198

Bratanov, P. I., & Bonek, E. (2003). Mobility model of vehicle-borne terminals in urban cellular systems. *IEEE Transactions on Vehicular Technology*, 52(4), 947–952. doi:10.1109/TVT.2003.808795

Braun, R. P. (1998, September). Tutorial: Fibre radio systems, applications and devices. *24th European Conference on Optical Communication*, (pp. 87–119).

Braun, R. P., & Grosskopf, G. (1998). Optical millimeter-wave systems for broadband mobile communications, devices and techniques. *Proceedings of the International Zurich Seminar on Broadband Communications, Accessing, Transmission, Networking*, (pp. 51-58).

Breyer, F., Hanik, N., Lee, S. C. J., & Randel, S. (2007). Getting the impulse response of SI-POF by solving the time-dependent power-flow equation using the Crank-Nicholson scheme . In Bunge, C. A., & Poisel, H. (Eds.), *POF modelling: Theory, measurement and application.* Norderstedt, Germany: Verlag Books on Demand GmbH.

Bronstein, N. I., & Semendjajew, K. A. (1981). *Taschenbuch der Mathematik.* Thun und Frankfurt, Germany: Verlag Harri Deutsch.

Bunge, C. A., Kruglov, R., & Poisel, H. (2006). Rayleigh and Mie scattering in polymer optical fibers. *Journal of Lightwave Technology*, 24(8), 3137–3146. doi:10.1109/JLT.2006.878077

Burden, R. L., & Faires, J. D. (2011). *Numerical analysis.* Brooks/Cole.

Burrus, C. A., & Miller, B. I. (1971). Small area double heterostructure AlGaAs electroluminescent diode sources for optical-fiber transmission lines. *Optics Communications*, 4(4), 307. doi:10.1016/0030-4018(71)90157-X

Camp, T., Boleng, J., & Davies, V. (2002). A survey of mobility models for ad hoc network research. *Proceedings of WCMC 2002: Special Issue on Mobile Ad Hoc Networking: Research, Trends and Applications*, 2(5), 483–502.

Campolo, C., Cozzetti, H. A., Molinaro, A., & Scopigno, R. (2012). *Overhauling ns-2 PHY/MAC simulations for IEEE 802.11p/WAVE vehicular networks.* IEEE ICC 2012, Workshop on Intelligent Vehicular Networking: V2V/V2I Communications and Applications, Ottawa, Canada

Campuzano, G., Aldaya, I., & Castañón, G. (December de 2009). Performance of digital modulation formats in radio over fiber systems based on the sideband injection locking technique. *ICTON Mediterranean Winter Conference ICTON-MW*, (pp. 1-5).

CAN in Automation. (2012). *About CAN in Automation.* Retrieved from http://www.cancia.org/index.php?id=aboutcia

Canbuskit 2012. (n.d.). Retrieved from http://www.canbuskit.com

Carr, H. (1992). Developments in capacitive transducers. *Non-Destructive Testing and Evaluation*, 10(1), 3. doi:10.1080/10589759208952778

Carruthers, J. B., & Kahn, J. M. (1998). Angle diversity for nondirected wireless infrared communication. *IEEE International Conference on Communications*, Vol. 3, (p. 1665).

Carruthers, J. B., & Kannan, P. (2002). Iterative site-based modelling for wireless infrared channels. *IEEE Transactions on Antennas and Propagation*, 50(5), 759–765. doi:10.1109/TAP.2002.1011244

Castañón, G., Aragón-Zavala, A., Ramirez-Velarde, R., Campuzano, G., & Tonguz, O. (2008). *An integrated availability analysis of RoF networks.* 2nd International Conference on Transparent Optical Networks ICTON.

Castañón, G., Campuzano, G., & Tonguz, O. (2008). High reliability and availability in radio over fiber networks. *OSA Journal of Optical Networking*, 7(6), 603–616. doi:10.1364/JON.7.000603

Chang, K. C., Saha, R. K., Bar-Shalom, Y., & Alford, M. (1996). Performance evaluation of multisensor track-to-track fusion. *Proceedings of the 1996 IEEE International Conference on Multisensor Fusion and Integration for Intelligent Systems*, (pp. 627-632).

Chang, C. H., Lu, H. H., Su, H. S., Shih, C. L., & Chen, C. L. (2009, November). A broadband ASE light source-based full-duplex FTTX/ROF transport system. *Optics Express*, 17.

Chang, G., Chowdhury, A., Jia, Z., Chien, H., Huang, M., Yu, J., & Ellinas, G. (2009, September). Key technologies of WDM-PON for future converged optical broadband access networks. *Journal of Optical Communications and Networking*, 1(4), C35–C50. doi:10.1364/JOCN.1.000C35

Chang, Q., Fu, H., & Su, Y. (2008, February). Simultaneous generation and transmission of downstream multiband signals and upstream data in a bidirectional radio-over-fiber system. *IEEE Photonics Technology Letters*, 20(3), 181–183. doi:10.1109/LPT.2007.912992

Chbat, M. W. (2000). Managing polarization mode dispersion. *Photonics Spectra*, 100.

Chen, Q., Schmidt-Eisenlohr, F., Jiang, D., Torrent-Moreno, M., Delgrossi, L., & Hartenstein, H. (2007). Overhaul of IEEE 802.11 modeling and simulation in NS-2. In *Proceedings of the 10th ACM International Symposium on Modeling, Analysis and Simulation of Wireless and Mobile Systems* (MSWiM), Chania, Crete Island, Greece, (pp. 159–168).

Cheng, L., Henty, B. E., Stancil, D. D., Bai, F., & Mudalige, P. (2007). Mobile vehicle-to-vehicle narrow-band channel measurement and characterization of the 5.9 GHz dedicated short range communication (DSRC) frequency band. *IEEE Journal on Selected Areas in Communications*, 25(8), 1501–1516. doi:10.1109/JSAC.2007.071002

Chen, H. H., & Wong, S. Y. (1997). Spectral efficiency analysis of new quadrature overlapped modulations over band-limited non-linear channels. *International Journal of Communication Systems*, 10(1), 1. doi:10.1002/(SICI)1099-1131(199701)10:1<1::AID-DAC319>3.0.CO;2-F

Chen, L., Yu, J. G., Wen, S., Lu, J., Dong, Z., Huang, M., & Chang, G. K. (2009, July). A novel scheme for seamless integration of ROF with centralized lightwave OFDM-WDM-PON system. *Journal of Lightwave Technology*, 27(14), 2786–2791. doi:10.1109/JLT.2009.2016984

Chen, T., Yang, Y., Zhang, H., Kim, H., & Horneman, K. (2011, October). Network energy saving technologies for green wireless access networks. *IEEE Wireless Communications*, 18, 30–38. doi:10.1109/MWC.2011.6056690

Chien, H., Hsueh, Y., Chowdhury, A., Yu, J., & Chang, G. (2010, February). On frequency-doubled optical millimeter-wave generation technique without carrier suppression for in-building wireless over fiber applications. *IEEE Photonics Technology Letters*, 22(3), 182–184. doi:10.1109/LPT.2009.2037333

Chiu, M., & Bassiouni, M. A. (2000). Predictive schemes for handoff prioritization in cellular network based on mobile positioning. *IEEE Journal on Selected Areas in Communications*, 18(3), 510–522. doi:10.1109/49.840208

Clarke, R. H. (1968). A statistical theory of mobile radio reception. *The Bell System Technical Journal*, 47(6), 957–1000.

Cliche, J., Allard, M., & Tétu, M. (2006). *Ultra-narrow linewidth and high frequency stability laser sources. Optical Amplifiers and their Applications/Coherent Optical Technologies and Applications.* Technical Digest.

Cochrane, P. (1989, September). The future symbiosis of optical fiber and microwave radio systems. *19th European Microwave Conference*, (pp. 72–86).

Cohen, L. G. (1982). Dispersion and bandwidth spectra in single-mode fibers. *IEEE Journal of Quantiatitve Electronics*, *18*(1), 49. doi:10.1109/JQE.1982.1071366

Consortium, L. I. N. (2012). LIN specification package. Retrieved from http://www.linsubbus.de/

Cooperation, M. O. S. T. (2007). *MOST brand book vol. 1.1*, Aug. 2007. Retrieved December 7, 2009, from www.mostcooperation.com

Cooperation, M. O. S. T. (2012). *MOST specifications.* Retrieved from http://www.mostcooperation.com/publications/specificationsorganizationalprocedures/index.html

Cozzetti, H. A., & Scopigno, R. (2011). Scalability and QoS in slotted VANETs: Forced slot re-use vs pre-emption. In *Proceedings of the 14th International IEEE Conference on Intelligent Transportation Systems (ITSC 2011)*, Washington, DC, USA, Oct. 2011.

Cozzetti, H. A., Campolo, C., Scopigno, R., & Molinaro, A. (2012). Urban VANETs and hidden terminals: Evaluation by RUG, a validated propagation model. In *Proceedings IEEE International Conference on Vehicular Electronics and Safety (ICVES'12)*, Istanbul, Turkey.

Cozzetti, H. A., Scopigno, R., Casone, L., & Barba, G. (2009), Comparative analysis of IEEE 802.11p and MS-Aloha in VANETs scenarios. In *Proceedings of the 2nd IEEE International Workshop on Vehicular Networking (VON 2009)*, Biopolis, Singapore.

Cozzetti, H. A., Vesco, A., Abrate, F., & Scopigno, R. (2010). Improving wireless simulation chain: Impact of two corrective models for VANETs. In *Proceedings of the 2nd IEEE Vehicular Networking Conference (VNC 2010)*, Jersey City, New Jersey, USA.

Crandall, I. B. (1918). Air-damped vibrating system – Theoretical calibration of the condenser transmitter. *Physical Review*, *11*, 449. doi:10.1103/PhysRev.11.449

Dana, K., Van-Ginneken, B., Nayer, S., & Koenderink, J. (1999). Reflectance and texture of real world surfaces. *ACM Transactions on Graphics*, *18*(1), 1–34. doi:10.1145/300776.300778

Daniels, R. C., & Heath, W. (2007, September). 60 GHz wireless communications: Emerging requirements and design recomendations. *IEEE Vehicular Technology Magazine*, 41–50.

De Jong, K. A., & Spears, W. M. (1989). Using genetic algorithm to solve NP-complete problems. *Proceeding of Third International Conference on Genetic Algorithms (ICGA)*.

De Pauw, P. (2008). *Melexis new fiber optic ttransceivers.* ITG VDE 5.4.1, Fachgruppentreffen Krems. Retrieved from http://www.pofac.de/downloads/itgfg/fgt25/FGT25_Krems_DePauw_Melexis-Transceivers.pdf

Delphi. (2003). *MOST Interconnectivity Conference, Frankfurt.*

Dintchev, P. A., Perez-Quikes, B., & Bonek, E. (2004). An improved mobility model for 2G and 3G cellular systems. In *International Conference on 3G Mobile Communication Technologies, 5th 3G 2004,* (pp. 402-406).

Dissanayake, S. D., Karunasekara, P. P. C., Lakmanaarachchi, D. D., Rathnayaka, A., & Samarasinghe, A. T. L. (2008). Zigbee wireless vehicular identification and authentication system. *Proceedings of ICIAFS*, *2008*, 257–260. doi:doi:10.1109/ICIAFS.2008.4783998

Djordjevich, A., & Savović, S. (2000). Investigation of mode coupling in step index plastic optical fibers using the power flow equation. *IEEE Photonics Technology Letters*, *12*(11), 1489–1491. doi:10.1109/68.887704

Djordjevich, A., & Savović, S. (2004). Numerical solution of the power flow equation in step-index plastic optical fibers. *Journal of the Optical Society of America. B, Optical Physics*, *21*, 1437–1438. doi:10.1364/JOSAB.21.001437

Djordjevich, A., & Savović, S. (2008). Coupling length as an algebraic function of the coupling coefficient in step-index plastic optical fibers. *Optical Engineering (Redondo Beach, Calif.)*, *47*(12), 125001. doi:10.1117/1.3049858

Dötzer, F., Kohlmayer, F., Kosch, T., & Strassberger, M. (2005). Secure communication for intersection assistance. In *Proceedings of the 2nd International Workshop on Intelligent Transportation*, Hamburg, Germany.

Drullmann, R., & Kammerer, W. (1980). Leitungscodierung und betriebliche Überwachung bei regenerativen Lichtleitkabelübertragungssystemen. *Frequenz, 34*(2), 45. doi:10.1515/FREQ.1980.34.2.45

Durana, G., Zubia, J., Arrué, J., Aldabaldetreku, G., & Mateo, J. (2003). Dependence of bending losses on cladding thickness in plastic optical fibers. *Applied Optics, 42*(6), 997–1002. doi:10.1364/AO.42.000997

Ebbinghaus, G. (1985). Small area ion implanted p+n Germanium avalanche photodiodes for a wavelength of 1.3 μm. *Siemens Research and Development Report, 14*(6), 284.

Eichler, S. (2007). Performance evaluation of the IEEE 802.11p WAVE communication standard. *Proceedings from VTC 07: 66th IEEE Vehicular Technology Conference,* (pp. 2199-2203).

ElBatt, T., Goel, S., Holland, G., Krishnan, H., & Parikh, J. (2006). *Cooperative collision warning using dedicated short range wireless communications.* 3rd ACM International Workshop on VANETs, Los Angeles, California, USA.

Elco, R. A., & Burke, B. (2003). *SPICE models with frequency dependent conductor and dielectric losses.* High Frequency Design SPICE Models.

Engel, A. (2005). *Tyco Electronics next generation MOST networking.* MOST All Members Meeting, Frankfurt, 5th of April 2005.

Erqun, A. S. (2003). Capacitive micromachined ultrasonic transducers: Theory and technology. *Journal of Aerospace Engineering, 16*(2), 76. doi:10.1061/(ASCE)0893-1321(2003)16:2(76)

ERSO - The European Road Safety Observatory. (n.d.). Retrieved from http://erso.swov.nl/

ETSI ES 202 663 V1.2.1. (2012-03). *Intelligent transport systems (ITS); European profile standard for the physical and medium access control layer of intelligent transport systems operating in the 5 GHz frequency band.*

ETSI TR 102 638 V1.1.1. (2009-06). *Intelligent transport systems (ITS); Vehicular communications; Basic set of applications; Definitions.*

ETSI TR 102 861. (2010). *Intelligent Transport Systems (ITS); On the Recommended Parameter Settings for Using STDMA for Cooperative ITS; Access Layer Part.*

ETSI TR 102 862. (2010). *Intelligent transport systems (ITS); Performance evaluation of self-organizing TDMA as medium access control method applied to ITS; Access layer part.*

ETSI TS 102 637-2. (2010). *Intelligent transport systems (ITS); Vehicular communications; Basic set of applications; Part 2: Specification of cooperative awareness basic service.*

ETSI TS 102 637-3. (2010). *Intelligent transport systems (ITS); Vehicular communications; Basic set of applications; Part 3: Specifications of decentralized environmental notification basic service.*

ETSI TS 102 687. (2010). *Intelligent transport systems (ITS); Decentralized congestion control mechanisms for intelligent transport systems operating in the 5 GHz range; Access layer part.*

ETSI TS 102 724. (2010). *Draft: Intelligent transport systems (ITS); Vehicular communications, channel specifications 5 GHz,* v 0.0.10 (2011/10).

ETSI. (2001). *ETSI, requirements and architectures for interworking between HIPERLAN/2 and 3rd generation cellular systems.* (Tech. rep. *ETSI TR, 101,* 957.

European Commission, Information Society and Media. (2011, October). *eCall – saving lives through in-vehicle communication technology.* Retrieved from http://ec.europa.eu/information_society/doc/factsheets/049-ecall-en.pdf

Fang, Y., Masaki, I., & Horn, B. (2002). Depth-based target segmentation for intelligent vehicles: Fusion of radar and binocular stereo. *IEEE Transactions on Intelligent Transportation Systems, 3*(3), 196–202. doi:10.1109/TITS.2002.802926

Faria, G., Henriksson, J. A., Stare, E., & Talmola, P. (2006). DVB-H: Digital broadcast services to handheld devices. *Proceedings of the IEEE, 94*(1), 194–209. doi:10.1109/JPROC.2005.861011

Farsi, M., Ratcliff, K., & Barbosa, M. (1999). An overview of controller area network. *Computing & Control Engineering Journal, 10*(3), 113–120. doi:10.1049/cce:19990304

Feng, F., & Reeves, D. S. (2004). Explicit proactive handoff with motion prediction for mobile IP. *Proceedings from WCNC 04: 6ᵗʰ IEEE Wireless Communications and Networking Conference,* Vol. 2, (pp. 855-860).

Ferguson, N., & Schneier, B. (2003). *Practical cryptography.* Hoboken, NJ: Wiley.

Fiore, M. (2006). *Mobility models in inter-vehicle communications literature.* Technical report. Retrieved from http://www.tlc-networks.polito.it/oldsite/fiore/papers/mobilityModels.pdf

Firecomms. (2003). *Firecomms launches RCLED-Based S200 solution for IEEE 1394 POF applications.* Press release. Retrieved from http://pofto.com/articles/FC200R%20S200%20PR.pdf

Fischer, H.-J. (2012). *Standards for C-ITS.* Retrieved from http://its-standards.info/Standards_for_C-ITS.pdf

Fitzek, F., Munari, M., Pastesini, V., Rossi, S., & Badia, L. (2003). Security and authentication concepts for UMTS/WLAN convergence. *Proceedings from VTC 03: 58ᵗʰ IEEE Vehicular Technology Conference,* Vol. 4, (pp. 2343-2347).

Flannery, D. (2001). Raman amplifiers: Powering up for ultra-long-haul. *Fiber Systems, 5*(7), 48.

Fleder, K. (1996). *The Bergeron method.* Texas Instruments.

FlexRay Consortium. (2008). *Home page.* Retrieved August 7, 2009, from www.FlexRay.com

Flexray. (2003). *Avidyne Corporation joins Flexray Consortium.* Retrieved from http://www.flexray.com/news/Avidyne_pr.pdf

Flexray. (2008). *FlexRay in the avionic context.* FlexRay Product Day, 2008. Retrieved from

Flexray. (2010). *Flexray specifications version 3.0.1.* Retrieved from http://www.flexray.com/index.php?sid=9c03bd01bb2a03269f772bb99935a742&pid=94&lang=de

Flexray. (2010). *Specifications version 3.0.1.* Retrieved from http://www.flexray.com/index.php?sid=9c03bd01bb2a03269f772bb99935a742&pid=94&lang=de

Friis, H. T. (1994). Noise figures of radio receivers. *Proceedings of the IRE,* (pp. 419-422).

Fu, W., et al. (2006). Design and performance evaluation of carrier lock detection in digital QPSK receiver. *IEEE International Conference on Communications,* Vol. 7, (p. 2941).

Fußgänger, K., & Roßberg, R. (1990). Uni and bidirectional 4λ x560 Mbit/s transmission systems using WDM devices and wavelength-selective fused single-mode fiber couplers. *IEEE Journal on Selected Areas in Communications, 8*(6), 1032. doi:10.1109/49.57806

Fuster, G., & Kalymnios, D. (1998). Passive components in POF Data Communications. *Proceedings of the 7th International Conference on Plastic Optical Fibres and Applications,* (pp. 75-80).

Gallager, R. G. (2001). Claude E. Shannon: A retrospective on his life, work, and impact. *IEEE Transactions on Information Theory, 47*(7), 2681. doi:10.1109/18.959253

Garito, A. F., Wang, J., & Gao, R. (1998). Effects of random perturbations in plastic optical fibers. *Science, 281,* 962–967. doi:10.1126/science.281.5379.962

Gerstel, O., & Ramasawami, R. (2000). Optical layer survivability- An implementation perspective. *IEEE Journal on Selected Areas in Communications, 18,* 1885–1899. doi:10.1109/49.887910

Gfeller, F. R., & Bapst, U. (1979). Wireless in-house data communication via diffuse infrared radiation. *Proceedings of the IEEE, 67*(11), 1474–1486. doi:10.1109/PROC.1979.11508

Ghafoor, S., & Hanzo, L. (2011). Sub-carrier-multiplexed duplex 64-QAM radio-over-fiber transmission for distributed antennas. *IEEE Communications Letters,* 1–4.

Ghani, N., Dixit, S., & Wang, T. S. (2003). On IP-WDM integration: A retrospective. *IEEE Communications Magazine,* 42–45. doi:10.1109/MCOM.2003.1232232

Giordano, E., Frank, R., Ghosh, A., Pau, G., & Gerla, M. (2009). Two ray or not two ray this is the price to pay. In *Proceedings of the 6th IEEE International Conference on Mobile Ad Hoc and Sensor Systems* (MASS), Macau SAR, China, (pp. 603–608).

Giordano, E., Frank, R., Pau, G., & Gerla, M. (2011). CORNER: A radio propagation model for VANETs in urban scenarios. *Proceedings of the IEEE, 99*(7), 1280–1294. doi:10.1109/JPROC.2011.2138110

Gloge, D. (1972). Optical power flow in multimode fibers. *The Bell System Technical Journal, 51*(8), 1767–1783.

Gloge, D. (1973). Impulse response of clad optical multimode fibers. *The Bell System Technical Journal, 52*, 801–816.

Gloge, D. (1973). Multimode theory of graded core fibers. *The Bell System Technical Journal, 52*, 1563.

Goldberg, D. E. (1989). *Genetic algorithms in search, optimization, and machine learning*. Addison-Wesley.

Gräfling, S., Mähönen, P., & Rihijärvi, J. (2010). Performance evaluation of IEEE 1609 WAVE and IEEE 802.11p for vehicular communications. *Proceedings of the Second International Conference on Ubiquitous and Future Networks*, (pp. 344–348).

Grag, S., et al. (2000). MAC scheduling polices for power optimisation in Bluetooth: A master driven TDD wireless system. *IEEE Vehicular Technology Conference*, Vol. 1, (p. 196).

Grau, G. P., Pusceddu, D., Rea, S., Brickley, O., Koubek, M., & Pesch, D. (2010). Vehicle-2-vehicle communication channel evaluation using the CVIS platform. In *Proceedings 7th International Symposium on Communication Systems Networks and Digital Signal Processing*, Newcastle upon Tyne, UK, (pp. 449–453).

Green, R. J. (2010). Optical wireless with application in automotives. *Proceedings of ICTON, 2010*, 1–4. doi:doi:10.1109/ICTON.2010.5549164

Green, R. J., Joshi, H., Higgins, M. D., & Leeson, M. S. (2008). Recent developments in indoor optical wireless systems. *IET Communications, 2*(1), 3–10. doi:10.1049/iet-com:20060475

Grilli, G. (2010). *Data dissemination in vehicular networks*. PhD thesis University of Rome "Tor Vergata", Rome, Italy.

Gronemeyer, S. A., & McBride, A. L. (1976). MSK and offset QPSK modulation. *IEEE Transactions on Communications, 24*(8), 809. doi:10.1109/TCOM.1976.1093392

Group, I. G. I. (2009). Plastic optical fiber (POF) selected reprint series, Vol. 1. Retrieved from http://igigroup.net/osc3/product_info.php?cPath=22&products_id=174

Grzemba, A. (2011). MOST: The automotive multimedia network. In *From MOST25 to MOST150*. Poing, Germany: Franzis Verlag GmbH. ISBN 9783645650618

Grzemba, A., & von der Wense, H. C. (2005). *LINBus –Systeme, Protokolle, Test von LINSystemen, Tools, Hardware, Applikationen*. Poing, Germany . *FranzisVerlag GmbH., ISBN13*, 9783772340093.

GSM. (2011). *GSM World*. Retrieved from www.gsmworld.com

Gunter, S. (2009). The green base station. *4th International Conference on Telecommunication - Energy Special Conference (TELESCON)*, (pp. 1-6).

Guo, L., Li, L., Cao, J., Yu, H., & Wei, X. (2007). On finding feasible solutions with shared backup resources for surviving double-link failures in path-protected WDM mesh networks. *Journal of Lightwave Technology, 25*(25), 287–296. doi:10.1109/JLT.2006.886721

Guo, L., Yu, H., & Li, L. (2005). Protection design for double-link failures in meshed WDM networks. *Acta Eletronica Sinica, 33*(5), 883–888.

Hall, D. L. (1992). *Mathematical techniques in multisensor data fusion*. Artech House.

Hall, D. L., & Llinas, J. (1997). An introduction to multisensory data fusion. *Proceedings of the IEEE, 85*(1), 6–23. doi:10.1109/5.554205

Hall, D. L., & Llinas, J. (Eds.). (2001). *Handbook of multisensor data fusion*. CRC Press. doi:10.1201/9781420038545

Haller, M. I., & Khuri-Yakub, B. T. (1994). Micromachined 1-3 composites for ultrasonic air transducers. *The Review of Scientific Instruments, 65*, 2095. doi:10.1063/1.1145231

HarmanBecker. (2007). *Migration from 1st to 2nd generation MOST systems*. MOST All members meeting, Frankfurt, 27th of March 2007.

Härri, J., Filali, F., & Bonnet, C. (2007). *Mobility models for vehicular ad hoc networks: A survey and taxonomy*. Research Report EURECOM. doi:10.1109/SURV.2009.090403

Hartenstein, H., & Labertaux, K. (2010). *VANET: Vehicular applications and inter-networking technologies*. Wiley 2010.

Hartmann, P., Qian, X., Wonfor, A., Penty, R., & White, I. (October de 2005). *1-20 GHz directly modulated radio over MMF link*. International Topical Meeting on Microwave Photonics.

Hassun, R., Flaherty, M., Matreci, R., & Taylor, M. (1997, August). Efffective evaluation of link quality using error vector magnitude techniques. *Proceedings of Wireless Communications Conference*, (pp. 89–94).

Haynes, H. D., et al. (2002). *Ultrasonic communication project phase 1 final report*. National Security Program, Oak Ridge, TN, Office Rep. Y/NSP-252.

Heddebaut, M., Deniau, V., & Adouane, K. (2004). In-vehicle WLAN radio-frequency communication characterization. *IEEE Intelligent Transportation Systems*, 5(2), 114–121. doi:10.1109/TITS.2004.828172

He, L., Botham, C. P., & O'Shea, C. D. (2004). An evolutionary design algorithm for ring-based SDH optical core networks. *BT Technology Journal*, 22, 135–144. doi:10.1023/B:BTTJ.0000015503.09815.48

Heller, C., & Reichel, R. (2009). *Enabling FlexRay for avionic data buses*. IEEE/AIAA 28th Digital Avionics Systems Conference, DASC '09, EADS Innovation Works, Munich, Germany.

Heller, C. (2010). FlexRay on aeronautic harnesses, 2010. *Institut für Luftfahrtsysteme.*, *ISBN3868537112*, 9783868537116.

He, R., Wen, H., & Li, L. (2004). Shared sub-path protection algorithm in traffic-grooming WDM mesh networks. *Photonic Network Communications*, 8(3), 239–249. doi:10.1023/B:PNET.0000041236.17592.b4

Heredia, P., Mateo, J., & Losada, M. A. (2007). Transmission capabilities of large core GI-POF based on BER measurements. *Proceedings of the 16th International Conference on Plastic Optical Fibres and Applications*, (pp. 307-310).

Hibbs-Brenner, M. K. (2009). VCSEL technology for medical diagnostics and therapeutics. *Proceedings of the Society for Photo-Instrumentation Engineers*, 7180, 71800–71810. doi:10.1117/12.815307

Higgins, M. D., Green, R. J., & Leeson, M. S. (2009). A genetic algorithm method for optical wireless channel control. *Journal of Lightwave Technology*, 27(6), 720–772. doi:10.1109/JLT.2008.928395

Higgins, M. D., Green, R. J., & Leeson, M. S. (2011). Channel viability of intra-vehicle optical wireless communications. *Proceedings of IEEE GLOBECOM Workshops, 2011*, 183–817. doi:doi:10.1109/GLOCOMW.2011.6162567

Higgins, M. D., Green, R. J., & Leeson, M. S. (2012). Optical wireless for intravehicle communications: A channel viability analysis. *IEEE Transactions on Vehicular Technology*, 61(1), 123–129. doi:10.1109/TVT.2011.2176764

Higgins, M. D., Green, R. J., Leeson, M. S., & Hines, E. L. (2011a). Multi-user indoor optical wireless communication system channel control using a genetic algorithm. *IET Communications*, 5(7), 937–944. doi:10.1049/iet-com.2010.0204

Hiroshi, O. (2004). Experiments on acoustic communication with quadrature amplitude modulation in multipath environment. *Japanese Journal of Applied Physics, Part 1: Regular Papers and Short Notes and Review Papers*, 43(5B), 3140. doi:10.1143/JJAP.43.3140

HIS iSupply. (2011). *LCD panel market growth slows in 2011*. Retrieved from http://www.isuppli.com/DisplayMaterialsandSystems/ MarketWatch/Pages/LCDPanelMarketGrowthSlowsin2011.aspx

Hoefer, E. E. E., & Nielinger, H. (1985). *SPICE*. Springer Verlag. ISBN 3-540-15160-5

Hogg, D. C., & Mumford, W. W. (1960). *The effective noise temperature of the sky* (pp. 80–84). The Microwave Journal.

Holm, S. (2005). Airborne ultrasound data communications: The core of an indoor positioning system. *Proceedings IEEE Ultrasonics Symposium*, Vol. 3, (p. 1801).

Holz, D. (2003). *Automatisierte Bewertung von Fusionsverfahren zur Fahrzeugumfelderfassung*. Master Thesis, FH Karlsruhe und Robert Bosch GmbH.

Ho, P.-H. (2004). State-of-the-art progress in developing survivable routing schemes in mesh WDM networks. *IEEE Communications*, 6(4), 2–16.

Ho, P.-H., & Mouftah, H. T. (2004). A novel survivable routing algorithm for shared segment protection in mesh WDM networks with partial wavelength conversion. *IEEE Journal on Selected Areas in Communications*, 22(8), 1548–1560. doi:10.1109/JSAC.2004.830475

Ho, P.-H., & Mouftah, H. T. (2004). On optimal diverse routing for shared protection in mesh WDM networks. *IEEE Transactions on Reliability*, 53(6), 216–225. doi:10.1109/TR.2004.829141

Ho, P.-H., Tapolcai, J., & Cinkler, T. (2004). Segment shared protection in mesh communications networks with bandwidth guaranteed tunnels. *IEEE/ACM Transactions on Networking*, 12(6), 1105–1118. doi:10.1109/TNET.2004.838592

Hossain, E., Chow, G., Leung, V., McLeod, B., Misic, J., Wong, V., & Yang, O. (2010). Vehicular telematics over heterogeneous wireless networks: A survey. *Computer Communications*, 33(7), 775–793. doi:10.1016/j.comcom.2009.12.010

http://arstechnica.com/gadgets/news/2011/07/2011will-hit420millionsmartphonesalessamsunghits433percent-growthsamsungsees300growthinsmartphonemarketshare.ars

http://www.eurtd.com/moet/PDF/EADSIW%20(Presentation)%20%20FlexRay%20Product%20Day.pdf

Huang, C. M. (Ed.). (2010). *Telematics communication technologies and vehicular networks: Wireless architectures and applications*. Hershey, PA: IGI Global.

Huang, D., & Letaief, K. B. (2006). Carrier frequency offset estimation for OFDM systems using null subcarriers. *IEEE Transactions on Communications*, 55(4), 813–822. doi:10.1109/TCOMM.2006.874001

IEEE 802.11 Working Group. (2007). *IEEE standard for information technology--Telecommunications and information exchange between systems Local and metropolitan area networks--Specific requirements Part 11: Wireless LAN medium access control (MAC) and physical layer (PHY) specifications.*

IEEE 802.11p Working Group. (2010). *IEEE standard 802.11p, Wireless LAN medium access control (MAC) and physical layer (PHY) specifications: Amendment 6, wireless access in vehicular environments.*

IEEE 802.11-Working Group. (2012). *IEEE standard for information technology--Telecommunications and information exchange between systems local and metropolitan area networks--Specific requirements part 11: Wireless LAN medium access control (MAC) and physical layer (PHY) specifications*

IEEE 802.15. (n.d.). *Task Group 3c (TG3c) Millimeter Wave Alternative PHY*. Recuperado el November de 2011, de http://ieee802.org/15/pub/TG3c.html

IEEE 802.15. (n.d.). *Wireless Personal Area Networks*. Recuperado el November de 2011, de http://www.ieee802.org/15/

IEEE 802.16. (n.d.). *Standard for Wireless Metropolitan Area Networks*. Recuperado el November de 2011, de http://ieee802.org/16/.

IEEE. (2006). *IEEE trial-use standard for wireless access in vehicular environments (WAVE)—Multi-channel operation*. Trial-Use Std. 1609.4.

IEEE. (2010). *Part 11: Wireless LAN medium access control (MAC) and physical layer (PHY) specifications amendment 6: Wireless access in vehicular environments.* IEEE 802.11p published standard.

Iizuka, N. (2008). *Some comments on merged draft from the viewpoint of the VL-ISC*. (IEEE 802.15-08-0759-02-0vlc, Project: IEEE P802.15Working Group for Wireless Personal Area Networks (WPANs).

Insua, I. G., Plettemeier, D., & Schffer, C. G. (2010, February). Simple remote heterodyne radio over fiber system for Gbps wireless access. *Journal of Lightwave Technology*, 28, 1–1. doi:10.1109/JLT.2010.2042426

International Organization of Motor Vehicle Manufacturers (OICA). (2011). Statistics on worldwide car production. Retrieved from http://www.worldometers.info/cars/

Ishimaru, A. (1991). Wave propagation and scattering in random media and rough surfaces. *Proceedings of the IEEE, 79*(10), 1359–1366. doi:10.1109/5.104210

Islam, A. R., & Bakaul, M. (2009, December). Simplified millimeter-wave radio-over-fiber system using optical heterodyning of low cost independent light sources and RF homodyning at the receiver. *International Topical Meeting on Microwave Photonics*, (pp. 1–4).

Ismail, T. M., & Seeds, A. (2005). Linearity enhancement of a directly modulated uncooled dfb laser in a multi-channel wireless-over-fibre system. *IEEE MTT-S International Microwave Symposium Digest*, (pp. 7-10).

Ito, H., Kodama, S., Muramoto, Y., Furuta, T., Nagatsuma, T., & Ishibashi, T. (August de 2004). High-speed and high-output InP-InGaAs unitraveling-carrier photodiodes. *IEEE Journal of Selected Topics in Quantum Electronics, 10*, 709–727.

ITU. (1988). *General characteristics of international telephone connections and international telephone circuits*. ITU-TG.114.

ITU. (2003). *Framework and overall objectives of the future development of IMT-2000 and systems beyond IMT-2000*. (Recommendation ITU-R M.1645).

ITU. (2003b). *One-way transmission time*. ITU G.114.

ITU-R. (n.d.). *Recommendation ITU-R M.1645*. Retrieved from http://www.itu.int/ITU-R.

Jakes, W. C. (Ed.). (1975). *Microwave mobile communications*. New York, NY: John Wiley & Sons Inc.

Jeruchim, M. C., Balaban, P., & Shanmugan, K. S. (2000). *Simulation of communication systems* (2nd ed.). New York, NY: Kluwer Academic.

Jeunhomme, L., Fraise, M., & Pocholle, J. P. (1976). Propagation model for long step-index optical fibers. *Applied Optics, 15*, 3040–3046. doi:10.1364/AO.15.003040

Jiang, T., Chen, H. H., Wu, H. C., & Yi, Y. (2010). Channel modeling and inter-carrier interference analysis for V2V communication systems in frequency-dispersive channels. *The Journal Mobile Networks and Applications, 15*(1), 4–12. doi:10.1007/s11036-009-0177-2

Johnson, D. B., Perkins, C. E., & Arkko, J. (2004). *Mobility support in IPv6*. Retrieved from http://www.ietf.org/rfc/rfc3775.txt.

Johnston, C. (2011). *Samsung sees 300% growth in smartphone market share: IMS Research finds that 420 million smartphone sales will have been made by the end of 2011*. Retrieved from.

Juang, R., Lin, H., & Lin, D. (2005). An improved location-based handover algorithm for GSM systems. *Proceedings from WCNC 05: 7th IEEE Wireless Communications and Networking Conference*, Vol. 3, (pp. 1371–1376).

Kaheel, A., Khattab, T., Mohamed, A., & Alnuweiri, H. (2002). Quality-of-service mechanisms in IP-over-WDM networks. *IEEE Communications Magazine, 40*(12), 38–43. doi:10.1109/MCOM.2002.1106157

Kahn, J. M., Krause, W. J., & Carruthers, J. B. (1995). Experimental characterization of non-directed indoor infrared channels. *IEEE Transactions on Communications, 43*(2,3,4), 1613-1623. doi: 10.1109/26.380210

Kahn, J. M., & Barry, J. R. (1997). Wireless infrared communications. *Proceedings of the IEEE, 85*(2), 265–298. doi:10.1109/5.554222

Kalman, R. E. (1960). A new approach to linear filtering and prediction problems. *Transactions ASME . Journal of Basic Engineering, 82*, 35–45. doi:10.1115/1.3662552

Kalman, R. E., & Bucy, R. S. (1961). New results in linear filtering and prediction theory. *Transactions ASME . Journal of Basic Engineering, 83*, 95–108. doi:10.1115/1.3658902

Kalymnios, D. (1999). Squeezing more bandwidth into high NA POF. *Proceedings of the 8th International Conference on Plastic Optical Fibres and Applications*, (pp. 18-24).

Kamra, V., & Kumar, V. (2011, January). Power penalty in multitone radio over-fibre system employing direct and external modulation with optical amplifiers. *International Journal for Light and Electron Optics, 34*, 44–48. doi:10.1016/j.ijleo.2010.02.001

Kao, C. K., & Hockham, G. A. (1966). Dielectric-fiber surface waveguides for optical frequencies. *Proceedings of the IEE, 113*(7), 1151.

Kapron, F. P. (1970). Radiation losses in glass optical waveguides. *Applied Physics Letters, 17*(10), 423. doi:10.1063/1.1653255

Kartalopoulos, S. V. (2000). *Introduction to DWDM technology: Data in a rainbow*. Piscataway, NJ: IEEE Press.

Kashima, N. (2006, August). Dynamic properties of FP-LD transmitters using side-mode injection locking for LANs and WDM-PONs. *Journal of Lightwave Technology, 24*(8), 3045–3058. doi:10.1109/JLT.2006.878056

Kastell, K. (2007). *Sichere, schnelle, ortsbasierte Handover in hybriden Netzen*. Darmstädter Dissertation. Aachen, Germany: Shaker-Verlag.

Kastell, K. (2011). Challenges for handovers in hybrid networks. *Proceedings from IWT 11: International Workshop on Telecommunications*.

Kastell, K. (2012). Novel mobility models and localization techniques to enhance location-based services in transportation systems. *Proceedings of ICTON 2012*.

Kastell, K., & Jakoby, R. (2006). *Fast handover with integrated authentication for hybrid networks*. IEEE-Vehicular Technology Conference (VTC) Fall 2006.

Kastell, K., Bug, S., Nazarov, A., & Jakoby, R. (2006). *Improvements in railway communication via GSM-R*. IEEE-Vehicular Technology Conference (VTC) Spring 2006, Melbourne, Australia.

Kastell, K., Fernandez-Pello, A., Fernandez, D., Meyer, U., & Jakoby, R. (2004). Performance advantage and use of a location based handover algorithm. *Proceedings from VTC 04: 60th IEEE Vehicular Technology Conference*, Vol. 7, (pp. 5260-5264).

Kastell, K., Meyer, U., & Jakoby, R. (2003). Secure handover procedures. In *Proceedings of the Conference on Cellular and Intelligent Technologies (CIC)*.

Kastell, K. (2009). Improved fast location-based handover with integrated security features. *Frequenz: Journal of RF/Microwave Engineering . Photonics and Communications, 63*, 1–2.

Kastell, K., & Jakoby, R. (2007). A location-based handover for use in secure hybrid networks. *Frequenz, 61*, 162–165. doi:10.1515/FREQ.2007.61.7-8.162

Katoh, K., Ibaraki, T., & Mine, H. (1982). An efficient algorithm for K shortest simple paths. *Networks, 12*, 411–427. doi:10.1002/net.3230120406

Kato, K. (1999). Ultrawide-band/high-frequency photodetectors. *IEEE Transactions on Microwave Theory and Techniques, 47*(7), 1265–1281. doi:10.1109/22.775466

Katz, E., & Sruber, G. L. (1995). Sequential sequence estimation for trellis-coded modulation on multipath fading ISI channels. *IEEE Transactions on Communications, 43*(12), 2882. doi:10.1109/26.477483

Kavian, Y. S., Rashvand, H. F., Ren, W., Leeson, M. S., Hines, E. L., & Naderi, M. (2007). RWA problem for designing DWDM networks delay against capacity optimization. *Electronics Letters, 43*(16), 892–893. doi:10.1049/el:20071219

Kavian, Y. S., Ren, W., Naderi, M., Leeson, M. S., & Hines, E. L. (2008). Survivable wavelength-routed optical network design using genetic algorithms. *European Transaction on Telecommunication, 19*(3), 247–255. doi:10.1002/ett.1263

Kawanishi, H., Yamauchi, Y., Mineo, N., Shibuya, Y., Murai, H., Yamada, K., & Wada, H. (2001, September). EAM-integrated DFB laser modules with more than 40-GHz bandwidth. *IEEE Photonics Technology Letters, 13*, 954–956. doi:10.1109/68.942658

Keiichi, M. (2007). Acoustic communication in air using differential biphase shift keying with influence of impulse response and background noise. *Japanese Journal of Applied Physics, 26*(Part 1), 4541.

Kennington, J., & Olinick, E. (2004). *A survey of mathematical programming models for mesh-based survivable networks*, (pp. 1-33). Southern Methodist University, Technical Report 04-EMIS-13.

Kenningtona, J. L., Olinick, E. V., & Spiride, G. (2006). Basic mathematical programming models for capacity allocation in mesh-based survivable networks. *International Journals of Management Sciences, 35*(6), 1–16.

Kennington, J. L., Lewis, K. R., & Olinick, E. V. (2003). Robust solutions for the WDM routing and provisioning problem: Models and algorithms. *Optical Networks Magazine, 4*(2), 74–84.

Kennington, J., Olinick, E., Ortynski, A., & Spiride, G. (2003). Wavelength routing and assignment in a survivable WDM mesh network. *Operations Research*, *51*(1), 67–79. doi:10.1287/opre.51.1.67.12805

Kenward, M. (2001). Plastic fiber homes in/on low-cost networks. *Fiber Systems*, *5*(1), 35.

Khawaja, B. A., & Cryan, M. J. (2009). A wireless hybrid mode locked laser for low cost millimetre wave radio-over-fiber systems. *International Topical Meeting on Microwave Photonics*, (pp. 1–4).

Khawaja, B. A., & Cryan, M. J. (2010, August). Wireless hybrid mode locked lasers for next generation radio-over-fiber systems. *Journal of Lightwave Technology*, *28*, 2268–2276. doi:10.1109/JLT.2010.2050461

Kibler, K., Poferl, S., Bock, G., Huber, H. P., & Zeeb, E. (2004). Optical data buses for automotive applications. *Journal of Lightwave Technology*, *22*(9), 2184–2199. doi:10.1109/JLT.2004.833784

Kibler, T. (2004). Optical data buses for automotive applications. *Journal of Lightwave Technology*, *22*, 2184–2199. doi:10.1109/JLT.2004.833784

Kim, H. S., Pham, T. T., Won, Y. Y., & Han, S. K. (November de 2009). Bidirectional WDM-RoF transmission for wired and wireless signals. *Communications and Photonics Conference and Exhibition*, (pp. 1–2).

Kitayama, K. (1998). Architectural considerations of radio-on-fiber millimeter-wave wireless access systems. *URSI International Symposium on Signals, Systems, and Electronics*, (pp. 378-383).

Kitayama, K. I., Stohr, A., Kuri, T., Heinzelmann, R., Jager, D., & Takahashi, Y. (2000, December). An approach to single optical component antenna base stations for broadband millimeter-wave fiberradio access systems. *IEEE Transactions on Microwave Theory and Techniques*, *48*, 2588–2595. doi:10.1109/22.899017

Kitayama, K., Kuri, T., Onohara, K., Kamisaka, T., & Murashima, K. (2002). Dispersion effects of FBG filter and optical SSB filtering in DWDM millimeter-wave fiber-radio systems. *Journal of Lightwave Technology*, *20*(8), 1397–1407. doi:10.1109/JLT.2002.800265

Klotz, A., Sparbert, J., & Hötzer, D. (2004). Lane data fusion for driver assistance systems. *Proceedings of the International Conference on Fusion 2004*.

Klueser, J. (2012). *CAN based protocols in avionics Version 1.1*. (20120412, Application Note ANION10104) Vector Informatik GmbH. Retrieved from https://www.vector.com/portal/medien/cmc/application_notes/ ANION10104_CANbased_protocols_in_Avionics.pdf

Klyucharev, M. Y., & Rastorguev, V. V. (1995). Elaboration of automobile radio systems on the radiometer base. In *Proceedings of Third Scientific Exchange Seminar "Radio Technical Systems and Devices of UHF", TUM*, Munich, Germany, (p. 7).

Koch, J., & Stämpfle, M. (2010). *Mathematik für das Ingenieurstudium*. Hanser. doi:10.3139/9783446425507

Kojima, Y., Yamada, K., & Ninomiya, Y. (2002). *Three-dimensional road structure estimation by fusion of a digital road map and an image*. (SAE 2002, 2002-01-0758).

Konak, A., & Smith, A. E. (1999). A hybrid genetic algorithm approach for backbone design of communication networks. *Proceedings of the Congress on Evolutionary Computation*, Vol. 1, (pp. 1817-1823).

Kopetz, H., & Bauer, G. (2003). The time-triggered architecture. *Proceedings of the IEEE*, *91*(1), 112–126. doi:10.1109/JPROC.2002.805821

Köster, W. (1983). Einfluss des Rückstreulichts auf die Nebensprechdämpfung in bidirektionalen Übertragungssystemen. *Frequenz*, *37*(4), 87. doi:10.1515/FREQ.1983.37.4.87

Krautkrämer, J., & Krautkrämer, H. (1990). *Ultrasonic testing of materials* (4th ed.). New York, NY: Springer-Verlag.

Kroh, R. (2006). *VANETs security requirements*. Deliverable: D1.1, Sevecom project. Retrieved from http://www.sevecom.org/Deliverables/Sevecom_Deliverable_D1.1_v2.0.pdf

Kuhl, W. (1954). Condenser transmitters and microphones with solid dielectric for airborne ultrasonics. *Acoustica*, *4*(5), 519.

Kukushkin, A. V., & Rastorguev, V. V. (1991). *The device for measurement of object speed.* (Patent of Russian Federation, N° 4910299/09/012782).

Küpfmüller, K. (1973). *Einführung in die theoretische Elektrotechnik.* Springer Verlag.

Kuri, T., Kitayama, K., & Takahashi, Y. (February de 2003). A single light source configuration for full-duplex 60-GHz-band radio-on fiber system. *IEEE Transactions on Microwave Theory and Technique, 51,* 431–439.

Kuri, T., Kitayama, K., & Takahashi, Y. (2000, April). 60-GHz-band full-duplex radio-on-fiber system using two-RF-port electroabsorption transceiver. *IEEE Photonics Technology Letters, 12,* 419–421. doi:10.1109/68.839038

Lages, U. S. (2001). *Untersuchungen zur aktiven Unfallvermeidung von Kraftfahrzeugen.* Fortschritt-Berichte VDI.

Laitinen, H., Lahteenmaki, J., & Nordstrom, T. (2001). Database correlation method for GSM location. *In Proceedings of the 53rd IEEE Vehicular Technology Conference,* Vol. 4, (pp. 2504–2508).

Lampropoulos, G., Passas, N., Merakos, L., & Kaloxylos, A. (2005). Handover management architectures in integrated WLAN/cellular networks. *IEEE Communications Surveys & Tutorials, 7*(4), 30–44. doi:10.1109/COMST.2005.1593278

Lan, K.-C., & Chou, C.-M. (2008). Realistic mobility models for vehicular ad hoc network (VANET) simulations. *Proceedings of ITST 2008,* Phuket, Thailand, (pp. 362-366).

Lannoo, B., Colle, D., Pickavet, M., & Demeester, P. (2007, February). Radio-over-fiber-based solution to provide broadband internet access to train passengers. *IEEE Communications Magazine, 45*(2), 56–62. doi:10.1109/MCOM.2007.313395

Lasowski, R., Scheuermann, C., Gschwandtner, F., & Linnhoff-Popien, C. (2011). Evaluation of adjacent channel interference in single radio vehicular ad-hoc networks. In *Proceedings of Consumer Communications and Networking Conference* (IEEE CCNC 2011), Las Vegas, NV, USA.

Lau E. K., Zhao X., Sung H., Parekh D., Chang-Hasnain C., & Wu M. C. (April de 2008). Strong optical injection-locked semiconductor lasers demonstrating > 100-GHz resonance frequencies and 80-GHz intrinsic bandwidths. *Optics Express, 16*(9).

Lawer, E. L. (1972). A procedure for computing the K best solutions to discrete optimization problems and its application to the shortest path problem. *Management Science, 18,* 401–405. doi:10.1287/mnsc.18.7.401

Lawrence, B. D., & Simmons, J. A. (1982). Measurements of atmospheric attenuation at ultrasonic frequencies and the significance for echolocation by bats. *The Journal of the Acoustical Society of America, 71*(3), 585. doi:10.1121/1.387529

Lecoche, F., Charbonnier, B., Frank, F., Dijk, F. V., Enard, A., & Blache, F. … Moodie, D. (2009, October). 60 GHz bidirectional optical signal distribution system at 3 Gbps for wireless home network. *International Topical Meeting on Microwave Photonics,* (pp. 1-3).

Lee, K. C., Lee, U., & Gerla, M. (2009). TO-GO: TOpology-assist geo-opportunistic routing in urban vehicular grids. In *Proceeding of the Sixth International Conference on Wireless On-Demand Network Systems and Services,* (pp. 11-18).

Lee, S. C. J., et al. (2008). *Low-cost and robust 1 Gbit/s plastic optical fiber link based on light-emitting diode technology.* Optical Fiber Conference (OFC), San Diego, CA, USA

Lee, T. P. (2001). *Prospects and challenges of optoelectronic components in optical network systems.* Seminar on Internat. Exchange & Technology Co-operation, Sept. 22 - 24, Wuhan, China.

Lee, C. H. (2007). *Microwave photonics.* Taylor & Francis Group.

Leen, G., & Heffernan, D. (2001). Vehicles without wires. *Computing & Control Engineering Journal, 12*(5), 205–211. doi:10.1049/cce:20010501

Leen, G., & Heffernan, D. (2002). Expanding automotive electronic systems. *Computer, 35*(1), 88–93. doi:10.1109/2.976923

Leen, G., & Heffernan, D. (2002a). TTCAN: A new time-triggered controller area network. *Microprocessors and Microsystems*, *26*(2), 77–94. doi:10.1016/S0141-9331(01)00148-X

Lee, T. C., & Chen, C. C. (2006). A missed-signal GFSK demodulator for Bluetooth. *IEEE Transactions on Circuits and Wystems. II, Express Briefs*, *53*(3), 197. doi:10.1109/TCSII.2005.858320

Leung, D., & Grover, W. D. (2005). Capacity planning of survivable mesh-based transport networks under demand uncertainty. *Photonic Network Communications*, *10*(2), 123–140. doi:10.1007/s11107-005-2479-z

Leung, P. S. K., & Feher, K. (1993). F-QPSK: Superior modulation technique for mobile and personal communications. *IEEE Transactions on Broadcasting*, *39*(2), 288. doi:10.1109/11.218001

Li, J., Xu, K., Wu, J., & Lin, J. (2007). A simple configuration for WDM full-duplex radio-over-fiber systems. *33rd European Conference and Ehxibition of Optical Communication (ECOC)*, (pp. 1-2).

Li, Z., Nirmalathas, A., Bakaul, M., Cheng, L., Wen, Y. J., & Lu, C. (2005). Application of distributed Raman amplifier for the performance improvement of WDM millimeter-wave fiber-radio network. *The 18th Annual Meeting of the IEEE Lasers and Electro-Optics Society*, (pp. 579–580).

Liberti, J. C., & Rappaport, T. S. (1996). A geometrically based model for line-of-sight multipath radio channels. In *Proceedings of the 46th IEEE Vehicular Technology Conference* (VTC '96), Vol. 2, (pp. 844–848). Atlanta, GA, USA.

Liburdi, F. (2005). *New IDB-1394 socket*. Retrieved from www.electronic-links.com

Li, C. (2008). Short-range ultrasonic digital communication in air. *IEEE Transactions in Ultrasonics, Ferroelectronics, and Frequency Control*, *55*, 908. doi:10.1109/TUFFC.2008.726

Lim, C., Nirmalathas, A., Attygalle, M., Novak, D., & Waterhouse, R. (October de 2003). On the merging of millimeter-wave fiber-radio backbone with 25-GHz WDM ring networks. *Journal of Lightwave Technology*, *21*(10), 2203-2210.

Li, M. J., Soulliere, M. J., Tebben, D. J., Nederlof, L., Vaughn, M. D., & Wagner, R. E. (2005, October). Transparent optical protection ring architectures and applications. *Journal of Lightwave Technology*, *23*, 3388–3403. doi:10.1109/JLT.2005.856240

Lim, C., Ampalavanapillai, N., Novak, D., Waterhouse, R., & Yoffe, G. (2004, October). Millimeter-wave broadband fiber-wireless system incorporating baseband data transmission over fiber and remote LO delivery. *Journal of Lightwave Technology*, *18*(10), 1355–1363. doi:10.1109/50.887186

Lim, C., Nirmalathas, A., Bakaul, M., Gamage, P., Lee, K., & Yang, Y. (2010, February). Fiber-wireless networks and subsystem technologies. *Journal of Lightwave Technology*, *28*(4), 390–405. doi:10.1109/JLT.2009.2031423

Lin, D. B., & Juang, R. T. (2005). Mobile location estimation based on differences of signal attenuations for GSM systems. *IEEE Transactions on Vehicular Technology*, *54*, 1447–1454. doi:10.1109/TVT.2005.851318

Lin, W. P. (2005, September). A robust fiber-radio architecture for wavelength division- multiplexing ring-access networks. *Journal of Lightwave Technology*, *23*, 2610–2620. doi:10.1109/JLT.2005.853150

Lin, X., Sun, X., Ho, P.-H., & Shen, X. (2007). GSIS: A secure and privacy-preserving protocol for vehicular communications. *IEEE Transactions on Vehicular Technology*, *56*(6), 3442–3456. doi:10.1109/TVT.2007.906878

Liu, Z., Sadeghi, M., de Valicourt, G., Brenot, R., & Violas, M. (February de 2011). Experimental validation of a reflective semiconductor optical amplifier model used as a modulator in radio over fiber systems. *IEEE Photonics Technology Letters*, *23*, 576– 578.

Liu, L., Wang, Y., Zhang, N., & Zhang, Y. (2010). UWB channel measurement and modeling for the intra-vehicle environments. *Proceedings of ICCT*, *2010*, 381–384. doi:doi:10.1109/ICCT.2010.5688815

Liu, Y., Tipper, D., & Siripongwutikorn, P. (2005). Approximating optimal spare capacity allocation by successive survivable routing. *IEEE/ACM Transactions on Networking*, *13*(1), 198–211. doi:10.1109/TNET.2004.842220

Lochert, C., Hartenstein, H., Tian, J., Fussler, H., Hermann, D., & Mauve, M. (2003). A routing strategy for vehicular ad hoc networks in city environments. *IEEE Proceedings of Intelligent Vehicles Symposium,* (pp. 156-161).

Lomba, C. R., Valada, R. T., & de Oliveira Duarte, A. M. (1998). Experimental characterisation and modelling of the reflection of infrared signals on indoor surfaces. *Proceedings IEE Optoelectronics, 145*(3), 191–197. doi:10.1049/ip-opt:19982020

Losada, M. A., Mateo, J., & Serena, L. (2007). Analysis of propagation properties of step index plastic optical fibers at non-stationary conditions. *Proceedings of the 16th International Conference on Plastic Optical Fibres and Applications,* (pp. 299-302).

Losada, M. A., Mateo, J., Martínez, J. J., & López, A. (2008). SI-POF frequency response obtained by solving the power flow equation. *Proceedings of the 17th International Conference on Plastic Optical Fibres and Applications.*

Losada, M. A., Garcés, I., Mateo, J., Salinas, I., Lou, J., & Zubía, J. (2002). Mode coupling contribution to radiation losses in curvatures for high and low numerical aperture plastic optical fibres. *Journal of Lightwave Technology, 20*(7), 1160–1164. doi:10.1109/JLT.2002.800377

Losada, M. A., Mateo, J., Garcés, I., Zubía, J., Casao, J. A., & Pérez-Vela, P. (2004). Analysis of strained plastic optical fibres. *IEEE Photonics Technology Letters, 16*(6), 1513–1515. doi:10.1109/LPT.2004.826780

Losada, M. A., Mateo, J., & Martínez-Muro, J. J. (2011). Assessment of the impact of localized disturbances on SI-POF transmission using a matrix propagation model. *Journal of Optics, 13,* 055406. doi:10.1088/2040-8978/13/5/055406

Lott, M., Siebert, M., Bonjour, S., von Hugo, D., & Weckerle, M. (2004). Interworking of WLAN and 3G systems. *Proceedings from 04. IEE Proceedings. Communications, 151,* 507–513. doi:10.1049/ip-com:20040600

Lubkoll, J. (2009). Optical data bus technologies for automotive applications. *The Mediterranean Journal of Electronics and Communications, 5*(4), 127. ISSN 1744-2400

Lueftner, T. (2003). Edge-position modulation for high-speed wireless infrared communications. *IEE Proceedings. Optoelectronics, 150*(5), 427. doi:10.1049/ip-opt:20030960

Lu, H. H., Patra, A., Ho, W. J., Lai, P. C., & Shiu, M. H. (2007, October). A full-duplex radio-over-fiber transport system based on FP laser diode with OBPF and optical circulator with fiber Bragg grating. *IEEE Photonics Technology Letters, 19,* 1652–1654. doi:10.1109/LPT.2007.905077

M'Edard, M., & Lumetta, S. S. (2002). *Network reliability and fault tolerance. Tech. Report., MIT.* Laboratory for Information and Decision Systems.

Mahajan, A., Potnis, N., Gopalan, K., & Wang, A. A. (2006). Urban mobility models for VANETs. In *Proceedings of the IEEE Workshop on Next Generation Wireless Networks (WoNGeN).*

Mahlke, G., & Gössing, P. (1987). *Fiber optic cables* (p. 77). Berlin, Germany: Siemens AG.

Maiman, T. H. (1960). Optical and microwave-optical experiments in Ruby. *Physical Review Letters, 4*(11), 564. doi:10.1103/PhysRevLett.4.564

Majkner, R. (2003). *Overview - Lightning protection of aircraft and avionics.* Sikorsky Corp. Retrieved December 7, 2009, from http://ewh.ieee.org/r1/ct/aess/aess_events.html

Malyon, D. J. (1991). Demonstration of optical pulse propagation over 10 000 km of fiber using recirculating loop. *Electronics Letters, 27*(2), 120. doi:10.1049/el:19910080

Mangel, T., Klemp, O., & Hartenstein, H. (2011). A validated 5.9 GHz Non-Line-of-Sight path-loss and fading model for inter-vehicle communication. *2011 11th International Conference on ITS Telecommunications (ITST),* (pp. 75-80).

Mao, S., et al. (2008). Modelling and measuring of weak frequency selectivity in OFDM modulator. *2008 4th IEEE International Conference on Circuits and Systems for Communications, ICCSC,* (p. 416).

Marcuse, D. (1979). Calculation of bandwidth from index profiles of optical fibers. *Theory Applied Optics, 18*(12), 2073. doi:10.1364/AO.18.002073

Markopoulos, A., Pissaris, P., Kyriazakos, S., Koutsorodi, A., & Sykas, E. D. (2004). Performance analysis of cellular networks by simulating location aided handover algorithms. *Proceedings from EWC 04: 5th European Wireless Conference on Mobile and Wireless Systems beyond 3G*, (pp. 605-611).

Masui, H., Kobayashi, T., & Akaike, M. (2002). Microwave path-loss modeling in urban line-of-sight environments. *IEEE Journal on Communications, 20*(6), 1151–1155.

Mateo, J., Losada, M. A., Garcés, I., Arrúe, J., Zubia, J., & Kalymnios, D. (2003). High NA POF dependence of bandwidth on fibre length. *Proceedings of the 12th International Conference on Plastic Optical Fibres and Applications*, (pp. 123–126).

Mateo, J., Losada, M. A., & Garcés, I. (2006). Global characterization of optical power propagation in step-index plastic optical fibers. *Optics Express, 14*, 9028–9035. doi:10.1364/OE.14.009028

Mateo, J., Losada, M. A., & Zubia, J. (2009). Frequency response in step index plastic optical fibers obtained from the generalized power flow equation. *Optics Express, 17*(4), 2850–2860. doi:10.1364/OE.17.002850

Matsuzawa, K. (1958). Condenser microphones with plastic diaphragms for airborne ultrasonics I. *Journal of the Physical Society of Japan, 13*, 1533. doi:10.1143/JPSJ.13.1533

Matsuzawa, K. (1960). Condenser microphones with plastic diaphragms for airborne ultrasonics II. *Journal of the Physical Society of Japan, 15*, 167. doi:10.1143/JPSJ.15.167

Mc Nair, J., Akyildiz, I. F., & Bender, M. D. (2000). An inter-system handoff technique for the IMT-2000 system. *Proceedings from Joint Conference of the IEEE Computer and Communications Societies,* (pp. 208-216).

McCarthy, D. C. (2001). Growing by design. *Photonics Spectra*, 88.

McLachlan, N. W. (1934). *Loudspeakers*. Oxford, UK: Clarendon Press.

Mecklenbräuker, C. F., Molisch, A. F., Karedal, J., Tufvesson, F., Paier, A., & Bernado, L. … Czink, N. (2011). Vehicular channel characterization and its implications for wireless system design and performance. *Proceedings of the IEEE, 99*(7), 1189–1212.

Medeiros, M., Laurncio, P., Correia, N., Barradas, A., da Silva, H., & Darwazeh, I. … Monteiro, P. (2007). Radio over fiber access network architecture employing reflective semiconductor optical amplifiers. *ICTON Mediterranean Winter Conference*, (pp. 1–5).

Meinke, H., & Gundlach, F. W. (1986). *Taschenbuch der Hochfrequenztechnik*. Springer Verlag.

Meireles, R., Boban, M., Steenkiste, P., Tonguz, O., & Barros, J. (2010). Experimental study on the impact of vehicular obstructions in VANETs. In *Proceedings of the 2nd IEEE Vehicular Networking Conference* (VNC 2010), Jersey City, New Jersey, USA.

Melchior, H. (1970). Photodetectors for optical communication systems. *Proceedings of the IEEE, 58*(10), 1466. doi:10.1109/PROC.1970.7972

Melexis. (2008). *Melexis new fiber optic ttransceivers*. ITG VDE 5.4.1, Fachgruppentreffen Krems.

Melexis. (2012). *TH8062 voltage regulator with LIN transceiver*. Datasheet of the LIN transceiver TH8062. Retrieved from http://www.melexis.com/LINSBC'sandBUSTransceivers/(general)/TH8062583.aspx

Menasce, D. A. (2002). QoS issues in web services. *IEEE Internet Computing, 6*(6), 72–75. doi:10.1109/MIC.2002.1067740

Miller, S. E. (1973). Research toward optical-fiber transmission systems. *Proceedings of the IEEE, 61*(12), 1703. doi:10.1109/PROC.1973.9360

Mitsubishi Rayon. (2009) *MOST All Members Meeting*, Frankfurt.

Mittal, S., & Mirchandani, P. (2004). *Implementation of K-Shortest path Dijkstra algorithm used in all-optical data communication networks*. SIE 546 Project Report.

Mohamed, N., Idrus, S., & Mohammad, A. (2008). *Review on system architectures for the millimeter-wave generation techniques for RoF communication link.* IEEE International RF and Microwave Conference.

Moreia, A. J. C. (1996). Performance of infrared transmission systems under ambient light interference. *IEE Proceedings. Optoelectronics, 143*(6), 339. doi:10.1049/ip-opt:19960696

MOST. (n.d.). *Cooperation.* Retrieved from www.MOST-Cooperation.com

MOST. (n.d.). *Specifications.* Retrieved from http://www.mostcooperation.com/publications/specifications-organizational-procedures/index.html

Mukherjee, B. (2000). WDM optical communication networks: Progress and challenges. *IEEE Journal on Selected Areas in Communications, 18*(10), 1810–1824. doi:10.1109/49.887904

Murthy, C. S. R., & Gurusamy, M. (2004). *WDM optical networks: Concepts, design, and algorithms.* Prentice Hall.

Naghian, S. (2003). Hybrid predictive handover in mobile networks. *Proceedings from VTC 03: 58th IEEE Vehicular Technology Conference*, Vol. 3, (pp. 1918-1922).

Nakasyotani, T., Toda, H., Kuri, T., & Kitayama, K. (January de 2006). Wavelength-division-multiplexed millimeter-waveband radio-on-fiber system using a supercontinuum light source. *Journal of Lightwave Technology, 24*, 404–410.

Narasimhan, R. (1996). Effect of electronic-ballast fluorescent lighting on wireless infrared links. *IEE Proceedings. Optoelectronics, 143*(6), 347. doi:10.1049/ip-opt:19960877

Navet, N., Song, Y., Simonot-Lion, F., & Wilwert, C. (2005). Trends in automotive communication systems. *Proceedings of the IEEE, 93*(6), 1204–1223. doi:10.1109/JPROC.2005.849725

Nelson, T. (2008). Near optimal common detection techniques for shaped offset QPSK and Feher's QPSK. *IEEE Transactions on Communications, 56*(5), 724. doi:10.1109/TCOMM.2008.060155

Ng'oma, A. (2005). *Radio-over-fibre technology for broadband wireless communication systems.* Doctoral Thesis.

Ng'oma, A., Fortusini, D., Parekh, D., Yang, W., Sauer, M., & Benjamin, S. (2010, August). Performance of a multi-Gbps 60 GHz radio over fiber system employing a directly modulated optically injection locked VCSEL. *Journal of Lightwave Technology, 28*(16), 2436–2444. doi:10.1109/JLT.2010.2046623

Niehsen, W., Garnitz, M., Weilkes, M., & Stämpfle, M. (2005). Informationsfusion für Fahrerassistenzsysteme. In Maurer, M., & Stiller, C. (Eds.), *Fahrerassistenzsysteme mit maschineller Wahrnehmung.* Springer. doi:10.1007/3-540-27137-6_3

Nikolopoulos, S. D., Pitsillides, A., & Tipper, D. (1997). Addressing network survivability issues by finding the K-best paths through a trellis graph. *Proceeding of IEEE Infocom*, Kobe.

Nirmalathas, A., Novak, D., Lim, C., & Waterhouse, R. B. (October de 2001). Wavelength re-use in the WDM optical interface of a millimeter-wave fiber wireless antenna base-station. *IEEE Transactions Microwave Theory and Techniques, 49*(10), 2006–2012.

Nolte, T., Hansson, H., & Bello, L. L. (2005). Automotive communications-past, current and future. *Proceedings of ETFA, 1*, 985–992. doi:doi:10.1109/ETFA.2005.1612631

Norris, J. A., & Nieto, J. W. (2008). Evaluation of a novel constant envelope spread-spectrum modulation technique. *Proceedings IEEE Military Communications Conference MILCOM 2008*, (p. 1).

NXP. (2011). *Datasheet TJA1027, LIN 2.2/SAE J2602 transceiver.* Retrieved from http://www.nxp.com/products/automotive/transceivers/lin_transceivers/series/TJA1027.html

Nyquist, H. (1928). Thermal agitation of electric charge in conductors. *Physical Review, 32*, 110–113. doi:10.1103/PhysRev.32.110

O'Reilly, J., & Lane, P. (1994, February). Remote delivery of video services using mm-waves and optics. *Journal of Lightwave Technology, 12*(2), 369–375. doi:10.1109/50.350584

Ochi, H., & Fukuchi, T. (2007). Development of titled Toroidal beam wideband transducer using quadrature phase shift keying for underwater acoustic communication. *Japanese Journal of Applied Physics*, *46*(7), 4961. doi:10.1143/JJAP.46.4961

Ogawa, H., Polifko, D., & Banba, S. (December de 1992). Millimeter-wave fiber optics systems for personal radio communication. *IEEE Transactions on Microwave Theory and Techniques, 40*, 2285–2293.

Ohno, S., Manasseh, E., & Nakamoto, M. (2011). Preamble and pilot symbol design for channel estimation in OFDM systems with null subcarriers. *EURASIP Journal on Wireless Communications and Networking*, *2011*, 2. doi:10.1186/1687-1499-2011-2

Olariu, S., & Weigle, M. C. (Eds.). (2009). *Vehicular networks: From theory to practice*. CRC Press. doi:10.1201/9781420085891

Olver, A. (1990). Potential and applications of millimetre-wave radar. *IEE Colloquium on Millimetre-Wave Radar*, (pp. 1-4).

O'Reilly, J. J., Lane, P. M., Heidemann, R., & Hofstetter, R. (1992, November). Optical generation of very narrow linewidth millimetre wave signals. *Electronics Letters*, *28*(25), 2309–2311.

Oren, M., & Nayer, S. (1994). Seeing beyond Lambert's law *Proceedings of ECCV*, (pp. 269-280). doi: 10.1007/BFb0028360

Othman, H. F., Aji, Y. R., Fakhreddin, F. T., & Al-Ali, A. R. (2006). Controller area networks: Evolution and applications. *Proceedings of ICTTA*, *2*, 3088–3093. doi:doi:10.1109/ICTTA.2006.1684909

Otnes, R., & Eqqen, T. H. (2008). Underwater acoustic communications: Long-term test of turbo equalisation in shallow water. *IEEE Journal of Oceanic Engineering*, *33*(3), 321. doi:10.1109/JOE.2008.925893

Oualkadi, A. E. (2011, November). *Obtenido de trends and challenges in CMOS design for emerging 60 GHz WPAN applications*. Intechopen. Retrieved from http://www.intechopen.com

Ou, C., Zang, H., Singhal, N. K., Zhu, K., Sahasrabuddhe, L. H., MacDonald, R. A., & Mukherjee, B. (2004). Sub-path protection for scalability and fast recovery in optical WDM mesh networks. *IEEE Journal on Selected Areas in Communications*, *22*(9), 1859–1875. doi:10.1109/JSAC.2004.830280

Ou, C., Zang, J., Zang, H., Sahasrabuddhe, L. H., & Mukherjee, B. (2004). New and improved approaches for shared-path protection in WDM mesh networks. *Journal of Lightwave Technology*, *22*(5), 1223–1232. doi:10.1109/JLT.2004.825346

Ozdemir, M. K. (2007). Channel estimation for wireless OFDM systems. *IEEE Communications Surveys & Tutorials*, 2nd Quarter.

Paajanen, M. (2000). Electromechanically film (EMFi) – A new multipurpose electret material. *Sensors and Actuators*, *84*, 95. doi:10.1016/S0924-4247(99)00269-1

Pakravan, M. R., & Kavehrad, M. (2001). Indoor wireless infrared channel chracterisation by measurements. *IEEE Transactions on Vehicular Technology*, *50*(5), 1053–1073. doi:10.1109/25.938580

Pan, J. J. (1988, May). 21 GHz wideband fiber optic link. *MTTS International Microwave Symposium Digest*, (pp. 977–978).

Panish, M. B. (1976). Heterostructure injection lasers. *Proceedings of the IEEE*, *64*(10), 1512. doi:10.1109/PROC.1976.10367

Papadimitratos, P., La Fortelle, A., Evenssen, K., Brignolo, R., & Cosenza, S. (2009). Vehicular communication systems: Enabling technologies, applications and future outlook on intelligent transportation. *IEEE Communications Magazine*, *47*(11), 84–95. doi:10.1109/MCOM.2009.5307471

Parekh, D., & Yang, W. Ng'Oma, A., Fortusini, D., Sauer, M., Benjamin, S., … Chang-Hasnain, C. (2010). *Multi-Gbps ask and QPSK-modulated 60 GHz RoF link using an optically injection locked VCSEL*. Optical Fiber Communication (OFC), Collocated National Fiber Optic Engineers Conference.

Parsons, J. D. (2001). *The mobile radio propagation channel*. New York, NY: John Wiley & Sons.

Pasupathy, S. (1979). Minimum shift keying: A spectrally efficient modulation. *IEEE Communications Magazine*, *17*(4), 14. doi:10.1109/MCOM.1979.1089999

Pato, S., Pedro, J., & Monteiro, P. (2009, July). *Comparative evaluation of fibre-optic architectures for next-generation distributed antenna systems*. 11th International Conference on Transparent Optical Networks.

Payne, D. N., et al. (1990). Fiber optical amplifiers. *Proceedings of OFC '90*, Tutorial, paper ThFl, (p. 335). San Francisco.

Pearsall, T. P. (1981). Photodetectors for optical communication. *Journal of Optical Communication*, *2*(2), 42.

Peng, P. C., Feng, K. M., Chiou, H. Y., Peng, W. R., Chen, J. J., & Kuo, H. C. (2007). Reliable architecture for high capacity fiber-radio systems. *Optical Fiber Technology*, *13*(3), 236–239. doi:10.1016/j.yofte.2007.02.001

Perez, S. R., Jimenex, R. P., Lopez-Hernandez, F. J., Hernandez, O. B. G., & Alfonso, A. J. A. (2002). Reflection model for calculation of the impulse response on IR wireless indoor channels using a ray-tracing algorithm. *Microwave and Optical Technology Letters*, *32*(4), 296–300. doi:10.1002/mop.10159

Pergola, L., Gindera, R., Jäger, D., & Vahldieck, R. (2007, March). An LTCC-based wireless transceiver for radio-over-fiber applications. *IEEE Transactions on Microwave Theory and Techniques*, *55*(3), 579–587. doi:10.1109/TMTT.2006.890527

Perkins, C. E. (2002). *IP mobility support for IPv4*. Retrieved from http://tools.ietf.org/html/rfc3344

Pesola, J., & Pönkänen, S. (2003). Location-aided handover in heterogeneous wireless networks. *Proceedings from MLW 03: Mobile Location Workshop*, (pp. 195–205).

Pfeiffer, T. (2001). Optical packet transmission system for metropolitan and access networks with more than 400 channels. *Journal of Lightwave Technology*, *18*(12), 1928. doi:10.1109/50.908792

Pilosu, L., Fileppo, F., & Scopigno, R. (2011). RADII: A computationally affordable method to summarize urban ray-tracing data for VANETs. In *Proceedings of the 7th International Conference on Wireless Communications, Networking and Mobile Computing* (IEEE WiCOM 2011), Wuhan, China.

Pleros, N., Vyrsokinos, K., Tsagkaris, K., & Tselikas, N. . (June de 2009). A 60 GHz radio-over-fiber network architecture for seamless communication with high mobility. *Journal of Lightwave Technology*, *27*(12), 1957-1967.

Porcino, D. (2001). Location of third generation mobile devices: a comparison between terrestrial and satellite positioning systems. In *Proceedings of the 53rd IEEE Vehicular Technology Conference*, Vol. 4, (pp. 2970–2974).

Porcino, D., & Hirt, W. (2003). Ultra-wideband radio technology: Potential and challenges ahead. *IEEE Communications Magazine*, *41*(7), 66–74. doi:10.1109/MCOM.2003.1215641

POSESKA. (n.d.). *Technical characteristics of POF*. Retrieved from http://www.pofeska.com/pofeskae/download/pdf/technical%20data_01.pdf

Potenza, M. (1996, August). Optical fiber amplifiers for telecommunication systems. *IEEE Communications Magazine*, *34*, 96–102. doi:10.1109/35.533926

Qi, G., Yao, J., Seregelyi, J., Paquet, S., & Blisle, C. (November de 2005). Generation and distribution of a wide-band continuously tunable millimeter-wave signal with an optical external modulation Technique. *IEEE Transactions on Microwave Theory and Techniques*, *53*, 3090–3097.

Qingyan, Y., Heng, W., & Jiuchun, G. (2008). Linking freeway and arterial data - Data archiving testing in supporting coordinated freeway and arterial operations. *11th International IEEE Conference on Intelligent Transportation Systems, ITSC*, pp. 259-264. ISBN: 978-1-4244-2111-4

Quist, T. M. (1962). Semiconductor Maser of GaAs. *Applied Physics Letters*, *1*(4), 91. doi:10.1063/1.1753710

Radziunas, M., Glitzky, A., Bandelow, U., Wolfrum, M., Troppenz, U., Kreissl, J., & Rehbein, W. (2007, January/February). Improving the modulation bandwidth in semiconductor lasers by passive feedback. *IEEE Journal on Selected Topics in Quantum Electronics*, *13*, 136–142. doi:10.1109/JSTQE.2006.885332

Rafiq, G., & Patzold, M. (2008). The influence of the severity of fading and shadowing on the statistical properties of the capacity of Nakagami-lognormal channels. *IEEE Global Telecommunications Conference, GLOBECOM 2008*, New Orleans, LA, (pp. 1-6).

Rai, V., Bai, F., Kenney, J., & Laberteaux, K. (2007). *Cross-channel interference test results: A report from VSC-A project.* (IEEE 802.11 11-07-2133-00-000p).

Rajagolapan, B. (2000). IP over optical networks: Architectural aspects. *IEEE Communications Magazine,* 94–102. doi:10.1109/35.868148

Rajamani, R. (2006). *Vehicle dynamics and control.* Springer.

Ramamurthy, S., Sahasrabuddhe, L., & Mukherjee, B. (2000). Survivable WDM mesh networks. *Journal of Lightwave Technology, 21*(4), 870–883. doi:10.1109/JLT.2002.806338

Ramamuthy, S., & Mukherjee, B. (1999). Survivable WDM mesh networks: Part II-Restoration. *Proceeding of IEEE ICC Conference,* (pp. 2023-2030).

Ramaswami, R., & Sivarajan, K. N. (1995). Routing and wavelength assignment in all-optical networks. *IEEE/ACM Transactions on Networking, 3,* 489–500. doi:10.1109/90.469957

Rappaport, T. S. (1996). *Wireless communications, principles and practice.* Prentice Hall.

Rastorguev, V., & Shnajder, V. (2010). Radiometric sensor of movement speed of vehicles. In *Proceeding of 12ᵗʰ International Conference on Transparent Optical Networks – ICTON'2010,* Munich, Germany, June 27 - July 1, 2010. ISBN: 978-1-4244-7797-5

Rastorguev, V., Betz, C., & Strobel, O. (2011). Optimizations of parameters of the radiometric sensor of movement speed of vehicles. *Proceeding of 13ᵗʰ International Conference on Transparent Optical Networks – ICTON'2011,* Stockholm, Sweden, June 26 – June 30, 2011. ISBN: 978-1-4577-0882-4/11

Raya, M., & Hubaux, J. P. (2007). Securing vehicular ad hoc networks. *Journal of Computer Security, 15*(1), 39–68.

Reiter, R., & Schramm, A. (2008). *Verfügbarkeitsrisiko senken – Neue Physical-Layer-Spezifikation für MOST150.* WEKA Fachmedien GmbH.

Ritter, M. B. (1993). Dispersion limits in large core fibers. *Proceedings of the International. Conference on Plastics Optical Fibres and Applications,* (pp. 31-34).

Robert Bosch Gmb, H. (1991). *CAN specification,* v 2.0. Stuttgart, Germany: Robert Bosch GmbH. Retrieved from http://www.semiconductors.bosch.de/media/pdf/canliteratur/can2spec.pdf

Robert Bosch Gmb, H. (2011). *Automotive handbook* (8th ed.). Plochingen, Germany: Robert Bosch GmbH.

Rousseau, M., & Jeunhomme, L. (1977). Numerical solution of the coupled-power equation in step-index optical fibers. *IEEE Transactions on Microwave Theory and Techniques, 25,* 577–585. doi:10.1109/TMTT.1977.1129162

Saarnisaari, H. (2007). A general receiver and constant envelope direct sequence signals. *Proceedings IEEE Military Communications Conference MILCOM 2007,* (p. 1).

Sabatini, A. M. (1997). Stochastic model of the time-of-flight noise in airborne sonar ranging systems. *IEEE Transactions on Ultrasonics, Ferroelectrics, and Frequency Control, 44*(3), 606. doi:10.1109/58.658313

SAE J2735. (2006). *Dedicated short range communications (DSRC) message set dictionary,* V1.0.

SAE. (2009). *Handbook for robustness validation of automotive electrical/electronic modules.*

Saha, D., & Birdsall, T. (1989). Quadrature-quadrature phase-shift keying. *IEEE Transactions on Communications, 37*(5), 437. doi:10.1109/26.24595

Sairam, S. (2002). Bluetooth in wireless communication. *IEEE Communications Magazine, 40*(6), 90. doi:10.1109/MCOM.2002.1007414

Salomon, D. (2006). *Curves and surfaces for computer graphics.* Springer-Verlag.

Samarasinghe, R., Friderikos, V., & Aghvami, A. H. (2003). *Analysis of intersystem handover: UMTS FDD & WLAN.* London Communications Symposium.

Sangiovanni-Vincentelli, A., & Di Natale, M. (2007). Embedded system design for automotive applications. *Computer, 40*(10), 42–51. doi:10.1109/MC.2007.344

Santi, P. (2012). *Mobility models for next generation wireless networks: Ad hoc, vehicular and mesh networks.* Wiley Series on Communications Networking & Distributed Systems. doi:10.1002/9781118344774

Saradhi, C. V., Gurusamy, M., & Zhou, L. (2004). Differentiated QoS for survivable WDM optical networks. *IEEE Communications Magazine, 42*(5), s8–s14. doi:10.1109/MCOM.2004.1299335

Schimmelpfennig, K. H., & Nackenhorst, U. (1985). Bedeutung der Querbeschleunigung in der Verkehrsunfallrekonstruktion - Sicherheitsgrenze des Normalfahrers. *Verkehrsunfall und Fahrzeugtechnik, 4,* 94–96.

Schindel, D., et al. (1993). Micromachined capacitance transducers for air-borne ultrasonics. *Proceedings of the Ultrasonics International Conference,* (p. 675).

Schindel, D. W. (1995). The design and characterization of micromachined air-coupled capacitance transducers. *IEEE Transactions on Ultrasonics, Ferroelectrics, and Frequency Control, 42,* 42. doi:10.1109/58.368314

Schmitt, G. (2009). The green base station. *4th International Conference on Telecommunication - Energy Special Conference (TELESCON),* (pp. 1-6). Frankfurt, Germany.

Schoch, E., Kargl, F., Weber, M., & Leinmuller, T. (2008). Communication patterns in VANETs. *Communications Magazine, 46*(11), 119-125. ISSN 0163-6804

Schott. (n.d.). *Spec sheet.* Retrieved from http://www.schott.com/korea/korean/download/specsheet_datacom_e&a_2_5.pdf

Scopigno, R. (2012). Physical phenomena affecting VANETs: Open issues in network simulations. In *Proceedings of the International Conference on Transparent Optical Networks (ICTON 2012),* Coventry, England.

Scopigno, R., & Cozzetti, H. A. (2009). Mobile slotted Aloha for VANETs. In *Proceedings of the IEEE 70th Vehicular Technology Conference* (VTC Fall 2009), Anchorage, Alaska, USA.

Scopigno, R., & Cozzetti, H. A. (2010a), Signal shadowing in simulation of urban vehicular communications. In *Proceedings of the 6th International Wireless Communications and Mobile Computing Conference* (IWCMC), Valencia, Spain.

Scopigno, R., & Cozzetti, H. A. (2010b). Evaluation of time-space efficiency in CSMA/CA and slotted VANETs. In *Proceedings of the IEEE 71st Vehicular Technology Conference* (VTC Fall 2010), Ottawa, Canada.

Scully, P., & Skehill, R. (2006). Demonstration of aggregate mobility models in Matlab. *IET Irish Signals and Systems Conference,* Dublin Institute of Technology, (pp. 271-276).

Seeds, A. J., & Williams, J. (2006, December). Microwave photonics. *Journal of Lightwave Technology, 24*(12), 4628–4641. doi:10.1109/JLT.2006.885787

Seibl, D. (2008). *Polymer-optical-fiber data bus technologies for MOST applications in vehicles. ICTON MW 2008.* Morocco: Marrakech.

Sell, H. (1937). Eine neue Methode zur Umwandlung mechanischer Schwingungen in elektrische und umgekehrt. *Biomedical Engineering, 18,* 3.

Sethna, F., Stipidis, E., & Ali, F. H. (2006). What lessons can controller area networks learn from FlexRay. *Proceedings of IEEE VPPC, 2006,* 1–4. doi:doi:10.1109/VPPC.2006.364320

Setiawan, G., Iskandar, S., Salil, K., & Lan, K. C. (2007). The effect of radio models on vehicular network simulations. *Proceedings of the 14th World Congress on Intelligent Transport Systems,* October 2007.

Shafik, R. A., et al. (2007). On the extended relationships among EVM, BER and SNR as performance metrics. *Proceedings of 4th International Conference on Electrical and Computer Engineering, (ICECE 2006),* (p. 408).

Sharkov, E. A. (2003). *Passive microwave remote sensing of the Earth: Physical foundations* (p. 613). Springer/PRAXIS.

Shen, A., Make, D., Poingt, F., Legouezigou, L., Pommereau, F., & Legouezigou, O. ... Duan, G. (2008). *Polarisation insensitive injection locked Fabry-Perot laser diodes for 2.5Gb/s WDM access applications.* European Conference on Optical Communications (ECOC).

Shenai, R., & Sivalingam, K. (2005). Hybrid survivability approaches for optical WDM mesh networks. *Journal of Lightwave Technology, 23*(10), 3046–3055. doi:10.1109/JLT.2005.856273

Shih, P., Lin, C., Huang, H., Jiang, W., Chen, J., & Ng'oma, A. ... Chi, S. (2009). 13.75-Gb/s OFDM signal generation for 60-GHz RoF system within 7-GHz license-free band via frequency sextupling. *35th European Conference on Optical Communication,* (pp. 1-2).

Shin, D. S., Kim, W. R., Woo, S. K., & Shim, J. I. (2008, June). Low-detuning operation of electroabsorption modulator as a zero-bias optical transceiver for Picocell radio-over-fiber applications. *IEEE Photonics Technology Letters, 20*(11), 951–953. doi:10.1109/LPT.2008.922916

Shiu, D. S., & Kahn, J. M. (1999). Differential pulse-position modulation for power-efficient optical communication. *IEEE Transactions on Communications, 47*(8), 1201–1210. doi:10.1109/26.780456

Shutko, A. M. (1986). *The UHF radiometer for a water surface, ground and terrestrial ground.* Moscow, Russia: Nauka. (in Russian)

Siebert, M., Schinnenburg, M., Lott, M., & Göbbels, S. (2004). Location aided handover support for next generation system integration. *Proceedings from EWC 04: 5th European Wireless Conference on Mobile and Wireless Systems beyond 3G,* (pp. 195-201).

Silventoinen, M. I., & Rantalainen, T. (1996). Mobile station emergency locating in GSM. In *Proceedings of the 1st IEEE International Conference on Personal Wireless Communications,* (pp. 232–238).

Simon, M. K., & Polydoros, A. (1981). Coherent detection of frequency-hopped quadrature modulations in the presence of jamming – 1. QPSK and QASK modulations. *IEEE Transactions on Communications, 29*(11), 1644. doi:10.1109/TCOM.1981.1094924

Sjöberg, K., Uhlemann, E., & Ström, E. G. (2011). How severe is the hidden terminal problem in VANETs when using CSMA and STDMA? In *Proceedings of the 4th IEEE Symposium on Wireless Vehicular Communications* (WiVEC), San Francisco, CA, US.

Sjöberg-Bilstrup, K., Uhlemann, E., Ström, E., & Bilstrup, U. (2009). On the ability of the IEEE 802.11p MAC method and STDMA to support real-time vehicle-to-vehicle communication. *EURASIP Journal on Wireless Communications and Networking, 13*. doi:doi:10.1155/2009/902414

Sklar, B. (1997). Rayleigh fading channels in mobile digital communication systems part I: Characterization. *IEEE Communications Magazine,* 136–146. doi:10.1109/35.620535

Sklarek, W. (2006). *High data rate capabilities of multicore glass optical fiber cables,* (p. 22). Fachgruppentreffen der ITG-FG 5.4.1, Optische Polymerfasern, Oct 2006. Retrieved from http://www.pofac.de/downloads/itgfg/fgt22/FGT22_Muenchen_Sklarek_GOF-Buendel.pdf

Smith, G., Novak, D., & Ahmed, Z. (1997, January). Technique for optical SSB generation to overcome dispersion penalties in fiber-radio systems. *Electronics Letters, 33,* 74–75. doi:10.1049/el:19970066

SMSC. (n.d.). Retrieved from www.smsc.com/

Sommer, C., Eckhoff, D., German, R., & Dressler, F. (2011). A computationally inexpensive empirical model of IEEE 802.11p radio shadowing in urban environments. In *Proceedings of the 8th International Conference on Wireless On-Demand Network Systems and Services* (WONS), Bardonecchia, Italy, (pp. 84-90).

Sommer, C., & Dressler, F. (2008). Progressing toward realistic mobility models in VANET simulations. *IEEE Communications Magazine,* 132–137. doi:10.1109/MCOM.2008.4689256

Son, M., & Kang, T.-G. (2008). *Project: IEEE P802.15 working group for wireless personal area networks* (WPANs). (doc.: IEEE 802.15-08-0759-02-0vlc).

Stämpfle, M., & Branz, W. (2008). *Kollisionsvermeidung im Längsverkehr - Die Vision vom unfallfreien Fahren rückt näher.* München, Germany: Tagung Aktive Sicherheit durch Fahrerassistenz.

Stämpfle, M., Holz, D., & Becker, J. C. (2005). Performance evaluation of automotive sensor data fusion. *8th International Conference on Intelligent Transportation Systems,* Wien, (pp. 590-595).

Stämpfle, M. (1996). Cellular automata and optimal path planning. *International Journal of Bifurcation and Chaos in Applied Sciences and Engineering, 6*(3), 603–610. doi:10.1142/S0218127496000291

Stämpfle, M. (1998). *Numerische Approximation des Flusses eines dynamischen Systems mit adaptiven Triangulierungsverfahren.* Shaker.

Stämpfle, M. (1999). Dynamical systems flow computation by adaptive triangulation methods. *Computing and Visualization in Science, 2*(1), 15–24. doi:10.1007/s007910050023

Stämpfle, M., Hunt, K. J., & Kalkkuhl, J. (2001). Efficient simulation of parameter-dependent vehicle dynamics. *International Journal for Numerical Methods in Engineering, 52,* 1273–1299. doi:10.1002/nme.254

Stöhr, A., Babiel, S., Cannard, P. J., Charbonnier, B., van Dijk, F., & Fedderwitz, S. (2010). Millimeter-wave photonic components for broadband wireless systems. *IEEE Transactions on Microwave Theory and Techniques, 58*(11), 3071–3082. doi:10.1109/TMTT.2010.2077470

Stohr, A., Kitayama, K., & Jager, D. (1999, July). Full-duplex fiberoptic RF subcarrier transmission using a dual-function modulator/ photodetector. *IEEE Transactions on Microwave Theory and Techniques, 47,* 1338–1341. doi:10.1109/22.775476

Street, A. M., Stavrinou, P. N., O'Brien, D. C., & Edwards, D. J. (1997). Indoor optical wireless systems: A review. *Optical and Quantum Electronics, 29,* 349–378. doi:10.1023/A:1018530828084

Streicher, A., et al. (2003). Broadband ultrasonic transducer for an artificial bat head. *Proceedings of the IEEE Ultrasonics Symposium,* Vol. 2, (p. 1364).

Streicher, A., et al. (2005). Broadband EMFi ultrasonic transducer for bat research. *Proceedings of the IEEE Ultrasonics Symposium,* Vol. 3, (p. 1629).

Strobel, O. (2007). *Optical data bus technologies for automotive applications. ICTON MW 2007, Sousse* (p. 1). Tunesia.

Strobel, O. A. (2013forthcoming). *Optical communication and related microwave techniques.* Chichester, UK: John Wiley & Sons.

Ström, E. G. (2011). On medium access and physical layer standards for cooperative intelligent transport systems in Europe. *Proceedings of the IEEE, 99*(7), 1183–1188. doi:10.1109/JPROC.2011.2136310

Suman, S., & Sudhir, D. (2007, November). Hybrid wireless-optical broadband-access network (WOBAN): A review of relevant challenges. *Journal of Lightwave Technology, 25*(11), 3329–3340. doi:10.1109/JLT.2007.906804

Sun, L., et al. (1997). Modelling and optimisation of micromachined air-coupled capacitance transducers. *IEEE Proceedings Ultrasonics Symposium,* Vol. 2, (p. 979).

Sundberg, C. (1986). Continnous phase modulation. *IEEE Communications Magazine, 24*(4), 25. doi:10.1109/MCOM.1986.1093063

Sun, X., Chan, C. K., Wang, Z., Lin, C., & Chen, L. K. (2007, April). A singlefiber bi-directional WDM self-healing ring network with bidirectional OADM for metro-access applications. *IEEE Journal on Selected Areas in Communications, 25,* 18–24. doi:10.1109/JSAC-OCN.2007.023305

Sykes, E. (2001). Modelling sheds light on next-generation networks. *Fiber Systems, 5*(3), 58.

Technologies, C. (n.d.). *60 GHz reflective electroabsorption modulator.* Obtenido. Retrieved from www.ciphotonics.com

Teshirogi, T., & Yoneyama, T. (2001). *Modern millimeter-wave technologies.* Ohmsha, Ltd.

Thakur, M. P., Quinlan, T., Ahmad Anas, S., Hunter, D. K., Walker, S. D., & Smith, D. W. … Moodie D. (2009). Triple-format, UWB-WiFi-WiMax, radio-over-fiber co-existence demonstration featuring low-cost 1308/1564 nm VCSELs and a reflective electro absorption transceiver. *Optical Fiber Communication (OFC), Collocated National Fiber Optic Engineers Conference (NFOEC),* (pp. 1-3).

Thakur, M. P., Quinlan, T., Bock, C., Walker, S. D., Toycan, M., & Dudley, S. (2009, February). 480-Mbps, bi-directional, ultra-wideband radio-over-fiber transmission using a 1308/1564-nm reflective electro-absorption transducer and commercially available VCSELs. *Journal of Lightwave Technology, 27*(3), 266–272. doi:10.1109/JLT.2008.2005644

Tietze, U., & Schenk, C. (1993). *Halbleiterschaltungstechnik.* Springer Verlag.

Toda, H., Nakasyotani, T., Kurit, T., & Kitayama, K. (2004). WDM mm-wave-band radio-on-fiber system using single supercontinuum light source in cooperation with photonic up-conversion. *IEEE International Topical Meeting on Microwave Photonics,* (pp. 161-164).

Tonguz, O. K., Tsai, H. M., Talty, T., Macdonald, A., & Saraydar, C. (2006). RFID technology for intra-car communications: A new paradigm. *Proceedings of IEEE VTC, 2006,* 1–6. doi:doi:10.1109/VTCF.2006.618

Tong, W., Tong, C., & Liu, Y. (2007). A data engine for controller area network. *Proceedings of Computational Intelligence and Security, 2007*, 1015–1019. doi:doi:10.1109/CIS.2007.137

Torrent-Moreno, M., Jiang, D., & Hartenstein, H. (2004). Broadcast reception rates and effects of priority access in 802.11-based vehicular ad-hoc networks. *Proceedings of the 1st ACM International Workshop on Vehicular Ad Hoc Networks*, Philadelphia, PA, USA.

Tsai, H. M., Tonguz, O. K., Saraydar, C., Talty, T., Ames, M., & Macdonald, A. (2007). Zigbee-based intra-car wireless sensor networks: A case study. *IEEE Transactions on Wireless Communications, 14*(6), 67–77. doi:10.1109/MWC.2007.4407229

Tsao, S., & Lin, C. (2002). Design and evaluation of UMTS-WLAN interworking strategies. *Proceedings from VTC 02: 56th IEEE Vehicular Technology Conference*, Vol. 2, (pp. 777-781).

Tsuzuki, K., Shibata, Y., Kikuchi, N., Ishikawa, M., Yasui, T., Ishii, H., & Yasaka, H. (2008). 10-Gbit/s, 200 km duobinary SMF transmission using a full C-band tunable DFB laser array co-packaged with InP Mach-Zehnder modulator. *IEEE 21st International Laser Conference*, (pp. 17-18).

Tuncer, T. E. (2002). ISI-free pulse shaping filters for receivers with or without a matched filter. *Proceedings IEEE International Conference on Acoustics, Speech and Signal Processing*, Vol. 3, (p. 2269).

Turton, H., & Moura, F. (2008), Vehicle-to-grid systems for sustainable development: An integrated energy analysis. *Technological Forecasting and Social Change, 75*(8), 1091-1108. ISSN: 00401625

Tyco Electronics. (2003). *MOST Interconnectivity Conference*, Frankfurt.

Ukkusuri, S. V., Mathew, T. V., & Waller, S. T. (2007). Robust transportation network design under demand uncertainty. *Computer-Aided Civil and Infrastructure Engineering, 22*, 6–18. doi:10.1111/j.1467-8667.2006.00465.x

UMTS. (2011). *UMTS Forum*. Retrieved from www.umts-forum.org

Ushiki, K., & Fukazawa, M. (1998). A new handover method for next generation mobile communication systems. *Proceedings from GTC 98: 41st IEEE Global Telecommunications Conference*, Vol. 5, (pp. 2560-2565).

Varma, V. K., Ramesh, S., Wong, K. D., Barton, M., Hayward, G., & Friedhoffer, J. A. (2003). Mobility management in integrated UMTS/WLAN networks. *Proceedings from ICC 03: IEEE International Conference on Communications,* Vol. 2, (pp. 1048-1053).

Wake, D., Lima, C. R., & Davies, P. A. (1995, September). Optical generation of millimeter- wave signals for fiber-radio systems using a dual-mode DFB semiconductor laser. *Microwave Theory and Techniques, 43*(9), 2270–2276. doi:10.1109/22.414575

Wake, D., Nkansah, A., & Gomes, N. (2010). Radio over fiber link design for next generation wireless systems. *Journal of Lightwave Technology, 28*, 2456–2464. doi:10.1109/JLT.2010.2045103

Wake, D., Nkansah, A., Gomes, N., de Valicourt, G., Brenot, R., & Violas, M. (2010, August). A comparison of radio over fiber link types for the support of wideband radio channels. *Journal of Lightwave Technology, 28*(16), 2416–2422. doi:10.1109/JLT.2010.2046136

Waltz, E., & Llinas, J. (Eds.). (1990). *Multisensor data fusion*. Artech House.

Wang, J., & Yan, W. (2009) RBM: A role based mobility model for VANET. *Proceedings of the International Conference on Communications and Mobile Computing*, Kunming, China.

Wang, S. S., Green, M., & Malkawi, M. (2001). Adaptive handoff method using mobile location information. *3rd IEEE Emerging Technologies Symposium on Broadband Communications for the Internet Era: Symposium Digest*, (pp. 97-101).

Wang, Y., Cheng, T. H., & Mukherjee, B. (2003). Dynamic routing and wavelength assignment scheme for protection against node failure. *Proceeding of IEEE GLOBECOM'03*, (pp. 2585-2589).

Wang, F., Gosh, A., Sankaran, C., Fleming, P. J., Hsieh, F., & Benes, S. J. (2008). Mobile WiMAX systems: Performance and evolution. *IEEE Communications Magazine, 46*(10), 41–49. doi:10.1109/MCOM.2008.4644118

Wang, J., & Yongacoglu, A. (1994). Performance of trellis coded-8PSK with cochannel interference. *IEEE Transactions on Communications, 42*(1), 6. doi:10.1109/26.275292

Ward's Auto. (2012). *World vehicle population.* Retrieved from http://wardsauto.com/ar/world_vehicle_population_110815/

Warrelmann, J. (2003). *Glasfaserbündel für Datenkommunikation,* (p. 15). Fachgruppentreffen der ITG-FG 5.4.1, Optische Polymerfasern, Offenburg 25- 26, März 2003. Retrieved from http://www.pofac.de/downloads/itgfg/fgt15/FGT15_Offbg_Warrelmann_GOF-Buendel.pdf

Watfa, M. (Ed.). (2010). *Advances in vehicular ad-hoc networks: Developments and challenges.* Hershey, PA: IGI Global. doi:10.4018/978-1-61520-913-2

Weiershausen, W., et al. (2000). Realization of next generation dynamic WDM networks by advanced OADM design. *Proceedings of the European Conference on Networks and Optical Communication,* 2000, (p. 199).

Weihs, M. (2006). Design issues for multimedia streaming gateways. *Proceedings of ICN/ICONS/MCL 2006.* doi: 10.1109/ICNICONSMCL.2006.74

Weinzierl, W., & Strobel, O. (2010). Simulation of throughput reduction of a WLAN over fibre system due to propagation constraints. *Proceedings of CriMiCo,* Sevastopol, Ukraine.

Wei, S., & Shong, C. (1999). A K-best paths algorithm for highly reliable communication networks. *LEICE Transaction on Communication . E (Norwalk, Conn.), 82-B*(4), 586–590.

Wen, H., Chen, L., He, J., & Wen, S. (2007, October). Simultaneously realizing optical millimeter-wave generation and photonic frequency down-conversion employing optical phase modulator and sidebands separation technique. *Asia Optical Fiber Communication and Optoelectronics Conference,* (pp. 427–429).

Wente, E. C. (1917). A condenser transmitter as a uniformly sensitive instrument for the absolute measurement of sound intensity. *Physical Review, 10,* 39. doi:10.1103/PhysRev.10.39

Wente, E. C. (1922). Sensitivity and precision of the electrostatic transmitter for measuring sound intensities. *Physical Review, 19,* 498. doi:10.1103/PhysRev.19.498

Wente, E. C. (1922). The thermophone. *Physical Review, 19,* 333. doi:10.1103/PhysRev.19.333

Wikins, G. D. (1996). Eye-safe free-space laser communications. *IEEE Proceedings of the National Aerospacce and Electronics Conference,* Vol. 2, (p. 710).

Wikipedia. (2012). *ARINC.* Retrieved from http://en.wikipedia.org/wiki/ARINC

Wikipedia. (2012). *CAN bus.* Retrieved from http://en.wikipedia.org/wiki/Controller_area_network

Wikipedia. (n.d.). *ARINC825.* Retrieved from http://en.wikipedia.org/wiki/ARINC_825

Wikipedia. (n.d.). *CANaerospace.* Retrieved from http://en.wikipedia.org/wiki/CANaerospace

Wikipedia. (n.d.). *Unshielded twisted pair.* Retrieved from http://en.wikipedia.org/wiki/Unshielded_twisted_pair#Unshielded_twisted_pair_.28UTP.29

Williams, S. K. (2000). IrDA: Past, present and future. *IEEE Personal Communications, 7*(1), 11. doi:10.1109/98.824566

Winner, H., Hakuli, S., & Wolf, G. (Eds.). (2009). *Handbuch Fahrerassistenzsysteme.* Vieweg Teubner. doi:10.1007/978-3-8348-9977-4

Wong, E., Prasanna, A. G., Lim, C., Lee, K. L., & Nirmalathas, A. (2009, April). Simple VCSEL base-station configuration for hybrid fiber-wireless access networks. *IEEE Photonics Technology Letters, 21,* 534–536. doi:10.1109/LPT.2009.2014393

Wong, G. S. K., & Embleton, A. F. W. (1995). *AIP handbook of condenser microphones: Theory, calibration and measurements.* New York, NY: AIP Press. doi:10.1121/1.413754

Xin, Y., & Rouskas, G. N. (2004). A study of path protection in large-scale optical networks. *Photonic Network Communications, 7*(3), 267–278. doi:10.1023/B:PNET.0000026891.50610.48

Xipeng, X., & Ni, L. M. (1999). Internet QoS: A big picture. *IEEE Network*, *13*(2), 8–18. doi:10.1109/65.768484

Xu, D., Xiong, Y., & Qiao, C. (2003). Novel algorithms for shared segment protection. *IEEE Journal on Selected Areas in Communications*, *21*(8), 1320–1331. doi:10.1109/JSAC.2003.816624

Yang, X., Shen, L., & Ramamurthy, B. (2005). Survivable lightpath provisioning in WDM mesh networks under shared path protection and signal quality constraints. *Journal of Lightwave Technology*, *23*(4), 1556–1564. doi:10.1109/JLT.2005.844495

Yao, J. (2009, February). Microwave photonics. *Journal of Lightwave Technology*, *27*(3), 314–335. doi:10.1109/JLT.2008.2009551

Zang, H., Jue, J., & Mukherjee, B. (2000). A review of routing and wavelength assignment approaches for wavelength-routed optical WDM networks. *SPIE Optical Networks Magazine*, *1*(1), 47–60.

Zang, H., Ou, C., & Mukherjee, B. (2003). Path-protection routing and wavelength-assignment (RWA) in WDM mesh networks under duct-layer constraints. *IEEE/ACM Transactions on Networking*, *11*, 248–258. doi:10.1109/TNET.2003.810313

Zhang, J., Zhu, K., & Mukherjee, B. (2004). A comprehensive study on backup reprovisioning to remedy the effect of double-link failures in WDM mesh networks. *Proceeding of IEEE ICC,* Vol. 27, (pp. 1654–1658).

Zhang, J., Zhu, K., Yoo, S. J. B., & Mukherjee, B. (2003). On the study of routing and wavelength assignment approaches for survivable wavelength-routed WDM mesh networks. *Optical Networks Magazine*, 16-27.

Zhang, J., & Mukherjee, B. (2004). Review of fault management in WDM mesh networks: Basic concepts and research challenges. *IEEE Network*, *18*, 41–48. doi:10.1109/MNET.2004.1276610

Zhang, J., Orlik, P. V., Sahinoglu, Z., Molisch, A. F., & Kinney, P. (2009). UWB systems for wireless sensor networks. *Proceedings of the IEEE*, *97*(2), 313–331. doi:10.1109/JPROC.2008.2008786

Zhang, Z. R., Zhong, W. D., & Mukherjee, B. (2004). A heuristic algorithm for design of survivable WDM networks. *IEEE Communications Letters*, *8*, 467–469. doi:10.1109/LCOMM.2004.832772

Zheng, Y. R., & Xiao, C. (2002). Improved models for the generation of multiple uncorrelated Rayleigh fading waveforms. *IEEE Communications Letters*, *6*(6), 256–258. doi:10.1109/LCOMM.2002.1010873

Zhensheng, J., Jianjun, Y., Georgios, E., & Chang, C. G.-K. (2007, November). Key enabling technologies for optical–wireless networks: Optical millimeter-wave generation, wavelength reuse, and architecture. *Journal of Lightwave Technology*, *25*(11), 3452–3471. doi:10.1109/JLT.2007.909201

Zhou, J., Xia, L., Cheng, X., Dong, X., & Shum, P. (February de 2008). Photonic generation of tunable microwave signals by beating a dual wavelength single longitudinal mode fiber ring laser. *Applied Physics*, *91*, 99–103.

Zhou, D., & Subramaniam, S. (2000). Survivability in optical networks. *IEEE Network*, *14*, 16–23. doi:10.1109/65.885666

Zhukovsky, A. P. (1992). *Radiothermal radiation of a terrestrial surface* (p. 76). Moscow, Russia: Publishing MAI. (in Russian)

Ziemann, O., Krauser, J., Zamzow, P., & Daum, W. (2008). *POF handbook: Optical short range transmission systems* (2nd ed.). Springer.

Zimmermann, W., & Schmidgall, R. (2011). Bussysteme in der Fahrzeugtechnik: Protokolle, Standards und Softwarearchitektur. ATZ/MTZ-Fachbuch. Auflage, Germany: Vieweg+Teubner Verlag. ISBN-13: 978-3834809070

Zimmermann, W., & Schmidgall, R. (2007). *Bussysteme in der Fahrzeugelektronik*. Vieweg Teubner. doi:10.1007/978-3-8348-9188-4

Zimmermann, W., & Schmidgall, R. (2011). Bussysteme in der Fahrzeugtechnik: Protokolle . *Standards und Softwarearchitektur.*, *ISBN13*, 9783834809070.

Zurek, R. A., et al. (2002). *Omnidirectional ultrasonic communication system*. (U.S. Patent 6363139).

About the Contributors

Otto A. Strobel is Head of Physics Institute and Director of Physics Laboratory, Faculty of Basic Sciences at Esslingen University of Applied Sciences, Germany. He passed an Apprenticeship in Electrical Engineering. He received his Dipl.-Phys. and Dr.-Ing. degree from Technical University of Berlin in 1980 and 1986 and his Dr. h.c. degree in 2005 from Moscow Aviation Institute, National Research University, Russia. In 2011 he was awarded as Honorary Professor by the Tecnológico de Monterrey, Mexico. He performed more than 30 visiting professor stays worldwide. He is author of about 90 publications in the field of fiber-optic technologies and optoelectronics, also author of the textbook (in German language) "Technology of Lightwave-Guides in Transmission and Sensing" (VDE 2012, 3rd edition), co-author of the text book (in German): "Photonics" (Springer, 2005) and co-author of the reference book "Resilient Optical Network Design: Advances in Fault-Tolerant Methodologies," IGI Global, Hershey, PA, USA 2011. Furthermore he is honorary workshop chair at the "International Conference on Transparent Optical Networks ICTON," chair member of the "International Workshop on Telecommunications IWT, Brazil," and also member of the Construction Consultative Committee of Wuhan Optics Valley of China. He has more than 10 years experience in companies' R&D, as consultant of Daimler, Bell Labs Germany (Alcatel-Lucent), HP, Agilent, Diehl Aerospace, and Siemens.

* * *

Ivan A. Aldaya was born in Spain in 1980. He received the B.E. from the Public University of Navarre (UPNA), Pamplona, Spain, in 2005. He joined Optimal Communication Ltd., London, United Kingdom, in 2006. Since 2008, he is a full time Ph. D. student at the School of Electronics and Information Technologies, Monterrey Institute of Technology and Higher Education (ITESM), Monterrey, Mexico, where he is part of the optical communications research group. In 2009, he spent a two-month research internship at the University of Bologna, Bologna, Italy. His main areas of research are radio over fiber networks, optical injection locking of semiconductor lasers, and physical modeling of optical routers.

Alejandro Aragón-Zavala graduated from Tecnológico de Monterrey, Campus Querétaro as Electronics and Communications Engineer in December 1991. In 1998 he received his MSc in Satellite Communication Engineering from the University of Surrey, and in 2003, his PhD in Antennas and Propagation at the same university. He has worked as an Engineer and Senior Consultant in the wireless communications industry for more than 15 years. Since 2003, he is working as a full-time Lecturer for Tecnológico de Monterrey, Campus Querétaro, and has been awarded as a National Researcher Level 1 for Sistema Nacional de Investigadores, in Mexico, in 2010. He is the author of more than 30 international papers and two books on wireless communications, both with Wiley. His research interests include: mobile communications, satellite systems, high-altitude platform systems, antenna design, and indoor radio propagation.

Joaquín Beas received the Bachelor of Science in Electronics Engineering from the Autonomous University of Nuevo León UANL Mexico in 2001. He received the Master of Science degree in Electronics and Telecommunications from the Ensenada Research Centre and Higher Education, Mexico in 2003. His thesis included the characterization and development of an all-optical wavelength converter using the Cross-Polarization Modulation (XPOL) phenomena in a Semiconductor Optical Amplifier (SOA). In 2004, he joined Motorola Mexico manufacturing facility working on new product introduction area, specifically focused on Hybrid Fiber Coax (HFC) and Passive Optical Network (PON) technologies. Since January 2010, he has been Ph.D. student of the Monterrey Institute of Technology and Higher Education (ITESM) in the program of Optical Communication Systems. His research interests include radio-over-fiber networks and milli-metric wave enabling technologies, distribution network topologies, and all-optical signal processing. Presently, he is working for Motorola at the Research and Innovation Technology Park, Monterrey, Nuevo Leon, Mexico on sustaining and optimization projects for optical fiber analog and digital transmitters and receivers and Erbium Doped Fiber Amplifiers (EDFAs) for HFC and RFoG networks.

Gabriel Campuzano is an Associate Professor. He received his undergraduate degree as an Industrial Physics Engineer from the same University (Tecnológico de Monterrey), Campus Monterrey, and holds a MSc (DEA) in Microwaves and Optoelectronics from the "Université de Pierre et Marie Curie," France. He then received the Ph.D. degree in Optical Communications from the "École Nationale Supérieure des Telécommunications" (Telecom ParisTech), France. His research experience includes participation in different projects in the information and communication technology fields with Alcatel Opto+, European Union project OPTIMIST thematic network, and the Government of the State of Nuevo Leon. His fields of interest range in radio over fiber systems for wireless access networks, including mobility in picocells, novel optoelectronic functions, and quantum cryptographic systems for telecommunication wavelengths. He is now working in the Department of Electrical and Computational Engineering at ITESM since September of 2003.

Gerardo Castañón is the head of the optical communications chair, an Associate Professor and member of the national research system in Mexico, and member of the Academy of Science in Mexico. Dr. Castañón received the Bachelor of Science in Physics Engineering from the Monterrey Institute of Technology and Higher Education (ITESM), México in 1987. He received the Master of Science degree in Physics (Optics) from the Ensenada Research Centre and Higher Education, México in 1989. He received the Master of Science degree in Electronics Engineering from ITESM in 1990. He also received the Master and Ph. D. degrees in Electrical and Computer Engineering from the State University of New York (SUNY) at Buffalo in 1995 and 1997, respectively. He was supported by the Fulbright scholarship through his Ph. D. studies. From January of 1998 to November of 2000 he was a Research Scientist working with Alcatel USA Corporate Research Center in Richardson TX. Where he was doing research on IP over WDM, dimensioning and routing strategies for next generation optical networks and the design of all-optical routers. From December of 2000 to August of 2002 he was a senior researcher with Fujitsu Network communications doing research on ultra high-speed transmission systems. He is now working in the Department of Electrical and Computer Engineering at ITESM since September of 2002. Dr. Castañón has over 60 publications in journals, book chapters and conferences, and 4 international patents. He is a senior member of the IEEE Communications and Photonics societies. He frequently acts as a reviewer for IEEE, IET, and Elsevier journals. Also he frequently participates as technical committee member of telecommunication conferences.

Roger J. Green became Professor of Electronic Communication Systems at Warwick in September 1999, and was Head of the Division of Electrical and Electronic Engineering from August 2003 for five years. He has published around 250 papers in the field of optical communications, optoelectronics, video and imaging, and has several patents. Over 50 Ph.D. research students have worked successfully under his supervision. He now leads the Communications Systems Laboratory in the School of Engineering at Warwick. His current interests include signal processing, optical wireless, and optical fibre communications. He is a Fellow of the UK IET and a Fellow of the Institute of Physics. He is a Senior Member of the IEEE, and currently serves on two IEEE committees concerned with communications and signal processing.

Matthew D. Higgins received his MEng in Electronic and Communications Engineering and PhD in Engineering from the School of Engineering at the University of Warwick in 2005 and 2009, respectively. Remaining at the University of Warwick, he then progressed through several Research Fellow positions with leading defence and telecommunications companies before undertaking two years as a Senior Teaching Fellow. Dr. Higgins now holds the position of Assistant Professor where his major research interests are the modelling of optical propagation characteristics in underwater, indoor and atmospheric conditions and how the channel can affect communications systems. Dr. Higgins is a Member of both the IEEE and IET.

David Hutchins joined the School of Engineering at the University of Warwick in 1988. He is a Processor of Electrical and Electronic Stream. His research interests are in the design and fabrication of ultrasonic systems, including transducers and sensors formed by micromachining techniques, infrared and other electromagnetic measurement techniques, security and cargo screening methods, communications, and imaging. Applications being investigated include materials characterisation, flow measurement, personal screening from a distance, and cargo screening of freight vehicles. He is an Associate Editor of the *IEEE Transactions on Ultrasonics, Ferroelectrics and Frequency Control.*

Kira Kastell is Director of Studies of the Electrical Engineering Study Programmes and Professor for Communication Theory of Frankfurt University of Applied Sciences, Germany. She received her first diploma degree in Electrical Engineering from Frankfurt University of Applied Sciences in 1998 and her second diploma degree in electrical engineering from Hagen University of distance education in 2002. She has been awarded a scholarship of the German Research Council's graduate school "System integration for ubiquitous computing in information technology" at Darmstadt University of Technology from where she also received her Dr.-Ing. degree in 2007. She also holds diploma degrees in Business Administration as well as Economics from Hagen University of Distance Education (both 2004), Germany. She has more than four years of experience in network planning for railways as project engineer for German Railways at Mannesmann Arcor. She is member of the editorial board of *Frequenz - Journal of RF-Engineering and Telecommunications*, reviewer and TPC member for several international conferences. Her major research interests include hybrid handovers, software defined radio, localization, and applied robotics.

Yousef S. Kavian received the B.Sc. (Hons) degree in Electronic Engineering from the Shahid Beheshti University, Tehran, Iran, in 2001, the M.Sc. degree in Control Engineering from the Amkabir University, Tehran, Iran, in 2003 and the Ph.D. degree in Electronic Engineering from the Iran University of Science and Technology, Tehran, Iran, in 2007. After one year appointment at Shahid Beheshti University, in 2008 he joined the Shahid Chamran University as an Assistant Professor for both research and teaching. He worked as a postdoctoral research fellow at Esslingen University and IAER, Germany, in 2010. His research interests include digital circuits, systems design, and optical and wireless networking. Dr. Kavian has over 80 technical publications including journal and conference papers and book chapters in these fields. He is a Senior Industrial Engineer and Trainer with more than 10 years industrial collaborations and experiences.

Mark S. Leeson received the degrees of BSc and BEng with First Class Honors in Electrical and Electronic Engineering from the University of Nottingham, UK, in 1986. He then obtained a PhD in Engineering from the University of Cambridge, UK, in 1990. From 1990 to 1992 he worked as a Network Analyst for National Westminster Bank in London. After holding academic posts in London and Manchester, in 2000 he joined the School of Engineering at Warwick, where he is now an Associate Professor (Reader). His major research interests are coding and modulation, nanoscale communications, and evolutionary optimization. To date, Dr. Leeson has over 220 publications and has supervised ten successful research students. He is a Senior Member of the IEEE, a Chartered Member of the UK Institute of Physics and a Fellow of the UK Higher Education Academy.

Chuan Li is a Research Assistant in the Faculty of Engineering at the University of Bristol, UK. He has his BEng, MSc, and PhD. He is part of the Solid Mechanics research group with the Department of Mechanical Engineering.

Alicia López received the M.Sc. degree in Communications Engineering and Ph.D. degree from the University of Zaragoza (UZ), Zaragoza, Spain, in 2002 and 2009, respectively. In 2002, she joined the Photonic Technologies Group (GTF), Aragon Institute of Engineering Research (i3A). Since 2004, she has been Assistant Professor in the Departamento de Ingeniería Electrónica y Comunicaciones, UZ. Her research interests include the use of plastic optical fibers in communications applications and the design and evaluation of optical networks.

M. A. Losada received her Ph.D. in Physics from the Universidad Complutense de Madrid (Spain) in 1990. She has worked at the McGill Vision Research Laboratories of McGill University in Montréal (Canada), in the Institute for Research in Optics of the Scientific Research Council of Spain (CSIC) and in Universidad Politécnica of Madrid. She has a tenured position as an Associate Professor of the Department of Electronics and Communications in Universidad de Zaragoza. She is a member of the Photonic Technologies Group (GTF) that is now a division of the Aragon Institute for Engineering Research (i3A). Now, her research interests are centered in optical communications based on plastic optical fibers and in the design of optical networks.

Jan Lubkoll received his Master of Science with honours from Friedrich-Alexander University Erlangen-Nuernberg, Germany, in the elite master program of Advanced Optical Technologies. His major research interests include technologies of micro systems, optoelectronics, and optical communication. In 2009 he completed his Bachelor of Engineering degree at Esslingen University of Applied Science, Germany. He has some years of experiences in fiber-optic data buses (FlexRay) in the aviation field, ultra short pulsed laser systems in material processing and computational simulation of optical systems.

Javier Mateo was born in Zaragoza, Spain, in 1964. He received the M.Sc. degree in Electrical Engineering from the Polytechnic University of Madrid in 1989 and the Ph.D. degree from the University of Zaragoza in 2000. From 1989 to 1993, he was with Cables de Comunicaciones S. A., where he worked on fiber optic sensors and optical communications. In 1993, he joined the Electronic Engineering and Communications Department of the University of Zaragoza, where he developed his Ph.D. degree dissertation. He is currently Professor of Optical Fiber Communications in the Department of Electronic Engineering and Communications at the University of Zaragoza. His professional research interests are in signal processing, in particular, applied to biomedical signals, fiber optic sensors, and optical communication systems.

Jürgen Minuth studied Electrical Engineering at the University of Stuttgart, Germany. He worked in automotive industry for eleven years in the field of electronic research, developing electronic control units, embedded operating systems, communication and network protocols, standardized software modules, fault tolerant electronic systems, communication hardware and electro-magnetic compatibility. At Esslingen, University of Applied Sciences, Germany, he has been a part-time Lecturer in transmission line theory and measurement engineering for 6 years, and since 2000 he is a Professor with the main focus on Electronics and Automotive Applications. In 2007 he overtook the coordination in Mechatronics at the CDHAW (Chinese German University of Applied Sciences), and since 2010, he is the head of the electronics lab, faculty of Mechatronics and Electrical Engineering. He was a working member in various FlexRay™ physical layer working groups from the very beginnings up to the final stage represented by the publishing of the FlexRay-specifications version 3.

Zaiton Abdul Mutalip received her BEng in Electronic and Telecommunications Engineering from Universiti Malaysia Sarawak, Malaysia (2000) and MSc Optical Communication System and Networks from University of Hertfordshire, UK (2008). She is attached with Universiti Teknikal Malaysia Melaka, Malaysia. She is now working on her PhD research in optical wireless in vehicles at University of Warwick.

Piet De Pauw was born in Dendermonde (Belgium) in 1956. He graduated as Electronic Engineer in 1979 and as Ph.D. in 1984 from the University of Leuven. Subject of Ph.D. thesis: Low Cost Solar Cells. He obtained an MBA from the Handels HogeSchool (now Flanders Business School) in Antwerpen in 1990. In 1984, after his Ph.D. research, he started as a Design Engineer and Project Leader of high voltage circuits in the company Mietec in Oudenaarde, Belgium. In 1986 he joined Bell Telephone Manufacturing Corp in Antwerpen, Belgium (now part of Alcatel Lucent) as head of the reliability laboratory, failure analysis laboratory, IC technology, and Product Engineering. In 1991, he joined Alcatel Microelectronics in Oudenaarde, Belgium, (now part of ON Semiconductor) as Reliability Manager, and later Quality and Reliability Manager. Between 1998 and 2002 he established and expanded the company Sipex Flanders Design Center in Zaventem Belgium, where he introduced new optoelectronic product lines. From 2000 till 2009 he was a member of the board of directors of Q-Star Test N.V. in Brugge. From April 2006 to 2009 he chaired the opto division of Melexis N.V., and from 2006 till 2011, he was site manager of Melexis Ieper. Starting from 2002 he is a consultant specialized in optoelectronic product development (image sensors, infrared sensors, Optical storage products, Opto datacom products). In this position, he headed the team developing the MOST25 and MOST150 products, resulting in a world record optical budget for these components. Currently he is working on fiber optic transceivers for avionics applications.

Zeina Rihawi was born in Syria in 1985. She received the B.Sc. degree in Electronic Engineering – Communication Department from Electric and Electrical Engineering Faculty, Aleppo University, Aleppo, Syria, in 2007, and the M.Sc. degree in Electronic Systems with Communications with distinction from University of Warwick, UK, in 2010. She is currently working toward the Ph.D degree at the School of Engineering, University of Warwick, UK. She worked as a Teaching Assistant for three semesters (2008-2009) in Communication Department, Electric and Electrical Engineering Faculty, University of Aleppo, Syria. Her main research interests are in the field of optical wireless communications, in-vehicle networks and multimedia communications.

Daniel Seibl completed his diploma studies in Electrical Engineering, in the field of Micro-Electro-Mechanical Systems and specialized in Micro Optics and Laser Technology in October 2008 at Esslingen University of Applied Sciences. His major research interests are in the fields of fiber-optic technologies. He already has experience in automotive fiber-optic data buses (MOST) and fiber-optic sensors. Now he is working as Research Associate at Esslingen University of Applied Sciences. His actual research work is dealing with optical 3D measurement, sensor systems, and high-power fluorescence illumination.

Riccardo Scopigno (M.Sc. 1995 *summa cum laude*, Ph.D. 2005) is Head of Multi-Layer Wireless Department of Istituto Superiore Mario Boella (ISMB). He has matured a 16-year working experience in the area of Telecommunications, obtaining, in the meantime his Ph.D. His skills cover telecommunication architectures, from theory to practice - as matured from his variegate working experience. He was a hardware designer for TLC systems in Italtel-Siemens (1997-1999); in Marconi (2000-2003), he achieved a good expertise in IP network design (as certified network engineer). He is also ISMB's representative in ETSI ITS (the working group for the standardization of intelligent transportation systems), in ERTICO, in the Car-to-Car Communication Consortium (C2C-cc) and in the European SmartCity Stakeholders' Platform. He is author of about 70 publications, including book chapters and papers at international IEEE conferences and international journals; he also acts as TPC co-chair at several conferences and is author of 5 patents and patent pending techniques for quality of service in wireless networks. His research interests include: scalability and QoS in wireless MAC, optimization of standard protocols and non-standard compliant techniques, vehicular communications, wireless automation, and cross-layer QoS for media over wireless.

Vladimir Rastorguev works with the State University Moscow Aviation Institute, Russia. He is a member of the scientific board of the Institute for Advanced Engineering and Research. He was on the CTS Technical Program Committee of the 13th International Conference on Transparent Optical Networks.

Martin Stämpfle is Professor and Director of the Mathematics Laboratory in the Faculty of Basic Sciences at Esslingen University of Applied Sciences, Germany. He was Lecturer for Mathematics in the Faculty of Basic Sciences at Esslingen University of Applied Sciences, Germany, from 2006 to 2008. In 1998 and 1999, he was Research Assistant at the Center of Systems and Control at Glasgow University, Scotland, UK. He received his Dr. rer. nat. from University of Ulm, Germany, in 1997, and the Promotionspreis of the Ulmer Universitätsgesellschaft for his Ph.D. thesis on the numerical solution of dynamical systems. In 994, he received his Diploma in Applied Mathematics also from University of Ulm, Germany. Martin Stämpfle is co-author of a comprehensive student textbook on mathematics.

His major research interests include numerical mathematics and scientific computing. In the field of applied research and development, he supervises a great variety of interdisciplinary projects in the context of driver assistance systems. He contributes to a large number of patents. He has 12 years of industrial experience as member, consultant, and qualification trainer with international automotive suppliers and manufacturers. He is member of many mathematical and engineering societies.

Bin Wang received the Ph.D. degree in Electrical Engineering from the Ohio State University. He is a full Professor of Computer Science and Engineering, Department of Computer and Engineering, Wright State University. Dr. Wang has led numerous research projects funded by DoD, NSF, DoE, AFOSR, AFRL, ODOD, and other sources. He is a recipient of US Department of Energy Career Award. Dr. Wang's research interests include cyber security, network security visualization, user behavior modeling, trust management, security and information assurance, game theory, layered sensing, wireless communication, wireless and mobile computing, wireless sensor networks, pervasive/ubiquitous computing, multimedia communication, real-time system and communication, quality of service provisioning, dense wavelength division multiplexing (DWDM) optical networks, optical burst switching, Grid computing, ultra-wide band, software defined radio, open spectrum access & cognitive radio networks, information theory, distributed signal processing, Semantic Web, RFID and medical/health informatics, organisms, and human cellular networks.

Index